Experimental Design and Data Analysis for Biologists

Applying statistical concepts to biological scenarios, this established textbook continues to be the go-to tool for advanced undergraduates and postgraduates studying biostatistics or experimental design in biology-related areas. Chapters cover linear models, common regression and ANOVA methods, mixed effects models, model selection, and multivariate methods used by biologists, requiring only introductory statistics and basic mathematics. Demystifying statistical concepts with clear, jargon-free explanations, this new edition takes a holistic approach to help students understand the relationship between statistics and experimental design. Each chapter contains further-reading recommendations and worked examples from today's biological literature. All examples reflect modern settings, methodology, and equipment, representing a wide range of biological research areas. These are supported by hands-on online resources including real-world datasets, full R code to help repeat analyses for all worked examples, and additional review questions and exercises for each chapter.

Gerry Quinn is an honorary professor in the School of Life and Environmental Sciences at Deakin University, having served as Chair in Marine Biology and Head of Warrnambool Campus during his academic career. He has extensive experience in teaching biostatistics at Deakin University and the University of Gothenburg.

Michael J. Keough is an ecologist, environmental scientist, and honorary professor in the School of Biosciences at University of Melbourne. He has taught classes in ecology, experimental design, and environmental science for many years at the University of Melbourne and in the United States.

"The new edition of this 'go-to' text, has been revised to take a more holistic, informative, and up-to-date approach to the subject. This remarkably accessible, practical and – at times – entertaining guide to how we can best translate our questions and ideas into informative experiments and analyses is highly recommended for everyone wanting to investigate, visualise, and analyse biological phenomena."

Professor Jonathon Havenhand,
University of Gothenburg

"The new book is an excellent resource for researchers, analysts and teachers. The text clearly outlines the important concepts of the tests and models. The examples are all based on published data which makes it easy to source the full manuscript, datasets and replicate the analyses, which is incredibly important and assists with interpretation."

Dr Victoria Goodall, *VLG Statistical Services*

"At last, a book for undergraduates and graduates that distills the complexities of biological data analysis into an easy-to-understand generalized linear model approach – with examples in R! The authors challenge the reader to think critically about data by providing important details on design, summary statistics, power analysis/effect size, model fit, and data visualization. This book will be used in my classes for years to come."

Professor Greg Moyer, *Mansfield University*

"I have been using *Experimental Design and Data Analysis* for teaching and research for 20 years and have been hoping for a second edition for 10. It was worth the wait. The new edition shares many attributes with the original. It is quite easy to read, the examples are varied and interesting, there are ample and revealing box examples and there remains an attitude of Statistics as a tool that will be used by many. There are a number of changes that reflect these attributes – two key ones are the emphasis on Generalized Models as a framework for a broad array of approaches and the conversion of examples to R, with code and additional material available online."

Professor Peter Raimondi,
University of California Santa Cruz

"I have taught in and coordinated a third-year design/ statistics paper for zoology and ecology students for 10 years – an enjoyable and sometimes challenging task. The first edition of Quinn and Keough has been immensely helpful for me in teaching this course. The second edition has been updated and expanded considerably. I'm sure it will continue to inspire my future teaching."

Professor Christoph Matthaei, *University of Otago*

"I was excited to see Quinn and Keough have updated their classic guide to experimental design and data analysis. I read the earlier edition of this book as a graduate student, and the advice it provides on experimental design is the foundation of my own studies, as well as my approach to training graduate students... This book is foundational reading for aspiring scientists. Not only does it teach you how to analyse your data, it also provides invaluable advice on how to communicate analyses and write up scientific studies. The book's advice will help give early career scientists the confidence they need to write-up and publish their first studies."

Professor Chris Brown, *University of Tasmania*

Experimental Design and Data Analysis for Biologists

Second Edition

Gerry P. Quinn
Deakin University

Michael J. Keough
University of Melbourne

CAMBRIDGE
UNIVERSITY PRESS

Shaftesbury Road, Cambridge CB2 8EA, United Kingdom

One Liberty Plaza, 20th Floor, New York, NY 10006, USA

477 Williamstown Road, Port Melbourne, VIC 3207, Australia

314–321, 3rd Floor, Plot 3, Splendor Forum, Jasola District Centre, New Delhi – 110025, India

103 Penang Road, #05-06/07, Visioncrest Commercial, Singapore 238467

Cambridge University Press is part of Cambridge University Press & Assessment,
a department of the University of Cambridge.

We share the University's mission to contribute to society through the pursuit of
education, learning and research at the highest international levels of excellence.

www.cambridge.org
Information on this title: www.cambridge.org/highereducation/ISBN/9781107036710

DOI: 10.1017/9781139568173

First published 2002
Second edition 2024

Printed in the United Kingdom by TJ Books Limited, Padstow, Cornwall

A catalogue record for this publication is available from the British Library.

A Cataloging-in-Publication data record for this book is available from the Library of Congress

ISBN 978-1-107-03671-0 Hardback
ISBN 978-1-107-68767-7 Paperback

Additional resources for this publication at www.cambridge.org/quinn-keough2.

Contents

Color plates can be found between pages 364 and 365.

Preface

Statistical analysis is at the core of most modern biological research, and many biological hypotheses, even deceptively simple ones, can require complex statistical models.

The landscape can be daunting, with a forest of acronyms, a bewildering array of terms – linear models, generalized linear models (GLMs), generalized linear mixed models (GLMMs), covariates, randomized blocks, mixed models, multilevel analysis, multivariate analysis, and so on, and apparently different statistical tribes. Adding to this complexity is a statistical package (*R*) that has become the standard but often offers several options for the same task and challenging online help. How can a biological researcher, particularly a beginner, make sense of this landscape?

Much of this complexity is unnecessary or illusory. Statistical analyses with different names are sometimes synonyms rather than new techniques to be learned. More importantly, many different analyses are better viewed as linear models built using a common framework rather than distinct tools. Understanding the framework prepares you to deal with a range of unfamiliar situations you may encounter.

As biologists, we try to explain natural phenomena as best we can, given our current knowledge. A good explanation has survived challenges from alternative explanations. We think of these explanations as models – simplified descriptions of nature – and we assess their adequacy by comparing them to data. Ideally, we challenge a particular explanation by finding or creating a novel situation in which our focal model will produce a pattern in the data that differs from the pattern produced by the alternatives.

We use statistical analysis to assess the fit of the data to a particular model. We look for a signal from that model against background noise in the data and estimate that signal's strength. The confrontation between models and data requires us to translate a broad, even qualitative model that is our biological explanation (or hypothesis) into a more precise statistical model. The specific way we compare the model to data depends on our statistical approach, but ultimately, we decide based on that comparison. The decisions may be formal, as in hypothesis tests, or informal but no less important – what do we do next in our research, or what do we advise end-users of the knowledge to do?

We can be led astray in many ways during this process:
- The sampling and experimental design doesn't match the biological question.
- The statistical model doesn't match the biological one.
- The data don't have properties assumed by the statistical model.
- The data aren't sufficient to distinguish signal from noise, between competing models or to allow you to estimate biological effects with confidence.
- You might have unrepresentative data because of poor design or bad luck.

These problems can cause you to expend lots of energy or resources, only to get an unclear or wrong answer to the original biological question. We aim to reduce the likelihood of this happening, and we use three simple principles:
- Think clearly about the biological problem, the different models or explanations in play, and what kind of data we need to distinguish these models unambiguously.
- Think in advance about the statistical model corresponding to each biological model and how the match between model and data will be assessed. Decide how much data you need to make confident decisions.
- Think before you analyze. Make sure that the data you've collected are consistent with assumptions of your statistical model(s); if not, transform the data or respecify the model.

These principles require most of the hard work before a single statistical analysis is attempted. They also require us to be our own harshest critics, to make sure that we have challenged each biological model severely.

We apply this approach to a wide range of statistical models that biologists use, from the simplest to some that are very complex. Most of these models are variants of GLMs. We start with simple models and gradually make them more complex. We show how different "methods" such as analysis of variance (ANOVA), multiple regression, logistic regression, etc. are closely related.

We need to:
- know the pitfalls and assumptions of particular statistical models;
- be able to identify the model appropriate for the sampling or experimental design and the data we plan to collect;
- be able to interpret the output of analyses using these models; and
- be able to design experiments and sampling programs optimally – that is, with the best possible use of our limited time and resources.

In This Book

Our approach encourages readers to understand the models underlying the most common designs in biology. We assume that our readers have completed some basic training in experimental design and data analysis, and we begin with reminders of some essential basic concepts. We then build the linear models common to so many biological situations. We begin with the structure of a linear model, the variables used in these models, and how we fit models to data with an emphasis on ANOVA. Because model-fitting is crucial, we outline the exploratory methods used to ensure an appropriate model is being used. The next few chapters provide detailed accounts of increasingly complex models. We start with simple models with a single predictor and a single response variable, and outline models for when these variables are continuous or categorical (ANOVA/regression/logistic regression/loglinear models). From there, we consider models with a continuous response variable and multiple predictors, starting with predictors that are only continuous or categorical. These chapters cover familiar approaches of factorial ANOVA and multiple regression and can be expanded to mixtures of continuous and categorical predictors. We then introduce models with fixed and random effects – linear mixed models – and models for correlated data, including nesting, multilevel modeling, and repeated measures. We extend these models to situations with categorical responses in generalized linear and mixed models. Our emphasis is on learning to build and fit linear models for particular situations rather than applying cookbook techniques.

After considering situations with single response variables, we briefly describe techniques, such as principal components, discriminant function analysis, and multidimensional scaling, used with multiple response variables. We also show how linear models can be applied to multivariate data.

Our emphasis for most of the book is on thinking clearly about collecting, analyzing, and interpreting data. We conclude with some thoughts about communicating this process clearly to our end-users, particularly those unfamiliar with the statistical approaches.

Each chapter includes a further reading section, where we list books or book chapters that provide background to or expand on particular statistical topics. These lists are clearly not exhaustive and simply represent books/authors that we, and more importantly our students, have found helpful. Other relevant books and papers from the primary statistical and biological literature are cited in all chapters.

Learning by Example

One of our strongest beliefs is that most biologists understand statistical principles much better when we see how they are applied to situations in our own discipline. Examples let us link statistical models and formal statistical terms (blocks, plots, etc.) or papers written in other fields and the biological situations we are dealing with. For example, how is our analysis and interpretation of an experiment repeated several times helped by reading literature about blocks of agricultural land? How does literature developed for psychological research help us measure changes in plants' physiological responses?

Throughout this book, we illustrate the statistical techniques with examples from the current biological literature. We describe how to analyze the data. We focus on specifying the appropriate model, assessing assumptions, and interpreting the statistical output. These examples appear as boxes throughout each chapter. We use examples where the raw data are available, mainly through public online repositories (e.g. individual journals, datadryad.org) but sometimes on our website, courtesy of the study's authors. Although we focus on data analysis rather than software, we have implemented all analyses in R, with the code available online.

The other value of published examples is that we can see how particular analyses can be described and reported. It is easy to allow the biology to be submerged beneath a mass of statistical output when fitting complex statistical models. We hope that the examples and our thoughts on this subject in the final chapter will help prevent this from happening.

This Book Is a Bridge

We assume that readers have some introductory training, but their research questions will usually require statistical models far beyond that introduction. Even a simple decision such as using each experimental animal more than once triggers an additional complexity in the analysis. Even if we use simple designs, we will be reading papers with more complex analyses. We have tried to cover the most common biological designs we have found students and colleagues using, and to provide readers with the tools to tackle more complex or unusual questions.

Biological data are often messy, and some readers will find that their research questions require more complex models than we describe here. The primary statistical literature provides ways of dealing with messy data or solutions to complex problems. We try to point the way to critical pieces of that statistical literature, providing the reader with the essential tools to deal with that literature or get help. The help might come from formally trained statisticians or more knowledgeable peers, but we need to speak a common language to describe the question and interpret the answer. If help comes from outside our specific discipline, we need to be aware of biological considerations that may cause statistical problems. We can't expect a statistician to know the biological idiosyncrasies of our particular study, but we may get misleading or incorrect advice if they lack that information. This book provides a bridge to these situations.

We hope this book will be used in two ways, as for the first edition. In formal classes, it is the base for a graduate, or perhaps advanced undergraduate subject. It should be preceded by an introductory-level class. There is too much material for a single-semester class, and instructors can choose subsets of material that are right for their students. This has been our approach in teaching. The book also functions as a reference source for individual researchers who need to deal with actual data collection. Researchers can use it for self-directed learning and get practical suggestions for dealing with data. They also have a launching point to delve more deeply into the relevant literature or to start conversations about more complex analyses.

Always remember that for biologists, *statistics is a tool* we use to illuminate and clarify biological problems. We aim to use these tools efficiently without losing sight of the biology that is the motivation for most of us entering this field.

What's Different This Time Around?

In many ways, our first edition was reactive. We were influenced by the experimental and sampling designs we encountered among colleagues and research students. We focused on the common, named analyses – regressions, ANOVA, analyses of covariance, etc. We did focus on mixed models, particularly those involving nesting, because designs for which these statistical models are appropriate were disproportionately common. While acknowledging a range of model-fitting approaches, we used ordinary least squares (OLS) approaches to illustrate the major analyses.

We also presented the hypothesis-testing approaches as the main criterion for deciding whether a hypothesis was supported or not, though we described others. We used this approach because it was (and remains) common and provides an easy link to planning studies with sufficient replication (through power analysis). Our view was that there are several approaches, and all have strengths and shortcomings. Careful statistical and philosophical issues underpin the different approaches, and we presented readers with an introduction to this literature. Our view was and continues to be that we should know these deeper issues. Whether or not we use *P*-values, confidence intervals, likelihoods, etc., we understand precisely what each method does and does not do. We have tried to encompass this diversity of views, using the signal/noise concept – confidence intervals, *P*-values, etc. are tools to avoid getting fooled by noise (Gelman & Loken 2014).

This Edition Differs from Its Predecessor in Several Important Ways

A holistic approach to linear models vs. a box of named tools. In recent years, we've lost count of how often it's been recommended (by advisors, reviewers, etc.) that a researcher use linear mixed models as an alternative to, for example, the planned partly nested ANOVA. After our initial confusion, given that a partly nested ANOVA is almost always a linear mixed model, we realized that three issues were conflated. First, just because two analyses have different names, they may not actually be different (see also partly nested vs. repeated measures designs in Chapters 11 and 12). Second, the linear mixed models were often GLMs, fitted using maximum likelihood (ML) estimation and emphasizing parameter estimation and model comparison. Third, sometimes, "new" approaches did not represent dramatic changes but could be something as simple as changing one predictor variable from categorical to continuous. We moved away from just describing the different named methods. Instead, the analyses done by most biologists share a common core of fitting a statistical model to data. It is usually a linear model. Biologists measure biological responses that may be discrete or continuous, and those variables may be distributed in many ways. Generally, they follow a distribution from the exponential family (Chapter 4), so we can fit a GLM. The form of the GLM depends on the distribution followed by the response variable and the nature of the predictor variable(s).

Suppose the response variable follows a normal distribution. Here, we can use least squares methods to fit the model, and it's a general linear model – but that's just a special case of a GLM. We can classify our predictors by whether they, too, are categorical or continuous and whether they are fixed or random (we'll explain this later!). If some of our predictors are fixed, and some are random, we magically have a GLMM. We focused the core of this book (Chapters 4–13) around building GLMs, using our taxonomy of predictors. We start by building simple models with single predictors, extending to more complex models with more complex relationships among the predictors. As we do this, we link the model to the "named" approaches you may find in papers and elsewhere. We've also pointed out how different named methods are related, and sometimes where different approaches have been used, despite the same underlying statistical model. To use a DIY metaphor, it's the change from a shed full of specific mains-powered tools to a core of standard batteries and chargers, matched with task-specific "skins" with the same shared power source.

We have brought the **estimation of effects** very much to the front. They were implicit through much of the first edition and explicit in parts (e.g. power analysis). In this edition, they are front and center – we need tools to identify signals from noise, but the nature of the signal (its strength and form) separates the trivial from the meaningful. We need to know we're not fooled by noise to start with, but we should be telling our audiences what the signal is like. We also need to acknowledge that detecting a signal may tell us little or nothing about its strength.

P-values remain one way of **identifying signals**, but not the only one, and we have tried to illustrate parallel approaches. Published biological papers often show strong effects (which may be why they were published!), and the different approaches often lead us to the same conclusion. While we have views on preferred approaches, it's possible to write a clear justification for several approaches. We encourage you to read thoroughly and critically and decide on your decision-making approach.

We finished the first edition with some thoughts about improving how we talk about our analyses because we felt this topic was underappreciated. We've expanded this section and tried to move it beyond our preliminary thoughts. We're delighted to see how science communication has increased in emphasis, bringing with it a broader diversity of target audiences. There are now many good books on the topic, and many students take a class as part of their training. There is still scope for improvement when reporting analyses, however.

Some Topics Have Been Reduced

We did try to provide an outline of the philosophical underpinnings of biological inference. Since then, there has been lots of published material, renewed attention to poor practices, debates about hypothesis testing, campaigns against using "statistical significance," and so on. Much of this discussion within the biological literature is a little simplistic, and it's the professional domain of others. Rather than presenting our summary, we've reduced this component and encourage you to read clearer accounts of the issues. Statisticians and philosophers write these accounts, but they are intended for practitioners.

We have assumed that you'll already have a basic statistical understanding in venturing into this book. That was the case in the first edition, but despite that, we embarked on at least two chapters of revisionary material. With our expanded coverage of linear models, we've compensated by reducing the recap into a single chapter. In that chapter, we highlight concepts we think you should already know. We'll leave it to you to check that you understand them clearly, and if not, or if by Chapter 4 you feel like you've jumped into the deep end, it may be time to revive your old course notes or do some more reading.

Models for Teaching

This book is a guide for researchers as well as a base for teaching, so it covers far more ground than can be covered in a single-semester subject. There are several options for using this book in class, depending on the course participants, their needs, and their backgrounds. We'll describe a few possibilities, based on our teaching experience.

A First Course for Graduate Students

The biggest challenge in teaching design and analysis to biology students is understanding just what they know as they enter the subject. Their backgrounds are diverse, ranging from recipes without background to math/stats majors. Even when we have an introductory statistics class as a prerequisite, little information may have been retained.

The first step is to assess the background of the students and the complexity of analyses they'll be expected to do. Our experience is that some of the more complex designs are linked to disciplines in which data analysis is seen as simple and straightforward, so the question is more "what do students need to know to analyze their data appropriately?"

• In an ideal world, with an earlier subject completed, we would assume Chapter 2 and most of 3, and ask students to revise this material before starting. Our aim would then be to use Chapters 4–13 and 17 as the core, selecting relevant sections to keep the content manageable. Split-plot and repeated measures designs are common in biology, so our aim is to cover their analysis at the end. This is challenging for many students, especially those with weaker quantitative backgrounds. Chapters 5 and 17 would not require much time, and Chapter 9 may or may not be required.

• If students are less well prepared, we need to spend more time on the earlier chapters. Early subject time could cover Chapters 2–5 in more depth and build toward Chapter 8. Some parts of Chapter 13 could be included, and Chapters 10–12 are optional. Chapter 17 provides a little light relief at the end.

Do We Teach Multivariate Methods?

In the past, the multivariate introduction in Chapters 14–16 was used if the audience included ecology students, particularly community ecologists. The chapters are a jumping-off point to an extensive literature. In recent years, we have seen multivariate methods embedded in "cookbooks" for genomics, proteomics, etc. We think it's valuable for students using these biological techniques to be aware of how data are being treated.

Ordinary Least Squares or Maximum Likelihood?

Some of the core chapters are somewhat bulky because we describe OLS and ML approaches to fitting and interpreting specific models. We do this because many traditional approaches taught to biologists are based around normal distributions and OLS, transforming data where necessary. Generalized linear model approaches rely heavily on ML estimation. In our "holistic" approach to linear models, the GLM/ML path is consistent and provides easier entry into GLMMs, additive models (GAMs), etc. However, for now, most students' backgrounds are more likely to be around OLS regression, ANOVA, etc.

We wouldn't teach these approaches in parallel. If students have a good background in OLS approaches, it can be prudent to use OLS, and introduce ML for advanced models where necessary. If we have a blank slate, it's tempting to use ML throughout. More advanced models can be challenging for students, and they can feel overwhelmed when trying to come to grips with ML (and wrestle with R) at the same time. We suggest being flexible, depending on how far down the path (e.g. to Chapter 8, 12, or 13) we're trying to get by the end of a course.

Advanced Graduate Students

Students in mid-to-late degree stages may already have completed some training and have experience handling data. They may be ready for more advanced work. In this case, Chapters 10–12, and parts of 13 and 9 could be the basis of an introduction to mixed models, with Chapters 1–8 as assumed knowledge. Other combinations of chapters can provide different emphases.

Acronyms

AIC	Akaike information criterion	GLMM	generalized linear mixed model; also GLME
AIC_C	small sample adjustment of AIC		
ALAN	artificial light at night	GLS	generalized least squares
ANCOVA	analysis of covariance	HSD	honestly significant difference
ANOSIM	analysis of similarities	i.i.d.	independently and identically distributed
ANOVA	analysis of variance		
AR	autoregressive	IRLS	iteratively reweighted least squares
ASE	asymptotic standard error	LAD	least absolute deviations
BENT	bad evidence no test	LASSO	least absolute shrinkage and selection operator
BF	Bayes factor		
BIC	Bayesian information criterion	LDA	linear discriminant analysis
BLUE	best linear unbiased estimator	LM	linear model
BLUP	best linear unbiased predictor	LME	linear mixed effect model; also LMM
BRT	boosted regression trees	LMG	Lindeman, Merenda, and Gold
CA	correspondence analysis	LOESS	locally weighted sums of squares; also sometimes LOWESS
CART	classification and regression trees		
CB	complete blocks	LR	likelihood ratio
CCA	canonical correspondence analysis	LSD	least significant difference
CI	confidence interval	MA	major axis
C-M-H	Cochran–Mantel–Haenszel (test)	MAD	mean absolute deviation
CR	completely randomized	MANOVA	multivariate analysis of variance
CV	coefficient of variation	MAR	missing at random
CWD	coarse woody debris	MCAR	missing completely at random
db-RDA	distance-based redundancy analysis	MCMC	Markov chain Monte Carlo
DC	deciles of risk	MDA	malondialdehyde
DCA	detrended correspondence analysis	MDES	minimum detectable effect size
df	degrees of freedom	MDS	multidimensional scaling
EM	expectation maximization	MI	multiple imputation
EMB	expectation maximization with bootstrapping	ML	maximum likelihood
		MNAR	missing not at random
EMS	expected mean squares	MRPP	multi-response permutation procedure
ES	effect size	MS	mean squares
FA	factor analysis	NB	negative binomial
FCS	fully conditional specification	NMDS	nonmetric multidimensional scaling
FDR	false discovery rate	N-P	Neyman–Pearson
GAM	generalized additive model	NR	Newton–Raphson
GAMM	generalized additive mixed model	OAW	ocean acidification and warming
GCV	generalized cross-validation	OLRE	observation-level random effect
GDM	generalized dissimilarity modeling	OLS	ordinary least squares
GLM	generalized linear model	OR	odds ratio

PC	principal component	SMA	standardized (or standard) major axis
PCA	principal components analysis	SNK	Student–Neuman–Keuls (test)
PCoA	principal coordinates analysis	SOC	soil organic carbon
PCR	principal components regression	SPLOM	scatterplot matrix
pdf	probability density function	SS	sum of squares
PERMANOVA	permutational analysis of variance	SSCP	sums of squares and cross products
PERMDISP	permutational comparison of dispersion	SVD	singular value decomposition
		SW	sum of (Akaike) weights
PEV	proportion of explained variance	TWINSPAN	two-way indicator species analysis
PLS	partial least squares	UBRE	unbiased risk estimator
PMVD	proportional marginal variance decomposition	UPGMA	unweighted pair-groups method using arithmetic averages
PSG	positively selected gene	UPGMC	unweighted pair-groups method using centroids
PUFAs	polyunsaturated fatty acids		
QDA	quadratic discriminant analysis	UPMC	unplanned pairwise multiple comparison
RA	reciprocal averaging		
RCB	randomized complete blocks	VC	variance component
RDA	redundancy analysis	VIF	variance inflation factor
REGW	Ryan–Einot–Gabriel–Welsch test	WHC	water-holding capacities
REML	restricted maximum likelihood	WLS	weighted least squares; see also GLS
RM	repeated measures	WPGMA	weighted pair-groups method using arithmetic averages
RMA	reduced major axis		
RMSE	root mean square error	ZA	zero-altered; also ZAB (binomial), ZAP (Poisson), and ZANB (negative binomial)
RT	rank transformation		
SD	standard deviation		
SE	standard error	ZI	zero-inflated; also ZIB (binomial), ZIP (Poisson), and ZINB (negative binomial)
SIC	Schwarz information criterion		
SIMPER	similarity percentages		

1 Introduction

As biologists, we are trying to explain natural phenomena. Whether in the lab or the field, at the level of a whole ecological community, or the metabolites within a cell, we want to know why things happen the way they do. We use these explanations for lots of things. They might help advise a government about how to conserve a threatened species or set levels of pollutants in the environment. They may be the jumping point for a new research proposal or scientific paper that will change the world. They may identify a problem that will be the basis of a spectacular thesis.

Biological explanations are not static or complete. They are our best attempts based on what we know, so they can always be improved with new information. They can, of course, also be wrong and need to be discarded for alternatives or modified dramatically. How hard we try to refine them will depend on how we're using the explanation. We usually decide whether a particular explanation or "model" is plausible by assessing how well it matches biological data. This step is often labeled statistical analysis and is commonly viewed as a standalone stage. Our approach is to step back and think about the entire process of what we do as biologists, and our thesis is that we can view the process as a series of models. Data analysis is one set of models in this series. It is integral but part of the process rather than a separate stage. We start by thinking about creating models to explain nature.

1.1 Almost Every Biological Theory or Hypothesis Is a Model of How Nature Works

Models are simplified descriptions of natural phenomena used to summarize our current understanding or to predict outcomes in new situations. They range from abstract verbal or qualitative descriptions to precise and complex mathematical descriptions. For example, we might have a model that estrogen plays a role in stimulating physical activity, that isoflurane is better than alpha chloralose for anesthetizing mice, or that aphid colonies with many members can afford to produce more soldiers. At the other extreme, advice around how many Atlantic cod should be caught this year comes from a complex set of mathematical models linking the expected number of cod remaining to fish, the life history of cod, and the function of the North Atlantic ecosystem.

Often, in the "classical" way science is seen as working, we start with a model or hypothesis. That model might come from previous knowledge, such as links between estrogen pulses and female activity or the physiological effects of anesthetics on mice, or it might come from observations of a pattern, such as variable proportions of soldier aphids. Our models are about processes that could produce the specific patterns in these examples. We want to know if those processes are important for understanding activity levels, use of anesthetic in experiments, or aphid biology. At other times, we don't have a particular model in mind, but we'd like one. For example, we might want to know what influences the diversity of birds in a forest fragment, but instead of a single specific cause, we might have a long list of potential influences, and our question is about the combination that best explains or lets us predict bird diversity. We have lots of candidate models, and we'd like to know the best or most useful one(s).

In the first situation, where we have pretty specific ideas, we can decide whether our model is credible by looking for or creating a situation that would expose the model's shortcomings:

• For mice and anesthetics, the best option is to expose mice to the current and the alternative anesthetics and compare their physiological responses. If our model is credible, we should see physiological differences. Changing the anesthetic should have little or no effect if it's wrong.

• For our aphids and soldiers model, we could observe aphid colonies of different sizes and see if colony size is related to soldiers, or, if possible, we could change colony size and see if the proportion of soldiers changes. To test this model, we also need to get more specific about how the proportion changes – is it linear, asymptotic, etc.?

• The estrogen model is a bit more complicated; if we just varied estrogen levels, there could be a range of physiological responses, so the experiment wouldn't challenge our explanation vigorously. Here, the model might be more detailed and linked to estrogen-sensitive neurons and a specific receptor (estrogen receptor-α). This more specific model could be assessed by disrupting this receptor. If the receptor knock-out has no effect, the model is wrong.

In the second situation, we have a list of potential influences (or, as we'll call them later, "predictors"), which might include the size of the forest patch when isolated, its degree of isolation, etc. That list might be long. We can devise many models, each with a combination of these influences, and we'd like to compare them.

1.2 We Use Data to Separate Wrong Models from (Possibly) Correct Ones (or Bad Models from Less Bad Ones)

There's nothing particularly scientific about the models we can conceive, but the scientific aspect is our efforts to separate poor models from better ones. We separate models by comparing them to data. Ideally, we collect data from a new situation with different conditions chosen so the alternative models vary in how well they would match the data. This confrontation between models and data is made by converting each biological model to a mathematical/statistical model and seeing how well it matches our data.

1.2.1 What Kind of Data Do We Need?
For each model, we can identify scenarios that could allow us to refute our suggested model:
• compare blood CO_2 in mice anesthetized in two ways (e.g. Low et al. 2016);
• choose different-sized aphid colonies and examine the proportion of soldiers (e.g. Shibao et al. 2004);
• compare nocturnal activity between control mice and those with estrogen receptor-α knocked out (e.g. Krause et al. 2021);
• examine bird diversity in forest patches where all our potential influences vary widely (e.g. Loyn 1987).

For simplicity, we'll focus on the first example. We're asking whether there's a difference in blood CO_2 between mice anesthetized with isoflurane and those with alpha chloralose, and you might have done a t test comparing the means for the two groups. We can also view our data analysis as comparing two models, one accounting for anesthetic type and one ignoring them. Looking at all our mice, we'll see variation in their blood CO_2 levels. That might come from genetic sources, their nutrition, their recent social interactions, and so on. For our experiment, this is unexplained variation – we're not controlling these sources completely or varying them in the experiment, nor are we trying to sort them out. We can estimate this background variation. Now, let's suppose that the anesthetic does matter. We can estimate how much it matters and see if knowing the anesthetic reduces the variation in blood CO_2. In the first model, the variation we see is just background "noise." In the second model, we still have background noise but also a potential "signal" from the anesthetic. We'd like to know if this first model, where there is no difference, is a good fit to the data, and we usually would also want to know if the second model fits the data better.

1.2.2 The Signal and the Noise
When we create our models, we're proposing that the biological processes that are the focus of our model generate a signal in the data, visible above the background noise.

We fit statistical models to decide if there is a strong signal among the noise:
• Is there evidence for a signal, or are we just describing noise?
• If we can detect a signal, how strong and important is it?

1.2.3 Random and Representative
We're making a jump here in translating our biological model into a mathematical description that describes what the data might look like if that model is the correct one. We're using the relationship between this statistical model and the data to decide whether our biological model might be close to correct or not. The data we collect are samples (see Chapter 3) from some broader "population" of possible observations. We take samples from these populations and use them to draw conclusions about the populations. Our samples must be truly representative of their populations, and we need to collect enough samples to allow us to distinguish between models. Being representative is best guaranteed by choosing or allocating randomly.

1.2.4 How Do We Decide If a Model Fits Well? What's the "Best" Model?

There are many ways to decide whether a model is good enough or better than its competitors. You'll have run across some of these approaches already, possibly with labels such as hypothesis testing, confidence intervals, and Bayesian. There is a long history of debate about different approaches. We don't have space to go into them, though we discuss them in a bit more depth in Chapter 2. Briefly, most have some merit and some weaknesses, and you may use one of them or perhaps even vary your approach with the biological task. You must understand what each approach involves, but one problem is that the same debate recurs, often surrounded by simplistic depictions of the opposing viewpoints. We recommend reading some of the more thoughtful literature on the topic. We like the explanations provided by Mayo (2018, 2021) and Gelman and Shalizi (2013a, b) as a starting point, but it would be easy to include many others (e.g. Efron & Hastie 2016; Hand 2021; Spanos 2019).

1.2.5 What Do I Do Next?

Regardless of the method you use, you must decide which models you'll retain and which you'll reject. That decision is profound, as it determines what you do next.

Suppose we decide that aphid colony size need not be in the model describing soldier production. There, we essentially throw away the original model. We go back to the original pattern again and look at one of the competing models of soldier production for our next step. But if our analysis leads us to decide that aphid colony size has an effect, we go down a different research path in which we might refine the model. We might, for example, propose some more sophisticated description of how colony size influences soldiers, perhaps focusing on social or environmental factors that might change the value of soldiers to a colony. Similarly, our decision about different anesthetics could influence future experimental practice and affect whether we discard the alternative anesthetic or refine its dosing regimen. The decision that we're making based on the data fundamentally changes the research we do next.

1.3 There Are Many Ways to Reach Wrong Conclusions!

"You can show anything with statistics." This statement trivializes the process, but analysts can vary in their conclusions. They may use different model-fitting approaches, which formulate the precise statistical version of the question differently. If you ask a different question of the data, don't be surprised if you get a different answer. This view is also a little dangerous because it encourages the notorious practice of "fishing" or P-hacking – trying lots of things until you get an answer you like (often a "statistically significant" result).

We find that, most often, you're misled because you have mismatches:
- Biological question and sampling or experimental design. You can't make the logical link between a pattern in your data and the biological model.
- Biological model and statistical model. Your statistical model does not describe the biological situation. You might get a clear answer from the analysis, but it won't answer your biological question.
- Statistical model and sampling or experimental design. You might have estimated model parameters and tested them, but you don't have confidence in those estimates.
- Statistical model and your data. Your statistical model has assumptions, but the results may be unreliable if the data do not meet them.

You might also not have enough data. You might, for example, decide that your idea is fine, but the data may have been so sparse and noisy that you can't tell whether the model matches the data or not. Rather than good/bad, your decision really should be good/don't know/bad.

You can also be unlucky: By chance, we get data that leads us to the wrong decision because our data are samples from populations, and often small samples.

You could also reach a wrong conclusion if the data have been modified. Fraud cases are relatively rare, but they can be influential.

These issues all lead to inferential problems – your data does not answer your question clearly.

Any of these missteps could cause our research recommendations or conclusions to be wrong. This error should not be a problem because science is supposed to be self-correcting – if we are sent down the wrong path, it's only until new data exposes that error. Any substantial claim would ideally have been challenged multiple times to survive. That's not the case in practice. You may have read recently of the "repeatability crisis," in which a substantial proportion of published papers across several disciplines did not have their conclusions supported when the work was replicated; for example, in psychology this was 33–47%, depending on how you measure it (Mayo 2021; Open Science Collaboration 2015). This is a worry; if we assume that the replicate studies were well designed, the original research likely reached the wrong conclusion – that is, the initial conclusions were misleading. As an aside, the errors were generally in favor of detecting "statistically significant" effects, not missing them. There's a second worrying finding. Despite results from these projects systematically checking conclusions, repeating studies to check their conclusions is frighteningly rare, well under 1% (e.g. Fidler et al. 2017), so it's not clear how often self-correction operates.

The solution is obvious: Do the best you can to avoid being misled in your own studies. Do everything possible in the lead-up to data collection to avoid being misled or getting an inconclusive answer.

1.4 There Are Also Many Ways to be Right

Just as there are many ways to be wrong or fool ourselves, there are often several ways to be right. Different statistical approaches lead to different formulations of statistical models and ways of confronting models with data. As we'll see repeatedly, even within the same statistical philosophy, there can be several ways to fit a particular statistical model to data.

In these cases there is not a right and a wrong approach, and often we finish at the same conclusion. There are common features to all approaches, such as clear logical links between biological models, structure of the data collection, and statistical models. It is critical that we decide on our approach and apply it critically and rigorously.

1.5 Our Philosophy in This Book

We've tried to sum up our approach in three simple statements. The elements aren't new, either in thinking about good ways to do science or good approaches to data analysis. However, formalizing the steps in the inferential process and being transparent about decisions reduces the chance of you becoming victim to some of the poor practices under renewed scrutiny. It also lessens the chance that you'll expend lots of your own time and resources but have little confidence in your conclusions.

1.5.1 Think Clearly

No data collection makes sense without a clear purpose. If it's to allow us to decide if a particular hypothesis should be refuted, be very clear about alternatives to that hypothesis. Try to produce an exhaustive list of alternative explanations and think about the situations, ideally experiments, that would allow us to separate the alternatives most unambiguously. Look specifically for alternatives that might produce the same pattern in the data as our main hypothesis. If they exist, we may need to

consider a different sampling design to separate them or use a multi-stage process whereby alternative explanations are whittled down. In experimental situations, this often involves careful thought about potential experimental artifacts, additional controls, etc. (see Chapter 3).

If data collection is to generate rather than explain patterns, the same careful thought is needed. What explanatory variables are plausible? Are they independent of each other? Can I measure variables where there might be a causal effect, or am I forced to measure "proxies" for these causal variables?

With experience, these steps may run in the background at the start of any project. As you learn these skills, it can help to formalize the process, stating the biological question, turning it into a hypothesis or model, listing the alternative hypotheses and how they can be distinguished, etc. Even for experienced researchers, this mental checklist is valuable. We aim to be our own harshest critics, trying to pick holes in our favored model or hypothesis.

1.5.2 Think in Advance

When we know what kind of sampling/experimental design will be necessary, it's time to think about the analysis workflow. This step involves decisions about sampling design and the appropriate response variables to measure. It's common to have several possible biological responses, each of which might give us a partial answer to our biological question. We should identify those that provide a clear answer and avoid generating a shopping list of possible responses – cherry-picking from a long list of variables to include in your paper is one of the unethical practices that have attracted renewed attention recently.

With a sampling design in mind, it's also time to formalize the statistical model that we'll use and the criteria we'll use to determine the match between model and data or between competing models. We need to carefully consider how well the statistical model's mathematical formulation matches the biological model. There will be assumptions to consider once we have the model and the model fitting in mind.

At this stage, we should also think about how much replication we need to get a clear answer. This step is likely to involve some calculation and may require preliminary data. It will also make us think about an important biological effect – we want to be sure that our data and analysis can detect meaningful effects.

By now, we have lots of the publication written. We have clear hypotheses, sampling designs, and we can also draft the data analysis section of the paper or the thesis's methods section. All we need are data!

1.5.3 Think before You Analyze

The planning stage involves some uncertain decisions. If the work is novel, we may not know how much background noise to expect, and we may not be sure whether any data will match the assumptions of our chosen statistical model. It's also possible that things may have gone wrong. Once we have the data, we need to look at them – look for mistakes, unusual values, check assumptions of planned analysis, etc. This is all done without considering the analysis outcomes. When you're satisfied, fit the statistical models. The pre-analysis can be lengthy, but it helps us make sensible decisions about the analysis when there are unexpected occurrences in the data.

Once we have the data analysis, we can draw some conclusions. It's important to stay faithful to the criteria we set up. Report the fit of the models you specified for the selected response variables. Tell the readers if you detected a signal among the noise and how strong and biologically important that signal was. Whether your biological idea was supported or refuted, tell us. Supply information about the uncertainty, so the audience knows that you have a conclusive answer, rather than retaining a particular idea because of weak evidence.

We can also compile a list of things not to do! Don't measure many things hoping that some will show a pattern, then report just those measures. Don't go back and run a different analysis or transform your data (i.e. change the model) in the hope of eventually finding a "statistically significant" result to report. Don't go back and restate your aims to match the results.

One way of avoiding some of these temptations is to preregister your study. Preregistration involves lodging the details above before the study (Nosek et al. 2018; Simmons et al. 2012) with your final paper or thesis being assessed against the preregistration details. Preregistration isn't the perfect solution – things will go wrong, and final studies may not match (Claesen et al. 2021; Goldacre et al. 2019) – but it does reduce many poor practices. It's a starting point where possible, rather than a prescription (Gelman & Loken 2014).

What we're doing is akin to Mayo's (2018) severe testing – making sure that we've looked hard for evidence for and against a particular idea and being open and transparent. Mayo emphasized three rather than two outcomes of our process – a claim (or hypothesis) might be refuted, it might be supported (for now), or it might have survived a challenge where the challenge itself is not severe. She defines a severely tested hypothesis as one

that "has been subjected to and passes a test that probably would have found flaws, were they present." Her third category is crucial, as it represents situations where we may have expended lots of energy, time, and resources but are left with uncertainty about the original questions because the test was not severe. She used the acronym BENT to describe situations of Bad Evidence, No Test; we find it useful. As mentioned earlier, these principles have been core elements of strong scientific practice for decades.

At its broadest, severe tests require us to cast a critical eye over every step of the process, from the biological idea to fitting a statistical model for the data, looking for places we can be misled or uncertain.

1.6 How This Book Is Structured

We've organized this book along the lines just outlined, with an emphasis on thinking clearly about the statistical models that we fit to data.

We start with a mini refresher (Chapter 2). We assume that you'll have taken an introductory statistics or data analysis course at some stage. That may have been some time ago, and it might have been when you couldn't quite see its point, so please start by making sure that you understand the basic concepts well. Be clear about the diverse kinds of data that biologists encounter (and the probability distributions of those data), how we fit models to data and estimate parameters, and how we decide. Some of these areas are contentious or frequently misunderstood. Even if you think you know this material, have a quick read through anyway.

Deciding from data requires several things. We need samples to be collected so they represent broader statistical populations (random sampling). When we design an experiment or sampling program to test whether a particular process is important, we need that sampling program to link clearly to the biological hypothesis and be capable of isolating it from competing explanations (confounding). We also need to understand the importance of replication, and the distinction between experimental or sampling units, which are true replicates, and observational units, which may not be independent replicates. We need enough replicates to be confident in our decisions – in Mayo's terminology, to make a severe test. Chapter 3 outlines the basic principles for sampling and experimental design.

We introduce you to most of the common statistical models biologists use. Traditionally, they have been taught under a bewildering array of names; each often had its technical requirements and language. We take the view that many are "flavors" of a single approach, which involves fitting linear models – using linear combinations of predictor variables to predict or explain values of a response variable. Chapter 4 introduces the concept of a linear model and describes the structure of several simple linear models with just a single predictor. These generalized linear models may be familiar to you already. Depending on whether the response and predictors are continuous or categorical, we can end up at linear and logistic regression, analysis of variance, or contingency table analyses.

Suppose we are using the fit of a statistical model to assess a biological one. We need to be sure that it's the right model for the data we collect and that any assumptions associated with that statistical model are satisfied. Chapter 5 describes various exploratory data techniques, which should be used routinely as a prelude to, and as part of, a full data analysis.

Chapters 6–13 build steadily through linear models of increasing complexity. We start by showing how we fit single predictor linear models to data, then expand to consider multiple predictors. Multiple predictors can be categorical, allowing factorial designs and assessing synergistic effects (interactions; Chapter 7), continuous, or a mixture of both (Chapter 8). As we increase the number of predictors, the number of possible models can be large. Chapter 9 describes approaches to finding the "best" model from a large range and determining how important each predictor is.

We increase complexity by considering the important distinction between fixed and random predictors, introducing hierarchical or multilevel designs where predictors are nested within other predictors (Chapter 10). This chapter introduces linear mixed models, where fixed and random predictors are present. Chapters 11 and 12 extend mixed models to common designs biologists use: splitplot (partly nested) and repeated measures designs. These designs are closely related and share features with nested designs of clustered observations, such as when a subject is recorded multiple times. Clustering in the data creates added complexity because observations within clusters may be correlated with each other. We use normally distributed data to illustrate most of these statistical models for simplicity. We finish this group of chapters by considering data that follow other distributions and response variables that are categorical. Chapter 13 introduces generalized linear mixed models, and we finish by showing the interested reader how to relax the linear requirement to use generalized additive models.

Biological data do not always have a single response variable, and Chapters 14–16 briefly introduce ways of handling multivariate data. While these methods are often used to identify patterns in the data, we also show how they can simplify large sets of predictor variables (Chapter 15). We also introduce a linear models approach to multivariate data (Chapter 16).

After all the hard work fitting models and deciding, we take a deep breath, relax a little, and think about how we can explain what we've done. Describing your analysis to close peers can be quite straightforward because of shared knowledge and experience, but we spend more and more of our time telling stories to different audiences. Chapter 17 has our recommendations for explaining complex analyses to these audiences, including practices we'd like to see more of and those we're happy not to see again.

1.7 A Bit of Housekeeping

In the first edition, we were agnostic concerning the choice of software packages. Biologists use a wide range of primarily commercial packages with different interfaces. We used one (SYSTAT), but we didn't provide code. The world has moved on, and most biologists now use the package R because it's free and supported by a massive user community. That community creates packages to do lots of statistical modeling tasks, which is wonderful, but it also means that several packages might be available for a single task. The user community is also active, so the core of R and its packages are updated many times per year. It is daunting for the average user to keep up.

We've decided to analyze all our worked examples using R. We've not presented that code in the book, as it would quickly become outdated (and we'd probably attract unwanted attention to our coding skills). We have, however, provided the code online, often with more detail than in the book. Throughout the book, you'll see the symbol ® when there's code available online and ® when the online version of a box contains additional material, including images. Keeping the R code online will also let us improve the code and update it for new R versions. Good introductions to using R for statistical analyses include Jones et al. (2022), Winter (2019), and Borcard et al. (2018), the latter focusing on multivariate analyses.

We've provided lots of worked examples taken from the biological literature. The practice of many scientific journals requiring data files to be provided is an important step in increasing the rigor of biological research, and it's great when looking for examples and raw data. For all our worked examples, we've run our own analysis. The website includes the code for all these examples, including lots of the exploratory steps. We also include a link to the data repository where the raw data are stored, or sometimes (often examples from the first edition) we include data files supplied to us by the authors. We've tried as much as possible to follow the original authors' decisions about which models to fit so that readers can see our vignette, and if interested, see the whole context for the study and how the authors interpreted the analyses. Sometimes we've used subsets of the data or done something slightly different, or there wasn't quite enough detail in the original paper to replicate the analyses. That's all detailed in the worked example boxes.

2 Things to Know before Proceeding

We assume that you've already learned something about statistical inference and data analysis. However, introductory classes vary, and we don't know when you had that training, what you learned, and what you've forgotten. In this chapter, we outline statistical concepts we'll rely on. Some of these concepts have been misunderstood or misused, so we suggest that you check you understand each one clearly. We've provided a brief summary and links to deeper coverage where necessary.

2.1 Samples, Populations, and Statistical Inference

Biologists usually wish to draw conclusions about statistical populations, defined as all the possible units and their associated observations of interest. A collection of observations from any population is a sample, and the number of observations in it is the sample size. Characteristics of the sample are called statistics (e.g. sample mean or sample regression slope), and characteristics of the population are called parameters (e.g. population mean or population regression slope).

The basic way of generating a sample is random sampling. Here, every observation has the same probability of being included in the sample. Further information on random and other types of sampling is provided in Chapter 3. Nearly all applied statistical procedures that use samples to draw conclusions about populations assume some form of random sampling. If the sampling is not random, we are never sure what broader population our sample represents.

Sample data in biology are usually arranged as:
• objects, which are sampling or experimental units (or smaller-scale observation units) and are commonly the cases or rows in our data file; and

• variables, which are the characteristics we measure or record from each object and are commonly the columns in our data file.

Variables can be continuous, taking any value within a range (such as weight) or discrete, taking only specific values (such as counts or presence/absence). Discrete variables can be ordered along a continuous scale (e.g. counts), with so many possible values that they are essentially continuous. They can also follow an ordinal scale (e.g. levels of agreement in questionnaires) or be unordered (nominal scale). A random variable is defined informally as one whose value we don't know until after our samples have been collected, but whose values can be assigned probabilities of occurrence.

The most important characteristic of biological data is their uncertainty. For example, let's take two samples, each consisting of the same number of observations, from a population and estimate the mean for some variable. The two means will almost certainly be different, despite the samples coming from the same population. This difference makes us uncertain about the population mean. The sample means can differ for two reasons (Hilborn & Mangel 1997). Process uncertainty results from the true population mean being different when each sample was taken. In some biological systems, such short-term variation can be common. Observation uncertainty results from sampling error; the mean value in a sample is an imperfect estimate of the population mean. Because of natural variability between observations, different samples nearly always produce different means. Observation uncertainty can also result from measurement error, where we use imperfect measuring devices. For many biological variables, natural variability is so great that we rarely worry about measurement error, but we should keep both types in mind.

Statistical inference deals with uncertainty, particularly drawing conclusions about populations from sample data and detecting any signal (i.e. biological effect) among the noise of variation. In the next section, we describe how quantifying uncertainty as probability and probability distributions is fundamental to statistical inference. Based on different interpretations of probability, we then cover the "classical" frequentist approach to statistical inference in some detail, focusing on estimating parameters and testing hypotheses, followed by a brief introduction to Bayesian inference.

2.2 Probability

In most statistical analyses, we view uncertainty in terms of probabilities, and understanding probability is crucial to understanding modern applied statistics. We will only briefly introduce probability here, but there are good general introductions (Antelman 1997; Kruschke 2011) from a biological (Hilborn & Mangel 1997; Nakajima 2013; Sokal & Rohlf 2012) or philosophical (Mayo 2018) perspective.

We usually talk about probabilities in terms of events; the probability of event A occurring is written $P(A)$. Probabilities can be between 0 and 1; if $P(A) = 0$, the event is impossible; if $P(A) = 1$, the event is inevitable. As a simple example and one used in nearly every introductory statistics book, imagine the toss of an unbiased coin. The sample space is the set of possible events or outcomes; with the coin-tossing example, the sample space is heads or tails (assuming a coin cannot land on its side). We can assign a probability to each event in the sample space, and most of us would state that $P(\text{heads})$ is 0.5 and $P(\text{tails})$ is 0.5, but what do we really mean by this statement?

There are two ways we can think about probability. The classical interpretation of probability is the relative frequency of an event we expect in the long run or in a long sequence of identical trials. In the coin-tossing example, the probability of heads being 0.5 is the expected proportion of heads in a long sequence of tosses. Problems with this long-run interpretation include defining what is meant by identical trials and the many situations in which uncertainty has no sensible long-run frequency interpretation – for example, the probability of a horse winning a particular race or rain tomorrow. The long-run frequency interpretation, termed frequentist or objective probability, is the more common one. It is how we must interpret confidence intervals and P-values from analyses.

The alternative way of interpreting probabilities is much more subjective and is based on a "degree of belief" about whether an event will occur. It is basically an attempt to quantify an opinion. It includes two slightly different approaches – logical probability, developed by Carnap and Jeffreys, and subjective probability, pioneered by Savage. The latter is a probability measure specific to the person deriving it (see Mayo 2018). The opinion on which the probability is based may come from previous observations, theoretical considerations, knowledge of the particular event under consideration, etc. This approach to probability has been criticized because of its subjective nature, but it has been widely applied in developing prior probabilities in the Bayesian approach to statistical analysis (Section 2.7).

Whether we are talking about frequentist or subjective interpretations, numerical probabilities used in statistical analyses must satisfy Komolgorov's properties (Kruschke 2011):
- They must be nonnegative.
- The sum of all probabilities in the sample space equals 1: $P(\text{heads}) + P(\text{tails}) = 1$.
- The probability that either of two mutually exclusive events occurs is the sum of their probabilities: $P(\text{heads or tails}) = P(\text{heads}) + P(\text{tails})$.

Another key concept is conditional probability, the probability of one event occurring given that a second event has occurred or is certain to occur. Conditional probability is expressed as $P(A|B)$, the probability of event A given event B. Conditional probability can be defined as:

$P(A|B) = P(A \text{ and } B)/P(B)$, that is, the joint probability of A and B occurring divided by the probability of B occurring.

Another way of formulating conditional probability is to use Bayes' theorem:

$P(A|B) = P(A) \times P(B|A)/P(B)$, that is, the probability of A times the probability of B given A divided by the probability of B.

When A is a hypothesized value of some parameter of interest and B is our sample data, then conditional probability and Bayes' theorem are fundamental to Bayesian statistical inference, introduced in Section 2.7.

2.3 Probability Distributions

A random variable (Y) will have an associated probability distribution that describes the relative probabilities of its possible values (the sample space). For discrete variables, the probability distribution will comprise a measurable probability for each outcome, such as 0.5 for heads and 0.5 for tails in a coin toss, or 0.167 for each one of the six sides of a fair die. Continuous variables are not restricted

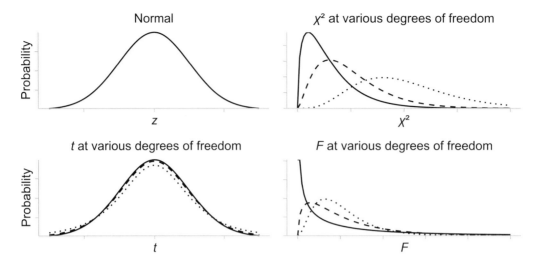

Figure 2.1 Probability distributions for four common statistics. For the t, χ^2, and F distributions, we show distributions for different degrees of freedom to show how the shapes of these distributions change.

to specific values, so there are infinite possible outcomes. The probability of any specific value of a continuous random variable is zero, so we usually talk about the probability associated with a range of values, represented by the area under the probability distribution curve between the two ends of this range. The probability density function (pdf) is the probability associated with a unit change at any specific value of the variable. The distribution is usually normalized so the total probability under the curve equals 1.

In many of the statistical analyses described in this book, we deal with two or more variables, and our statistical models will often have more than one parameter. Then we may need to switch from single probability distributions to joint probability distributions, where probabilities are measured not as areas under a single curve, but volumes under a more complex distribution. A common joint pdf is the bivariate normal distribution.

Probability distributions nearly always refer to the distribution of variables in populations. The expected value of a random variable $[E(Y)]$ is simply the mean (μ) of its probability distribution. The expected value is an important concept – many statistical modeling procedures are trying to model the expected value of a random response variable. The mean is a measure of the center of a probability distribution. Other measures include the median (the middle value) and the mode (the most common value). It is also important to measure the spread of a distribution. The most common measures are based on deviations from the center; for example, the variance is measured as the sum of squared deviations from the mean.

2.3.1 Distributions for Variables

Most statistical procedures rely on knowing the probability distribution of a variable we are analyzing. We can define many probability distributions mathematically and some are important for common biological situations. We will introduce three common distributions from the exponential family, a group of distributions that are mathematically convenient for statistical modeling but also are realistic for many biological variables; we will expand on the exponential family in Chapter 4.

Let's consider continuous variables first. The normal (also termed Gaussian) distribution is a symmetrical probability distribution with a characteristic bell shape (Figure 2.1). It is defined as:

$$f(y) = \frac{1}{\sqrt{2\pi\sigma^2}} e^{-(y-\mu)^2/2\sigma^2},$$

where $f(y)$ is the probability density of any value y. The normal distribution can be defined simply by its mean (μ) and variance (σ^2), which are independent of each other. A normal distribution is often abbreviated to $N(Y: \mu, \sigma^2)$ or $N(\mu, \sigma^2)$. There are infinitely many possible combinations of mean and variance, so there is an infinite number of possible normal distributions. The standard normal distribution (z distribution) is a normal distribution with a mean of 0 and a variance of 1. The normal distribution is one of the most important probability distributions for data analysis. Many common statistical procedures in biology (e.g. linear models) assume an underlying normal distribution.

The normal distribution is symmetrical, but continuous variables can have asymmetrical distributions. Variables limited by zero but having no upper limit are often positively skewed with long right-hand tails. Transformations to logarithms are often used in this situation, and other distributions that might be useful are presented in Box 4.2.

There are also probability distributions for discrete variables. There are two possible outcomes if we toss a coin – heads or tails. Processes with only two possible outcomes are common in biology (e.g. animals live or die, or a particular tree species can be present or absent from samples from a forest). A process that can have only one of two outcomes is sometimes called a Bernoulli trial, and we often call the two outcomes success and failure. We will only consider a stationary Bernoulli trial, one where the probability of success is the same for each trial – that is, the trials are independent.

The probability distribution of the number of successes in n independent Bernoulli trials is called the binomial distribution, another important distribution in biology:

$$P(y = r) = \frac{n!}{r!(n-r)!} \pi^r (1-\pi)^{(\pi-r)},$$

where $P(y = r)$ is the probability of a particular value (y) of the random variable (Y) being r successes out of n trials, n is the number of trials, and π is the probability of a success. Note that n is fixed, and therefore the value of a binomial random variable cannot exceed n. The binomial distribution can be used to calculate probabilities for different numbers of successes out of n trials, given a known probability of success on any individual trial. It is also an important distribution for generalized linear modeling variables with binary outcomes. A generalization of the binomial distribution to when there are more than two possible outcomes is the multinomial distribution, which is the joint probability distribution of multiple outcomes from n fixed trials.

Another very important probability distribution for discrete variables is the Poisson distribution. It usually describes variables representing the (typically small) number of occurrences of a particular event, such as counts of organisms in a plot, cells in a microscope field of view, or seeds taken by a bird per minute. The pdf of a Poisson variable is:

$$P(y = r) = \frac{e^{-\mu}\mu^r}{r!},$$

where $P(y = r)$ is the probability that the number of occurrences (y) of an event equals an integer value (r)

and μ is the mean (and variance) of the number of occurrences. A Poisson variable can take any integer value between zero and infinity because the number of trials is not fixed, in contrast to the binomial and the multinomial. One of the characteristics of a Poisson distribution is that the mean (μ) equals the variance (σ^2). For small values of μ, the Poisson distribution is positively skewed, but once μ is greater than about five, the distribution is symmetrical.

2.3.2 Distributions for Statistics

The remaining probability distributions to examine are used for determining probabilities of statistics derived from our sample data. These are termed sampling distributions and are used extensively for confidence intervals (CIs) and hypothesis testing. Four particularly important ones (Figure 2.1) are:

1. The z (standard normal distribution) represents the probability distribution of a random variable with a mean of 0 and a variance of 1, as described above; it is the ratio of the difference between a sample statistic and its population value to its population standard deviation.

2. Student's t distribution represents the probability distribution of a random variable that is the ratio of the difference between a sample statistic and its population value to the sample standard deviation. The t distribution is symmetrical, like a normal distribution. Its shape becomes more similar with increasing sample size. We can often convert a single sample statistic to a t value and use the t distribution to determine the probability of obtaining that t (or one smaller or larger) for a specified population parameter value.

3. χ^2 (chi-square) distribution represents the probability distribution of a variable that is the square of values from a standard normal distribution and so can't be negative. Variances have a χ^2 distribution, so this distribution can be used for interval estimation of population variances.

4. F distribution represents the probability distribution of a variable that is the ratio of two independent χ^2 variables, each divided by its degrees of freedom (df). Because variances are distributed as χ^2, the F distribution is used for comparing variances.

Different versions of the latter three distributions are used, depending on the degrees of freedom associated with the sample or samples (see Box 2.1).

2.4 Frequentist ("Classical") Estimation

Estimating population parameters from sample data is, or at least should be, the key aspect of statistical analyses in biology. What are we after when we estimate population

Box 2.1 Degrees of Freedom

Degrees of freedom is a term that biologists use all the time in statistical analyses, but few probably really understand. We will attempt to make it a little clearer. The degrees of freedom are the number of observations in our sample(s) that are "free to vary" when we are estimating a parameter. For example, we use $n - 1$ as the df to estimate the standard deviation because we have used 1 df to estimate the mean, so there are $n - 1$ remaining. More generally, when fitting statistical models, we need to adjust the df to avoid misrepresenting the error associated with predictions from the model (Gelman et al. 2020). For a model with p parameters to be estimated based on a sample size of n, the df is usually $n - p$.

parameters? A good estimator should be unbiased (it doesn't consistently over- or underestimate the parameter), consistent (so as the sample size increases the estimator will get closer to the population parameter), and efficient (meaning it has the lowest variance among competing estimators).

There are two broad types of estimation:
• Point estimates provide a single value that estimates a population parameter.
• Interval estimates provide a range of values that might include the parameter with a known probability. The most common interval estimate is the CI, a range of values within which we have some level of confidence that the true value of the parameter might lie. The CI is derived from the standard error (SE), a measure of how close our point estimate is likely to be to the true population parameter – that is, the precision of our point estimate. Standard errors and CIs are based on extrapolating from our single sample of observations to what we would expect if we took repeated samples from the population.

2.4.1 Methods for Estimation

There are numerous methods for estimating parameters of populations. Imagine we wish to estimate the average value of a continuous variable in a population (i.e. the population mean (μ)), based on a random sample of observations. Two common estimation methods are:
• Ordinary least squares (OLS). The OLS estimator for a parameter is the one that minimizes the sum of the squared differences between each sample observation and the estimate of μ. OLS point estimation doesn't have any distributional assumptions but assumes a normal distribution for CIs.
• Maximum likelihood (ML). The ML estimator for a parameter is the one that maximizes the likelihood of observing the sample data. To determine an ML estimator for μ, we need to generate the likelihood of observing our sample data for different possible values of μ, termed the likelihood function (Figure 2.2). Maximum likelihood

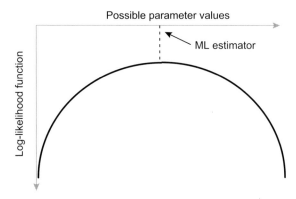

Figure 2.2 Generalized log-likelihood function for estimating a parameter.

point and interval estimation assumes a known distribution for the data, but in contrast to OLS the distribution doesn't have to be normal. The principles of ML, especially the likelihood function, also have a fundamental role in Bayesian inference (Section 2.7).

We will provide more details on OLS and ML estimation in the context of linear models in Section 4.4 and Box 4.3.

The OLS and ML estimates are sensitive to unusual observations, especially with small sample sizes, and robust methods have been developed for dealing with these situations. These methods are commonly based on downweighting more extreme observations, so they are less influential. Some of these robust methods are described in the next section. Ordinary least squares and ML interval estimation (e.g. CIs) assumes we know the underlying probability distributions for the population from which our data arose. This is not always possible and there are methods for estimation of parameters that don't assume any distribution based on resampling from the original sample – that is, we use the original sample to generate the repeated samples from which our frequentist inference is based. These resampling estimation methods are described in Section 2.4.7.

Table 2.1 Common population parameters and sample statistics

Parameter	Statistic	Formula
Mean (μ)	$\bar{y}$	$\dfrac{\sum_{i=1}^{n} y_i}{n}$
Median	Sample median	$y_{(n+1)/2}$ if n odd $\left(y_{n/2} + y_{(n/2)+1}\right)/2$ if n even
Variance (σ^2)	s^2	$\dfrac{\sum_{i=1}^{n} (y_i - \bar{y})^2}{n-1}$
Standard deviation (σ)	s	$\sqrt{\dfrac{\sum_{i=1}^{n} (y_i - \bar{y})^2}{n-1}}$
Median absolute deviation (MAD)	Sample MAD	$\text{Median}(|y_i - \text{median}|)$
Coefficient of variation (CV)	Sample CV	$\frac{s}{\bar{y}} \times 100$
Standard error of $\bar{y}$ ($\sigma_{\bar{y}}$)	$s_{\bar{y}}$	$\frac{s}{\sqrt{n}}$
95% confidence interval for μ		$\bar{y} - t_{0.025(n-1)} \frac{s}{\sqrt{n}} \le \mu \le \bar{y} + t_{0.025(n-1)} \frac{s}{\sqrt{n}}$

2.4.2 Simple Parameters and Statistics

Consider a population of observations of the variable Y recorded from all N sampling units in the population. We take a random sample of n observations $(y_1, \ldots y_i, \ldots y_n)$ from the population. We usually would like information on the population's location or central tendency (i.e. where is its middle?) and some measure of spread (i.e. how different are observations in the population?). Common estimates of location and spread parameters are given in Table 2.1 and illustrated in Box 2.2.

2.4.2.1 *Center (Location) of Distribution*

The most common estimators for the center of a distribution are based on the sample data being ordered from smallest to largest (order statistics) and then forming a linear combination of weighted order statistics. The best known is the sample mean ($\bar{y}$), which is an unbiased estimator of the population mean (μ), where each observation is weighted by $1/n$ (Table 2.1). Assuming a normal distribution, the sample mean is the OLS and ML estimator of the population mean. It is a reliable measure of location for symmetrical distributions, but less so for skewed distributions or when unusually small or large observations (outliers; see Section 5.2) are present. Alternatives include:

- The median is the middle measurement of a set of data sorted in order of magnitude (i.e. ranks). The median is an unbiased estimator of the population mean for normal distributions, is a better estimator of the center of skewed distributions, and is more resistant to outliers.

- The trimmed mean is calculated after omitting a proportion (commonly 5%) of the highest (and lowest) observations, usually to deal with outliers.
- The Winsorized mean is like the trimmed mean, except the omitted observations are replaced by the nearest remaining value.

Another approach is to use M-estimators. The weightings given to the different observations change gradually from the middle of the sample and incorporate a measure of variability in the estimation procedure. They include the Huber M-estimator and the Hampel M-estimator, which use different functions to weight the observations. They may be useful when outliers are present because they downweight extreme values. Wilcox (2022) provides full details.

Finally, we can use R-estimators, based on the observations' ranks rather than the observations themselves. They form the basis for many rank-based "nonparametric" tests. The main one is the Hodges–Lehmann estimator, which is the median of the averages of all possible pairs of observations.

Our experience is that biologists mostly focus on estimating the mean of a population or use the median for nonsymmetrical distributions.

2.4.2.2 *Spread or Variability*

Table 2.1 shows common measures of sample spread. The range, the difference between the largest and smallest observation, is the simplest, but there is no clear link between the sample range and the population range. In general, the sample range rises with sample size. The

Box 2.2 Ⓡ Ⓔ Worked Example of Traditional Inference for Two-Group Design: Anesthetic Effects on Mice

Low et al. (2016) examined the effects of two different anesthetics on aspects of mouse physiology. Twelve mice were anesthetized with isoflurane. Eleven mice were anesthetized with alpha chloralose. The two anesthetics represent our categorical predictor variable or factor, and blood CO_2 levels were recorded after 120 minutes as our response variable. Exploration of the data was based on boxplots of the response for the two groups. They indicated some skewness in the distribution of CO_2 and a wider spread for alpha chloralose, one unusually large value for isoflurane, and blood CO_2 levels were generally higher under alpha chloralose (see Figure 5.5).

With this type of study design there are at least three aims of the statistical analysis. First, we assume that these two groups of mice represent random samples from two populations of mice, one anesthetized with isoflurane and one anesthetized with alpha chloralose. We want to estimate the characteristics of the two populations. The common sample statistics are as given:

Anesthetic	n	Mean	Median	SD	Variance	SE	95% CI for mean
Isoflurane	12	50.0	50.5	11.4	129.8	3.3	42.8–57.2
Alpha chloralose	11	70.9	64.0	20.2	408.1	6.1	57.3–84.5

Note the interpretation of CIs. For isoflurane, we say that 95% of all CIs resulting from repeated sampling from this population with this sample size will contain the population mean and 5% won't. The frequentist extension of this logic is that there is a 95% probability that the interval 42.8–57.2 will contain the population CO_2 mean.

Second, we would be interested in the size of the effect of the anesthetic treatment on blood CO_2, an estimate of the difference between population group means. The raw difference between the means is 20.9 (SE: 6.76; 95% CI: 35.0–6.9), a 42% increase in CO_2 concentration by using alpha chloralose over isoflurane. The effect size (ES), standardized by the standard deviation pooled across groups, is 1.29 (95% CI = 0.33–2.25). The ES could also be standardized by a pooled SE, resulting in a t statistic.

Third, we might want to test null hypotheses formally with these data. A formal test of equal population variances for the two groups did not suggest strong evidence against the null (Levene median test: $F_{1,21} = 2.604, P = 0.122$; Section 5.3.2), even though one variance was three times greater than the other. If we base our conclusion on statistical "significance" with an $\alpha = 0.05$, we would not reject H_0.

The hypothesis of main interest is about group means and a pooled variance t test of the H_0 of no difference between the anesthetics in population mean blood CO_2 levels indicated strong evidence against H_0 ($t = -3.09, \text{df} = 21, P = 0.006$). This P-value is the probability from repeated sampling from these two populations with these sample sizes of getting a t value (and therefore a mean difference) of this magnitude or greater. We would reject the H_0 based on a threshold of $\alpha = 0.05$.

We noted that the variance in CO_2 for alpha chloralose was three times larger than for isoflurane. While our formal test of equality of variances did not suggest strong evidence for a difference, a separate variance t test might be more robust; our conclusions are not affected ($t = -3.02, \text{df} = 15.49, P = 0.008$).

There is little justification for using a nonparametric (rank-based) test of H_0 that there was no difference in a more general measure of location, as only the distribution for alpha chloralose suggested asymmetry. For illustration, the Mann–Whitney–Wilcoxon test resulted in a U statistic of 114 and $P = 0.003$.

sample variance, which estimates the population variance, is an important measure. The numerator of the formula is called the sum of squares (SS, the sum of squared deviations of each observation from the sample mean) and the variance is the average of these squared deviations. Note we might expect to divide by n to calculate an average (this is the ML estimator of the variance), but then s^2

consistently underestimates σ^2, that is, it is biased. Instead, we divide by $n - 1$ (the degrees of freedom; Box 2.1) to make it an unbiased estimator; it is now a restricted ML (REML) estimator; we will describe REML estimation for variances in more detail in Chapter 10.

The one difficulty with s^2 is that its units are the square of the original observations; for example, if the observations are lengths in mm, then the variance is in mm², an area, not a length. The sample standard deviation s, which estimates σ, the population SD, is the square root of the variance. In contrast to the variance, the SD is in the same units as the original observations.

The coefficient of variation (CV) is used to compare standard deviations between populations with different means, and it provides a measure of variation independent of the measurement units. The sample CV describes the SD as a percentage of the mean; it estimates the population CV.

Some measures of spread are more robust to skewed distributions and those with unusual observations, including:
- The median absolute deviation (MAD) is the sensible measure of spread to present in association with medians.
- The interquartile range is the difference between the first quartile (the observation which has 0.25 or 25% of the observations below it) and the third quartile (the observation which has 0.25 or 25% of the observations above it). It is used in constructing boxplots (Chapter 5).

Rowland et al. (2021) offer a nice summary of these measures, along with a flowchart to guide their use.

2.4.3 Sampling Distribution of the Mean

Having a point estimate is only the first estimation step. We also need to know our estimate's precision. If its value varies widely under repeated sampling, it will not be especially useful for frequentist inference. If repeated sampling produces a very consistent estimator, it is precise, and we can be more confident that it is close to the population parameter (if it is unbiased). The logic for determining the precision of estimators is well covered in most introductory texts, so we will describe it only briefly, using normally distributed variables as an example.

Assume that our sample has come from a normally distributed population. We can standardize this distribution using μ and σ:

$$z = \frac{y_i - \mu}{\sigma}.$$

For a standard normal distribution, half of the population falls between -0.674 and 0.674, 95% falls between -1.96 and 1.96, and 99% between -2.576 and 2.576. The z distribution can be used to calculate any sized interval; for example, in our sample, 95% of the population falls between $\mu - 1.960\sigma$ and $\mu + 1.960\sigma$ (Figure 2.3).

Usually, we only deal with a single sample (with n observations) from a population. If we took many samples from a population and calculated all their sample means,

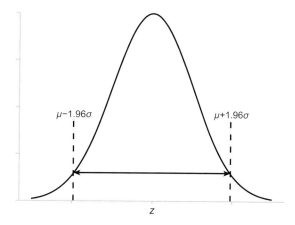

Figure 2.3 Plot of normal probability distribution, showing points between which values 95% of all values occur.

we could plot the frequency (probability) distribution of the sample means, because the sample mean is itself a random variable. This sampling distribution of the mean has three important characteristics.
- The probability distribution of means of samples from a normal distribution is also normally distributed.
- As the sample size increases, the probability distribution of means of samples from any distribution will approach a normal distribution. This result is the basis of the central limit theorem.
- The expected value or mean of the probability distribution of sample means is the mean of the population (μ) from which the samples were taken.

2.4.4 Standard Error of the Sample Mean

Let's consider the sample means to have a normal probability distribution. We can calculate their variance and standard deviation, just like we could calculate the variance of the observations in a single sample. The expected value of the SD of the sample means is:

$$\sigma_{\bar{y}} = \frac{\sigma}{\sqrt{n}},$$

where σ is the SD of the original population from which the repeated samples were taken, and n is the size of each sample.

We rarely have many samples from the same population, so we estimate the SD of the sample means from our single sample. The SD of the sample means is called the standard error of the mean:

$$s_{\bar{y}} = \frac{s}{\sqrt{n}},$$

where s is the sample estimate of the SD of the original population and n is the sample size.

The SE of the mean is a measure of the precision of our sample mean. It is termed "error" because it tells us about the error in using $\bar{y}$ to estimate μ. If the SE is large, repeated samples would likely produce quite different means, and any single sample mean might not be close to the true population mean. If the SE is small, repeated samples would likely produce similar means, and we would be more confident that any specific sample mean is a good estimate of the population mean.

2.4.5 Confidence Intervals for Population Mean

In Section 2.4.3, we converted any value from a normal distribution into its equivalent value from a standard normal distribution, the z score. Equivalently, we can convert any sample mean into its equivalent value from a standard normal distribution of means using:

$$z = \frac{(\bar{y} - \mu)}{\sigma_{\bar{y}}}.$$

The denominator is simply the SD of the mean, $\sigma/\sqrt{n}$, or SE. Because this z score has a normal distribution, we can determine how confident we are in the sample mean, that is, how reliable it is as an estimate of the true population mean. We simply calculate values in our distribution of sample means between which a percentage (often 95% by convention) of means occurs. As we showed above, 95% of sample means fall in the range $\mu \pm 1.96\sigma_{\bar{y}}$.

Now we can combine this information to make a CI for μ:

$$P(\bar{y} - 1.96\sigma_{\bar{y}} \leq \mu \leq \bar{y} + 1.96\sigma_{\bar{y}}) = 0.95.$$

This CI is an interval estimate for the population mean, although the probability statement is actually about the interval, not about the population parameter, which is fixed. We will discuss the interpretation, and common misinterpretation, of CIs below. The only problem with this equation is that we usually don't know σ, so we also don't know $\sigma_{\bar{y}}$. We can only estimate the SE from s. Our standard normal distribution of sample means is now the distribution of $(\bar{y} - \mu)/s_{\bar{y}}$. This is the random variable t (Section 2.3.2). A t distribution is flatter and more spread than a normal distribution (Figure 2.1). We use it to calculate CIs for the population mean (and commonly parameters from linear models) in the usual situation of not knowing the population SD.

The shape of the t distribution varies with sample size, or, more accurately, the degrees of freedom, $n-1$ (see Box 2.1). This is because s provides an imprecise estimate of σ if the sample size is small, increasing in precision as the sample size increases. When n is large

(say, >30), the t distribution is very similar to a normal distribution (because our estimate of the SE based on s will be close to the real SE). Remember, the z distribution is simply the probability distribution of $(y - \mu)/\sigma$ or $(\bar{y} - \mu)/\sigma_{\bar{y}}$ if we are dealing with sample means.

The CI (in this case 95% or 0.95) for the population mean then is:

$$P(\bar{y} - t_{0.025(n-1)} s_{\bar{y}} \leq \mu \leq \bar{y} + t_{0.025(n-1)} s_{\bar{y}}) = 0.95,$$

where $\pm t_{0.025(n-1)}$ is the value from the t distribution with $n-1$ df between which 95% of all t values lie, and $s_{\bar{y}}$ is the SE of the mean. The interval size will depend on the sample size and SD, both of which are used to calculate the SE, and the required level of confidence.

Using 95% as a threshold for CIs is common practice, providing a direct connection to the standard use of the 5% threshold for statistical significance when testing null hypotheses (Section 2.5). We can use this equation to determine CIs for different levels of confidence – for example, for 99% CIs, simply use the t value between which 99% of all values lie. The 99% CI will be wider than the 95% CI.

2.4.5.1 Interpretation of Confidence Intervals for Population Mean

We usually do not consider μ a random variable but a fixed, albeit unknown, parameter. Therefore, the CI is actually not a probability statement about the population mean. We are not saying there is a 95% probability that μ falls within this specific interval we have determined; μ is fixed, so the CI we have calculated for a single sample contains μ or it doesn't. The probability associated with CIs is interpreted as a long-run frequency. Different random samples from the same population will give different CIs. If we took 100 samples of size n and calculated the 95% CI from each sample, 95 of the intervals would contain μ and five wouldn't. Antelman (1997) summarized a CI succinctly as "one interval generated by a procedure that will give correct intervals 95% of the time."

2.4.6 Standard Errors and Confidence Intervals for Other Statistics

The SE is simply the SD of the probability distribution of a specific statistic, so we can calculate SEs for statistics besides the mean. Sokal and Rohlf (2012) listed the SE formulae for many statistics. We can use the methods just described to reliably determine SEs for statistics (and CIs for the associated parameters) that assume normality. When divided by their SE, these statistics follow a t distribution, and we can calculate CIs.

Box 2.3 ⓡ Worked Example of Resampling Inference for Two-Group Design: Anesthetic Effects on Mice

We will illustrate resampling estimation and randomization tests with the mouse anesthetic data and compare with the methods from Box 2.2. If you run the bootstrap and randomization test yourself, you won't get precisely the same answer, as is usually the case for resampling/randomization procedures.

Anesthetic	n	Mean	Jack SE	Boot SE	95% CI Boot perc	95% CI Boot BCA
Isoflurane	12	50.0	3.29	3.11	43.58–57.87	44.42–59.96
Alpha chloralose	11	70.9	6.09	5.75	59.09–85.36	60.18–87.34

The jackknife SEs of the mean are essentially the same as in Box 2.2. The bootstrap SEs based on 9,999 resamples are slightly smaller. The bootstrap percentile CIs are narrower than the bias-corrected and accelerated intervals, although the latter are more reliable.

We can also use randomization to test the H_0 that there is no difference in mean CO_2 between anesthetics, so that any possible allocation of observations to the two groups is equally likely. This test minimizes the effect of the underlying assumptions of the t test from Box 2.2. The resultant z statistic was 2.62 with $P = 0.009$ for a difference as large or more extreme than observed, based on 9,999 randomizations.

One important exception is the sample variance, which has a known distribution that is not normal. To derive CIs for the population variance, we can use the chi-square (χ^2) distribution, which is the distribution of the random variable:

$$\chi^2 = \frac{(y-\mu)^2}{\sigma^2}.$$

This is simply the square of the standard z score discussed above. The χ^2 distribution is a sampling distribution, so, like the random variable t, there are different probability distributions for χ^2 for different sample sizes; this is reflected in the df ($n-1$). The probability distribution is skewed for small df (Figure 2.1), but it approaches normality as df increases.

Now back to the sample variance. It turns out that the probability distribution of the sample variance is a chi-square distribution. Strictly speaking,

$$\frac{(n-1)s^2}{\sigma^2}$$

is distributed as χ^2 with $n-1$ df. We can rearrange this equation, using the χ^2 distribution, to determine a 95% CI for the variance:

$$P\left(\frac{s^2(n-1)}{\chi^2_{a/2,n-1}} \leq \sigma^2 \leq \frac{s^2(n-1)}{\chi^2_{1-a/2,n-1}}\right) = 0.95,$$

where the lower bound uses the χ^2 value below which 2.5% of all χ^2 values fall and the upper bound uses the χ^2

value above which 2.5% of all χ^2 values fall (i.e. $\alpha = 0.05$). This interval is sensitive to deviations from normality (McCulloch et al. 2008), and there are better ways of estimating variances in linear models (Chapters 10–12).

2.4.7 Resampling Methods for Frequentist Estimation

The methods described above for calculating SEs for a statistic and CIs for a parameter rely on the statistic's sampling distribution following a normal or a t distribution. For some other statistics, like the median, the sampling distributions are not simple. We must rely on alternative, computer-intensive resampling methods to measure the precision of our estimates. Without other information, the best guess for the distribution of the population is the observations in our sample.

Two common resampling methods are the bootstrap and the jackknife. We illustrate them in Box 2.3, and good recent introductions include Sprenger (2011) and Efron and Hastie (2016).

2.4.7.1 Bootstrap

The bootstrap estimator's sampling distribution is determined empirically by randomly resampling, with replacement, from the original sample, usually with the same original sample size. Because sampling is with replacement, repeated bootstrap samples will vary. The desired statistic can be determined from each bootstrapped sample. The bootstrap estimate of the parameter is the mean of these statistics. The SE of the bootstrap estimate is simply their SD.

Confidence intervals for the unknown population parameter can also be calculated using the bootstrap samples. There are at least three methods (Efron & Hastie 2016):

• The percentile method, where CIs are calculated directly from the frequency distribution of bootstrap statistics. For example, if we resampled 1,000 times, we would sort the bootstrap statistics in ascending order, and the lower limit of the 95% CI would be the 25th value, and the upper limit would be the 975th value. Unfortunately, the distribution of bootstrap statistics is often skewed, especially for statistics other than the mean. The CIs calculated using the percentile method will not be symmetrical around the bootstrap estimate of the parameter, so will be biased.

• The bias-corrected method first determines the percentage of bootstrap samples with statistics lower than the bootstrap estimate. This is transformed to its equivalent value from the inverse cumulative normal distribution (z_0). We determine the percentiles for the values $(2z_0 + 1.96)$ and $(2z_0 - 1.96)$ from the normal cumulative distribution function and use them as the percentiles for our CI.

• The accelerated bootstrap further corrects for bias based on the influence each bootstrap statistic has on the final estimate.

More bootstrap samples are required to calculate CIs than SEs; Efron and Hastie (2016) recommend at least 2,000 for CIs, whereas sizes as small as 200 were suggested initially for SEs.

2.4.7.2 Jackknife

The jackknife is an earlier alternative to the bootstrap that is less computer-intensive for calculating SEs. The statistic (θ^*) is calculated from the full sample of n observations, then from the sample with the first data point removed (θ^*_{-1}), the second data point removed (θ^*_{-2}), etc. Pseudovalues for each observation in the original sample are calculated as:

$$\overline{\theta}_i = n\theta^* - (n-1)\theta^*_{-i},$$

where θ^*_{-i} is the statistic calculated from the sample with observation i omitted. Each pseudovalue is simply a combination of two estimates of the statistic, one based on the whole sample and one based on the removal of that particular observation.

The jackknife estimate of the parameter is the mean of the pseudovalues $(\tilde{\theta})$. The SE of the jackknife estimate is:

$$\sqrt{\frac{n-1}{n}\sum\left(\theta^*_{-i} - \overline{\theta}\right)^2}.$$

The jackknife SE is usually biased upwards compared to the true SE. The jackknife is not usually used for CIs because so few samples are available when the original sample size was small. Barber et al. (2021) suggested a modified version, jackknife+, which can provide CIs for regression predictions.

2.5 Hypothesis Testing

The other main component of statistical inference, besides estimating population parameters, and one that has dominated the application of statistics in the biological sciences, is testing hypotheses about those parameters.

Mayo (2018) summarizes this process as having several components:

(A) A hypothesis, typically about the value of some parameter (θ) that reflects some biological process affecting the data. If we make certain assumptions about the process, we can specify the data expected if this hypothesis is correct.

(B) A "distance function," calculated from the data, that will allow us to assess how well the data fit the hypothesis. When we can calculate its expected distribution under the hypothesis, this distance function is called a test statistic.

(C) A test rule, by which we decide whether the observed value of the test statistic is consistent with the hypothesis, e.g. if it exceeds a particular value.

Mayo includes an additional step as part of her error-statistical approach to consider how well the process has interrogated the hypothesis.

2.5.1 Frequentist Statistical Hypothesis Testing

The early development of statistical hypothesis testing was led primarily by Fisher and Neyman and Pearson; see Chapter 13 of Spanos (2019) for a recent detailed summary of their approaches. Let's start with Fisher, and use one of the simple examples from Chapter 1 – whether the size of an aphid colony affects the proportion of soldiers. As part of taking this question to a model that can be tested statistically, we need to get more specific. We'll suppose that the proportion of soldiers (y) rises linearly with aphid density (x), so any relationship would be of the form $y = \beta_0 + \beta_1 x$. This is a linear regression model (see Chapters 4 and 6) where β_0 is the y-intercept, and our parameter of interest (θ) is the value of the slope, β_1.

> **Box 2.4** Biological versus Statistical Significance
>
> It is essential to distinguish between biological and statistical significance. If we take larger and larger samples, we can detect very small differences. Whenever we get a (statistically) significant result, we must still decide whether the effects we observe are biologically meaningful. For example, we might measure 100 snails in each of two populations, and we would almost certainly find that the two populations were different in size. However, if the mean size differed by $\approx 1\%$, we may struggle to explain the biological meaning of such a slight difference.
>
> What is biologically significant? The answer has nothing to do with statistics but with our biological judgment, and the answer will vary with the questions being asked. Small effects of experimental treatments may be biologically significant when dealing with rates of gene flow, selection, or some physiological measurements, because small differences can have important repercussion. For example, small changes in the concentration of a toxin in body tissues may be enough to cause mortality. In contrast, small effects may be less important for ecological processes at larger spatial scales, especially under field conditions.
>
> Biologists need to think carefully about how large an effect must be before it is biologically meaningful. In particular, setting biologically important ESs is crucial for ensuring that our statistical test has adequate power (Section 3.3).

Using Mayo's steps above, one of Fisher's followers would:

A. Construct a null hypothesis (H_0). If colony size has no effect, $\beta_1 = 0$.

B. Choose a test statistic that will compare our sample slope value (b_1) to that predicted and generate an SE from the data. Our test statistic, more generally $(b_1 - \theta)/s_b$, and in this case $(b_1 - 0)/s_b$, will follow a t distribution; if H_0 is correct, that distribution will be centered on zero, and will have $n - 2$ df. The sampling distribution of the t statistic when the H_0 is true is also called the central t distribution. For the aphid example (described fully in Box 6.2), $n = 16$, our sample slope is 29.3, and the SE is 4.68, so $t = 6.255$.

C. Fisher's test rule compares the test statistic from the data to values expected from such samples when H_0 is correct. For convenience, he proposed possible thresholds corresponding to 5, 1, and 0.1% of test statistic values but suggested that researchers be free to nominate alternative thresholds (before any data analysis!). A test statistic value beyond the threshold is evidence for rejecting H_0. In the aphid example, the critical value for the t distribution is 1.76 (with a 5% threshold) – under H_0, only 5% of sample t statistics should exceed this value with repeated sampling. Therefore, we reject H_0. Our test statistic is large, with $P = 0.000011$. If H_0 is true, it's very unlikely that we'd see a value so large.

A key component of this framework is the P-value, the probability of obtaining our sample data or data more extreme if the null hypothesis is true (i.e. $P(\text{data}|H_0)$). We will consider issues associated with P-values more in Section 2.6. A result where the test

statistic exceeds the threshold value – that is, the P-value is below the probability associated with the threshold – is often called a statistically significant result. This does not imply a biologically significant or a biologically meaningful effect (see Box 2.4). The threshold is called the critical value and its probability given the symbol α or α_c. Using a threshold for determining statistical significance has been an issue of strong debate and polarized opinions among statisticians and biologists (Section 2.6).

More strictly, Fisher viewed a test statistic above the threshold as either a rare event or evidence that H_0 is incorrect (Spanos 2019). This rare event is important. Under H_0, we would expect some datasets to produce a test statistic above the threshold or critical value by chance. For $\alpha = 0.05$, those rare events will occur 5% of the time. These rare events would mislead us, leading to rejection of a correct H_0. This is labeled a Type I error, and we'll return to it. Importantly, Fisher's approach controls the risk of these errors.

Jerzy Neyman and Egon Pearson extended Fisher's work, introducing an alternative hypothesis (H_1), and their process involved a comparison of H_0 and H_1. In the aphid example, $H_0: \beta_1 = 0$ vs. $H_1: \beta_1 \neq 0$. A sample whose test statistic led to the rejection of H_0 provided support for H_1. In their process:

A. The specification of a hypothesis (H_0, which they labeled a test hypothesis) proceeded the same way, but with the specification of H_1. The alternative is a compound hypothesis; unlike H_0, it is not a specific value but a range of values outside H_0.

B. Identify a test statistic, as above.

C. They were explicit about identifying critical values of the test statistic – for example, a value of t that the test statistic would only exceed in a proportion α of cases. Rather than the data just providing evidence against H_0, rejection of H_0 also implies acceptance of H_1.

The Neyman–Pearson (N-P) framework introduced an important idea. A particular value of the test statistic could arise in two ways. The process underlying H_0 might have generated it, or H_0 might be false, and the data arose from processes associated with H_1. Fisher's approach recognized the chance that data leading us to reject H_0 might result from bad luck (i.e. a Type I error). In the N-P approach a value of the test statistic that resulted in acceptance of H_0 might have been obtained because H_0 is true or resulted from H_1 and by chance have been less than the critical value. They labeled this error Type II (given the symbol β).

They made one additional step. Once we recognize two kinds of errors, we can try to minimize or control them. If we are choosing a threshold, we control the Type I errors automatically. Neyman and Pearson showed that if we specify the alternative hypothesis, we can calculate the probability of our value of the test statistic being a Type II error. The risk of this error can be controlled by varying the sample size when planning to collect data, as we'll describe in Section 3.3.

Under the N-P scheme, the P-value provides no additional information beyond indicating whether we should reject the H_0 at our specified significance level. They emphasized making a dichotomous decision about the H_0 (reject or not reject) and the possible errors associated with that decision, whereas Fisher was more concerned with measuring evidence against the H_0. There is some logical justification for providing P-values (Section 2.6), as it allows readers to use their own thresholds to decide whether to reject the H_0. This P-value is also important for those using Mayo's severity as a measure of how strongly a hypothesis has been challenged.

Note the relationship between the hypothesis test illustrated here and CIs described in Section 2.4.5. A CI for our test statistic is also constructed using the same t distribution. Then, a test of this H_0 with a 0.05 significance level is the equivalent of seeing whether the 95% (0.95) CI for the parameter overlaps zero; if it does, we have no evidence to reject H_0.

We will use the term statistical "test" in this chapter in the sense of generating a P-value as a measure of evidence against the H_0, although most biologists use the term in the context of whether the probability associated with a test statistic generated by sample data is below some predefined threshold or significance level, such as 0.05 (see also Section 2.6).

Population situation	Statistical conclusion: Reject	Statistical conclusion: Retain
Effect	Correct decision Effect detected	Type II error (β) Effect missed
No effect	Type I error (α) Effect claimed; none exists	Correct decision No effect detected; none exists

Figure 2.4 Statistical decisions and errors when testing null hypotheses.

2.5.2 Decision Errors

When we use the N-P protocol to test H_0 based on a statistical significance threshold, there are four possible outcomes based on whether H_0 was true (no effect) or not (real effect) for the population (Figure 2.4). A rejection of H_0 implies that some alternative hypothesis (H_1) is true. Clearly, two outcomes result in the right decision being made; we correctly reject a false H_0 or retain a true H_0. What about the two errors?

Examine Figure 2.5, which shows the probability sampling distribution of t when the H_0 is true (left curve) and the probability sampling distribution of t when a particular H_1 is true (right curve). We never know what this latter distribution looks like in practice because if H_0 is false we don't know the real H_1. Initially, H_1 is a compound hypothesis, and there will be a different distribution for each specific H_1 but only one sampling distribution for H_0. To estimate the Type II error, we need to specify an H_1.

Traditionally, biologists have been most concerned with Type I errors. This is probably because "statistically significant" results imply falsification of a null hypothesis and therefore "progress," and maybe because some have wrongly equated statistical significance with biological significance (see Box 2.4). Therefore, we think we protect ourselves (and our discipline) from false significant results by using a conservative threshold (e.g. 0.05), controlling our Type I error rate to α. Why don't we use a lower significance level to protect ourselves even more? Mainly because for most statistical tests, for a given sample size and level of variation, lowering the Type I error rate (the significance level) results in more Type II errors (imagine moving the vertical line to the right in Figure 2.5) if H_1 is true. A cynic might claim that it also becomes harder to achieve a "statistically significant" result!

For some activities, especially environmental assessment and experiments involving human health, Type II errors may be of much greater concern than Type I. Consider an environmental monitoring program. A Type I error results in an erroneous claim of an unacceptable

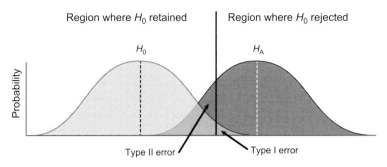

Figure 2.5 Graphical representation of Type I and Type II error probabilities, using a *t* test as an example.

impact, triggering unnecessary and possibly costly remedial measures. A Type II error, on the other hand, is a failure to detect a change that has occurred. The verdict of "no impact" results in harmful activities continuing. There is no immediate cost, but sometime in the future the environmental change may become apparent. By then, environmental degradation may have increased or become more widespread, and mitigation or rehabilitation may be harder or impossible, even at significant cost. In some cases, we might want to protect against Type II errors, even at the expense of Type I errors.

A good sampling design should protect against both errors, primarily by taking large enough samples to be confident in our decisions and parameter estimates. Sampling is often constrained by time or resources, raising error rates. If we opt to fix α or β, scaling back increases the risk of the other error, by implication prioritizing avoidance of one type of error.

Mapstone (1995) proposed that neither error rate should be fixed. Neyman and Pearson also did not see these rates as fixed and required only that decision criteria be determined in advance. Mapstone suggested fixing the ratio of Type I and Type II errors *a priori*, based on the relative consequences of each error (scalable decision criteria). As the number of samples changes, the risks of both errors change, while maintaining their relationship to each other. Downes et al. (2002) showed how to implement this approach when designing environmental monitoring.

The complement of Type II error is power. Suppose the Type II error is the probability that a value of our test statistic is less than the critical value under H_1. Power, $1 - \beta$, is the probability of H_1 generating a test statistic above the critical value (and leading to H_0 being rejected correctly). Power is a useful concept for planning a study with low error rates and to help interpret a test that retains H_0 and be confident that the hypothesis has been challenged. We will consider power in more detail as part of experimental design in Section 3.3.

Scrutinizing a decision not to reject H_0 is crucial. Underwood (1997), in describing the logical structure of hypothesis testing, clearly indicates how Type II errors can misdirect research programs completely.

Mayo's (2018) severity approach also uses similar calculations. Power focuses on calculating the probability of the test statistic being $< \alpha_c$ under H_1, and her severity calculation will essentially give the same result.

2.5.3 One- and Two-Tailed Tests

An important question in planning a study and specifying a H_0 is whether we expect any difference to occur in a particular direction. In our aphid example, do we expect any effect of colony density to be positive or negative? Are differences in either direction plausible and interesting? Perhaps colony defense is more important for small (low-density) colonies. Alternatively, soldiers might be expensive to produce, and only larger colonies might have "spare" resources. In this case, H_1 is that $\beta_1 \neq 0$; this is a two-tailed test and is the most common among biologists (Figure 2.6). We are interested in large positive or negative values of our test statistic. If our biological explanation could only result in a difference in one direction, our H_1 is that $\beta_1 > 0$ (or < 0). This test is one-tailed, and we are only interested in large values of the test statistic in one direction. Two-tailed tests (and CIs) are typically symmetrical, but this is not required. One could test a hypothesis where the critical values in each direction are different (Cox 2020).

We should test one-tailed hypotheses with care because we are obliged to ignore differences in the other direction, no matter how tempting it may be to deal with them after the fact. Statistical software usually defaults to two-tailed tests.

2.5.4 Multiple Hypothesis Testing

One of the most challenging issues when making decisions based on statistical significance thresholds is the

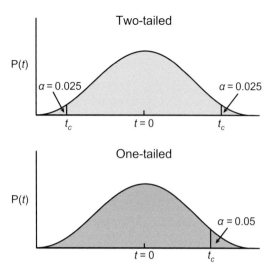

Figure 2.6 Probability distributions of t for (a) two-tailed and (b) one-tailed tests, showing critical t values (t_c).

potential accumulation of Type I errors under circumstances of multiple testing. With more statistical tests, the probability of making at least one Type I error rises. The probability of making one or more Type I errors in a set (or family) of tests is called the family-wise Type I error rate, although Day and Quinn (1989) and others have termed it experiment-wise Type I error rate because it is often used in the context of multiple comparisons within a single experiment. The problem potentially occurs whenever there are multiple simultaneous statistical tests that use thresholds to manage Type I error.

If the tests are independent of each other, the family-wise Type I error rate can be calculated as $1 - (1 - \alpha)^c$ for c tests at significance level α. For $\alpha = 0.05$ and $c = 3$, the Type I rate is 0.14, and when $c = 10$, that rate rises to 0.40.

Generating spurious "statistically significant" results after doing many tests (and possibly reporting only on those significant ones) is at the core of the dubious statistical practice of P-hacking – using methods to search for and/or increase the likelihood of statistically significant results rather than focusing on strong tests of specific, biologically relevant hypotheses (Head et al. 2015).

The issue of multiple testing has become more critical recently when hypothesis tests are used to screen large numbers of response variables, such as in metabolomics, proteomics, and genome-wide association studies. In these cases, there may be thousands of variables and we are identifying candidates for further research. We don't want to follow the path of false "significances," nor do we wish to miss important candidates.

The different approaches for dealing with multiple testing are mostly based on keeping the family-wise

Type I error rate low by reducing each test's Type I error rate (sometimes termed the comparison-wise error rate when comparing group means). Each test will then have a more stringent threshold and, consequently, much reduced power if the H_0 is false. The traditional priority of recommendations for multiple testing has been Type I rather than Type II error rates.

The other issue is what level to set for family-wise error rate. It is common practice for biologists to set the family-wise Type I error rate to the same threshold level as they would use for individual comparisons. This is not easy to justify, especially as it reduces the comparison-wise Type I error rate to very low levels, increasing the probability of Type II errors. This strategy is very conservative, and we should consider alternatives. One option may be to control the family-wise error rate but set a threshold level >0.05 to maintain the power of the individual tests.

Our general approach is not to adjust for multiple testing when our clearly specified biological and statistical hypotheses are part of our original study design. Adjustments might be justifiable in situations where we are screening large numbers of response variables.

2.5.4.1 *Adjusting Threshold Levels or P-values*

Methods to achieve this include:

• The Bonferroni procedure tests each comparison at α/c, where α is the nominated significance level (e.g. 0.05) and c is the number of comparisons in the family. It provides great control over Type I error but is very conservative when there are many comparisons. The big advantage is that it can be applied to any situation with a family of tests.

• The Dunn–Sidak procedure is a modification of the Bonferroni procedure that slightly improves power for each comparison, tested at $1 - (1 - \alpha)^{1/c}$.

• The sequential Bonferroni is a major improvement on the Bonferroni procedure, where the c test statistics or P-values are ranked from largest to smallest, and the smallest is tested at α/c, the next at $\alpha/(c - 1)$, the next at $\alpha/(c - 2)$, etc. Testing stops when a statistically non-significant result occurs. This procedure provides more power for individual tests and is recommended for any situation where the Bonferroni adjustment is applicable.

• Resampling-based adjusted P-values. Westfall and Young (1993a, b) developed a resampling approach that measures how extreme any particular P-value is out of a list from multiple tests, assuming all H_0s are true.

• There are other methods of controlling the family-wise Type I error rate specifically for multiple comparisons among three or more group means (Section 6.2.7.2).

2.5.4.2 *False Discovery Rates*
Another option is to focus on controlling the false discovery rate (FDR) rather than family-wise Type I error rate. The FDR is the expected proportion of Type I errors among the rejected (statistically significant) hypotheses. This approach provides more powerful individual tests and is particularly useful when there are many tests with a substantial proportion likely to be statistically significant.

2.5.5 Testing Hypotheses about Means and Variances for Two Populations
The *t* test we described above tests a hypothesis about a single population parameter, in that example a regression slope. The same logic applies if we want to compare parameters between two populations. For example, for the anesthetic example introduced in Chapter 1, we have samples from two "populations" of mice, each receiving a different anesthetic (Box 2.2). We then test the H_0 that $\mu_1 = \mu_2$:

$$t = \frac{\bar{y}_1 - \bar{y}_2}{s_{\bar{y}_1 - \bar{y}_2}},$$

where

$$s_{\bar{y}_1 - \bar{y}_2} = \sqrt{\frac{(n_1 - 1)s_1^2 + (n_2 - 1)s_2^2}{n_1 + n_2 - 2}\left(\frac{1}{n_1} + \frac{1}{n_2}\right)}.$$

This is the SE of the difference between the two means. It is like the one-parameter *t* test except the single sample statistic is replaced by the difference between two sample statistics. The population parameter specified in the H_0 is replaced by the difference between the parameters of the two populations specified in the H_0, and the SE of that difference replaces the SE of the statistic:

$$t = \frac{(\bar{y}_1 - \bar{y}_2) - (\mu_1 - \mu_2)}{s_{\bar{y}_1 - \bar{y}_2}}.$$

We follow the steps in Section 2.5.1 and compare *t* to the *t* distribution with $n_1 + n_2 - 2$ df in the usual manner, or we could use an ANOVA *F*-ratio test, which is often labeled an *F* test. We'll use the plain "*F*" throughout.

You may have already encountered a modified version of this comparison, where observations are paired, rather than two independent sets. In this case, a paired *t* test is used, focusing on the differences for each pair of observations and testing the H_0 of zero mean paired differences.

We can also test whether the variances of two populations are the same. Recall that variances are distributed as χ^2, and the ratio of two χ^2 distributions is an *F*

distribution. To test the H_0 that $\sigma_1^2 = \sigma_2^2$ (comparing two population variances), we calculate:

$$F = \frac{s_1^2}{s_2^2}$$

where s_1^2 is the larger sample variance, and s_2^2 is the smaller sample variance. We compare this *F* with an *F* distribution with $n_1 - 1$ df for the numerator (sample one) and $n_2 - 1$ for the denominator (sample two). We will consider *F* tests on variances in more detail in Chapters 4 and 6–12.

2.5.6 Parametric Tests and Their Assumptions
The tests we have just described are classified as parametric tests because we can specify a probability distribution for the variable from which our samples came and the test statistic. All statistical tests have some assumptions. If these assumptions are not met, the test may be less reliable because the test statistic may no longer be distributed as expected. *P*-values and CIs may also be unreliable.

For example, the two-group comparison outlined above assumes that the observations are from normally distributed populations, with equal variances, independent of each other, and the observations are sampled randomly from clearly defined populations. We should consider these assumptions at the design stage and try to ensure that they will be satisfied (Chapters 4 and 5). If this is not possible or the subsequent data have unexpected properties, we should consider the implications. If our testing procedure is robust to violations of the assumptions, we might proceed, but otherwise consider alternative procedures.

Three well-known alternatives are robust tests, randomization procedures, and rank-based methods.

2.5.6.1 *Robust Parametric Tests*
Several tests have been developed to compare two means without assuming equal variances. For example, most software offers approximate versions of the *t* test, called variously the Welch, Welch–Aspin, Satterthwaite-adjusted, Behrens–Fisher, and separate variances *t* tests. The most common version of the test recalculates the df for the *t* test as:

$$\frac{\left(s_1/\sqrt{n_1} + s_2/\sqrt{n_2}\right)^2}{\left(\dfrac{\left(s_1/\sqrt{n_1}\right)^2}{n_1 + 1} + \dfrac{\left(s_2/\sqrt{n_2}\right)^2}{n_2 + 1}\right)} - 2.$$

This results in fewer df and is therefore more conservative. Such a test is more reliable than the traditional *t* test when variances are very unequal, particularly when sample sizes are unequal.

2.5.6.2 *Randomization (Permutation) Tests*

These tests resample or reshuffle the original data many times to directly generate the sampling distribution of a test statistic (Box 2.3). The steps in the randomization test for a two-sample comparison are (Manly & Navarro Alberto 2022):

1. Calculate the difference between the means of the two groups (D_0).
2. Randomly reassign the observations, keeping the group sizes constant, and calculate the difference between the means of the two groups (D_1).
3. Repeat this step many times, each time calculating the D_i. Typically, $>5{,}000$ randomizations would be used.
4. Calculate the proportion of the D_is $\geq D_0$. This is the "*P*-value," and it is evaluated as described in Section 2.5.1.

Randomization tests can be used for a range of other analyses, including linear models for univariate and multivariate data, as we describe in Chapter 16 (see also Anderson & Robinson 2001). They are handy when analyzing data for which the distribution is unknown and when random sampling from populations is not possible (e.g. we are using data that occurred opportunistically, such as museum specimens) or perhaps when other assumptions such as independence of observations are questionable (Manly & Navarro Alberto 2022). Results are more difficult to relate to any larger population. Manly and Navarro Alberto (2022) did not see this as a serious problem and pointed out that a big advantage of randomization tests is when a population is not relevant or the whole population is effectively measured. We illustrate a randomization test in Box 2.3. Manly and Navarro Alberto (2022) is an excellent introduction to randomization tests from a biological perspective.

2.5.6.3 *Rank-Based Nonparametric Tests*

Statisticians have long appreciated the logic behind randomization tests, but the computations involved were prohibitive without powerful computers. An early solution to this problem was to rank the observations and then randomize the ranks to develop probability distributions of a rank-based test statistic. Using the ranks of the observations removes the assumption of normality of the underlying distribution(s) in each group, although other assumptions (e.g. equality of spread) may still apply.

Although there are several rank-based nonparametric tests (Hollander et al. 2013), we will only illustrate the alternative to the two-population *t* test here. The Mann–Whitney–Wilcoxon test assesses whether two samples come from identical populations. The procedure is:

1. Rank all the observations, ignoring the groups. Tied observations get the average of their ranks.
2. Calculate the sum of the ranks for both samples.
3. Compare the smaller rank sum to the probability distribution of rank sums, based on repeated randomization of observations to groups, and test in the usual manner.
4. For larger sample sizes, the probability distribution of rank sums approximates a normal distribution, and the *z* statistic can be used, or use a randomization test.

Another nonparametric approach is rank transformation tests. The data are transformed to ranks, which are then analyzed using the equivalent parametric analysis. They are less used, with their role subsumed by randomization tests.

These nonparametric tests generally have lower power than the analogous parametric tests when parametric assumptions are met, but the difference in power can be surprisingly small – for example, $<5\%$ for the Mann–Whitney–Wilcoxon test versus *t* test (Hollander et al. 2013).

2.6 Comments on Frequentist Inference

Frequentist inference has long been the dominant approach to statistical analysis in the biological sciences and elsewhere. More recently, aspects of this frequentist paradigm, especially null hypothesis significance testing, have been intensely debated and criticized. This has prompted the American Statistical Association to release a clarification statement on *P*-values' context, process, and purpose (Wasserstein & Lazar 2016). The statement was followed by a set of papers edited by Wasserstein et al. (2019) examining *P*-values and alternatives (and see, e.g. Ellison et al. 2014 for an ecologically oriented discussion). Some concerns are warranted, but most statistical procedures have shortcomings when subjected to scrutiny. Statistical hypothesis testing is likely to persist for some time, at least in modified form. We feel it is essential for biologists to understand the logical and philosophical basis when listening to these debates and the appropriate ways to use thresholds (Mayo & Hand 2022). We'd use the historical reviews contained in Spanos (2019), Mayo (2019, 2021), and Efron and Hastie (2016) as starting points.

A thorough review of this literature is beyond the scope of our book, but we will highlight a few important issues.

• Much of the criticism has been about the frequentist interpretation of probability and its impact on the interpretation of *P*-values and CIs. We have already pointed out the long-run frequency interpretation of CIs. It has been argued that researchers would prefer an interval that covered 95% of possible parameter values; Bayesian credible intervals do this if we are willing to treat the parameter as a random variable (Section 2.7). Similarly, *P*-values do not provide a probability that a particular hypothesis such as H_0 is true, only the probability of getting our sample data or data more extreme from repeated sampling from the relevant populations if H_0 is true. Additionally, a frequentist *P*-value is based not only on our observed data but also on unobserved more extreme data. This is one reason why it is difficult to reconcile *P*-values with Bayesian posterior probabilities or Bayes factors for a specific hypothesis, independently of any prior knowledge (Section 2.7).

• Related to the first point is the issue of whether *P*-values are useful measures of evidence against a hypothesis, such as H_0. While some statisticians have argued that the *P*-value should be discarded entirely, most of the papers in the discussions edited by Ellison et al. (2014) and Wasserstein et al. (2019) look for common ground and recommend that the *P*-value, if interpreted correctly and used in conjunction with measures such as ES, can remain as part of an analytical toolbox. We agree with Wasserstein and Lazar (2016) that the *P*-value, presented as a continuous variable, provides one measure of incompatibility of data with a hypothesis – that is, evidence against the H_0.

• Irrespective of how one interprets *P*-values, another issue has been using thresholds for statistical significance. Most biological researchers, especially those using experimental methods, still use statistical significance thresholds to make decisions about H_0s. Setting a statistical significance threshold usually falls under the error-statistical framework around Type I and Type II errors and is an integral part of power analysis to determine sample size. We also note that thresholds are often used in other areas of statistical analyses, such as differences in information criteria to select models (Section 9.2.3) or for deciding on the strength of evidence using Bayes factors (Section 2.7.4). Nonetheless, many statisticians are moving away from statistical significance thresholds, even those that still see value in *P*-values. The choice of threshold is often seen as arbitrary, especially given that *P*-values are sample-size dependent (but see the detailed defense of statistical significance by Mayo & Hand 2022).

• Finally, the frequentist approach is not well suited for incorporating prior knowledge into our analyses. While prior information should always influence the biological question and design of a study, the choice of response and predictor variables, and desirable sample sizes, it isn't formally included in our model-fitting and analyses. When estimating parameters, the Bayesian approach explicitly incorporates prior knowledge (Section 2.7.1), expressed as a probability distribution.

Our frequentist approach in this book is to provide readers with as much statistical information as possible in the worked examples so they can apply their own strategy to the interpretation of the analyses. We provide *P*-values as measures of how incompatible the data are with the relevant hypotheses and use them in addition to parameter and/or ES estimates with CIs, and appropriate graphical and tabular presentations, to draw conclusions.

2.7 Bayesian Statistical Inference

Bayesian statistical inference represents a fundamentally different philosophical and statistical approach from frequentist methods. For Bayesian inference, we make two major adjustments to how we think about parameters and probabilities. First, we now consider the parameters of interest to be random variables that can take a range of possible values, each with different probabilities or degrees-of-belief of being true. This contrasts with the classical approach where the parameter was considered fixed but unknown. Second, we must abandon our frequentist view of probability. Our interest is now only in the sample data we have, not in some long-run hypothetical set of identical experiments (or samples). In Bayesian methods, probabilities can incorporate subjective degrees-of-belief (Section 2.2), although such opinions can still be quantified using probability distributions.

Bayesian inference is derived from Bayes' rule or theorem, published in 1764, for relating conditional probabilities (see brief history and background in Link & Barker 2009; McCarthy 2007). Bayes' rule tells us how probabilities of an event might change based on previous evidence. In the context of applied Bayesian statistics, Bayes' rule has been simplified as:

$$P(\theta|\text{data}) = \frac{P(\text{data}|\theta)P(\theta)}{P(\text{data})},$$

where θ is a parameter to be estimated or a hypothesis to be evaluated, $P(\theta)$ is the "unconditional" prior probability of θ being a particular value, $P(\text{data}|\theta)$ is the likelihood of observing the data if θ is that value, and $P(\text{data})$ is the "unconditional" probability of observing the data. It is used to ensure the area under the probability distribution of θ equals 1 (termed "normalization"), and $P(\theta|\text{data})$ is the posterior probability of θ conditional on the data being observed. This formula can be re-expressed as:

posterior probability $\propto$ likelihood $\times$ prior probability.

Bayesian analyses focus on the posterior distribution for a parameter, usually expressed as a probability distribution of parameter values, given the observed data (which enters the Bayesian analysis via the sample data) and any prior knowledge about the parameter.

There are multiple "schools" of Bayesian statistics, summarized by Spiegelhalter et al. (2000):
• Empirical Bayes uses data from previous studies to derive priors and often interprets analyses in a frequentist context.
• Reference Bayes is based on objective or reference prior distributions.
• Proper Bayes incorporates a range of prior knowledge in prior distributions and draws conclusions without using loss or utility functions.

• Decision-theoretic Bayes focuses on decision-making that maximizes the expected utility or minimizes the expected loss.
Our brief summary only considers the proper school.

We illustrate a Bayesian analysis of the Low et al. (2016) dataset in Box 2.5. The parameter of interest, in this case, is the difference in population mean blood CO_2 concentrations between the two anesthetics. The posterior probability becomes the posterior probability distribution of the difference between means given the sample data. The prior probability is the probability distribution of the difference between means based on any previous knowledge. The likelihood is the usual likelihood function that we would use with ML estimation. It is important to note that this example was fitted as a simple linear model with three parameters. A full Bayesian analysis could include prior information about all three: the intercept (overall

---

Box 2.5 🅡 🅔 Worked Example of Bayesian Inference for Two-Group Design: Anesthetic Effects on Mice

We reanalyzed the mouse anesthetic data from Box 2.2 from a Bayesian perspective. We could do this in a few ways, including coding the analysis as a t test or single-factor linear model (ANOVA) directly in programs like WinBUGS (e.g. Kery 2010, McCarthy 2007). We chose the R package *rstanarm* (Goodrich et al. 2022) as it simplifies the coding for common analyses, especially linear models.

We set the analysis up as a simple linear effects model rather than a t test, as the former illustrates a more general modeling approach (Chapter 4). Our model has CO_2 concentration as the response and an intercept (mean of chloralose) and a regression coefficient representing the difference in population means between the two anesthetics, the parameter of interest.

Recall that the estimated mean difference from our standard analysis was 20.9 with a 95% CI of 6.9–35.0. Our first Bayesian analysis will use "weakly informative" normal (we could also have used a t distribution with 21 df) priors for the intercept and the mean difference, rather than flat (uniform) priors which can be problematical; we also assume an exponential distribution for the residual (unexplained) SD. The posterior distribution for the mean difference is presented the online version of this box; the median difference is 20.68 with a 95% credible interval of 5.98–34.52; these results are similar to those from Box 2.2. The Bayes factor for the effect of anesthetic type against a model that assumes no effect is 8.04, indicating eight times more support for the model that includes an effect of anesthetic type.

We will now look at the impact of specifying more informative prior normal distributions for the mean difference. The first prior will assume a mean CO_2 difference of 25 with relatively high precision (SD = 5); that is, we are reasonably confident in our prior. The second will have the same mean difference but less precision (SD = 20). The third will have a much bigger difference between means (50) but with relatively high precision (SD = 5).

Prior (mean, scale)	Median of posterior	95% credible interval
Mean = 25, SD = 5	23.66	15.63–31.39
Mean = 25, SD = 20	21.27	7.72–34.08
Mean = 50, SD = 5	41.66	32.41–50.63

The effect of the priors is apparent in these results. Specifying a prior difference of 25 moved the median of the posterior from 20.68 to 23.66, although this change was much less if our prior had low precision. Doubling the expected difference from 25 to 50 resulted in a marked increase in the median of the posterior.

population mean), the regression coefficient (the difference between population means measuring the effect of the anesthetics) and the model error term (unexplained variation).

Let's examine the components of Bayes' rule in a bit more detail.

2.7.1 Prior Knowledge and Probability

Prior probability distributions measure the relative "strength of belief" in possible values of the parameter of interest without taking the observed data into account. Banner et al. (2020) provide a recent review of the types of prior distributions, emphasizing the distinction between (1) informative priors that incorporate subjective and objective prior information, and (2) noninformative priors that represent the absence of any prior information.

For much biological research, it is probably unrealistic to assume that we have absolutely no previous knowledge about the parameter(s) of interest. For example, Low et al. (2016) cite other studies that examined the effects of anesthetics, including those in their study, on mouse physiology. Evidence from previous studies can be used in several ways to develop priors (Spiegelhalter et al. 2000), such as results from a single, highly relevant study, results from a formal meta-analysis synthesizing previous research, or weighting ("discounting") previous studies differently depending on their relevance. Informative priors can also be based on more subjective opinions. McCarthy (2007) and Spiegelhalter et al. (2000) provide excellent overviews of how opinion, differences of opinion, and expert consensus can be incorporated into priors. Both emphasize that even experts can be biased and are likely to be overly optimistic about the effects of new interventions.

Informative priors can also be considered strong. The probability distribution for the parameter has an obvious peak and relatively small variance (i.e. high precision), indicating that we are confident in our prior prediction for the values a parameter might take. In these circumstances, the prior may have more influence on the posterior than the data. For example, Dennis (1996) and Mayo (1996; see also 2018) highlighted some potential practical and philosophical issues associated with using strong subjective prior information, expressing concern that subjective biases could drive the outcome of analysis.

Like Banner et al. (2020), we suspect this concern about the potential influence of subjective priors on the outcome of an analysis has led many biologists using Bayesian methods to be conservative and use a noninformative prior. There are different types of noninformative prior. Commonly, default priors (e.g. a normal distribution with a very large variance) or vague/flat/diffuse priors (e.g. a uniform distribution where nearly all possible values are considered equally likely) are used. These priors have been criticized as not being truly noninformative and may contradict what is actually known about the behavior of a parameter (Banner et al. 2020; Gelman et al. 2017) – for example, a flat prior for sex ignores the fact that sex ratios tend not to vary much. An alternative is the weakly informative prior that reflects minimal but realistic prior information about the parameter of interest; it is now the default in some Bayesian software (Goodrich et al. 2022). When we use a noninformative prior such as a normal distribution with a very large variance, only the data influence the posterior distribution of the parameter.

It is very important that biologists justify their priors and explain their derivation when using Bayesian methods. This applies to how previous knowledge is incorporated into an informative prior and the basis for using a noninformative prior. Banner et al.'s (2020) review of Bayesian analyses in ecological research found that this was rarely the case. Additionally, a sensitivity analysis showing that the posterior was not greatly affected by or sensitive to relatively small changes in prior distributions would provide more confidence in the conclusions.

2.7.2 Likelihood Function

The likelihood function $P(\text{data}|\theta)$, standardized by the expected value (mean) of likelihood function $[P(\text{data})]$, is how the sample data enter Bayesian calculations. The likelihood function is not strictly a probability distribution, although we refer to it as the probability of observing the data for different parameter values. The likelihood function for linear models is generally based on the distribution used for modeling the predicted response, such as normal for continuous responses, Poisson for count responses, etc.

2.7.3 Posterior Probability

Most conclusions from Bayesian inference are based on the posterior probability distribution of the parameter. This posterior distribution represents our prior probability distribution modified by the likelihood function. The sample data only enter Bayesian inference through the likelihood function. The posterior distribution cannot usually be derived analytically except for very simple models because of the complexities imposed by the denominator (the normalizing constant) in the Bayes equation. Instead, we use computer-intensive iterative (Markov chain Monte Carlo; MCMC) methods to take draws (samples or replicates) from the posterior distribution for a given prior, likelihood, and dataset. The most common MCMC method used to obtain posterior distributions is the Gibbs sampler. If we obtain enough replicate draws from the posterior, we can describe the shape of the distribution and its relevant parameters. Consecutive random draws from a distribution using MCMC will be correlated (the Markov chain component of MCMC). This means that MCMC methods can require large numbers of draws before they converge and become stable, and with most MCMC software these initial draws before convergence are not used to derive the posterior, in a process called "burn-in." McCarthy (2007), Kery (2010), and Zuur et al. (2012) provide concise and readable introductions from a biological/ ecological perspective.

Bayesian inference is usually based on the shape of the posterior distribution, particularly the range of values over which most of the probability mass occurs. The best point estimate of the parameter is determined from the mean of the posterior distribution, or sometimes the median or mode if we have a nonsymmetrical posterior. The Bayesian analogs of frequentist CIs are termed Bayesian credible or probability intervals (sometimes highest density intervals), and 95% intervals are commonly reported. These 95% credible interval boundaries are the two values between which 95% of the posterior distribution lies. These credible intervals are interpreted differently from the frequentist CIs. Because the parameter is treated as a random variable in Bayesian inference, the credible interval is telling us directly that there is a 95% probability that the value of the parameter falls within this range, based on the sample data; the probability and interval are directly about the parameter, in contrast to frequentist CIs.

With a "noninformative" (flat) or weakly informative prior distribution, the Bayesian credible interval will be similar to the classical, frequentist CI, albeit with a different interpretation. If we have a more informative prior distribution, then the Bayesian credible interval would be narrower than the classical CI.

A wide range of probability distributions can be used as priors and posteriors, depending on the parameter of interest (see Appendix B in McCarthy 2007), including any of the exponential family of distributions described in Box 4.2. For parameters of linear models, such as means and regression coefficients, normal, lognormal, or t distributions are commonly used. When the prior and the posterior are of the same form for the same model, then they are said to be conjugate. There are other useful conjugate distributions for some models and parameters, such as the beta distribution which is conjugate to the parameter of a binomial distribution, and the gamma which is conjugate to the parameter of a Poisson

distribution (see Chapter 13) and it is often used for the precision parameter of a normal distribution.

2.7.4 Model Comparison and Bayes Factors
The Bayesian approach can evaluate a particular hypothesis about a parameter by seeing if the credible interval we obtain from the posterior distribution contains the value we specify in our hypothesis – for example, does the 95% credible interval for a regression coefficient or difference between means include zero? We are essentially asking the probability of getting a parameter value of zero, given our sample data. This is a more direct test of a null hypothesis than the frequentist approach, which provides the probability of getting our data or data more extreme if the parameter value is zero.

The Bayesian framework can also be used to choose between alternative models, such as one representing the null hypothesis versus one representing an alternative hypothesis. The ratio of posterior probabilities of two competing models can be expressed as the ratio of the prior probabilities of the two models multiplied by a term called the Bayes factor (BF). The BF is the ratio of the prior probabilities of the data, given each model, or the ratio of the (marginal) likelihoods of the observed data, averaged over the prior distribution of the parameters, for each of the two models (Bolker 2008). The ratio of the posterior probabilities of the two models is an odds ratio, and the BF measures how much the odds of one model over the other change when the data are considered (McCarthy 2007). A BF > 1 indicates that the first model is better supported by the data, and vice-versa for a BF < 1. The BF can also be considered a relative measure of evidence for the two competing models. Bayes factor thresholds have been proposed to distinguish different strengths of evidence in favor of one model over the other (Kass & Raftery 1995).

Bayes factors are not straightforward. They can be challenging to interpret when using noninformative (e.g. uniform) priors. Their derivation, especially the marginal likelihood calculation (Bolker 2008), is tricky. There are statistical links between BFs and information criteria (e.g. Akaike information criteria; AICs) that we introduce as measures of fit that penalize model complexity in Section 4.5.3 and describe in more detail in Section 9.2.1. Bolker (2008) showed how BFs based on strong priors converge to AICs.

2.7.5 Final Comments
We do not take a Bayesian approach to the analyses and worked examples. Most of the statistical models we describe have relatively few parameters, especially models used for analyzing designed experiments, and the

response variables generally follow common distributions from the exponential family. Without strongly informative prior distributions, the Bayesian point estimates and credible intervals for the model parameters will be close to their frequentist counterparts, keeping in mind the different interpretations of credible intervals for parameters as a random variable and CIs for fixed parameters (McCarthy 2015). We do recognize the limitation of not being able to incorporate prior knowledge formally into our analyses, but our analyses of the worked examples followed those of the original authors, which were not in a Bayesian framework. We also see prior information as most valuable when used to frame, rather than answer, questions.

Further Reading

This chapter is intended as a refresher of basic content that you should already be familiar with and, therefore, we assume in the following chapters. You might want further reading for two reasons.

You need more than a brief refresher:

There are many introductory texts available, and we generally recommend those directed at biologists. Depending on the depth of knowledge wanted, there are "classic" texts such as Sokal and Rohlf (2012), or, more simply, McKillup (2012), both of which take a hypothesis-testing approach. More recent texts include Whitlock and Schluter (2020) and Crawley (2014), who provide excellent introductions with detailed R code and numerous examples, and Gelman et al.'s (2020) book focusing on regression models.

You're not satisfied with brief summaries of what can be complex issues:

Probability: Nakajima (2013) or Mayo (2018) for a more philosophical perspective.

Hypothesis testing: This is a complex area, often mixing philosophical concepts, thresholds and decision-making, and statistical "significance." Detailed introductions are provided by Spanos (2019), Efron and Hastie (2016), and Mayo (start with her 2018 book).

Bayesian approaches: For biologists, we like McCarthy's (2007) book, and his more recent discussion of different statistical approaches (McCarthy 2015).

3 Sampling and Experimental Design

This chapter continues with the building blocks of design and analysis. In the opening chapter, we established the basis for challenging biological ideas strongly, with robust answers. The second chapter reminded you of how we fit statistical models to data and make decisions based on those analyses. The other part of good practice is making sure that we collect appropriate data. Our samples should match the biological questions and statistical models. We should be alert to other factors that could confound our interpretations of the analysis and collect enough data to answer our questions as unambiguously as possible.

3.1 Sampling Design

The design of the sampling regime is fundamental. We need to define a statistical population of interest and take samples[1] of the right type and size to characterize it. Populations[2] should be defined at the start of any study and this definition should include their spatial and temporal limits and hence the limits to our sampling and conclusions. For example, if we sample from animals at a certain location in December 2021, our inference is restricted to that location in December 2021. We cannot formally infer what the population might be like at any other time or in any other place, although we can speculate or possibly use other information to extrapolate our conclusions/predictions to other locations or times.

Samples are collections of sampling or experimental units, depending on the study. Units may be natural (e.g. cells, individual organisms, ponds, etc.) or artificially delineated (e.g. plots or quadrats, aliquots of water, aquaria, etc.). Sampling error arises because any sample will be an imperfect representation of its population, so different samples will likely produce different estimates of population parameters. Sampling error is measured by the standard error – the standard deviation of a sample statistic based on repeated sampling. We record data from each unit by measuring it completely or by taking a subsample from it. That measurement is an observational unit. Observational units might differ from sampling/experimental units, and when they do, the distinction can be very important, as we'll see in following chapters.

We aim to design a cost-effective sampling program that provides precise and unbiased estimates of population parameters. In this chapter, our introduction to sampling is also based on sampling without replacement, which provides better estimates of population parameters than sampling with replacement (Manly & Navarro 2014) and is more common in biological research.

Methods for selecting sampling units can be classified into two types, probability and nonprobability sampling.

3.1.1 Probability Sampling

Probability samples are where the probabilities of selecting every possible sampling unit are known. Probability sampling results in samples where population parameters can be estimated reliably and the variance (and precision) of the estimates calculated.

---

[1] Biologists and environmental scientists often use the term "sample" to describe a single experimental or sampling unit, such as a sample of mud from an estuary or a sample of water from a lake. In contrast, a statistical sample is a collection of one or more of these units from some defined population. We will only use the term sample to represent a statistical sample, unless there are no obvious alternative words for a biological sample, as in this case.

[2] Biologists also use "population" to mean a group of organisms of the same species living in the same place. We'll use "biological population" if necessary to distinguish it from the statistical population.

3.1.1.1 *Simple Random Sampling*

Simple random sampling is where, for a given sample size (n), all the possible combinations or sets of n sampling units in our population have an equal chance of being selected (Alf & Lohr 2007). Strictly, random sampling should be done by giving all possible sampling units an identifier and then choosing units for the sample using a random selection process (e.g. using a random number generator). This method is often difficult, especially in field biology. The sampling units may not represent natural distinct units (e.g. they are quadrats or plots) or we may be unable to number them easily in advance (e.g. free-roaming animals). For spatial sampling, one solution is to overlay a spatial grid and randomly choose the location of sampling units.

Formulae for estimating population means and variances, SEs of the estimates, and CIs for parameters assume simple random sampling. Simple random sampling should always be our goal. Its downside is that it may be more expensive, with higher cost of locating randomly placed units. When the population is heterogenous or we wish to estimate parameters at a range of spatial or temporal scales, it may be less efficient than other sampling designs.

3.1.1.2 *Stratified Sampling*

Stratified sampling is where the population is partitioned into levels or strata that represent clearly defined groups of units within the population and we sample independently (and randomly) from each group.

Estimating population means and variances from stratified sampling requires modification of the formulae provided in Chapter 2 for simple random sampling. The estimate of each stratum's population mean is as with simple random sampling. The estimate ($\bar{y}$) of the overall population mean is:

$$\sum_{h=1}^{l} \left(\frac{N_h}{N} \right) \bar{y}_h,$$

where there are $h = 1$ to l strata, N_h is the total units in stratum h, N is the total units in the population, and $\bar{y}_h$ is the sample mean for stratum h. The SE of this mean ($s_{\bar{y}}$) is:

$$\sqrt{\sum_{h=1}^{l} \left(\frac{N_h}{N} \right)^2 \frac{s_{\bar{y}_h}^2}{n_h}},$$

where $s_{\bar{y}_h}^2$ is the sample variance of the mean for stratum h and n_h is the sample size for stratum h.

Many biological populations are essentially infinite, so N and N_h are unknown. Under these circumstances, the proportion of the population in stratum h (e.g. measured by area) can be used instead of the ratio N_h/N (Manly & Navarro 2014). An approximate $1 - \alpha$ CI can be determined as usual using the t distribution:

$$\bar{y} \pm t_{\alpha/2}\, s_{\bar{y}} \bar{y} \pm t_{\alpha/2} s_{\bar{y}}.$$

When statistical models are fitted to data from stratified sampling designs, the strata should be included as a predictor variable in the model. The observations from the different strata cannot be simply pooled and considered a single random sample except maybe when we have evidence that the strata are not different in terms of our response variable – for example, from a preliminary test between strata.

Total sampling effort can be spread among the strata in several ways:

1. Equal allocation, with the same number of sampling units from each stratum. This might be appropriate if a statistical comparison between strata is planned, although it may not be the most efficient method.

2. Proportional allocation, where the number of units sampled from each stratum is proportional to the total possible units in each stratum. One downside of this approach is that parameter estimates in rarer strata may not be very precise because of the smaller allocated sample size.

3. Optimal allocation, where the number of sampling units in each stratum is allocated to minimize the variance (i.e. SE) of an estimated population parameter. This method requires knowledge of the variance within each stratum. A more sophisticated approach is to include the cost (in dollars or time) of sampling units in each stratum. These approaches are discussed in Section 10.5.3.

3.1.1.3 *Cluster Sampling*

Cluster sampling also modifies basic random sampling to account for heterogeneity in the population. Imagine we can identify primary sampling units (clusters) in a population (e.g. individual trees). For each primary unit (tree), we then sample all secondary units (e.g. branches on each tree). Simple cluster sampling is where we record all secondary units within each primary unit. Two-stage cluster sampling is where we take a random sample of secondary units within each primary unit. Three-stage cluster sampling is where we take a random sample of tertiary units (e.g. leaves) within each secondary unit (e.g. branches) within each primary unit (e.g. trees). Simple random sampling is usually applied at each stage, although proportional sampling can also be used.

3.1.1.4 Systematic Sampling

Systematic sampling is where we choose sampling units that are equally spaced, spatially or temporally. For example, we might choose plots at 5-m intervals along a transect or we might choose weekly sampling dates.

Systematic sampling can have a single random starting point, where the first unit is chosen randomly and then the remainder evenly spaced. Alternatively, a cluster design could be used, where clusters are chosen at random and then there is systematic selection on secondary sampling units within each cluster.

There are two major limitations of systematic sampling. With no random repetition of sampling units, an unbiased estimate of sampling variance and SEs for estimates of population parameters are not available (Gruitjer et al. 2006). Aune-Lundberg and Strand (2014) suggested that treating systematic spatial sampling as simple random sampling when spatial autocorrelation is unknown or absent at least provides conservative estimates of sampling variance. They also suggest, as does Fewster (2011), that post-sampling stratification (not influenced by the observations), with either overlapping or nonoverlapping strata, and then using stratified sampling estimators (see above) usually performs better than simple random sampling. Fewster (2011) also proposed the "striplet" estimator, which has excellent properties for systematic spatial sampling.

The second limitation is the risk that the regular spacing may coincide with an unknown environmental gradient and so any inference to the whole population would be biased. This situation is probably more likely in field biology (e.g. ecology), where environmental gradients can occur at a range of different spatial scales, although Manly and Navarro (2014) argued that such matching is unlikely if sampling units are far enough apart to be essentially independent. Systematic sampling through time, in contrast, may be more likely to match some underlying temporal cycle.

3.1.1.5 Unequal Probability Sampling

Unequal probability sampling is where different sampling units in the population have different probabilities of inclusion in a sample (Thompson 2012). Characteristics that might influence a sampling unit's probability of being part of a sample include its size or prior information about the response variable for a unit.

Horvitz–Thompson estimators provide unbiased estimates of population parameters from unequal probability sampling because they weight the observations differently depending on the probability of that sampling unit being selected. The estimate of the population mean is:

$$\frac{\sum_{i=1}^{n} y_i / p_i}{\sum_{i=1}^{n} 1 / p_i},$$

where p_i is the probability of observation i being selected and $\sum(1/p_i)$ is an estimate of the population size. Standard errors (and CIs) can be estimated, but the formulae are a bit messy (see Thompson 2012 for details). They depend on the individual probabilities of units being included and the joint probabilities of any two units being included. An alternative estimator, the Hansen–Hurwitz estimator, can be used when sampling is with replacement. It is particularly useful when the probabilities of units being included are proportional to the size of the units (probability-proportional-to-size method – again see Thompson 2012).

3.1.1.6 Adaptive Sampling

Finally, we should briefly mention adaptive sampling, which is where the sampling design is modified based on initial observations (e.g. preliminary estimates of parameters) from the sampling program. For example, we might change our sample size based on preliminary estimates of variance or we might even change to a stratified design if the initial simple random sampling indicates clear strata in the population that were not detected early on.

3.1.2 Sample Size for Random Sampling

If we have some idea of the variability between sampling units in our population, we can estimate the sample size required to be confident that any sample mean will be within some specified bounds around the true mean amount under repeated sampling. The calculations are relatively straightforward for simple random sampling and are based on the central limit theorem. For example, to determine the necessary minimum sample size to estimate a population mean,

$$n \geq \frac{z^2 s^2}{d^2},$$

where z is the value from a standard normal distribution for a given confidence level. For 95%, $z = 1.96$. s^2 is the estimate of the variance of the population (usually from some pilot sample or earlier information), and d is the maximum allowable absolute difference between the estimated and true population means. The reliability of estimated sample size depends on the variance estimate from the pilot study being similar to the population variance when we sample. Comparable formulae are available for other population parameters.

Sample size determination is more problematic for other sampling designs. For stratified random sampling, we have already discussed different options for distributing a predetermined overall sample size among strata, including allocating more effort to strata with larger variances. With systematic sampling, sample size determination is particularly tricky. We suspect most biologists simply use methods based on simple random sampling.

Another approach for figuring out required sample sizes when testing null hypotheses is to use *a priori* power analysis, which is covered in Section 3.3.

3.1.3 Nonprobability Sampling

When we can't define the probability of each possible sampling unit being included in our sample, we have nonprobability sampling, which includes convenience, haphazard, and purposive (or judgment) sampling.

3.1.3.1 *Convenience Sampling*

This approach chooses sampling units based on convenience or availability. For example, freshwater stream surveys might only examine sites near bridges or other access points, wildlife samples might only come from road kills, and bird counts sometimes rely on volunteers who only search their local neighborhood. In biomedical research, sampling people with particular conditions might rely on those that volunteer or report for other reasons. Genetic analyses, particularly those used for historical biogeography, may use herbarium and museum samples, which may be a combination of planned collection and serendipitous lodging of material.

Convenience samples are not easy to analyze and interpret because they generally have poor statistical properties and result in biased estimates of population parameters (Gruitjer et al. 2006). It is often difficult to determine what larger population they represent, although they can represent a random sample from a narrowly defined population. For example, stream surveys might represent the population of all stream sites near bridges in a particular area, but we don't know how that population relates to all sites along streams.

3.1.3.2 *Haphazard Sampling*

Haphazard sampling attempts to mimic true random sampling by selecting sampling units haphazardly, trying to avoid any bias. It is sometimes used where random sampling is technically difficult, such as ecological sampling using plots or quadrats over a large area. Making inferences about populations from haphazard sampling assumes that the sample is the equivalent of a random sample, and there are reasons this assumption may not be met. For example, researchers may introduce subconscious biases that minimize sampling effort or make the sample "representative" of what they perceive the population to be.

3.1.3.3 *Purposive Sampling*

Purposive sampling is where sampling units are selected for a specific study purpose. It is most often used where sampling units are people selected because researchers judge them most representative of the population (Levy & Lemeshow 2008). Like convenience sampling, purposive samples generally have poor statistical properties and are difficult to use for population inference.

3.2 Experimental Design

The value of experiments depends critically on them being designed properly.

Experimental design has its own terminology, although as Hurlbert (2009) pointed out, this terminology is sometimes used inconsistently. Manipulative experiments have a treatment structure, the set of treatments (the simplest case being a treatment and a control) being applied. A set of related treatments is sometimes termed a factor and experiments can have multiple factors. The smallest units to which an experimental treatment is applied are experimental units (Casler 2015; Hurlbert 2009). The design structure describes how the factor groups (treatments) are applied to experimental units.

One of the most important constraints on the unambiguous interpretation of an experiment is confounding. While the term confounding has several definitions in statistics (see Greenland 2014), it is commonly used to mean that differences due to experimental treatments – that is, the contrast specified in your hypothesis – cannot be separated from other extraneous factors that might be causing the observed differences. A simple, albeit trivial, example will illustrate the problem. Imagine you wished to test the effect of a particular hormone on some behavioral response of crayfish. You create two groups of crayfish, males and females, and inject the hormone into the male crayfish and leave the females as the control group. Even if other aspects of the design are fine, differences between the means of the two groups cannot be unambiguously attributed to effects of the hormone. The two groups are also different sexes, which may influence the behavior of the crayfish. In this example, the effects of hormone are obviously confounded with any sex effects, and it is impossible to determine the cause of any difference. The obvious solution is to randomize the allocation of crayfish, so the two groups are just as likely to have males and females or, even better, include sex as an additional factor. Unfortunately, confounding is rarely this obvious, and it can sneak into an experimental design in many ways, especially through inappropriate replication, lack of proper controls and lack of randomized allocation of experimental units to treatments. These pathways will be our focus in this section.

Confounding is sometimes intentional. When we have too many treatment combinations for the available experimental units, we might confound some interactions so we can test main effects (Section 7.2.2). This should be done only with great care.

3.2.1 Replication of Experimental Units

Replication means having multiple experimental or sampling units at the scale of the experimental treatments. Most biological systems are noisy and replication allows us to estimate this noise, so we can separate treatment effects from noise.

Replication at the scale of the experimental treatments also helps minimize the risk of confounding treatment differences with other systematic differences between experimental units. For example, imagine an experiment to examine the effects of two feed types on the growth of a farmed fish species (see Thorarensen et al. 2015 for an overview of correct designs for this type of experiment). Suppose there are only two large tanks available, and food is added to whole tanks, so they are the experimental units. One tank with multiple fish might be given feed type A and the other given feed type B. Unfortunately, with this design, the effect of feed type is completely confounded with inherent differences between the two tanks; these differences might be related to water supply, position of tank within the facility, what the tank has been previously used for, which fish are in each, etc. In contrast, if we had two or more replicate tanks for each of the two feed type treatments, we can be more confident in attributing differences between the two groups of tanks to feed type rather than inherent tank differences. Replication of experimental units does *not* guarantee protection from confounding because it is still possible that, by chance, all our tanks receiving feed type A differ from those receiving feed type B in ways other than the feed. However, the risk of confounding is reduced by increasing replication of experimental units, especially when combined with randomization (Section 3.2.3).

While most biologists know of the need for replication, they sometimes attempt to do this by increasing the number of observational, rather than experimental, units. Probably no other aspect of experimental design causes more problems for biologists (Hurlbert 1984). In our fish growth study example with a single tank for each feed type, there were multiple fish within each tank. When measuring growth, these fish are the observation units (Figure 3.1) and we might measure the growth rate of 10 fish per tank. We could compare the mean growth of fish between the two tanks with something simple, like a *t* test. However, this test does *not* evaluate the effect of feed type; it only evaluates whether we can detect a statistical difference between the two samples of fish. The fish are not the experimental units because the treatment (feeding) was applied at the scale of a tank. The fish are the observation units and represent subsampling units from each tank; they do not independently assess the effect of feed type. Using the fish to statistically test for differences between tanks and inferring differences between feed types is what Hurlbert (1984) termed simple pseudoreplication. Essentially, simple pseudoreplication is a confusion between experimental and observation units. It exaggerates the evidence for treatment differences (Hurlbert 2009) because the df for a test (or CI) will be based on many observation units rather than fewer, and often solitary, experimental units.

Confounding because of inappropriate replication is not restricted to manipulative experiments. Say, as

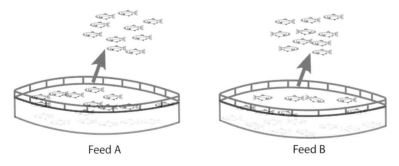

Feed A   Feed B

Figure 3.1 Example of an inappropriately replicated study on the effect of feed type on fish growth in an aquaculture setting. A single tank for each feed type is subsampled with 10 replicate fish.

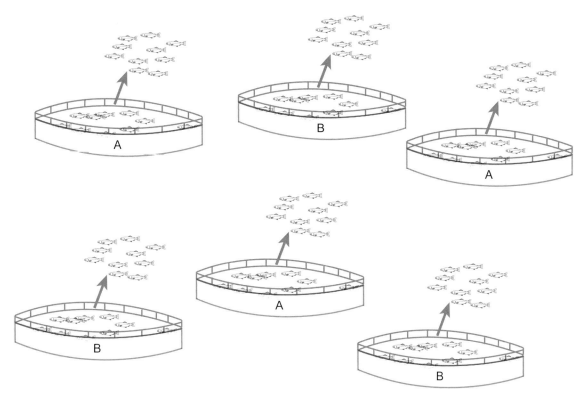

Figure 3.2 Example of an appropriately replicated study on the effect of feed type on fish growth in an aquaculture setting. Multiple tanks for each feed type are subsampled with 10 replicate fish.

ecologists, we wish to examine the effects of fire on the species richness of soil invertebrates. Fire is difficult to manipulate in the field, so investigators often make use of a natural wildfire. In our example, a burnt area might be located and compared to a nearby unburnt area. Within each area, replicate cores of soil are collected, and the species richness of invertebrates determined for each core. We have the same problem as with the fish growth experiment. The appropriate "experimental" units, noting this is not a true manipulative experiment but a study making use of a natural disturbance, for testing the effects of fire are the areas, the scale at which the fire "treatment" was applied. The effects of fire are completely confounded with other inherent differences between the two areas, and cores are just subsamples or observational units that tell us about variation within a single area.

How do we overcome this confounding? The starting point is being very clear about the experimental units. We do this by thinking about what a treatment is and how we produce an instance of this treatment. For the fish feed experiment, the treatment is to place a group of fish in a tank and then feed them, so we need replicate tanks for each feed type (Figure 3.2). Similarly, we need replicate burnt and unburnt areas. Such designs

with correct replication provide the greatest protection against confounding.

Hurlbert (2009) described temporal pseudoreplication, where repeated measurements through time on each experimental unit are treated as if they were independent units and independent assessments of treatment effects.

It is possible to include observational units in our statistical models, but we recommend avoiding this step where possible. A good rule of thumb is to work from a data file that has only data for experimental units. Any data from subsampling are aggregated to provide a single value for each experimental unit. This approach allows us to fit a simpler statistical model than if subsamples are included. It also reduces the chance of an incorrectly specified model. This aggregation can be made at the data preparation stage, or it could be made before the subsamples are processed. Earlier aggregation can be an advantage when the cost of processing each single observational unit is high. Processing each observational unit, then combining data into an experimental unit, will be much more expensive than pooling and processing a single experimental unit sample (Carey & Keough 2002). If you must include the subsamples or observational units because you are interested in relative variability at different spatial or temporal scales, you must construct a statistical model that includes the hierarchical (Chapter 10) or correlated (Chapters 11 and 12) nature of the data.

Sometimes, though, replication is difficult or impossible. For example, we might have an experiment in which constant-temperature rooms are the experimental units, but because of their cost and availability within a research institution, only two or three are available. Experiments at large spatial scales, such as ecosystem manipulations (Carpenter et al. 1995), often cannot have replication because replicate units simply don't exist in nature or for ethical reasons. There are a few options for designing and analyzing such experiments:

• If the experiment doesn't last long, we could run it several times, switching the treatments between the experimental units each time. For example, run the fish growth experiment once with a single tank for each feed type, and then repeat it after reversing which tank receives which food type. Repeating the experiment several times will reduce the likelihood that any tank effects will confound the effects of feed type. The different "runs" of the experiment would still be incorporated into the statistical model that is fitted.

• We could try to measure all factors that could influence our response variable (e.g. fish growth) and see if they vary between experimental units. If they do not differ, we are more confident that the only difference between the two tanks in our unreplicated experiment is food type. We can never be sure that we have accounted for all the relevant variables, so this is far from an ideal solution.

• We might also draw on other studies that have explicitly addressed the confounding factors and shown that their effects are small. These additional lines of evidence might let us argue that the confounding factors are unlikely to account for the patterns in the data. This is risky and depends on how easily readers can be convinced.

• Sometimes, especially in nonmanipulative studies, we may be able to replicate experimental units for one treatment but not another. This situation often occurs in environmental impact assessment, where there may be a single impact site but many potential control locations. Downes et al. (2002) provide a critical evaluation of this option.

3.2.2 Controls

Experiments involve changing some potentially influential process and recording a biological response. They provide the strongest biological evidence, but it is not enough to just observe the response to change. We need evidence that the response would not have occurred anyway. For example, Hairston (1989) described a test of whether the abundance of the salamander *Plethodon jordani* was affected by its congener *P. glutinosus*. He removed *P. glutinosus* experimentally from plots and the population of *P. jordani* increased during the following years. However, *P. jordani* showed the same pattern on control plots with *P. glutinosus* not removed. Without the control plots, the increase in *P. jordani* might have been incorrectly attributed to *P. glutinosus*.

We also need evidence that the response comes from the biological process that changed, rather than the things we did to change that process. We use controls to eliminate as many artifacts as possible introduced by our experimental procedure – that is, to prevent confounding. For example, Pursall and Rolff (2011) measured immune response of beetles by injecting a little dead bacteria culture. Immune responses might be to the bacteria, but the treatment also involved handling beetles, injecting them, and introducing the bacteria in a solution. A minimal control was a control with beetles injected with the sterile insect ringer solution in which bacteria were suspended. This is a procedural control. We might also want a control treatment of beetles that were not touched, so we can measure any effects of handling. An equivalent ecological example might be using cages to exclude predators, where the cages might also alter shading, air or water movement, etc. Isolating the effects of predation needs a control that accounts for these artifacts.

Controls should be viewed as treatments logically necessary to make a strong test of our hypothesis. They can be complex, and where several controls are needed they can increase the overall resources needed for an experiment. This can be a concern, particularly for situations where there are ethical issues. Kramer and Font (2017) suggested that by using "historical controls," (data from control animals from previous similar experiments) the sample size required for an animal experiment could be reduced. This step should be subjected to very close scrutiny before adopting it.

3.2.3 Randomization

We have emphasized the importance of random sampling from populations for statistical inference. With experiments, the experimental units within each treatment must represent a random sample from an appropriate real or theoretical population of experimental units. This ensures that our estimates of population parameters (means, treatment effects) are unbiased, and our statistical inferences (conclusions from the statistical modeling) are reliable.

For example, our experimental animals that received a substance in a treatment should represent a random sample of all possible animals we could have given the substance, and about which we wish to draw conclusions. In the fish growth experiment, our aquaculture tanks must be a random sample of all possible tanks for that treatment. We must clearly define our treatment populations when we design the experiment. We can only draw conclusions about the population from which we have taken a random sample. Of course, this shouldn't prevent us discussing the broader implications of our experiment as part of publishing the results.

Random sampling is not the only way randomness is incorporated into experimental design. A fundamental principle of experimental design is that the experimental units be randomly assigned to treatment groups, resulting in a completely randomized (CR) design. This minimizes the risk of systematic differences between experimental units that might confound our interpretation of treatment effects.

An artificial example, analogous to one described by Underwood (1997), involves an experiment looking at the difference in growth rates of newly hatched garden snails, fed either the flowers or leaves of a particular plant. The flowers are only available briefly because the plant blooms soon after rain. When the flowers are available, we feed them to any snails that hatch over that period. Snails that hatch after the flowering period are given the leaves of the plant. The obvious problem here is that the two groups of snails may be inherently different because they hatched at different times. Snails that hatch earlier may be genetically different, have different yolk levels in their eggs, etc. Our results may reflect the effect of diet, or they may reflect differences in the snails that hatch at different times, and these two sources of variation are confounded. Clearly, we should take all the snails that hatch over a given period, say the flowering period, and give some flowers and others leaves to eat.

The allocation of experimental units to treatments raises the difficult issue of randomization versus interspersion (Hurlbert 1984). Consider an experiment examining the effects of increased nutrients on algal biomass on stones in a freshwater stream. We randomly choose 10 stones in a stream section and randomly allocate 5 as nutrient-enhanced (N) stones and five as controls (C). What if, by chance, all the C stones end up upstream from all the N stones (Figure 3.3)? This arrangement would concern us because we want our N and C stones to be interspersed to avoid confounding nutrient effects with upstream–downstream differences.

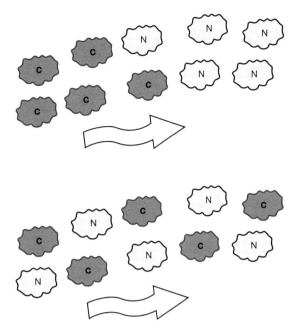

Figure 3.3 Possible result of random allocation of 10 stones in a stream to two treatments – nutrient enhancement (N) and control (C) – that would be unacceptable, along with regular positioning of stones combined with systematic allocation to two treatments to guarantee interspersion. The arrows indicate the direction of stream flow.

The simplest solution if we end up with such a clumped pattern is to re-randomize – any other pattern (except the complete reverse, with all treatment stones upstream) will incorporate some spatial interspersion of treatments and controls. However, we must decide *a priori* what degree of spatial clumping of treatments is unacceptable; re-randomizing until we get a particular pattern of interspersion is not really randomization.

Why not guarantee interspersion by arranging our stones regularly spaced along the stream and alternating the treatments (Figure 3.3)? With this design, stones within each treatment group no longer represent a random sample of possible stones, so it is difficult to decide to which population of stones our inferences refer. The regular spacing might also coincide with an unanticipated periodic environmental process that could confound our interpretation of nutrient effects (see also Section 3.1.1). A compromise might be to select stones randomly but then ensure interspersion by alternating N and C. At least we have chosen our stones randomly to start with so the probability of our treatments coinciding with some unknown gradient along the stream won't change compared to a CR design. However, there is still a problem because, once we have allocated an N, the next stone must be a C, and it is still a bit difficult to define our populations.

3.2.4 Independence

Lack of independence between experimental units can arise in several ways and complicate the analysis and interpretation. At the most practical level, experimental units must be separated enough that the outcome of one unit does not influence the outcome of another. Using the stream example from Section 3.2.3, the stones should be far enough apart that nutrients do not leach onto control stones, and the movement of herbivores, for example, to one stone doesn't automatically cause a corresponding decline in a neighboring stone.

Lack of independence can be included deliberately into the design structure of an experiment. For example, we might be interested in how pollutants move through an animal. We could expose many animals and then take a random sample for assay at each of several times, but we could also use fewer animals and monitor them through time. This second design option uses fewer animals, but measurements from the same animal may be correlated. We deal with these situations in Chapters 11 and 12.

3.2.5 Reducing Unexplained Variance

Any experiment attempts to characterize the signal from a particular biological process (explained variation) or hypothesis against a background of noise (the unexplained variation). We can see a signal more clearly and estimate its strength more precisely if there is less noise. Good experimental design will include considering how we might reduce the unexplained variation. Options include:

• Including additional predictor variables to explain some of the noise. These predictors may be unrelated to the biological question but capable of explaining some variation in the response variable. Adding them to the model shifts some variation from unexplained to explained. We discuss these options starting in Chapters 8 (e.g. analysis of covariance in Section 8.2) and 10.

• Narrowing the scope of the populations from which our samples are drawn, so the units in the population are more similar. This option may reduce noise but it also reduces our ability to generalize any results. As long as we interpret the results carefully, this is not a serious problem, and we can make the generalizing of results our next research challenge.

3.2.6 Limitations of Manipulative Experiments

Manipulative experiments involve the investigators changing biological systems. These manipulations can introduce artifacts, which we try to eliminate with appropriate controls (Section 3.2.2), but controls are not always perfect. Experimental manipulations are also often somewhat artificial settings, from which we wish to make inferences about the natural world. Extending inferences outside these artificial settings requires caution, and researchers should think carefully about differences that might constrain our interpretation.

Logistic constraints on experimental manipulations, especially in fields like ecology, mean that large-scale processes can be difficult to study experimentally. Even if a manipulation can be implemented at a large spatial scale, replication may be very difficult and suitable controls may not be available. Even when replication is feasible, we may be unable to allocate experimental units randomly, making it difficult to rule out preexisting systematic differences between units. There are ways to strengthen causal inference for such experiments. Collecting data from before and after the manipulation (e.g. Before–After–Control–Impact designs) is used widely in environmental impact assessment (Downes et al. 2002). Other methods for dealing with confounding (Section 3.2.1) can also be used. Unreplicated experiments have been particularly influential for our ecological understanding of whole ecosystems (e.g. Carpenter et al. 1995). The lack of replication is not, strictly speaking, a statistical problem, but a logical one about the evidence needed to confidently infer a causal relationship.

Experiments can also be limited in duration by grant cycles (typically around three years), thesis timelines, etc. When we fix the duration of an experiment, that decision may influence our conclusions. Is the lack of an observed effect simply because we did not run the experiment for long enough? Would the size of the effect change through time? To illustrate the importance of experimental duration, we used field experiments to examine the effects of pedestrians trampling on rocky intertidal algal beds over summer (Keough and Quinn 1998). We ran the experiment over seven years. At one rocky shore, the pattern in the first year showed an effect of trampling but with rapid recovery by the end of summer. After the second year, no recovery occurred from the most intense trampling. Stopping the experiment after one year would have produced conclusions very different from those resulting from the multi-year experiment. Experimental duration should be influenced by investigator's prior knowledge of the specific biological system. When external factors limit that duration, we should be cautious in our interpretations.

3.3 Sample Size for Detecting Differences: Power Analysis

Experiments are designed to separate competing explanations, but we need enough data to distinguish confidently between alternatives. In Section 3.1.3, we showed how to determine the sample size to estimate a population parameter with confidence. This rationale of determining sample size can be extended to experiments. If we specify the kind of change that would be important biologically, we can calculate how many units we need to distinguish this change from background noise. The most common tool for determining sample size is power analysis, which we introduced in Section 2.5.2 as part of controlling error rates in statistical hypothesis testing. Recall from Section 2.5.2 that the complement to a Type II error is the concept of power – the long-run probability of detecting a given effect with our sample(s) if it occurs in the population(s). More usefully, statistical power is a measure of our confidence we could detect a specified, biologically relevant, effect if one occurs. Power analysis is not just appropriate for manipulative experiments. We can apply it to any sampling or experimental situation where we can specify a biologically important effect and have a measure of variance in the population from which we are sampling.

Power analysis is most useful for determining acceptable sample sizes. Supposing there *is* a change of a particular size, how much sampling would be needed to detect that change with reasonable certainty? We could also ask, given limited resources, what kind of change could we reasonably expect to detect? Power analysis has also been used to assess a study after completion, although this approach has been criticized. If there was an effect of a particular size, would we have detected it with our samples?

To determine the power of a statistical hypothesis test, we need to specify the alternative hypothesis (H_A), or effect size, that we wish to detect. Power depends on:

• Effect size (ES) – how big a change or response is of biological interest. We are more likely to detect large effects.

• Sample size (n) – a given effect is easier to detect with a larger sample size.

• Variation (σ or σ^2) between sampling or experimental units – it is harder to detect an effect if the populations are more variable.

• Significance level (α) to be used for rejecting the null hypothesis (H_0) – power varies with α, and as mentioned in Chapter 2, by convention most biologists use a value of $\alpha = 0.05$. This may also be expressed as a Type I error rate we wish to accept.

Sometimes, you may see power described as a function of sample size, significance level, and ES. Here, the background variation has been incorporated into a standardized ES. We generally prefer to keep ES and σ separate, because usually in biology we think of an important effect as an absolute change in our response variable, rather than a change beyond patterns of normal variability (e.g. 2 or 3σ).

More formally,

$$\text{Power} \propto \frac{\text{ES}\alpha\sqrt{n}}{\sigma}.$$

Exactly how we link these parameters to power depends on the particular statistical model being fitted (hence the proportional sign in the equation). For individual cases, we construct a specific equation, usually using the relevant noncentral statistical distribution, which in turn requires precise knowledge of the statistical model that will be used (see Box 3.1 and Figure 3.4).

We will first consider *a priori* power analysis for determining sample size, and then look at the pros and cons of *a posteriori* power analysis.

3.3.1 Using Power to Plan Experiments (*a priori* Power Analysis)

Power analysis can be used in two ways to design an experiment or sampling program.

3.3.1.1 Sample Size Calculation (Power, σ, α, ES Known)

The most common use of power analysis during the planning of an experiment is to decide how much replication is necessary. We can then decide whether it is feasible to use this many replicate units. To do these calculations, we need to specify the ES and have an estimate of variance. The most challenging step will be specifying the ES (Section 3.3.3).

3.3.1.2 Effect Size Calculation (Power, n, σ, Known)

If external factors will restrict the number of observations to relatively low levels, the alternative approach is to calculate the constraints of the experiment. With the available number of units and the expected background variation, what is the smallest change we could expect to identify confidently? This situation is common when the sampling itself is expensive, such as laboratory analyses for chemicals at very low concentrations or oceanographic sampling requiring large ships. Ethical considerations often constrain the number of units, and research students often operate with small budgets.

Box 3.1 Worked Example of Power Analysis for Simple Two-Group Design: Anesthetic Effects on Mice

We will use the example from Low et al. (2016), comparing different means of anesthetizing or sedating mice to improve the results of imaging (Box 2.2). They concluded that the commonly used alpha chloralose caused unacceptable elevations of blood CO_2 levels, raising them from around 50 mm Hg to 70. Researchers have a range of candidates that might replace alpha chloralose, and we will plan a study to evaluate one of these candidates, compared to a standard, isoflurane. The experimental subjects are mice, so a formal ethics approval will be required, and most animal ethics committees require a rationale, often using power analysis, for the number of animals.

We'll follow the steps outlined in Section 3.3 to design this new study.

Effect size. Ideally, our new sedative candidate will have the same effect on blood physiology as isoflurane, and we want to detect any harmful effect. For our example, the effect of alpha chloralose of 20 mm Hg was considered very unacceptable, so we will set a lower threshold and aim to detect an elevation of 15 mm Hg.

Statistical model and desired power. With a single categorical predictor variable with two groups, this design can use a two-sample t test as the statistical model, and we will set 0.8 as the target power, consistent with common practice. We left α at its conventional 0.05 value, consistent with the value chosen in the original study.

Assumptions. The t test assumes normal distributions, with equal variances among treatments. We can use the Low et al. data as a guide, and Figure 5.5 does not indicate any strong deviation from normality. We will plan to analyze data with no transformations.

Variation. We are fortunate in this case that we'll follow the protocol of Low et al. and use CO_2 after 120 min. Our experimental units are mice, so we need to know how much variation there is for CO_2 of individual mice. We can use the Low et al. data to do this. Our "standard" is isoflurane, and in the original study the SD was 11.4.

When we plug this information into a power calculator, we find that 11 mice per group will make us confident of detecting a deleterious effect (Figure 3.4).

As part of the planning process, we might also explore possible scenarios for the experiment:

• We have used the isoflurane group of mice to estimate the variance, but inspection of the data from Low et al. shows a higher variance for the alpha chloralose mice (in this case, SD of 20.2). What if this is the real variation, or a better approximation of it, than the isoflurane group? It might be helpful to calculate the sample size under this scenario as the worst case. From Figure 3.4, we see that we'd need 30 mice per group.

The major decision at this point in crafting an ethics application is whether to be optimistic or pessimistic about the variation that will occur.

• The earlier work allowed us to identify an appropriate ES, but in the absence of this information it can be useful to explore the potential of our study using a range of sample sizes and calculate the kind of change we'd detect for a given size. We have plotted this relationship for the expected data pattern and note that when the sample size is 11, we can detect an effect slightly less than 15 mm Hg.

• We could also combine these approaches and plot minimum detectable effect size (MDES) values for the worst-case variance as well. Note that very small samples of a few mice can only detect gross changes, such as a doubling of blood CO_2.

This approach has been termed "reverse power analysis" by Cohen (1988), and the ES we calculate is the MDES. Calculating the MDES may be an option if you can't specify an ES.

For example, from surveys of intertidal mollusks in protected and unprotected areas near Melbourne, Australia, we found changes of 15–25% in the mean size of species collected by humans (Keough et al. 1993). Because we did not determine which areas were protected, we also measured sizes of species not harvested in great numbers to be confident that the patterns for collected species did not reflect some unmeasured environmental variable. We detected no difference for noncollected species, but we needed to be confident that we could have detected a pattern like that shown by collected species. We used power analysis to show that our sampling program would have detected a change as small as 10% for some of these species.

This option is not a complete solution to being unsure about what constitutes an important change. You must

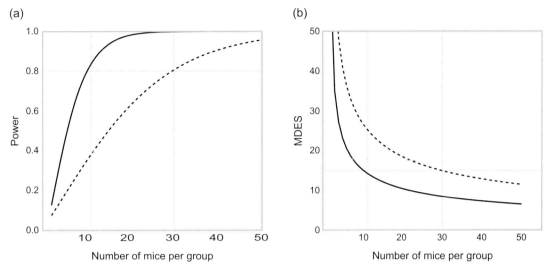

Figure 3.4 Visualizations of the relationships between power, detectable effect size, and sampling effort, using the mouse example from Low et al. (2016). Sampling effort is reflected by the number of mice used in each treatment group. (a) The relationship between sample size and power for the designated effect size (15). (b) The relationship between minimum detectable effect size (MDES) and sample size. The solid lines show the relationship using an optimistic estimate of variance on each panel, and the dashed line shows the relationship for a more pessimistic estimated variance. Limit lines show target power and effect sizes, respectively, and drop-down lines link to the relevant sample size.

still decide whether detecting that effect is worthwhile. Is the MDES so large it's hardly worth pursuing? There's no escaping a discussion of what constitutes an important change!

3.3.1.3 Sequence for Using Power Analysis to Design Experiments

The design stage of any experiment or sampling program should include these steps.

1. State clearly the patterns to be expected in the data if little or no effect occurs, and the patterns expected if there are biologically meaningful changes. In formal inferential statistical terms, this corresponds to clear formulations of the null hypothesis and its alternative.

2. Identify the statistical model to be applied to the data and state the desired power and the decision criteria to be used.

3. Identify the assumptions of that statistical model. If possible, use existing or comparable data as a rough guide to whether those assumptions are likely to be satisfied. Consider the relevance of possible data transformations. If you expect to use transformed data, express the ES on that transformed scale. For example, suppose you are interested in doubling numbers of a particular organism and will analyze log-transformed data. There, your ES will be 0.301 when converted to a log scale.

4. Estimate the variation in the variable to be analyzed – that is, the background noise. This is generally the variation between replicate units. If you have no information of your own, look at similar studies from other places or research groups or consider a pilot study. The estimate of variability must be on scales of space and time similar to your planned data. There is no reason to expect that variation on one scale will be a good predictor of variation on a different scale.

If you have complex experimental designs, which we cover in Chapters 7 and 10–12, particularly those with fixed and random effects (mixed models), you need to think about which source of variation is used to test a particular hypothesis, because there may be several candidates.

5. The next step depends on whether logistical constraints will limit your design.

    a. If we aim to design the best experiment, we specify the ES we wish to detect. The implication here is that detecting smaller changes has lower priority. In practice, this decision is tough, but it is critical. Using our desired ES, an estimate of σ, and the chosen value of σ, we can estimate the number of replicate units needed to detect that ES with power $(1 - \beta)$.

    b. If we have constraints on the number of replicate units, we can use σ, the chosen values of α and β, and our maximum number of units to determine the

MDES. It can be helpful to calculate MDES values for a range of replicate numbers and represent the results as a plot of MDES versus sample size. This relationship can then show how much return we would get for an increase in sampling effort or the sample size necessary to reach a particular MDES.

3.3.2 *Post Hoc* Power Calculation

Power analysis might be used in two ways after completion of an experiment or sampling program. The first is to calculate "observed power" (Hoenig & Heisey 2001). In this approach, the ES and sample variance are estimated from the data and used in the power equation. For example, in a two-sample *t* test we would use the observed difference between the two means as the ES. This observed ES is unrelated to a biologically important one. Perhaps most important, the observed power calculated in this way has a direct relationship with the *P*-value; higher *P* means lower power. This calculation of observed power tells us nothing new (Hoenig & Heisey 2001; Greenland 2012) and is a complete waste of time. Some software offers this option as part of the default output and a good rule of thumb is to disregard any power value for which you have not been asked to specify an ES of interest. Mayo (2018) suggests that "observed" is not a helpful description, because it is not linked to the effect of interest. She labels it "shpower" – a form of power we shouldn't talk about!

The second, and only useful, application of *post hoc* power analysis is to calculate power to detect a specified effect or to calculate the MDES for a given power. Calculating *post hoc* power requires that we define the ES we wished to detect. We have collected the data, so we know *n* and have an estimate of σ. It is obviously useful to demonstrate that our test had high power to detect a biologically important and pre-specified ES, and, therefore, any effects are smaller. The downside is that if power is low, you have only demonstrated your inability to design a very good experiment or, more charitably, your bad luck in having more variable data than expected. Given the uncertainty associated with most variance estimates used in power calculations, this latter point is not trivial. Many *a priori* sample size calculations are done using quite uncertain estimates of variance, and the final dataset will usually provide better estimates.

Hoenig and Heisey (2001) and Greenland (2012) both argued that higher observed power does not imply stronger evidence for a nonrejected null hypothesis, and they did not recommend this approach. Mayo (2018) laid out the reasons why this is so, but also identified *post hoc* analysis as an important calculation as part of her severity framework.

Power analysis is much more useful for planning a study than trying to interpret statistically nonsignificant hypothesis tests. Nonetheless, using power analysis to help interpret statistically nonsignificant hypothesis tests can be useful if done correctly. Providing measures of observed ES and showing you had good power to detect pre-specified ESs of biological interest will make non-significant results much more valuable and can help to get them published.

3.3.3 Effect Size

The most critical step of power analysis is deciding an appropriate ES. An ES is usually a measure of differences between treatment groups, the strength (steepness) of a linear (or nonlinear) relationship, or the odds of one variable changing in response to another variable. Some common ES measures are described by Nakagawa and Cuthill (2007) and Ialongo (2016). The more complex the study design and statistical model, the more difficult it is to define an ES.

How do we decide on an important effect? The decision is not statistical, but usually uses biological judgment by the researcher, who must understand the broad context of the study. Most research is not self-contained but extends or challenges existing knowledge. We might want to:

• compare results for our species to those for other species;
• compare the role of a particular biological process to other processes acting on a particular species or population; or
• contrast the physiological responses to a chemical, gas mixture, exercise regime, etc., to other such environmental changes.

In these cases, we should be guided by two questions:
1. Can we identify a change in the response variable that is important for the organism? For example, what change in a physiological variable would likely impair an organism's function or what change in population density would affect the risk of local extinction?
2. What were the levels of response observed in the related studies that we intend to compare to our own?

These questions sound simple but are often difficult. It is especially the case in whole-organism biology, where we are often dealing with little-studied biological systems. Here, we may be unable to predict critical levels of population depletion, changes in reproductive performance, etc. The available information gets richer as we move to sub-organismal measurements, where work is often done on broadly distributed species, standard laboratory organisms, or on systems that are relatively consistent across a wide range of animals or plants. In any case, we must decide what kind of change is important to us.

3.3.3.1 *What if We Can't Confidently Identify an Effect Size?*

Often, we cannot select an ES we could defend easily. Here, three options are available:

1. Use an arbitrary value as a negotiating point. In many environmental impact studies the target is 50% or 100%, relative to a control group. These values seem accepted as being "large" and with the potential to be important. The major value of this approach is in environmental monitoring, where a sampling program may be contentious and require a negotiated target as a basis for increases and decreases in the monitoring program.

2. Use a standard definition of large, medium, and small effects. This approach came from a review of the behavioral and psychological literature by Cohen (1988), who suggested standardized effects (ES/σ) of 0.8, 0.5, and 0.2, respectively. These values might have been appropriate within his discipline, but they are too crude for the range of biological situations we encounter. For example, the critical change in migration rates between geographically separated populations for genetic differentiation will be very different from the value that alters population dynamics. Any broad recommendation such as Cohen's must be tempered by biological judgment.

3. Rather than use a single ES, plot detectable ES versus sampling effort or power versus ES. While we don't have a formal criterion for deciding whether to proceed, this approach is useful for giving an idea of the potential of the experimental or sampling program.

The critical feature is that a target ES must be specified by the researcher, depending on the change in the response variable that is biologically important. This decision is specific to the research question, and there are no generic answers.

3.3.4 Using Power Analyses

The importance of *a priori* power calculations is that the proposed experiment or sampling program can be assessed to decide whether the MDES, power, or sample size values are acceptable. For example, if the variable of interest is the areal extent of seagrass beds in a shallow embayment, and a given sampling program would detect only a thousand-fold reduction over 10 years, it would be of little value. Such a reduction would be blindingly obvious without an expensive monitoring program and public pressure would stimulate action before that time anyway.

There are two possible outcomes from an *a priori* power analysis. The first is that the analysis has enough sensitivity to detect a biologically relevant ES for your planned sample size, and you're good to go. The second outcome is that you have inadequate power for the desired ES and replication should be increased. If you proceed without increasing sample size, you must know the real limitations of the data collection. Proceeding is a major gamble – if an important effect does occur, the chance of you detecting it may be very low – often less than 20%, rather than the conventional target of at least 80%. There is a high probability you'll not detect an effect, and it will be a result in which you have little confidence. Your resources will have been wasted. You may be lucky and the effect of your treatments much larger than the one you aimed to detect, but that outcome is unlikely. Large and obvious effects are often associated with trivial or self-evident questions.

When should you gamble? Again, there's no simple answer, as we are dealing with a continuum rather than a clear cut-off. If the power is close to your target (e.g. 0.80), you wouldn't be too worried about proceeding, but what of 0.60 or 0.50? The decision will most often result from a suite of considerations:

- How exciting would detecting this effect be?
- How important is it to get some information, such as an ES estimate, even if it's not conclusive?
- Will some other people add to my data, so we'll eventually be able to get a clear answer to the question?
- Would an unpublishable nonsignificant result be a career impediment? The answer to this question depends on your goals, career stage, how strong your scientific record is, etc.

If you aren't willing to gamble, you have only a couple of options. The first is to look hard at the experimental design. Are there ways to make the experiment more efficient, so you need less time or money to deal with each replicate unit? Decreasing the resources needed for each replicate may allow you to increase the sample size. Alternatively, are there other variables that could be incorporated into the design to reduce the background noise?

The second option for those using inferential statistics is intermediate between a calculated gamble and rethinking the analysis, and is the approach introduced in Section 2.5.2. One conventional approach would be to use a less stringent criterion for statistical significance – that is, increase α – producing an increase in power. This solution isn't satisfactory, as the Type II error rate still varies with sampling effort. Mapstone (1995) proposed that, when we must compromise a sampling program, we preserve the relative rates of the two errors rather than fixing the absolute rate of one error. During the design phase, we determine desirable error rates that reflect the consequences of each error. Any compromises should preserve those relative rates while allowing each rate to rise or fall. Downes et al. (2002) provided worked examples of this approach for environmental monitoring.

Occasionally, the calculations may show that the MDES is much less than the desirable ES, suggesting that the experimental/sampling program is more sensitive than expected. Here, you could consider reducing the replication, with the possibility of using "spare" resources for further studies. Our experience suggests this latter situation is uncommon.

Power calculations as part of an experimental or sampling program design are approximate. We often use estimates of variation that tend to have large CIs, so our power estimates would also have considerable imprecision. However, uncertainty in our estimates of variance and desired ES are not formally considered as part of power analysis. Wang et al. (2005) suggested that a Bayesian approach, combined with simulation, could incorporate this uncertainty by using Bayesian estimates (based on noninformative priors) in the usual sample size formulae. Other Bayesian approaches to sample size determination are described by Inoue et al. (2005), who connect frequentist and Bayesian methods for sample size determination, and Kruschke (2011).

A priori power analysis should be a routine part of planning any experiment, even if researchers do not plan on making inferences based on "statistical significance" or prefer to work in a Bayesian context. Our initial power estimates may be uncertain, especially when we have a poor estimate of the variation present in our data. However, we will at least know whether "important" effects are likely to be detected, given our resources.

3.4 Key Points

• Getting clear answers to biological questions has three important components: designing data collection programs that allow us to separate competing explanations, collecting data that are representative of broader statistical populations, and collecting enough data to be confident about our conclusions.

• Random sampling from (statistical) populations should be the default approach, but some biological situations require more complex sampling.

• Confounding describes the situation where several explanations for patterns in the data cannot be separated, most commonly where we can't distinguish our biological question of interest from other factors.

• In experiments, confounding can occur when appropriate controls are missing.

• Confounding also occurs because of inappropriate replication. Experimental or sampling units represent independent instances, particularly of experimental treatments, and they let us measure background variation. Observational units are the actual measurements, and there can be multiple observations for each replicate experimental or sampling unit. Confounding can occur when observational units are used as the measure of variation against which we assess effects.

• When there are multiple observational units for each replicate sampling or experimental unit, we recommend summarizing them as a mean (or other measure) before data analysis.

• We also strongly recommend determining an appropriate number of units before collecting data. This can be framed as the number necessary to estimate a parameter with a target confidence or precision or to detect a biological change of a given size.

• Power analysis is used for the second task. To be effective, the statistical models and their assumptions need to be specified clearly, and we need to identify a biological change we wish to detect – the effect size.

Further Reading

Sampling design: Levy and Lemeshow (2008) and Thompson (2012) are recommended general books on sampling, with Gruijter et al. (2006) and Manly and Navarro (2014) having a more ecological emphasis. Thompson (2012) provides an excellent introduction to adaptive sampling, but a more detailed text is Thompson and Seber (1996).

Experimental design: Most introductory and linear models textbooks will include sections on experimental design and power analysis, and we recommend Mead et al. (2012) for a comprehensive overview with emphasis on block and other more complex designs. Underwood (1997), Hairston (1989), and Resetarits and Fauth (1998) are also good.

4 Introduction to Linear Models

We introduced the concept of models in Chapter 1, particularly how we test biological hypotheses by creating a statistical model that is contrasted with data. Now we want to introduce these statistical models, primarily linear models that are used very widely in biology. We will:

- introduce linear models and the kinds of data to which they apply;
- describe how different components of models are put together;
- outline how we fit these models to data;
- show how to decide whether a model is a good fit to the data; and
- list common assumptions made when fitting linear models.

Linear models include many statistical analyses familiar to biologists, such as linear regression, ANOVA, analysis of covariance (ANCOVA), linear mixed models, logistic and Poisson regression, and loglinear models. In this chapter, we show how they are variations on a common theme.

4.1 What Is a Linear Model?

We can assess the straight-line relationship between two random variables with the familiar (Pearson) correlation coefficient (r). Where we can specify a response or dependent variable and one or more predictor variables, also termed the independent variables or covariates, we use linear models.

We expect the predictor variables to explain some of the pattern in the response variable.

A key aspect of most linear models is that, for statistical inference such as CIs and evaluating hypotheses, we must assume some form of probability distribution for the response variable. A model defined in terms of a probability distribution with estimable parameters is a

parametric model and will be the focus of much of this book. Nonparametric models are, however, covered as part of several later chapters.

Consider a population of N units, from which we take a sample of n. For each unit i, there is a value for each response and predictor variable. Linear statistical models take the general form:

$$(\text{response variable})_i = (\text{linear predictors})_i \\ + (\text{random error})_i.$$

We are modeling or predicting the value of the response variable for any unit from the linear relationship with the predictor variable(s). The response variable can be continuous (e.g. measurement) or discrete (e.g. counts, presence/absence, categories). The predictor variables, and their parameters, are included as a linear combination (Box 4.1). They can be continuous (e.g. classical "regression" models), categorical (e.g. classical ANOVA models), or both. Predictors can also be fixed (inference restricted to the values used) or random (inference generalized to a broader population). The random error component represents the part of the response not explained by the model.

For example, we introduced in earlier chapters the Shibao et al. (2004) study of factors affecting the production of soldiers in a gall-forming aphid. They had a sample of galls (from a population of possible galls) and recorded the proportion of soldiers, and aphid density. The proportion of soldiers is our response variable (Y) and aphid density is the predictor (X). A suitable linear model would be:

$$y_i = \beta_0 + \beta_1 x_i + \varepsilon_i$$

$$(\text{soldier proportion})_i = \beta_0 + \beta_1 (\text{aphid density})_i + \varepsilon_i,$$

where y_i and x_i are the values of the response variable (soldier proportion) and predictor variable (aphid density), respectively, for unit i, β_0 (regression intercept,

Box 4.1 What Does Linear Mean?

"Linear model" has been used in two distinct ways. It can refer to a model of a straight-line relationship between two variables. This is often the interpretation biologists are familiar with and can be termed "linear in form" (Boldina & Beninger 2016). The second meaning is that a linear model is simply one in which the response variable is described by a linear combination of parameters (regression slopes, intercept), and "no parameter appears as an exponent or is multiplied or divided by another parameter" (Kutner et al. 2005). Here, "linear" refers to the combination of parameters, not necessarily the shape of the relationship. Under this definition, linear models with a single predictor variable can represent straight-line relationships and curvilinear relationships, such as models with polynomial terms.

proportion of soldiers when aphid density is zero) and β_1 (regression slope, the change in soldier proportion for an increase in aphid density of one unit) are relevant parameters of the model to be estimated, and ε_i is a random error component for each unit. Parameter β_1 provides a measure of the "effect" of the predictor variable on the response, although effect here is a statistical term. A causal relationship can only be inferred if the study design allows this, for example via experimental manipulation. This simple linear regression model will be familiar to many biologists. Our statistical focus is on estimating the model parameters with CIs, assessing how well the model fits our sample data, and possibly evaluating hypotheses about the parameters.

Gelman and Hill (2006) emphasized that linear models should be valid – the response variable should reflect the biological phenomenon or outcome of interest. All predictor variables biologically relevant to the question (and measurable) should be included. The model should be generalizable to the population from which the observations came. The first two criteria combine researchers' judgment based on their knowledge of their study system and specifying clear research questions and hypotheses. The third criterion should be met via an appropriate sampling regime.

4.2 Components of Linear Models

The first step is to consider the nature of the response and predictor variables, as they determine the specific nature of the model.

4.2.1 Types of Response Variables

Modern linear models can accommodate a broad range of response variables and underlying probability distributions. While most biologists are familiar with normally distributed continuous responses, linear models can handle other types of responses, including discrete variables such as binary data and counts. The probability

distributions for response variables (and error terms) in linear models usually come from the exponential (dispersion) family of distributions (see Box 4.2).

4.2.1.1 Continuous Response Variables

The normal (Gaussian) distribution applies to a range of measurement-type variables and underlies much statistical inference using OLS. Commonly, however, biological measurements have a positively skewed distribution. If the response variable follows a lognormal distribution, it is common practice to transform it to a logarithmic scale so it follows a normal distribution. Two other distributions are sometimes used for skewed continuous variables in linear models. The gamma distribution may be applicable when the response is a time interval, such as waiting times between events or arrival times, and the inverse Gaussian has been applied to modeling lifetimes.

4.2.1.2 Discrete Response Variables

The two commonest discrete response variables in biology are (1) those that can take only one of two values, for example presence/absence, alive/dead, or more generally success/failure; and (2) those representing counts, which can only take nonnegative integer values. Binary responses are usually modeled using a binomial distribution, the distribution of the number of successes from a series of independent trials. Count response variables are usually modeled using a Poisson distribution, in which the mean equals the variance. Biological count variables, however, are often overdispersed, with variance > mean, especially for larger counts like numbers of organisms in sampling units. Strategies for modeling overdispersed data will be considered in Section 13.6.2.

Discrete variables can also have more than two states. They may be ordered or ordinal – such as predator choices between several increasing size classes of prey from the same species – or unordered or nominal – such as predator choices between prey of several species. We

Box 4.2 The Exponential and Exponential Dispersion Family of Distributions

Distributions from the exponential family are characterized by the relationship between the expected value or mean (μ) and the variance (σ^2). For each distribution, there is a natural or canonical link function that transforms the mean of the response variable to ensure a linear relationship with the predictors. This transformed mean is termed the natural or canonical parameter (θ). The variance is also commonly expressed as a function of the mean, termed the variance function. Two of the most common exponential distributions are:

• The binomial distribution for binary or proportion responses. The mean is the probability of one outcome (e.g. success), sometimes given the symbol π, and the variance is the odds of one outcome. The canonical parameter is termed the logit, the log of the odds of one outcome.
• The Poisson distribution for counts. The mean equals the variance, and the canonical parameter is the log of the mean response.

The exponential family can be extended to allow more complex (and flexible) relationships between the mean and the variance by including an additional parameter related to the variance called the dispersion parameter (ϕ). Now we have the exponential dispersion family. For binomial and Poisson distributions, ϕ is set to 1. When ϕ is unknown and estimated from the data, a range of other distributions are possible, including:

• The normal (Gaussian) distribution is used for continuous response variables and the mean and the variance are unrelated. The canonical parameter is simply the mean and the dispersion parameter is the variance (σ^2). While we can model normally distributed responses with standard linear models using OLS, the normal distribution is part of the exponential dispersion family and can also be modeled as a generalized linear model (GLM) using ML.
• The gamma and inverse Gaussian distributions can, like the normal, be used for continuous response variables, but are particularly applicable when the response variable is strongly positively skewed. Their link functions and canonical parameters are inverse functions of the mean, and the dispersion parameter is estimated from the data. The gamma distribution may be applicable when the response is a time interval, such as waiting times between events or arrival times, and the inverse Gaussian has been applied to modeling lifetimes.
• The negative binomial is a mixture of two exponential distributions, a Poisson distribution with an overlying gamma distribution and a parameter k that specifies how much the distribution differs from a Poisson; the bigger k, the greater the variance is compared to the mean. It is useful for modeling count data where the variance is greater than the mean (i.e. overdispersed data).
• The quasi-Poisson distribution is a Poisson distribution where ϕ is no longer set to 1 but is estimated from the data, so the variance is not constrained to equal the mean. Like the negative binomial, it is used for modeling count data where the variance is greater than the mean (Section 13.6.2.1). There is an equivalent modification of the binomial called the quasi-binomial (Section 13.6.5).

The statistical characteristics of these distributions are summarized in Table 4.1.

Table 4.1 Statistical characteristics of distributions

Distribution	Common response	Canonical parameter (θ)	Variance function	Canonical link
Binomial	Discrete binary or proportions	$\log[\mu(1-\mu)]$	$\mu(1-\mu)$	Logit
Poisson	Discrete count	$\log(\mu)$	μ	Log
Normal (Gaussian)	Continuous symmetrical	μ	1	Identity
Gamma	Continuous positively skewed	$-1/\mu$	μ^2	Inverse
Inverse Gaussian	Continuous positively skewed	$-1/2\mu^2$	μ^3	Inverse square
Negative binomial	Discrete count	$\log[\mu/(\mu+k)]$	$\mu+\mu^2/k$	Log

Box 4.2 (cont.)

The other distribution important in biological research is the lognormal distribution, applicable for continuous positively skewed response variables. The lognormal is characterized by the log-transformed response having a normal distribution. The lognormal distribution is not part of the family of exponential distributions, so it is not usually modeled directly in linear models. There are several methods for modeling lognormal distributions (Gustavsson et al. 2014), but commonly the response variable is transformed to a logarithmic scale and modeling proceeds using a normal distribution.

For good summaries of these and other probability distributions used in statistical modeling, see Bolker (2008), Dunn and Smyth (2018), or Pekár and Brabec (2016).

will use binary response variables to illustrate how to analyze such data, but the methods can be extended to multi-state variables.

4.2.2 Types of Predictor Variables

Predictors may be continuous, with many possible values, or categorical (discrete), where the units are allocated into groups or categories, or the predictor has only a few states.

4.2.2.1 Categorical (Discrete) vs. Continuous

Categorical predictors are sometimes called factors. Continuous predictors are quantitative, while categorical predictors can be quantitative (e.g. different temperature or drug dose or toxicant concentration groups) or qualitative (e.g. locations or different drugs or toxicants). For example, the predictor variable "anesthetic" from the study on mouse physiology (Low et al. 2016; see also Box 2.2) is categorical and qualitative, being two anesthetics. In contrast, for the study of soldier production by gall-forming aphids, the predictor is continuous and quantitative, as density is measured numerically for each gall.

4.2.2.2 Fixed vs. Random

Another important classification is whether a predictor is fixed or random. A fixed predictor is where all the values of the predictor (e.g. all the groups of a categorical factor) that are of interest are included in our sampling/experimental program and model. Conclusions for a fixed predictor are restricted to those groups we used. We cannot extrapolate our statistical inference beyond them to other values or groups not in the study. If we repeated the study, we would usually use the same values of the fixed predictor again. When all predictors are fixed, we have fixed effects models.

A random predictor is where we have a population of possible predictor values, and we are only using a random selection of them. Our inference is about the population of possible predictor values. If we repeated our study, we

would usually take another random sample of predictor values from the population of possible values. Random categorical predictors in biology often represent replication (e.g. sites, blocks, subjects) to measure variability. When all predictors are random, we have random effects models.

In the mouse physiology study, anesthetic is clearly a fixed effect, as the research question is about two specific anesthetics – inference could not be extrapolated to other anesthetics. If Low et al. (2016) had chosen anesthetic types at random from a population of possible anesthetics, then anesthetic type could be considered a random effect, although the study would not make much biological sense.

As sampling or experimental designs (and biological questions) become more complex, it is not unusual to include fixed and random predictors, producing mixed effects models.

4.3 Assembling our Linear Model

A (generalized) linear model has three components. The first two are:

- Random component – the response variable and its probability distribution. The probability distribution must be from the exponential (dispersion) family of distributions, which includes the normal distribution. These distributions are summarized in Box 4.2.
- Systematic component (or linear predictor as per Agresti 2015) – the predictors in the model and their associated parameters (model coefficients such as regression slopes).

Once we have chosen our response variable and its probability distribution and specified our predictors, we consider how the predictors link to response values. We do this with the third component:

- Link function, which links the random and the systematic components. The link function is essentially a transformation of the response variable, so the transformed response now has a linear relationship with the predictors. In practice, a link function, designated $g(\mu_i)$, links the expected value of $y_i (\mu_i)$ to the predictors. The same link function is used for all observations.

The three commonest link functions used by biologists in linear modeling are:
- Identity, $g(\mu_i) = \mu_i$, models the mean or expected value of y_i. It is used in well-known linear models based on a normal distribution such as linear regression.
- Log, $g(\mu_i) = \log(\mu_i)$, models the log of the mean (or expected value) of y_i. It is used for count data in Poisson regression and loglinear models.
- Logit, $g(\mu_i) = \log[\mu_i/(1 - \mu_i)]$, used for binary and proportion responses in logistic regression or, more generally, logit models. It models the log of the odds of one outcome (e.g. success) against the alternative (e.g. failure).

The link function converts the expected value of y_i to the natural parameter. It allows a linear model form to be fitted for a range of continuous and discrete data types that would otherwise be nonlinear.

These broad models are termed generalized linear models. Linear models can be expressed in two ways. We can specify the actual value of any observation or its expected or mean value. Standard linear regression models relate the individual values (y_i) to the predictors (and their associated parameters) and they explicitly include an error term (ε_i) :

$$y_i = \beta_0 + \beta_1 x_i + \varepsilon_i.$$

For each value of x_i, the error term represents the difference between the observed value and its expected value.

The simple linear regression model above can be re-expressed to model the expected value (mean) of y_i at each x_i:

$$E(y_i) = \beta_0 + \beta_1 x_i.$$

As we are now modeling the expected value, the error term is no longer included. This is the standard way of expressing GLMs, and common GLMs usually don't include an explicit error term (Agresti 2015). Nonetheless, the concept of "error" in fitting these models is still important.

The parameter(s) of most interest in linear models are those associated with the predictor variables, as they measure the "effect" of the predictors on the response. The parameter β_1, the regression slope measuring the change in response for a unit change in the predictor, is of most interest in the linear regression model.

Linear models commonly involve multiple predictors:

$$y_i = \beta_0 + \beta_1 x_{1i} + \beta_2 x_{2i} + \cdots + \varepsilon_i.$$

As an example, Ercit et al. (2014) examined the factors affecting the distance jumped (response variable) by laboratory-reared tree crickets (see Box 8.1). They had three predictor variables measured for each cricket: pronotum length, leg mass, and number of eggs. The linear model they used was:

$$\begin{aligned}(\text{jump distance})_i = {}&\beta_0 + \beta_1(\text{pronotum length})_i \\ &+ \beta_2(\text{leg mass})_i + \beta_3(\text{egg number})_i \\ &+ \varepsilon_i.\end{aligned}$$

Each regression coefficient or slope measures the change in the response for a single unit change in that predictor, independent of the others.

4.4 Estimation for Linear Models

Our primary aim is to fit our model to our data, estimate model parameters, and, if relevant, evaluate hypotheses about them. We introduced the principles underlying frequentist estimation in Section 2.4. Our estimators should be best linear unbiased estimators (BLUE) – that is, unbiased and have minimum variance (be the most precise possible). We also introduced two estimation methods, OLS and ML, that are most commonly used to estimate the parameters of linear models. Here we will illustrate OLS and ML estimation for a simple linear model with a continuous response and a continuous predictor recorded on each of n sampling units from a population of possible units, using the aphid study (see also Box 6.2); more statistical detail, including the common computer algorithms used for both, are given in Box 4.3. We can fit a linear regression model with proportion of soldiers as the response and aphid density as the predictor recorded from each gall in our sample. While we have a sample of units from a population and have recorded the random response and predictor variables for each unit, the statistical basis of regression models is a little different. The predictor values are treated as fixed and there is a probability distribution of values of the response (soldier proportion) for each value of the predictor (aphid density) – see Figure 4.1. The regression model predicts the expected value for the response (i.e. the mean of each of these probability distributions) at each x_i. This regression model is valid even though both our response and predictor are random variables in practice and our sample data only has a single value of y_i for each x_i; this issue is discussed in more detail in Section 6.1.1.

In the simple linear regression model our interest is estimating β_0 and β_1.

Box 4.3 OLS and ML Estimation for Linear Models

We briefly introduced OLS and ML estimation in Section 2.4.1. Now we'll extend these methods to estimating parameters of a linear model. Consider the linear regression model $y_i = \beta_0 + \beta_1 x_i + \varepsilon_i$, illustrated with the data on the proportion of soldiers (response) and aphid density (predictor) from a sample of galls (Shibao et al. 2004):

$$(\text{soldier proportion})_i = \beta_0 + \beta_1 (\text{aphid density})_i + \varepsilon_i$$

Our aim is to estimate the parameters β_0 (intercept), β_1 (slope), and σ_ϵ^2 (variance of the error terms).

Ordinary Least Squares

The OLS estimators for β_0 and β_1 (b_0 and b_1) are the values that minimize the SS between each sample observation of the response variable (y_i) and the value predicted by our model at each value of x_i:

$$\sum_{i=1}^{n} (y_i - \widehat{y}_i)^2 = \sum_{i=1}^{n} [y_i - (\beta_0 + \beta_1 x_i)]^2,$$

where $\widehat{y}_i$ is the predicted (or fitted) y_i. This is termed the residual or error sum-of-squares ($\text{SS}_{\text{Residual}}$) and the OLS estimators for β_0 and β_1 minimize this $\text{SS}_{\text{Residual}}$ (Figure 4.2).

To find the estimates of β_0 and β_1 that minimize the $\text{SS}_{\text{Residual}}$, we set the derivatives of this SS with respect to β_0 and β_1 (the changes in the SS for a change in β_0 and β_1) to zero to produce a set of "normal" equations:

$$\sum_{i=1}^{n} y_i = nb_0 + b_1 \sum_{i=1}^{n} x_i,$$

$$\sum_{i=1}^{n} x_i y_i = b_0 \sum_{i=1}^{n} x_i + b_1 \sum_{i=1}^{n} x_i^2.$$

Solving these equations gives the OLS estimates b_0 and b_1 (see Section 6.1.2). For models with multiple predictors, this process is best represented with some matrix algebra and involves taking the inverse of the matrix of predictor variable values multiplied by its transpose $\left[(\mathbf{X}' \mathbf{X})^{-1} \right]$. Unfortunately, this inversion can introduce inaccuracies, so most modern linear models software uses a different approach called QR decomposition of the $\mathbf{X}$ matrix. Hilbe and Robinson (2013) provide a good explanation.

Note that the residual variance $\left(\sigma_\epsilon^2 \right)$ is separately estimated by the $\text{MS}_{\text{Residual}}$ ($\text{SS}_{\text{Residual}}/\text{df}$) from the ANOVA of the fit of our model (Section 4.5.1).

Maximum Likelihood

Maximum likelihood estimators for β_0 and β_1 (b_0 and b_1) are those that maximize the likelihood of observing our sample data. To determine ML estimators, we need the likelihood function, which provides the likelihood of the observed data for all possible values of the parameters we are estimating. The likelihood function is derived from the product of the probability densities for each observation, given the parameters:

$$L(\theta) = \ln \left(\prod_{i=1}^{n} f(y_i; \theta) \right) = \sum_{i=1}^{n} \ln[f(y_i; \theta)]$$

The likelihood function is basically the joint probability distribution of the data and the parameters. The f represents a specific probability distribution; for a regression model like this, we would use the normal distribution, whereas we could use others from the exponential family such as the binomial for binary responses or the Poisson for counts. Working with products (Π) makes computation more difficult so we usually convert the likelihood function to a log-likelihood function:

$$LL\left(y | \beta_0, \beta_1, \sigma^2\right) = \sum_{i=1}^{n} \ln \left[f\left(y_i | x_i, \beta_0, \beta_1, \sigma^2\right) \right],$$

Box 4.3 (cont.)

where ln is the natural logarithm, and we are now working with sums (Σ). Note that in contrast to OLS, where σ_ϵ^2 was estimated separately from the ANOVA, the residual variance is usually part of the likelihood function, although for a simple regression model like this we could still estimate β_0 and β_1 without estimating σ_ϵ^2. Generally, likelihood functions include all the model parameters.

The likelihood and log-likelihood functions for a simple regression model, focusing just on the intercept and slope parameters, is represented graphically in Figure 4.3. To find the maximum of the log-likelihood function, we need to find the values of the parameters where the slope of the log-likelihood function is zero. To do this, we set the derivative (i.e. the slope or the change in the log-likelihood for a given change in the values of the parameters; termed the first derivative) to zero and solve for β_0 and β_1 (and potentially σ_ϵ^2). With normal-based models like this simple regression model, the ML estimates b_0 and b_1 are the same as the OLS estimates above.

Maximum likelihood estimation for linear models based on nonnormal distributions requires some sort of optimization procedure to find the parameter estimates that maximize the log-likelihood function. Two common algorithms are the Newton–Raphson and the Fisher scoring methods. Most common GLMs use the canonical link function (e.g. identity link for normal distributions, log-link for Poisson distributions) and these two algorithms are then identical and are often referred to as iteratively reweighted least squares (IRLS; a form of GLS). Iteratively reweighted least squares converges quite quickly to reliable estimates, although when sample sizes are small, and/or when we include random effects and their associated variances in the models, convergence can be problematic. Increasing the number of iterations or reducing the number of parameters in the model, and hence in the likelihood function, can help.

Standard errors are related to the curvature of the log-likelihood function with greater (steeper) curvature indicating more precise parameter estimates. Standard errors of ML estimates for normal-based models are derived in the usual way, assuming the sampling distribution of the estimate is also normally distributed. For ML estimates derived using IRLS for nonnormal GLMs, SEs are trickier to determine. We use second derivatives, measures of how quickly the first derivative or slope of the log-likelihood function changes. The collection of these observed second derivatives, one for each parameter we are estimating, is termed the Hessian matrix. Alternatively, and sometimes more easily, we can use the expected values of the second derivatives, derived from our model, and termed the information matrix. Standard errors are determined from the inverse of the square root of the elements in the information matrix. The details are not critical, just appreciate that we are measuring the steepness of the peak of the log-likelihood function for each parameter – the steeper the peak, the smaller the standard error.

One final piece of complexity for ML estimation is the derivation of CIs. While we can use the traditional (Wald) method of combining the SE with the standard normal z distribution if the parameter estimate is normally distributed, this may not be suitable for nonnormal GLMs when the estimate is not normally distributed, especially if the sample sizes are small. An alternative is to derive profile CIs based on profile log-likelihoods, likelihood functions for a single parameter maximized over (with respect to) the remaining parameters. This is an iterative process where maximizing the likelihood over the other parameters is done for a range of values of the parameter of interest. The profile likelihood for a parameter can then be used to construct a CI based on the χ^2 distribution.

Good introductions to OLS and ML estimation can be found in Bolker (2008) and Hilbe and Robinson (2013), although most linear models textbooks will provide coverage of both approaches.

4.4.1 Ordinary Least Squares

Ordinary least squares estimators minimize the sum of squared deviations, called residual or error sums-of-squares (or $SS_{Residual}$), between each sample observation of the response variable (y_i) and the value predicted by our model ($\hat{y}_i$) for its value of x_i:

$$\sum_{i=1}^{n} (y_i - \hat{y})^2.$$

The differences between the individual observed and predicted values, $y_i - \hat{y}$, are residuals. They can be viewed as estimates of the model error terms since our best estimate of the expected value or mean of y_i for each x_i is $\hat{y}$. Residuals are particularly important for assessing the fit of most linear models.

In addition to the intercept (β_0) and the regression slope (β_1), we also wish to estimate σ_ϵ^2, the variance of the error terms (sometimes called the error variance). This error variance is often abbreviated to σ^2, referring to assumed constant unexplained variance from the model. We need an estimate of σ^2 to evaluate the model assumptions and measure how well the model fits our sample data. We also use it to calculate SE for the other parameters and assess confidence in our predictions from the model. For relatively simple models with few parameters, we can use the ANOVA (Section 4.5.1) to estimate the error variance – the $SS_{Residual}$ divided by its degrees of freedom, termed the mean square ($MS_{Residual}$). However, this ANOVA approach is often not effective for estimating variances in more complex models, particularly with multiple random effects, and methods based on ML are preferred.

Point estimates of model parameters are of limited use, and statistical inference requires some measure of uncertainty. Standard errors for the estimated model coefficients are determined in the same way as for population means. They rely on our estimate of σ^2 and are standard output from statistical software. Confidence intervals are usually determined using the t or z (for sample sizes >30) distribution and the SE.

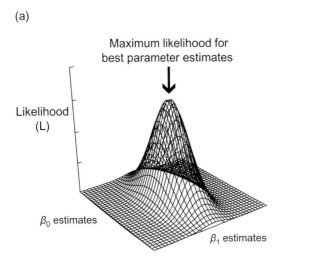

Figure 4.1 Diagrammatic representation of a linear regression model (based on an example from Shibao et al. 2004) showing the population of y_i at two values of x_i. Note that the population regression model relates the mean of Y at each X-value (x_i) to $\beta_0 + \beta_1 x_i$.

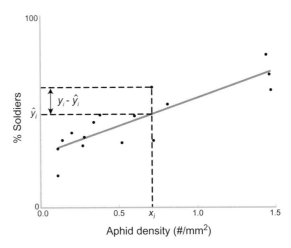

Figure 4.2 Illustration of the sum of squared deviations minimized in OLS estimation for the linear regression model relating proportion of soldiers to aphid density (data from Shibao et al. 2004).

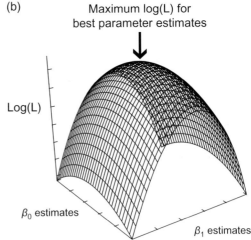

Figure 4.3 Illustration of a log-likelihood function used in ML estimation for a linear regression model, conditional on known σ^2 (to simplify the likelihood function to two dimensions). Panel (a) shows likelihood, and (b) shows the broad shape of the log-likelihood function.

We are often interested in evaluating hypotheses about these parameters, especially the hypothesis that the regression slope is some specific value, often 0. If $\beta_1 = 0$, there is no relationship between the predictor and the response. Individual linear model parameters can be evaluated in one of two ways:

• Using a t test for the model parameter, where the t statistic is calculated as the estimated parameter minus the value specified in the hypothesis, divided by its SE. The probability (P-value) of obtaining this t statistic or one more extreme under the specified hypothesis is determined by comparing it to a t distribution with the appropriate degrees of freedom.

• Comparing the fit of the model with the parameter included (termed the full model) to a reduced model omitting that parameter. For a full model with a single predictor, as in the gall example, we partition the total variation in the response into that explained by the model and unexplained (residual or error variance) using an ANOVA. The ratio of two variances is termed an F statistic (Section 2.5.5); in this case the ratio of explained to unexplained variance is equivalent to comparing the explained (or unexplained) variances of the full and reduced models. The probability of obtaining this F or one more extreme under the hypothesis (in this case, $\beta_1 = 0$) is determined by comparing it to an F distribution with the degrees of freedom associated with the explained and unexplained variances.

The term "ordinary" is used with OLS because there are other types of least squares estimation. Generalized least squares (GLS; also termed weighted least squares, WLS) modifies OLS so unequal variance, and even non-independence, of the error terms can be modeled (Gelman & Hill 2006). One difficulty with GLS is having to use sample estimates of the variance (and covariance) matrix for the weighting, when these estimates may not be close to the true population values. Penalized least squares minimizes the residual sum-of-squares like OLS, but includes a term that penalizes unnecessarily complex models, analogous to the way information criteria for model fit, such as AIC and BIC (Bayesian information criterion) do (Section 4.5.3). Both are used by some software packages for linear mixed models and will be discussed further in Chapter 10.

OLS estimators are usually more straightforward to calculate than ML estimators. While they have no distributional assumptions for point estimates, they do have assumptions for CIs and hypothesis tests (Section 4.6). The major application of OLS is estimating parameters (of fixed effects) in linear models based on normal distributions.

4.4.2 Maximum Likelihood

Maximum likelihood parameter estimates maximize the likelihood of observing our sample data (Section 2.4.1). We determine them by deriving the likelihood function, which provides the likelihood of observing the sample data as a function of the parameter values. The specific likelihood function for a linear model depends on the underlying probability distribution used to represent the response variable. The ML estimates for a linear regression model are the values of b_0, b_1, (and potentially s^2) that maximize the relevant likelihood function based on a normal distribution.

Graphically, likelihood (or log-likelihood) functions will have as many dimensions as there are parameters to be estimated, and the functions are generally "hill-shaped" (Gelman & Hill 2006), with ideally a single optimum (maximum) for the set of parameters (Figure 4.3). Maximum likelihood estimates are usually obtained using an iterative optimization method.

Estimation of the residual variance in linear regression models was done separately from the other parameters using OLS. With ML, the variance is usually incorporated as part of the likelihood function. The ML estimator is $SS_{Residual}/n$, but this is biased because it doesn't consider that we have also estimated the intercept and slope (β_0, β_1). An alternative estimator, termed REML is $SS_{Residual}/(n-2)$; this is the same as that derived using the ANOVA method and is unbiased because it accounts for the fixed parameters. We will address ANOVA, ML, and REML for variance estimation in more detail in Chapter 10.

Derivation of SEs is based on the estimated parameter having an approximate normal distribution for large sample sizes. The CIs are usually calculated using the z or t distribution and the SE. Profile CIs based on profile likelihood functions are more robust for nonnormal sampling distributions and small sample sizes.

Evaluating hypotheses about the model parameters (e.g. β_1) can be done in at least two ways:

• The Wald (z test) test, where z is calculated as the estimated parameter minus the value specified in the hypothesis (usually zero for null hypotheses) divided by its SE. The probability of obtaining this z or one more extreme under the null hypothesis is determined by comparing to a z distribution (or z^2 is compared to a χ^2 distribution with df $= 1$).

• Likelihood-ratio test using the ratio (Λ) of the maximized likelihood (l_0) of the reduced model under the hypothesis (e.g. $\beta_1 = 0$) to the maximized likelihood (l_1) of the full model including the parameter of interest. The test is actually based on $-2\log$(likelihood ratio), which follows a χ^2 distribution if sample sizes are not too small:

$$-2 \log \varLambda = -2 \log(l_0/l_1) = -2(L_0 - L_1)$$

where L_0 and L_1 are the maximized log-likelihoods of the reduced and full models, respectively. This approach is the likelihood equivalent of comparing models using explained variance with OLS. Generally, likelihood-ratio tests are preferred to Wald tests, especially with smaller sample sizes.

While ML estimators have distributional assumptions for both point estimates and CIs and hypothesis tests, the distributions don't have to be normal, so they are preferred for GLMs based on other distributions from the exponential family.

4.4.3 Robust Estimation Methods for Linear Models

Ordinary least squares and ML are widely used but they can have two drawbacks:

1. We must assume specific relationships between the mean and the variance for the response variable and/or error terms.

2. Standard errors and CIs based on ML methods are usually reliable for large sample sizes, but less reliable for small samples. Small sample sizes also make it more difficult to assess the distributional assumptions.

There are methods that are more robust to these assumptions (Sections 2.4.7 and 2.5.6), and the main ones used for linear models include:

• Rank-based methods, mostly focusing on evaluating hypotheses about parameters – that is, producing P-values. They are commonly termed nonparametric or distribution-free.

• M-estimators, for example a parameter estimate that minimizes the least absolute deviations (LAD), the absolute values of the model residuals.

• Resampling and permutation procedures. These methods will be described for specific models in later chapters.

4.5 How Well Does a Model Fit?

A key component of analyzing linear models is assessing how well our model fits the sample data. Measures of fit allow us to distinguish between competing models as descriptors/explanations for our observed data and choose a single "best" model if needed (but see Chapter 9). Some measures of fit, such as explained variance or deviance (see below) also provide the basic information needed for statistical inference (CIs and hypothesis tests). Fit is commonly assessed with OLS or ML, but there is a class of measures based on information criteria that are increasingly important when fitting complex models.

4.5.1 OLS Measures of Fit: ANOVA

Ordinary least squares estimation lets us partition the total variation in our response variable into two components: (1) variation explained by our predictors, and (2) variation not explained by them. Partitioning the variation like this is called analysis of variance, or ANOVA (Table 4.2). The term ANOVA is commonly used by biologists (and some statistical textbooks) to refer to a specific class of linear models with only categorical predictors and a normally distributed response variable. This is somewhat of a misnomer – while ANOVA is used to analyze such models, it broadly applies to all linear models based on OLS estimation (including classical "regression") and it has a role in ML models (Gelman 2016).

For example, the variation in proportion of soldiers in galls is partitioned into the variation explained by our linear predictor, aphid density, and the variation not explained. Formulae to illustrate calculating SS for simple models will be provided in Chapter 6. A widely used summary measure of fit for individual models is the coefficient of determination, or R-squared (r^2 or R^2), expressing $SS_{Explained}$ as a proportion of SS_{Total}.

Sum-of-squares are useful summaries of variation because, for many linear models, they are additive: $SS_{Explained} + SS_{Residual} = SS_{Total}$. However, they are difficult to compare directly because they don't consider the number of components making up each source of variation. We can convert the SS into a variance (also called a mean square) by dividing by the appropriate df; the df are the number of components contributing to the SS, minus the number of parameters already estimated (see Box 2.1).

While MS are no longer additive $\left(MS_{Regression} + MS_{Residual} \neq MS_{Total}\right)$, the ratio of two

Table 4.2 Generalized ANOVA table for linear models

Source of variation	Sum-of-squares	Degrees of freedom	Mean square
Explained by linear combination of predictor(s)	$SS_{Explained}$	$df_{Explained}$	$MS_{Explained}$
Unexplained or residual or error	$SS_{Residual}$	$df_{Residual}$	$MS_{Residual}$
Total	SS_{Total}	df_{Total}	

MS is an F (Section 2.5.5), and when the model's assumptions are met, it follows an F distribution with df associated with the two variances. We can evaluate a null hypothesis that the population variances estimated by the two mean squares (e.g. $MS_{Regression}$ and $MS_{Residual}$) are the same. This is the equivalent of comparing the fit of the full model (with β_0 and β_1) to a reduced model with only β_0. This model-comparison approach, based on the ANOVA, can compare any two nested (i.e. where one model is a subset of the other) linear models and, therefore, test a wide variety of hypotheses about predictors.

4.5.2 ML Measures of Fit: Log-Likelihood and Deviance

Maximum likelihood takes a different approach. The usual practice (Hosmer et al. 2013) is to compare the maximized likelihood of our model (l_1) to the maximized likelihood of what's called the saturated model (l_S), a model with as many parameters as there are observations (like a linear regression model with intercept and slope parameters for two observations). The saturated model will fit the data perfectly, but is not that useful. Our aim with model fitting is usually to find the most parsimonious model, the one that fits best with the fewest parameters. The saturated model does, however, provide a consistent basis to assess the fit of competing models.

The log-likelihood ratio statistic comparing our model to the saturated model is termed the deviance – the larger the deviance, the worse the fit. For any model, maximizing the (log) likelihood is the same as minimizing the deviance. This is a generalization of minimizing the $SS_{Residual}$ that happens with OLS. Indeed, if we have a linear model with a normally distributed response (and error terms), the $SS_{Residual}$ equals the deviance (Agresti 2015). Comparing the fit of two nested models fitted using ML is a matter of calculating the difference between the deviances, which itself is a deviance, with df being the difference in the two models' df. It follows a χ^2 distribution if sample sizes are not too small.

4.5.3 Information Criteria

While measures of fit based on analyzing variance or deviance are an integral part of linear model analyses, they are not necessarily ideal for comparing models with varying numbers of parameters. Adding extra parameters (e.g. extra predictors) means the r^2 rises and the deviance will decrease, even if those parameters/predictors contribute little additional information. The model with the most parameters will always fit better than models with subsets of these parameters. There are modifications to account for the number of model parameters (e.g. adjusted r^2), but

an alternative is to use information criteria that consider the fit (e.g. log-likelihood) and impose a penalty for overfitting (i.e. too many predictors). The two most familiar measures are the AIC and BIC. A general expression for the AIC is:

$$-2\,(\text{log-likelihood} - p),$$

where p is the number of parameters in the model. The BIC is similar but also considers sample size. Smaller values indicate a better fit. Information criteria can be used for comparing nonnested models, such as models that are not subsets of another model. We will examine information criteria in Chapter 9 and later chapters, where we are interested in selecting the "best" models from a set of candidates.

4.6 Assumptions for Linear Model Inference

In fitting linear models and estimating their parameters, we make assumptions about our data and model structure. We have already emphasized the importance of a valid, biologically relevant model based on a robust sampling or experimental design (Gelman & Hill 2006). The other assumptions relate to the model's characteristics or the error terms from our model. We will describe the assumptions underlying OLS-based inference and then broaden this to consider ML.

4.6.1 Assumptions for OLS

For our OLS estimators to be BLUE, the Gauss–Markov theorem states that certain assumptions about the model error terms must hold (Weisberg 2013). One is that the model is linear in the parameters (Box 4.1).

When the linear model is written as in Box 4.3 with an explicit error term for each x_i, the remaining assumptions are often expressed as $\varepsilon \sim$ i.i.d.N(0, σ^2), that is, the error terms should be **i**ndependently and **i**dentically **d**istributed with zero mean and constant variance for different values of the predictor. The **N** denotes normally distributed. This is not a requirement for BLUE or a Gauss–Markov condition, but it is required for further inference on the estimators. The Gauss–Markov conditions are often equated to assumptions about the response variable (for each x-value) because for many simple linear models the error terms and the response variable are the only random terms, and the linear model can be expressed without an error term. However, sometimes, the distribution of error terms is normal despite the response variable having a skewed or even multimodal distribution (see Gelman & Hill 2006: 46; Kery & Hatfield 2003), so focusing on the error terms using model residuals is better.

Violations of the assumptions can result in unreliable CIs and P-values, although OLS estimation is reasonably robust, at least to nonnormality and unequal variances (see, e.g. Boldina & Beninger 2016; Nimon 2012). Nonetheless, checks of the assumptions before using our model for statistical inference are important because (1) they may suggest alternative approaches such as transformation of the response or fitting a GLM based on a nonnormal distribution, (2) they also provide diagnostic measures to identify outliers and influential observations, and (3) nonnormal distributions or unequal variances may tell us something about the biology of the system we are studying.

A key to checking assumptions and model fit is inspecting patterns in the residuals, because they are our best estimate of the unknown model error terms. Residuals measure how far each observation is from the value predicted by our model and the most common plot is of residuals against X-values or predicted Y-values; the latter is especially useful as it means a single residual plot can be used despite how many predictors are in the model. Patterns of residuals represent patterns in the error terms from the linear model.

Some statistical assumptions can also be checked with formal statistical tests (e.g. Levene's test for homogeneity of variances among groups, the Shapiro–Wilk test for normality), although the tests often focus on the response variable rather than the model errors. We mention some of these methods in later chapters, but we don't recommend this approach for two main reasons. First, these methods are usually applied as tests of statistical significance, which have the problems we outlined in Section 2.6. They are sample-size dependent, so they could reject the H_0 of normality or equal variances with large sample sizes even if differences were small and our model inference would still be reliable. Equally, with small sample sizes, they might miss problems that might compromise our inference. P-values from these methods should be used only with other diagnostics, including the magnitude of any deviations. Second, statistical tests of normality or homogeneity of variances by themselves provide little information on the underlying cause of the problem and little help in determining appropriate remedies. We much prefer evaluating the assumptions using techniques that help identify appropriate corrective action (Section 5.3).

4.6.1.1 Zero Conditional Mean of Errors
This assumption is that the error terms' mean (i.e. expected value) for any value of x_i equals zero. It implies that the error terms are uncorrelated with the predictors. Violation can occur if an important predictor is omitted

from the model or the functional relationship is misspecified (Boldina & Beninger 2016) – for example, a linear relationship is modeled when the true relationship is nonlinear. It can cause biased estimates of model parameters. Plotting residuals against the predictors is about the only realistic check for this assumption.

4.6.1.2 Independence of Errors
We assume that the model error terms are independent (uncorrelated). While the assumption strictly refers to model error terms, it is often described as independence of observations, as correlated observations usually result in correlated error terms. Correlated errors commonly arise because of the study design. For example, repeated observations through time on the same units are likely to result in a correlation (autocorrelation) among error terms, with observations closer together being more similar than those further apart in time; this can cause underestimation of the true residual variance and SEs of model parameters, and unreliable P-values, such as inflated Type I error rates (Boldina & Beninger 2016; Kenny & Judd 1986). For these designs, we can fit models incorporating these correlations (Chapters 10–12).

It is always worth checking for correlated errors, especially with data that may not be derived from completely random sampling (e.g. systematic or haphazard sampling) such as ecologists sampling along transects using contiguous or systematic quadrats. Correlated errors may reveal an underlying gradient of biological interest. The degree of autocorrelation is measured by the correlation coefficient between successive error terms. It can be detected in plots of residuals against the predictors by an obvious positive, negative, or cyclical trend and explored using variograms, a common graphical method in spatial statistics (Dale & Fortin 2009). The Durbin–Watson test evaluates the H_0 that the autocorrelation parameter $= 0$ and is mostly relevant for regression models with a continuous predictor. Because we expect positive autocorrelation, this test is often one-tailed. The Durbin–Watson test is specifically designed for first-order autocorrelations and may not detect other patterns of nonindependence (Kutner et al. 2005).

If autocorrelation is identified, it may be possible to fit linear models using GLS, allowing for particular correlation structures (and different variances) – see Chapter 12. For more extended time series, more sophisticated modeling procedures are commonly used.

4.6.1.3 Homogeneity of Error Variances
The variance of the model error terms is assumed to be constant for different values of the predictor, sometimes

termed homoskedasticity. Heterogeneous variances (heteroskedasticity) often result from a nonnormal underlying distribution. Heterogeneous variances can also be due to unusual observations (i.e. outliers; Section 5.2). They affect the SEs for the estimated model coefficients and, therefore, the reliability of inference about model parameters.

The best way of assessing heteroskedasticity is with residual plots, usually residuals against predicted values, looking for an unequal spread of residuals. For categorical predictors with multiple Y-values in each group, this results in a plot of residuals against group means. Boxplots of residuals or observations within each group are also useful if the sample size is large enough.

4.6.1.4 Normality of Errors

Populations of error terms are assumed to be distributed normally for each value or group of the predictor variable x_i. This usually implies that the Y-values are also normally distributed. Inference for linear model parameters is generally robust to this assumption and some statisticians don't consider it a critical issue (Gelman & Hill 2006). However, there are two situations to watch out for:

• very nonnormal distributions, such as binomial, resulting from a response variable with only two outcomes or multimodal distributions;

• very skewed distributions where the mean and the variance are related (usually positively with right-skewed distributions common with biological variables).

Boxplots or probability plots of residuals or Y-values are the best ways of checking normality (Section 5.1.1).

4.6.1.5 Solutions

For variance and normality issues, we have several options:

• Appropriate transformations can normalize the distributions and reduce the mean–variance relationship (Sections 5.4 and 6.5.5).

• Fit a GLM with another distribution from the exponential family (Box 4.2 and Chapter 13).

• Generalized least squares (Section 6.6.2) can be applied if there is a consistent pattern of unequal variance.

• Robust or nonparametric approaches can help with nonnormality.

• Rely on the robustness of inference using OLS if the deviations from normality are not severe.

4.6.2 Assumptions for ML

We will consider ML assumptions in the context of GLMs. Because the link function transforms the response variable, GLMs usually do not include a simple additive error term. The model is usually expressed in terms of the expected or mean value of the transformed response. The assumptions underlying ML are commonly applied to the response variable.

Maximum likelihood estimation does have distributional assumptions, even for point estimation of parameters. While this may appear to be a limitation, its big advantage is that many other distributions besides normal can be used. The probability distribution of the response variable enters the likelihood equation via the mean and the variance, so the relationship between the mean and the variance is a key assumption to be evaluated. For example, modeling count response variables is commonly done using the Poisson distribution, where we assume the variance equals the mean. Using probability distributions where the variance is functionally related to the mean makes the homogeneity of variance assumption irrelevant. The assumption is now that the patterns of variance match those expected from the relevant probability distribution.

Checking the mean–variance relationship for GLMs can be done in several ways. Examining it for groups of observations can be useful. More commonly, as with OLS-based models, we examine patterns in the residuals. Plotting residuals against predicted values is still the best option, although it is complicated for binomial response variables, for example.

The other key assumption of ML estimation in GLMs is that of independence – that is, observations are uncorrelated with each other. The same comments as for OLS estimation apply here.

4.6.3 Model Diagnostics

Residuals are also important for evaluating how well the model fits the data and identifying which models (i.e. combinations of predictors) fit better or worse. Model diagnostics also include measures of influence – how much each observation influences the estimated model coefficients. They will be examined in more detail as we work through different linear models.

4.7 Types of Linear Models

We will now look at the most common linear models used by biologists. Our aim here is to emphasize that they essentially use the same framework, even though their statistical analyses might have different names, such as linear regression, ANOVA, logistic regression, and so on.

4.7.1 General Linear Models

General linear models, commonly just called linear models, are restricted to situations with a continuous response variable and an underlying normal distribution. The predictors can be continuous or categorical. These models have a random component, the response variable Y, a systematic component being the predictor variable X (or variables X_1, X_2, etc.), and an identity link function, as we are modeling the expected or mean value of y_i at each value of x_i. They are usually fitted using OLS, although ML could also be used. These models are commonly separated into linear regression models (continuous predictor) and ANOVA models (categorical predictors), although this distinction is artificial. The underlying linear model is essentially the same, and the ANOVA technique is a component of both.

4.7.1.1 Continuous Response and Predictor(s): "Regression" Models

Consider a random sample of observations, with values for a continuous response variable and a continuous predictor recorded for each observation. The linear model that describes the relationship between the response and the predictor variables in such situations is sometimes called a "linear regression" model. For the gall aphid example it is:

$$(\text{soldier proportion})_i = \beta_0 + \beta_1 (\text{aphid density})_i + \varepsilon_i.$$

Conceptually there is a probability distribution of possible Y-values at each x_i and we wish to estimate the expected value or mean y_i at each x_i, so we can rewrite this model as:

$$\mathrm{E}\big[(\text{soldier proportion})_i\big] = \beta_0 + \beta_1 (\text{aphid density})_i.$$

We end up with our fitted regression function:

$$\hat{y}_i = b_0 + b_1 x_i,$$

where $\hat{y}$ is the fitted or predicted Y for x_i, and b_0 and b_1 are the estimates of β_0 and β_1.

4.7.1.2 Continuous (Normal) Response and Categorical Predictors: "ANOVA" Models

Now consider a random sample of observations, with values for a continuous response and a categorical predictor recorded for each observation. Most designs that generate data like this would have multiple observations from several categories (groups) of the predictor variable. We now have a categorical predictor – a factor – with multiple sample observations for each group – for example, multiple mice, each with a CO_2 reading, for two anesthetic types.

What linear model might be appropriate here? The parameter β_1 in the linear model for a continuous predictor is usually considered a regression slope. When the predictor is categorical, especially in our example where anesthetic type is qualitative, a regression slope is not that helpful. Our focus is on estimating category or group means and measuring effects of the predictor by differences in group means. There are three equivalent ways of specifying the model.

One common way identifies an "effect" of each predictor and is called the effects model:

$$y_{ij} = \mu + \alpha_i + \varepsilon_{ij},$$

where y_{ij} is the value of Y for observation j in group i. For our mouse physiology example, there are two groups and the model would be:

$$(\text{blood } CO_2 \text{ level})_{ij} = \mu + (\text{effect of anesthetic type})_i + \varepsilon_{ij}.$$

This model has three parameters. First, μ is the overall mean blood CO_2 level, ignoring the group structure in our data. The overall mean is equivalent to the Y-intercept from the regression model. Second, α_i is the effect of anesthetic i on blood CO_2 level, measured as the difference between the overall mean (μ) and the mean of group i: ($\mu_i - \mu$). This parameter measures the effect of each anesthetic on mean blood CO_2 level and is equivalent to the regression slope for our continuous predictor. Third, ε_{ij} is the random error for each observation, being the difference between each y_{ij} and the mean of group i.

We usually estimate μ and α_i (and therefore μ_i) using OLS. We end up with our fitted model function:

$$\hat{y}_i = \bar{y} + (\bar{y}_i - \bar{y}),$$

where $\hat{y}$ is the fitted or predicted Y-value group i and $\bar{y}$ and $(\bar{y}_i - \bar{y})$ are the estimates of μ and α_i. This fitted function is telling us that the best estimate or predictor for any observation is the mean of the group.

Rather than expressing each group effect as a deviation from the overall mean, we could model the group means directly. This is a means model:

$$y_{ij} = \mu_i + \varepsilon_{ij}.$$

We could also fit a regression model by converting our categorical predictor into several continuous (therefore quantitative) dummy variables that take only two values (0/1 or +1/−1). We need one fewer dummy variable than groups, and the model is:

$$y_{ij} = \mu + \beta_1 (\text{dummy}_1)_{ij} + \beta_2 (\text{dummy}_2)_{ij} + \cdots \\ + \beta_{p-1} \left(\text{dummy}_{p-1}\right)_{ij} + \varepsilon_{ij}.$$

The regression coefficients for multiple dummy variables where there are three or more groups can be converted into an overall measure of effect of the predictor or the effects of specific groups. This is the model fitted by most software, even when you use an effects model, and emphasizes that regression and ANOVA "techniques" are essentially the same.

4.7.2 Generalized Linear Models

The most common probability distributions used with GLMs by biologists are normal and gamma distributions for continuous responses, and binomial, Poisson, and negative binomial for discrete responses (Box 4.2). Three scenarios are particularly common.

4.7.2.1 *Binary Response, Continuous Predictor(s): Logistic Models*

Response variables that can have only two outcomes (e.g. alive vs. dead, present vs. absent, success vs. failure) are common in biological research. For example, Polis et al. (1998) studied the factors that control spider populations on islands in the Gulf of California. Potential predators included lizards of the genus *Uta* and scorpions. They were interested in how the presence/absence of lizards changed with the shape of each island (perimeter:area, PA). The data are displayed in Figure 4.4. The response variable of interest (π_i) is the probability that $y_i = 1$, that is, any island in our population has lizards, and the predictor is the PA ratio. While we could fit a nonlinear model to these data, it is better to use the GLM framework.

The random component is the probability that an island has lizards and the appropriate probability distribution for such a binary response is the binomial distribution. The systematic component is the predictor variable

(PA). The link function for a binomial distribution is the logit link (see Box 4.2), the log of the odds that $y_i = 1$ (i.e. an island has lizards).

Our GLM models the expected value of the log of the odds of having lizards against the PA ratio:

$$\text{logit}(\pi_i) = \ln\left(\frac{\pi_i}{1 - \pi_i}\right) = \beta_0 + \beta_1 x_i,$$

$$\text{logit (odds of island having lizards)} = \beta_0 + \beta_1(\text{PA}).$$

This model has two parameters. β_1 is the regression coefficient or slope, measuring the change in the log of the odds of lizards being present for an increase in one unit of PA. β_0 is the intercept, measuring the log of the odds of lizards being present for an island with PA = 0.

This model should look familiar. It's similar to the linear regression model, but with three main differences. First, we are now modeling the log of the odds of an outcome occurring rather than the mean of a continuous normally distributed response variable. Second, the error term from the linear regression model is no longer present as we focus on modeling the expected value. Third, because our response variable clearly does not follow a normal distribution, OLS is not appropriate. We fit the model and estimate the parameters with ML.

Categorical response variables can have more than two categories, and the logistic regression model can be generalized to a multi-category logit model (Agresti 2013, 2019). Our GLM is now based on a multinomial distribution for the response variable rather than a binomial distribution, but still uses the logit link.

Data with a categorical predictor can be modeled with a logistic regression model using dummy variables for the predictors as described in Section 4.7.1.

4.7.2.2 *Poisson Response, Continuous Predictor(s)*

Count variables cannot be negative and usually follow a Poisson distribution, especially if the counts are small (means < 10). As an example, Fill et al. (2021) studied the post-fire ecology of wiregrass in a section of pine savanna. They burnt the area and counted the number of culms (aerial stems) on each plant five months later, along with its basal area.

The random component is the number of culms on each plant, with most values <10, so a Poisson distribution is appropriate. The systematic component is the predictor variable (basal area). The link function (and natural parameter) for a Poisson distribution is the log-link (see Box 4.2), so we model the log of the mean of the response variable (culms) for each value of the predictor:

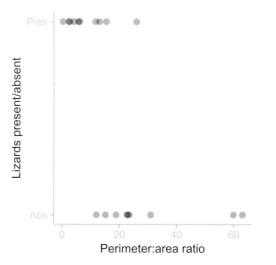

Figure 4.4 Illustration of data with a binary response variable and continuous predictor (data from Polis et al. 1998).

Table 4.3 Contingency table data for grizzly bears from Morehouse et al. (2016)

Offspring were cross-classified as to whether they were problem bears (negative interactions with other bears or humans) and whether their mothers were problem bears.

		Mother a problem?		
		Yes	No	Row total
Offspring	Yes	5	18	23
a problem?	No	3	50	53
	Col. total	8	68	76

$$\log(\mu_i) = \beta_0 + \beta_1 x_i,$$

$$\log \text{(mean culms)} = \beta_0 + \beta_1 \text{(basal area)}.$$

β_1 is the regression coefficient or slope, measuring the change in the log of the mean number of culms for a one-unit increase in plant basal area.

This model is usually called a Poisson regression model, but the broader term for GLMs that use the Poisson distribution and the log-link is a loglinear model, commonly used for analyzing contingency tables.

4.7.2.3 *Contingency Tables: Loglinear Model*

The final common GLM used by biologists is for analyzing contingency tables, where sampling or experimental units are cross-classified by two or more categorical variables. Consider the study of Morehouse et al. (2016), who examined whether grizzly bears from parents involved in conflict (with either other bears or humans) were themselves more likely to be involved in conflicts (Table 4.3).

The biological question of interest here is one of association or independence – are problem offspring associated with problem parents, or is an offspring's status independent of its mother? This kind of example might be familiar to you as a 2×2 χ^2 test for independence. We can also approach this analysis as a GLM where we model the expected counts in each cell of the contingency table, treating them as a Poisson variable, as the response against the row and column variables as "predictors." In the study of Morehouse et al. (2016), the two variables are whether the mother was a problem bear (Mother: two categories) and whether the cub was a problem bear or not (Cub: two categories), resulting in a 2×2 table. As with a Poisson regression model, the model for a contingency table analysis uses the log-link function, so it is a loglinear model modeling the log of the expected frequencies (f_{ij}) in each cell of the table. Calling row variable A and column variable B:

$$\log(f_{ij}) = \beta_0 + \beta_i^A + \beta_j^B + \beta_{ij}^{A \times B},$$

although loglinear model parameters for contingency tables are often expressed as λs (see Section 13.7.1.5):

$$\log(\text{expected number of bears})$$
$$= \lambda + \lambda_i^{(\text{Cub})} + \lambda_j^{(\text{Mother})} + \lambda_{ij}^{(\text{Cub} \times \text{Mother})}.$$

The key term of biological interest is the last one, the interaction between row and column variables on log number of bears in each cell of the table, measuring whether problem offspring are associated with problem mothers and vice-versa. This model is compared to the fit of a model without the interaction, a model representing independence of rows and columns. Lack of independence is identified by the fit of the interaction model being better than the independence model.

There is a key difference between this loglinear model and the other GLMs introduced in this chapter. This model is not directly modeling a response variable for each sampling/experimental unit, but uses the counts (frequencies) in each cell as the response. We'll explore these models in more detail in Chapter 13.

4.7.3 Linear Mixed Models (General and Generalized)

As outlined in Section 4.2.2, predictor variables can be considered fixed – where all the groups or values of interest of the predictor have been used and the statistical focus is on estimating mean values – or random – where the groups or values of the predictor are a random sample from a population of possible values and the focus is on estimating the variances. It is common for biological study designs to include both predictor types, and the linear models we fit are linear mixed models.

There is a long history of ANOVA approaches to mixed models, particularly in agriculture and psychology. Some widely used designs (randomized blocks, repeated measures, nested designs) are linear mixed models. The ANOVA models were fitted by OLS, making use of assumed underlying normal distributions. Modern statistical approaches use linear mixed effects models that emphasize estimating the random effects with ML or REML. These approaches also accommodate nonnormal distributions and are then labeled generalized linear mixed models (GLMMs). We will consider them in Chapters 10–13.

4.8 Key Points

• Statistical analyses are often presented as one of a series of named "techniques," such as regression, ANOVA, and ANCOVA, which are taught in separate ways. It is more helpful to see them as variants on a broad theme of linear models.

• Linear models specify a response variable as a linear function of one or more predictor variables. The exact form of a linear model depends on the type of response variable and the number and types of predictors.

• Predictors can be continuous or discrete (called factors) and can be classified as fixed or random. This leads to fixed, random, and mixed models.

• GLMs use a link function to relate predictors to responses. The three commonest link functions model the mean (identity), log of the mean (log), or log of odds (logit), depending on the properties of the response variable.

• We fit linear models to the data using two main methods, OLS and ML, although there are other options. Maximum likelihood is used when the response variable is not distributed normally and there is a relationship between the mean and the variance. For normal distributions and homogeneous variances, OLS has been used commonly, but ML and OLS produce the same result.

• The fit of a model to the data is assessed by ANOVA for OLS models and log-likelihood, deviance, and information criteria for ML models.

• Linear models make assumptions about the response variable and model structure. When these assumptions are violated, our conclusions can become unreliable, so it's particularly important to assess these assumptions before running the full analysis.

• The assumptions vary between OLS and ML models, but looking for patterns in the residuals is valuable in almost all cases. Examining relationships between means and variances is useful when there are categorical predictors.

• Several kinds of linear models are particularly common in biology:
  – A continuous (normal) response and
    ▪ one or more continuous predictors, often called regression;
    ▪ one or more categorical predictors, often called ANOVA – we can use dummy variable coding to produce a regression, with identical results;
    ▪ continuous and categorical predictors, often called ANCOVA.
  – A binary response and continuous predictors, called logistic regression.
  – A response variable with a Poisson distribution and continuous predictors, called Poisson regression. This situation is common where the response is counts.

Further Reading

Overall introduction to linear models: Most linear models textbooks provide good introductions, including classics such as Sokal and Rohlf (2012) for biologists. We also recommend Kutner et al. (2005) for a readable and comprehensive overview, and Agresti (2015) provides more statistical detail and an emphasis on GLMs (see also Dunn & Smyth 2018). We also really like the Gelman et al. regression books (Gelman & Hill 2006; Gelman et al. 2020).

Analysis of variance: Underwood (1997) provides an excellent overview of the logic and application of classical ANOVA from an ecological perspective.

5 Exploratory Data Analysis

When you finally have some data, the first step is to look at them before any formal statistical analysis. Do preliminary checks of your data to:
- reassure yourself that you do have some meaningful data;
- detect any errors in data entry;
- detect patterns in the data that may not be revealed by the statistical analysis you will use;
- ensure that the planned analysis is appropriate:
  - check that assumptions of the analysis are met, and if necessary, interpret departures from the assumptions;
  - ensure that you fit the appropriate relationship between continuous variables;
- detect unusual values, termed outliers.

This stage is labeled *exploratory data analysis*.

5.1 Basic Graphical Tools

Simply examining the raw data in rows and columns is possible for small datasets. Graphical techniques are much better for large samples, especially with multiple variables. Graphic displays are very important in the analysis of data, and they have four main functions (Snee & Pfeifer 1983):
- exploration, which involves checking data for unusual values, making sure the data meet the assumptions of the chosen analysis, and occasionally deciding what analysis (or model) to use;
- analysis, which includes checking assumptions but primarily ensuring that the chosen model is a realistic fit to the data;
- presentation and communication of results, particularly summarizing quantitative information; and
- graphical aids, which are visual displays for specific statistical purposes, e.g. power curves for determining sample sizes

We describe graphical displays for the first two functions here and the third in the final chapter. However, some graphs are helpful for more than one function – for example, scatterplots of Y against X are important exploratory tools and often the best way of communicating such data to readers. The fourth function is spread through several chapters.

5.1.1 Some Common Basic Graphs

The most important thing we want to know about our sample data, and therefore about the population from which our data came, is the shape of the distribution. Many of the statistical procedures we describe in this book assume, to some extent, that the error terms from our models (and usually the response variables) have normal or some other specified exponential distributions. The simplest way of examining the distribution of values of a variable is with a density plot, where the frequencies ("densities") of different values, or categories, are represented. We describe four standard density plots.

5.1.1.1 *Histogram*

A straightforward way of examining the distribution of a variable in a sample is to plot a histogram, a graphical representation of a frequency (or density) distribution. A histogram is a type of bar graph (see Chapter 17) grouping the observations into *a priori* defined classes or bins on the horizontal axis and their frequency on the vertical axis (Figure 5.1). If the variable is continuous, the size (width) of the classes will depend on the number of observations: more observations mean that more bins can be used. The values of a discrete variable usually determine the bins. Histograms are very useful for examining the shape of a distribution (Figure 5.1). For example, is the distribution symmetrical or skewed? Is it unimodal or

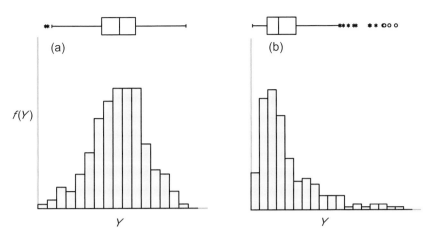

Figure 5.1 Histograms and boxplots for (a) normal and (b) positively skewed data ($n = 200$).

multimodal? The vertical axis can also be relative frequency (proportions), cumulative frequency, or cumulative relative frequency. Unfortunately, histograms are not always particularly informative in biology, especially experimental work, because we often deal with small (<20) sample sizes.

A helpful addition to a histogram is superimposing a more formal probability density function (pdf). For example, we could include a normal pdf, based on our sample mean and variance, or any of a range of other common distributions. An alternative approach is to not stipulate a specific distribution for the sample but to use the observed data to generate a probability density curve. This is nonparametric estimation because we are not assuming a specific underlying population distribution for our variable. Our estimation procedure may produce probability density curves that are symmetrical, asymmetrical, or multimodal, depending on the density pattern in the observed data. The standard reference to nonparametric density estimation is Silverman (1986), and the most common method is kernel estimation. We construct a window of a certain width for each observation, like the categories in a histogram. We then fit a symmetric pdf (called the kernel) to the observations in each window; commonly, we use the normal distribution. The estimated density for any value of our variable is simply the sum of the estimates from the density functions in each window. Even when the kernel is a normal distribution, the calculations are tedious, but kernel density estimators are now standard options in statistical software.

The window width is sometimes termed the smoothing parameter because it influences the shape of the final estimated density function. For standard kernel density

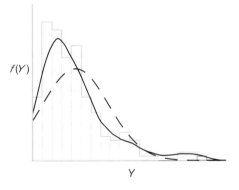

Figure 5.2 Histogram with normal density function (dashed line) and kernel density curve or smooth (solid line) for a positively skewed distribution ($n = 200$). The smoothing parameter for the kernel curve equals 1.

estimation, the smoothing parameter is constant for all observations; other approaches allow the smoothing parameter to vary depending on the local density of data (Silverman 1986). If the smoothing parameter is low (narrow windows), the density function can have numerous modes, although many will be artificial with small sample sizes. The density function will be much smoother if the smoothing parameter is high (wide windows), but important details might be missed, such as real modes. Kernel estimation requires a large sample size to fit a pdf reliably in each window and enough windows to represent the detail present in the data. The user determines the pdf fitted in each window. Symmetrical distributions such as normal are most common, although others are possible (Silverman 1986).

For the positively skewed distribution plotted in Figure 5.2, a normal distribution function based on the

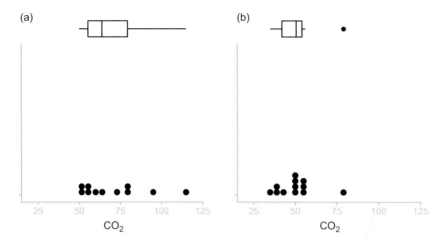

Figure 5.3 Dotplots and boxplots of blood CO_2 concentrations of mice anesthetized with alpha chloralose (a) or isoflurane (b). Data from Low et al. 2016; see Box 2.2.

sample mean and variance would fit the data poorly. The nonparametric kernel smoothing curve is a much more realistic representation of the data distribution. The kernel density estimator is particularly useful if we have a reasonable sample size. It may suggest which more formal parametric distribution should be used in modeling. Density estimation is also useful for bivariate distributions.

5.1.1.2 *Dotplot*

A dotplot is a plot where each observation is represented by a single dot or symbol, with the variable's value along the horizontal axis. Multiple observations with the same value are stacked above each other. Univariate dotplots can effectively represent single samples because skewness and unusually large or small values are easy to detect (Figure 5.3). For large samples, graphical techniques such as jittering, where individual data points are shifted slightly (and randomly), can reduce the overlap of points and make the plots easier to interpret.

5.1.1.3 *Boxplot*

A boxplot, also called a box-and-whiskers plot, is a good alternative for displaying the sample observations of a single variable (Figure 5.4; see also Figures 5.3 and 5.5), as long as we have a reasonable sample size (roughly 10 or more). The boxplot uses the median to identify location and 25% quartiles for the hinges (ends of the box). The difference between the values of the two hinges is called the spread (and used to calculate the interquartile range). The lines (or whiskers) extend to the extreme values of the data (encompassing upper and lower quartiles) unless there are unusually large or small values. These values, called sample outliers (Section 5.2), are highlighted and plotted individually (Figure 5.4), and the

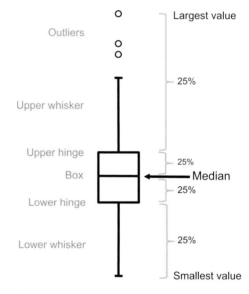

Figure 5.4 Components of a boxplot.

whiskers typically show the range of values not considered outliers. In conventional boxplots, outliers are points falling more than 1.5 times the spread beyond the hinges, and the whisker is drawn to cover this range. The value of 1.5 was present in Tukey's (1977) original formulation of the boxplot, but the actual formulae for identifying outliers can vary between different textbooks and software packages. Boxplots efficiently indicate several aspects of the sample:

• The middle of the sample is identified by the median, which is resistant (robust) to unusual values.

• The sample variability is indicated by the distance between the whiskers (with or without the outliers).

• The shape of the sample, especially whether it is symmetrical or skewed.

• Sample outliers – extreme values that are very different from the rest of the sample.

Because boxplots are based on medians and quartiles, they are very resistant to extreme values, which don't affect the basic shape of the plot very much. The boxplots and dotplots for blood CO_2 concentration in mice given two anesthetics (Low et al. 2016) are presented in Figure 5.3 (see Box 2.2). The positive skewness for the alpha chloralose mice is apparent, whereas the isoflurane distribution is slightly negatively skewed, with one sample outlier in the other direction. Boxplots can also be used in research publications (Chapter 17) instead of the more traditional means (±SD or similar). This is particularly the case when we use analyses not based on normal distributions because the mean might not be particularly appropriate as a measure of the center of a distribution.

More elaborate equivalents to boxplots are also available. Boxplots were developed as a tool that required little computation, but the information that could be shown using quartiles was limited. They have been extended in many useful directions, showing more of the actual distribution of the points. Vase plots show detail of the spread, while violin (Hintze & Nelson 1998) and bean (Kampstra 2008) plots show the entire distribution. Rousseeuw et al. (1999) also described the bagplot, a bivariate version of the boxplot. We find violin plots especially useful for large sample sizes, as they show more subtle changes to distributions than are possible using just a boxplot. A basic violin plot is produced by generating a kernel pdf for a particular sample, then making a reflected copy of the function to generate the "violin" shape. However, data with particularly unusual distributions may not resemble any stringed instrument, let alone a violin! We recommend the added step of showing a conventional boxplot within the violin area (Figure 5.5). When using these more elaborate techniques, be wary of results from small sample sizes because density kernels may be quite uncertain.

It is also possible to produce asymmetrical plots, where the upper and lower whiskers are calculated independently, using the upper and lower quartiles that make up the spread (Rousseeuw et al. 1999), rather than assuming a distribution roughly symmetrical about the median.

Boxplots can be harder to interpret with large samples because the absolute number of outliers increases, and there have been modifications to visualize outliers effectively at all sample sizes (Carter et al. 2009).

5.1.1.4 Probability Plot

Probability plots examine a cumulative frequency distribution of the data and compare the shape of that

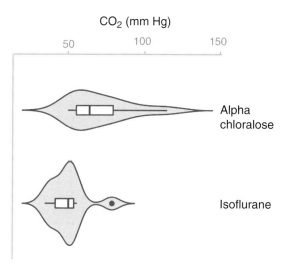

Figure 5.5 Violin and boxplot combination for blood CO_2 concentrations of mice anesthetized with alpha chloralose or isoflurane (data from Low et al. 2016; see Box 2.2). The violin component is a kernel probability distribution of the data.

distribution to that expected of a specified distribution having the same shape and relevant parameters (for a normal distribution, mean and variance). If your data match the specified distribution, the plot will be a straight line, and various kinds of skewness, multimodality, etc., will show as a kinked line. There are two common probability plots, p–p and q–q. A p–p plot compares cumulative distribution functions – typically, one axis shows the proportion of your data falling below a particular value. The other shows the expected proportions for a normal distribution with the same mean and standard deviation as the data. For our purposes in this chapter, the related q–q plot compares quartiles of data between quartiles from a specified distribution and quartiles of the data distribution. For a normal distribution, data are denser around the mean, so a p–p plot shows deviations around the middle of the distribution. In contrast, a q–q plot reveals more about the edges of the distribution. A q–q plot is shown in Figure 5.6 for a normal and a skewed (lognormal) distribution.

5.1.1.5 Scatterplot

We are often interested in the relationship between two or more continuous variables, one of which may be the dependent or response variable. A fundamental graphical technique is the scatterplot. The vertical axis represents one variable (the response, if defined), the horizontal axis represents the other variable, and the points on the plot are the individual observations. Scatterplots are very informative, especially when bordered by boxplots for each variable (Figure 5.7). Nonlinearity and outliers can be

(a)

(b)

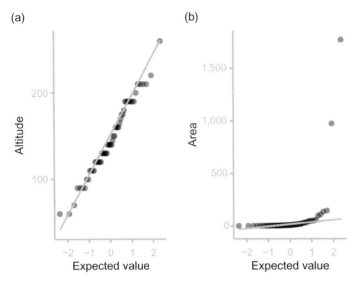

Figure 5.6 Probability plots for (a) normally (altitude) distributed and (b) strongly skewed (area) variables from 56 remnant forest patches in southeast Australia (data from Loyn 1987; see Box 8.2).

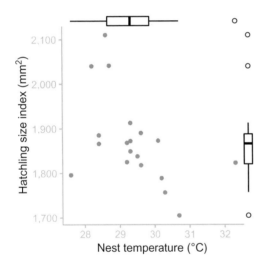

Figure 5.7 Basic scatterplot to explore and visualize the relationship between a pair of continuous variables. The data show the relationship between hatchling size and nest temperature for turtles (Booth & Evans 2011).

identified, as can departures from fitted linear models, particularly when we add some kind of smoothing function to the scatterplot (e.g. Figure 6.1).

5.1.1.6 *Scatterplot Matrix*

An extension of the scatterplot to three or more variables is the scatterplot matrix (SPLOM). Each panel in the matrix represents a scatterplot between two variables. The panels along the diagonal can indicate which variable forms the horizontal and vertical axes, or show other univariate displays such as boxplots or frequency distributions (see Figures 8.1 and 8.3). Scatterplot matrices are particularly useful for examining potential collinearity among predictor variables in multiple regression models (Chapter 8).

5.1.2 Smoothing

Sometimes we know that a straight-line model is an inappropriate description of the relationship between Y and X because a scatterplot shows pronounced nonlinearity or because we know theoretically that some other model should apply. Other times we simply have no preconceived model, linear or nonlinear, to fit to the data. We merely want to investigate the nature of the relationship between Y and X, often to decide whether fitting a linear model as part of our formal analysis is appropriate. In both situations, we require a method for fitting a curve to the relationship between Y and X not restricted to a specific model structure (such as linear). Smoothers are a broad class of techniques that describe the relationship between Y and X, etc., with few constraints on the form the relationship might take. Because of the absence of a specified form of the relationship between Y and X, they are sometimes called nonparametric regression analyses.

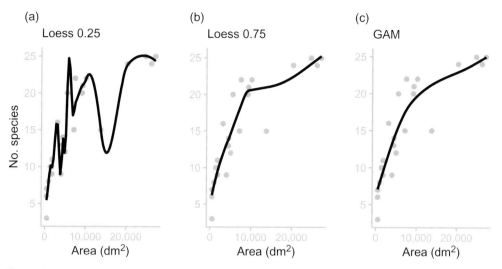

Figure 5.8 Smoothing functions through scatterplots showing the relationship between numbers of species of invertebrates and area of clumps of mussels on a rocky intertidal shore (Peake & Quinn 1993). (a) LOESS smoother with smoothing parameter of 0.25, (b) LOESS smoother with smoothing parameter of 0.75, and (c) a GAM smoother (cubic spline; see Section 13.9).

However, this term also includes linear models based on ranks (Section 6.6.1). The usual linear model aims to separate the data into two components (Chapter 4):

model + residual.

Smoothing also separates data into two components:

smooth + rough.

The rough component should have as little information or structure as possible.

In this section, we will introduce smoothing functions as exploratory tools to describe the relationship between two continuous variables informally, but they are also part of generalized additive models (GAMs) developing predictive relationships between a response and one or more continuous predictors (Section 13.9).

A smoothing function involves each observation being replaced by a predicted value from some form of regression model fitted through the surrounding observations. These surrounding observations are those within a window (sometimes termed a band or a neighborhood) that covers a range of observations along the horizontal axis. The X-value on which the window is centered is termed the target. The size of the window, that is the number of observations it includes, is determined by a smoothing parameter for most smoothers. Successive windows overlap, so the resulting fitted line is "smooth," although the smoothing parameter determines smoothness.

The simplest smoothing function is a running (moving) means (averages) smoother based on the mean of all the observations in a window. Each window is centered on the target X-value. The remaining X-values in the window can be determined by either including a fixed number on both sides of the target X-value or a set number of nearest observations to the target regardless of which side of the target they occur. The latter tend to perform better (Hastie & Tibshirani 1990), especially for locally weighted smoothers (see Cleveland's LOESS below). Using running medians instead of means makes the smoothing more resistant to extreme observations (i.e. more robust).

The most common smoothing function used for exploring the relationship between continuous variables is locally weighted regression scatterplot smoothing (LOESS or LOWESS; Cleveland 1994 and see Figure 5.8). A linear or sometimes quadratic polynomial regression model is fitted within each window, and each observed Y-value is replaced with that predicted by the local regression for the target X-value. The observations in a window are also weighted differently depending on how far they are from the target X-value using a tri-cube weight function. Observations further from the target X-value are downweighted compared with values close to the target X-value. The final LOESS smooth is often an excellent representation of the relationship between Y and X. However, the choice of smoothing parameter (window size) can be important for interpretation (see below).

More complex smoothers include splines that approach the smoothing problem by fitting polynomial regressions (Section 8.1.11), usually cubic polynomials, in each window. They are most commonly used in GAMs and will be described in Section 13.9.

There are two important practical issues when applying smoothers. First, whichever smoothing method is used, an important decision for the user is the value for the smoothing parameter – that is, how many observations to include in each window (Figure 5.8). Increasing the number of observations in each window (larger smoothing parameter) produces a flatter and "smoother" smooth with less variability, but is less likely to represent the real relationship between Y and X well (the smooth is probably biased). In contrast, fewer observations in each window (smaller smoothing parameter) produces a "jerkier," more variable, smooth but which may better match the pattern in the data (less biased). In an exploratory context, we recommend trying different smoothing parameter values as part of the phase of examining patterns in data before formal analyses. Second, there are difficulties when the endpoints (the smallest and largest X-values) are the targets, because their windows will

usually exceed the range of the data. Commonly, the weighting function tends toward zero approaching the endpoints.

5.1.3 Residual Plots

One of the most informative graphical tools for statistical modeling is a residual plot, which can take several common forms. The residuals from our fitted linear model, which may be standardized or studentized to constant variance to improve interpretation (see Section 6.5.1; Table 6.4) are plotted against the predictor (independent) variable(s) or more commonly the predicted values from our model. The most common residual plots are scatterplots, but boxplots can also be used when categorical predictors or groups are involved (Figure 5.9).

These residuals can be used in a diagnostic fashion to assess the model's fit to the data. In particular, residual plots can tell us whether the model's assumptions are met and whether there are unusual observations that do not match the model very well, termed model outliers (Section 5.2); these observations will have large residuals and could be influential for our model fit.

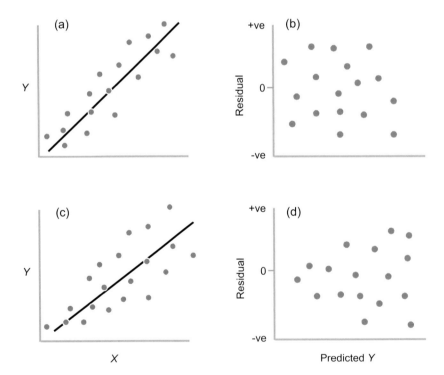

Figure 5.9 Diagrammatic representation of residual plots from linear regression: (a) regression showing even spread around the line; (b) associated residual plot; (c) regression showing increasing spread around the line; and (d) associated residual plot showing characteristic wedge shape typical of skewed distribution.

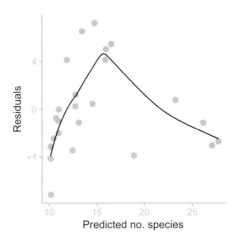

Figure 5.10 Residual plot for data in Figure 5.8, with residuals calculated from a linear regression. The LOESS smoother shows a strong pattern in the residuals and highlights the nonlinear relationship.

If the distribution of Y-values for each x_i is positively skewed (e.g. lognormal, Poisson), we would expect larger $\hat{y}_i$ (an estimate of the population mean of y_i) to be associated with bigger residuals. A wedge-shaped pattern of residuals, with a broader spread of residuals for larger x_i or $\hat{y}_i$ as shown in Figure 5.9, indicates increasing variance in ε_i (and y_i) with increasing x_i associated with nonnormality in Y-values and violating the assumption of homogeneity of variance of OLS models (Section 5.3.2). Transformation of Y to a different scale, such as logarithms, will usually help if the transformed variable makes biological sense. The ideal pattern in the residual plot is a scatter of points with no obvious pattern of increasing or decreasing variance in the residuals (Figure 5.9). Nonlinearity can be detected by a curved pattern in the residuals (Figure 5.10), and sometimes fitting a smoother to the residuals can help highlight patterns. Model outliers also stand out as having large residuals (Section 5.2).

5.2 Outliers

Outliers (or unusual values) are observations very different from the rest of the sample, termed sample outliers, or very different from those predicted by the fitted model, termed model outliers. These two types of outliers can be different; observations far from the fitted model may not necessarily be unusual values for Y. Outliers can seriously influence the results of analyses. There are two aspects to dealing with outliers: (1) identifying them and (2) handling them.

There are formal tests for detecting outliers, which assume that the observations are normally distributed. Dixon's Q test examines the difference between the

outlier and the next closest observation relative to the overall range of the data (Sokal & Rohlf 2012). However, such tests have difficulties when there are multiple outliers. As we have argued for other aspects of checking data, outliers are best detected with exploratory data analysis rather than statistical tests. For example, boxplots will highlight unusually large or small values, and plots of residuals from linear models reveal observations a long way from the fitted model, as will scatterplots with an appropriate smoothing function. For some linear models (e.g. linear regression), Cook's D statistic indicates the influence of each observation on the analysis result (Section 6.5.3).

Once you identify an outlier, first check to make sure it's not a mistake, such as a data entry error or equipment malfunction. These outliers often show up as impossible values (e.g. a 3 m ant, a blood pressure that would make an animal explode, etc.). If you can classify an outlier unambiguously as a mistake, delete it.

An outlier can result when something unusual happened to that observation. Perhaps the tissue preparation took longer than usual, or an experimental enclosure was placed in an unusual physical location. Here, you may have had *a priori* cause to suspect that value. It is essential to keep detailed notes of your experiments to identify potential outliers (and, of course, for many other reasons). If you suspected this observation *a priori*, you might justify deleting it.

In other cases, you may simply have an anomalous value. Although evolutionary biologists might make their reputations from rare variants, they are an unfortunate fact of life for most of us. If you have no reason to suspect an outlier as a mistake, there are three options. First, run the analysis with and without the outlier(s) to see how much the outcome of the analysis is affected; Cook's D statistic measures how much an observation influences our estimated parameters from a linear model. If the conclusions are altered, you are in trouble and should determine why those values are so different. Perhaps you are unwittingly counting two very similar species or have a batch of laboratory animals that came from very different sources. Sometimes thinking about why particular observations are outliers can stimulate new research questions. Second, use statistical techniques robust to outliers; for example, for simple analyses rank-based tests can provide some protection (Section 2.5.6). Third, remember that outliers may occur because we do not fit the right model, such as fitting an OLS model to data not normally distributed. Transformations, described in the following two sections, and fitting a more appropriate model might result in these points no longer being unusual.

It is crucial that outliers only be deleted when you have *a priori* reasons to do so – dropping observations just because they are messy or reduce the chance of getting a statistically significant result is unethical. The other unacceptable behavior is to run the analysis and then go back and look for outliers to remove if the analysis is not significant.

5.3 Am I Fitting the Right Model?

Many analyses in this book are based on linear models. They have important assumptions we must assess before applying linear models and assessing their fit formally. We discuss these assumptions in the chapters but briefly introduce them here in exploratory data analysis. Sometimes, these assumptions are not critical because the analysis is robust to violations of assumptions being violated. Other assumptions are critical because the statistical analysis may give unreliable results when assumptions are violated.

5.3.1 The Underlying Probability Distribution

The most commonly used linear models are based on the error terms (and usually also the response variable) coming from a population with a distribution from the exponential family of distributions (see Box 4.2). The normal (Gaussian) distribution is the basis for many linear model analyses. Most OLS analyses are robust to the assumption of normality to some degree, and some statisticians don't consider this assumption particularly important (e.g. Gelman & Hill 2006). Despite this robustness, the symmetry (roughly equal spreads on each side of each sample's mean or median) should be checked with a graphical procedure like boxplots or p–p plots. We don't suggest you do any formal analyses of these plots, but just look for major asymmetries.

The most common asymmetry in biological data is positive skewness: populations with a long right tail (Figure 5.1). Positive skewness in biological data is often because the variables have a lognormal (measurement variables) or a Poisson (counts) distribution. In our experience, skewed distributions are more common than symmetrical distributions. This makes sense when you realize that most variables cannot have values less than zero (lengths, weights, counts, etc.) but have no mathematical upper limit, although there may be a biological limit. Their distributions are usually truncated at zero, resulting in skewness in the other direction. Transforming skewed variables to a different scale (e.g. log or power transformations) often improves their normality, although a GLM

based on the Poisson (or negative binomial) distribution is probably more applicable for count data.

The other problematic distribution is multimodal, with two or more distinct peaks. The best option is to treat each peak of the distribution as representing a different "population" and to split your analyses into separate populations. In ecological studies you might get such a problem with different cohorts in a population of plants or animals and be forced to ask questions about the mean size of the first, second, etc., cohorts. In physiological or genetic studies, you might get such a result from using animals or plants of different genotypes. For example, allozymes with "fast" and "slow" alleles might produce two classes of physiological response, and you could analyze the response of fast and slow tissues as an added factor in your experiment.

One final distribution that often causes problems in biological data is when we have many zeroes and a few nonzero points. Such datasets are termed zero-inflated, and the distribution is so skewed that no transformation will normalize the distribution; whatever we do to these zeroes, they will remain a peak in our distribution. Rank-based approaches will not fare much better, as these values will all be assigned the same (tied) rank. In this situation, your data probably reflect two processes, such as whether a particular experimental or sampling unit has a response or not and the level of response when it occurs. Options for analysis are discussed in Section 13.6.3.

5.3.2 Homogeneity of Variances

The choice of a probability distribution for model error terms (or response variable) in a linear model also implies a particular relationship between the mean and the variance (see Box 4.2). When using a normal distribution, our linear model analysis also assumes that the variance in the response variable is independent of the predictor variables. This assumption is more important than normality. If the response variable is normal, unequal variances will probably be due to a few unusual values, especially with small sample sizes. Suppose the response variable has a lognormal or Poisson distribution. There, we expect a relationship between the means (expected or predicted values from the linear model) and the variances related to the underlying distribution. Transformations that improve normality will also usually improve the homogeneity of variances, although again using generalized linear models that allow positive mean–variance relationships can be a better approach, especially for counts.

There are formal tests for variance homogeneity, especially when the predictor variable is categorical. Levene's test is considered the most robust and is based on absolute deviations of each observation from its respective group mean or, preferably, median (see Box 6.5). Our reluctance to recommend such tests for evaluating linear model assumptions has been discussed in Section 4.6.1, although they may be useful if the central hypothesis is about variances. Less formal but much more helpful checks include side-by-side boxplots for multiple groups, which allow a check of homogeneity of the spread of samples (Figures 5.3 and 5.5). Note that plots of residuals from the model against predicted values are also valuable exploratory checks.

5.3.3 Is My Linear Model "Linear"?

Parametric correlation and linear regression analyses are based on straight-line relationships between variables. The simplest way of checking whether your data are likely to meet this assumption is to examine a scatterplot of the two variables or a SPLOM for more than two variables. Figure 5.8 illustrates how a scatterplot showed a nonlinear relationship between the number of invertebrate species inhabiting clumps of mussels of different areas on a rocky intertidal shore (data from Peake & Quinn 1993). While lots could be written on the value of scatterplots as a preliminary check of the data, the best illustration remains the classic example datasets from Anscombe (1973), available widely online, which show how easy it is to fit inappropriate linear models without preliminary scatterplots.

5.4 Is It Normal to Transform Data?

We stated in the previous section that transforming data to a different measurement scale can solve data not distributed in the way assumed by the proposed model. In this section, we will elaborate on data transformations.

The justification for transforming data to different scales before data analysis is based, at least in part, on the appreciation that the scales of measurement we use are often arbitrary. For example, we might measure the length of an object in centimeters, but we could just as easily measure the length in logarithm units, such as log centimeters. We could do so directly by altering the scale on our measuring device.

Transformations are common for measurements we encounter in everyday life. Sometimes, these transformations simply change the zero value – that is, adding a constant. Slightly more complex transformations may change the zero value and rescale the measurements by a constant value, such as the change in temperature units from Fahrenheit to Celsius. Such transformations are linear in that the relationship between the original variable

and the transformed variable is a straight line. Conclusions from our statistical analyses will be identical usually for the untransformed and the transformed data.

In data analysis, particularly in biology, transformations that change the data nonlinearly are more common. There are at least five aims of data transformations for statistical analyses, especially for linear models based on normal distributions:

- to make the model error terms and the response variable closer to the assumed distribution (e.g. to make the distribution of the data symmetrical);
- to reduce any relationship between the mean and the variance (i.e. to improve homogeneity of variances);
- to reduce the influence of outliers, especially when they are at one end of a distribution;
- to improve linearity in regression analyses; and
- to make multiplicative effects on the raw scale additive on a transformed scale, reducing the size of interaction effects (Chapters 7 and 8).

The most common use of transformations in biology is to help positively skewed data meet the distributional and variance assumptions required for linear models based on the normal distribution. Sokal and Rohlf (2012) and Tabachnick and Fidell (2019) provide excellent descriptions and justifications of transformations. These authors are reassuring to those who are uncomfortable about transforming their data, feeling they are "fiddling" the data to increase the chance of getting a meaningful result from their analysis. However, a decision to transform is always made before the analysis is done and without looking at the results.

A transformation changes your response variable and, therefore, the research question addressed by the analysis. You might hypothesize that the growth of plants varies with density. If you decide to log-transform your response, the hypothesis becomes "log-growth varies with density," or perhaps in the first case growth is defined as mg of weight gained, whereas after log-transforming growth is the log-mg weight gained. There is an argument that your results should be presented on the transformed scale, although many biologists analyze transformed data but present results such as means on the original scale or after back-transforming from the transformed scale. However, doing so without careful thought can give the reader a misleading impression of the effects of the predictor(s).

After any transformation aimed at meeting distributional assumptions, you should re-check your data to ensure the transformation improved the data distribution. Sometimes, log or square-root transformations can skew data just as severely in the opposite direction and produce new outliers.

5.4.1 Transformations and Distributional Assumptions

Continuous variables with skewed distributions can sometimes be analyzed with GLMs based on the gamma or inverse Gaussian distributions (Box 4.2); they often have a lognormal distribution, defined as a distribution for which the log-transformed variable has a normal distribution. In biology, it has been common practice to transform variables with lognormal distributions, so the transformed data are simply the logs (to any base) of the original data. They will then have a symmetrical distribution (e.g. Figure 6.2). A log-transformed scale is often the default scale for commonly used measurements, especially those covering a wide range of values (e.g. one or more orders of magnitude). For example, pH is simply the log of the concentration of H^+ ions, and most cameras represent aperture as f-stops, with each increase in f representing a halving of light reaching the sensor (a $\log_2$ scale).

While log transformations can be effective at normalizing positively skewed variables, this is not always the case (Feng et al. 2013a). There can also be issues with back-transforming (by taking the anti-log) key outputs from analyses of log-transformed data (Bland et al. 2013; Feng et al. 2013a, b). There is also an issue using log-transformation if the variable has zero values, such as count data, because you can't take the log of zero. Traditionally, biologists have used $\log(Y + c)$ where c is an appropriate constant. Some researchers use the smallest possible value for their variable as a constant; others use an arbitrarily small number, such as 0.001 or, most commonly, 1. If models based on a normal distribution are required, power transformations described below are preferred, as zeroes are not affected.

Another common type of transformation useful for biological data (especially counts or measurements) is the power transformation, which transforms Y to Y^p, where $p > 0$. The square-root transformation, where $p = 0.5$, applies for data with right skew, particularly for counts (Poisson-distributed). Cube roots ($p = 0.33$) and fourth roots ($p = 0.25$) will be more effective for increasingly skewed data; fourth-root transformations are commonly used for abundance data in ecology when there are lots of zeroes and a few large values. In practice, count data are better handled using GLMs, based on a Poisson or, if the variance is greater than the mean, a negative binomial distribution (O'Hara & Kotze 2010; but see Ives 2015).

If skewness is negative – that is, the distribution has a long left tail – Tabachnick and Fidell (2019) suggested reflecting the variable before transforming. Reflection simply involves creating a constant by adding 1 to the largest value in the sample and then subtracting each observation from this constant. Left or negatively skewed distributions are uncommon for biological variables.

Log and power transformations can be considered part of the Box-Cox family of transformations:

$$(y^\lambda - 1)/\lambda \text{ when } \lambda \neq 0 \text{ and } \log(y) \text{ when } \lambda = 0$$

When $\lambda = 1$, we have no change to the distribution; when $\lambda = 0.5$, we have the square-root transformation, and when $\lambda = -1$, we have the reciprocal transformation, etc. The Box-Cox family of transformations can also find the best transformation, in terms of normality and homogeneity of variance, by an iterative process that selects a value of λ that maximizes a log-likelihood function (Sokal & Rohlf 2012).

When data are percentages or proportions, they are bounded at 0% and 100%. Log and power transformations don't work very well for these data because they change each end of the distribution differently. One common approach is to use the arcsin transformation $\sin^{-1}(\sqrt{y_i})$, where y_i is a proportion. It is most effective if y_i is close to 0 or 1 and has little effect on mid-range proportions. While this transformation has been common, it has a drawback: It produces values in radians or degrees, which cannot be linked easily to the original scale. This transformation has been questioned by Warton and Hui (2011), who argue that it has been used in two situations: data that follow a binomial distribution and data recorded as percentages but which are not necessarily binomial (they use percentage nitrogen content as an example). Warton and Hui (2011) recommend using logistic regression (GLM with binomial link function) for binomial data (Chapter 13), and the logit transformation $\log(y_i/(1 - y_i))$ for percentages as an alternative transformation much easier to interpret biologically.

Finally, we should mention the rank transformation, which converts the observations to ranks. The rank transformation differs from the other transformations discussed here because it is bounded by 1 and n, where n is the sample size. This transformation is extreme, as it results in equal differences (one unit, except for ties) between every pair of observations in this ranked set, regardless of their absolute difference. It therefore results in the greatest loss of information of all the monotonic transformations.

5.4.2 Transformations and Linearity

Transformations can also improve the linearity of relationships between two variables and thus make linear regression models more appropriate. For example, allometric

Table 5.1 Means for treatment and control groups for a factorial experiment with two factors (control/treatment and Time 1/Time 2) Artificial data and arbitrary units were used.

	Untransformed		Log-transformed	
	Time 1	Time 2	Time 1	Time 2
Control	10	50	1.000	1.699
Treatment	5	25	0.699	1.398

relationships with body size nearly always have a better linear fit after one or both variables are log-transformed. Note that nonlinear relationships can also be investigated with a nonlinear model.

5.4.3 Transformations and Additivity

Transformations also influence how we measure effects in linear models. For example, let's say we measured the effect of an experimental treatment compared to a control at two times. If the means of our control groups are different at each time, how we measure the effect of the treatment is important. Some very artificial data are provided in Table 5.1 to illustrate the point. At Time 1, the treatment changes the mean value of our response variable from 10 to 5 units, a decrease of 5. At Time 2, the change is from 50 to 25, a change of 25. On the raw scale of measurement, the effects of the treatments are very different, but in percentage terms they are identical, both showing a 50% reduction. Biologically, which is the most meaningful measure of effect, a change in raw scale or a percentage scale? In many cases, the percentage change might be more biologically relevant, and we would want our analysis to conclude that the treatment effects are the same at the two times. Transforming the data to a log scale achieves this (Table 5.1).

The measurement scale can also affect the interpretation of interaction terms in more complex linear models (Section 7.1.9).

5.4.4 Do We Really Need a Data Transformation?

The transformation approach, in part, has been based on creating a path to a model with a normally distributed response variable, which can be fitted using OLS estimation. Historically, this occurred because OLS is computationally simpler than ML. The widespread availability of ML estimation in statistical software means we can now use GLMs (and GLMMs), where we only need the data to follow one of the exponential

distributions and we can directly model the response variable. We address GLMs in Chapter 13.

5.5 Standardizations

Another change we can make to the values of our variable is to standardize them in relation to each other. If we include two or more continuous variables in an analysis, such as regression analysis or a more complex multivariate analysis, converting all the variables to a similar scale is often important before they are included in the one analysis. Several standardizations are possible. Centering a variable simply changes it so it has a mean of zero:

$$y_i = y_i - \hat{y}.$$

This is sometimes called translation (Legendre & Legendre 2012).

Variables can also be altered to range from 0 (minimum) to 1 (maximum). Legendre and Legendre (2012) describe two ways of achieving this:

$$y_i = \frac{y_i}{y_{max}} \quad \text{and} \quad y_i = \frac{y_i - y_{min}}{y_{max} - y_{min}}.$$

The latter is called ranging, and both methods are particularly useful as standardizations of abundance data before multivariate analyses that examine dissimilarities between sampling units in terms of species composition (Section 14.5).

Changing a variable so it has a mean of zero and an SD of 1 is often termed standardization:

$$y_i = \frac{y_i - \bar{y}}{s}.$$

The standardized values are also called z scores and represent the values of the variable from a normal distribution with a mean of 0 and an SD of 1.

5.6 Missing Data

It is not uncommon in biological research for datasets to have missing values. Unbalanced sample sizes for analyses with categorical predictors do not pose particular problems, as long as all categories have observations. However, when missing values cause whole categories to be lost, or when datasets involve repeated measurements of sampling units through time, they can be more challenging to deal with. Before we look at other ways of handling missing values, it is worth considering the different mechanisms by which missing data can arise (Little & Rubin 2019; McKnight et al. 2007; Nakagawa 2015; Nakagawa & Freckleton 2008).

5.6.1 Missing Data Mechanisms

If the probability that an observation is missing is independent of the other variables, the missing observations are termed missing completely at random (MCAR). This implies that the missing observations are a random subset of the data. Missing observations that are not MCAR can be of two types. First, the probability that an observation is missing might depend on the values of the other variables. For example, the pattern of missing data may depend on the group in which the object occurs, where another variable classifies objects into groups. This is termed missing at random (MAR). MCAR and MAR are considered ignorable missing values. Second, the missing values might be nonignorable (MNAR) because whether an observation is missing depends on its value (e.g. only large values of the variable are missing) or some additional unobserved variable. It is often difficult to distinguish MAR and MNAR because we don't have information about the missing values (Nakagawa & Freckleton 2008).

Consider a study that sampled numerous water quality and environmental variables from a set of streams. Imagine that one stream was missing a value for one variable (e.g. pH). Suppose the value is missing because of a random malfunction of a meter or a mistake by a researcher who forgot to record the value. This observation might be MCAR because it is not related to any of the variables in the dataset. If the value is missing because the stream was at a high altitude and weather conditions precluded access, then the observation might be MAR because the value of another measured variable (elevation), but not the unobserved pH value, determined the probability of it being missing. Finally, suppose the value is missing because the original pH reading was unusual enough that either the meter could not measure it or the researcher assumed it was a mistake and omitted it. There, the missing value depends on the variable itself and is nonignorable. This is more common when the observations depend on responses from subjects, such as in marketing surveys or clinical trials. However, studies on animal behavior may suffer from this nonresponse. Most methods for dealing with missing data assume at least MAR.

5.6.2 Detecting Missing Data

For smaller datasets, missing values can be easily seen in the raw data matrix. However, it is often useful to visualize the pattern of missing data across the complete set of variables. Nakagawa (2015) provided an excellent introduction to some of these methods, including missingness maps and paired panel plots available in specialist software. He also pointed out there is no straightforward way of distinguishing between MAR and MNAR. Researchers need to use their knowledge of the biological system under investigation to decide whether MAR is a reasonable assumption.

5.6.3 Methods for Missing Data
5.6.3.1 *Deletions*

The simplest way of dealing with missing observations is to delete sampling or experimental units with the missing values; this is casewise or listwise deletion. It is the default for most statistical software routines. An alternative when the focus is on examining patterns of associations in the dataset, such as pairwise correlations among the variables, is pairwise deletion, where units are deleted only where one of the pair of variables has missing values but not the rest. Unfortunately, both methods involve the loss of information, especially if the number of units is small, and resulting parameter estimates are likely to be biased (Nakagawa 2015).

5.6.3.2 *Single Imputation*

Imputation involves replacing (substituting) the missing values with some estimate of the values. We could, for example, replace the missing observation with the mean value of the variable calculated from the nonmissing observations or use a regression model to predict the imputed observation from other variables in the data. These methods are of limited value because (1) the imputed values are not independent of the observed data for a variable and the precision of the estimates of parameters based on these imputed values is generally underestimated; and (2) imputing a single value provides no measure of uncertainty.

5.6.3.3 *Multiple Imputation*

Rubin (1987) developed a method termed multiple imputation as a solution for the problems with single imputation, especially the lack of any measure of uncertainty (see also Little & Rubin 2019). Multiple imputation comprises three steps. First, missing data are imputed multiple times using a particular imputation model based on a specific distribution for the missing values (see below). Rubin (1987) recommended 3–10 imputations, but modern computer power means there is no real limit to the number and it should depend on the extent of missing values relative to the size of the dataset (Dong & Peng 2013). Second, the complete datasets (observed and imputed values) are then analyzed as usual. Third, model parameters are estimated as the mean of estimates from the analyses of the imputed datasets. The SE of this average estimate includes the variance between imputations and the variance within each dataset using a set of equations known as Rubin's rules. Multiple imputation is a considerable improvement over single imputation, indicating how different imputed values affect the outcome of our analysis.

The first step of determining the imputed values is the trickiest, and the two approaches we briefly describe are termed *data augmentation methods* (McKnight et al. 2007). They can be used in missing data analysis themselves (they deal with the missing values and the subsequent analysis as a single procedure), and they are often used as the first step in MI (Dong & Peng 2013; Nakagawa 2015).

First are expectation maximization (EM) algorithms that are a way of obtaining ML estimates of means and variances of, and covariances between, variables when there are missing data. Some function of the missing data, like a predictive distribution, is incorporated into the likelihood function.

Second is to take a Bayesian approach and use Markov chain Monte Carlo (MCMC; Section 2.7.3) procedures to create a posterior distribution of missing values. Both methods require starting values, such as those produced by a stochastic version of regression imputation (Nakagawa 2015). The MCMC approach can handle a variety of probability distributions, whereas the EM/ML approach assumes multivariate normality of the variables; transformations can help to achieve the latter. Uncertainty estimates, and therefore SEs of parameter estimates, are derived directly from the posterior distribution with the MCMC approach. The EM approach needs to use bootstrapping (and becomes the EMB [expectation maximization with bootstrapping] algorithm). Finally, whether you use EMB or MCMC for the initial step, check that the algorithms have reached convergence in the estimates of means and variances, and covariances (Nakagawa 2015).

The multiple imputation methods, and the data augmentation procedures on which the initial imputation step is based, assume the missing data are at least MAR. However, McKnight et al. (2007) point out that multiple imputation can also be reliable under some MNAR conditions, which is useful given the challenge for biologists distinguishing between MAR and MNAR missing data. The other point to remember is that while the computational details for multiple imputation, EM, and MCMC are complex, the availability of software for missing data analyses has improved greatly in recent years. As always, it's important to know what your preferred software is doing.

5.7 Key Points

- Exploratory data analysis is an essential first step to ensure that you are fitting the right kind of statistical model to your data.
- Boxplots and probability plots are valuable for assessing the distribution of the response variable, but residual plots are the most valuable display.
- Scatterplots are important to check for linear vs. curvilinear relationships.
- Initial graphical analysis is also very valuable for identifying outliers, which can have a great influence on your analysis. They should be examined to decide if they are errors or legitimate observations.
- Transformations are routinely used to improve the fit of biological data to the assumptions of the planned statistical models. They can be used to allow OLS estimation, but there is increasing use of GLMs for data that are not normally distributed.
- Missing data are common in biology, and are considered at the exploratory stage. It is important to decide why observations are missing before deleting observations or imputing substitute values.

Further Reading

Exploratory data analysis: Modern statistical analysis texts usually include sections on exploratory data analysis, although the term was coined by Tukey (1977), and his book remains valuable, along with Hoaglin et al. (1983) and Cleveland (1993), with Ieno and Zuur (2015) providing a recent broad introduction.

Smoothing: Good, detailed descriptions can be found in Hastie and Tibshirani (1990), Fox and Weisberg (2018), and Ieno et al. (2015).

6 Simple Linear Models with One Predictor

Chapter 4 introduced the framework for fitting linear models. In this chapter, we will explain the analyses of two common single-predictor models. We outline the study designs to which these models are applied, parameter estimation, evaluation of hypotheses about relevant parameters, and assessing model fit. We will also consider alternatives to the OLS methods commonly used to fit these models. In this chapter, we focus on continuous response variables with an underlying normal distribution. We cover other response variables in Chapter 13.

6.1 Linear Model for a Single Continuous Predictor: Linear Regression

Simple linear regression, where we assume a straight-line relationship between a continuous response variable and a single, usually continuous, predictor variable, is one of biology's most widely applied statistical techniques. Linear regressions apply commonly in two scenarios:

1. Values of the continuous response and predictor variables are recorded in an observational sampling study from sampling units selected randomly from a larger population. Usually, a single observation of the response variable is recorded for each predictor value, the single observation being from a population of possible values of the response variable for each predictor value. Essentially, the response and predictor variables are random.

2. The investigator determines the predictor values and can include multiple units (and therefore multiple observations of the response variable) for each value of the predictor. This is sometimes termed fixed-X regression.

Fitting a linear regression model has three primary purposes:

- Describe the linear relationship between Y and X, that is, estimate the parameters of the linear model and evaluate them.

- Determine how much variation (uncertainty) in Y can be explained by the linear relationship with X, and how much remains unexplained.

- Predict new values of Y from new observations of X.

We will use three examples from the biological literature to illustrate linear regression analysis.

Coarse Woody Debris in Lakes

Humans' land use has altered the input of coarse woody debris (CWD), loose woody material that provides habitat for freshwater organisms and affects hydrological processes in rivers and lakes. Christensen et al. (1996) studied the possible dependence of CWD on shoreline vegetation and lake development using a sample of North American lakes. We will use their data to model the relationships between the response variable CWD basal area and the predictor riparian tree density or density of cabins. These analyses are presented in Box 6.1.

Soldier Production in Aphids

We first presented in Chapters 2 and 4 the study by Shibao et al. (2004), who examined factors affecting the production of soldiers in a gall-forming aphid. Their interest was a linear relationship between soldier production (proportion of soldiers) as the response and aphid density as the predictor. These data are analyzed in Box 6.2.

The third example about species of invertebrates inhabiting patches of mussels on rocky intertidal shores will be introduced later in the chapter (Section 6.5), when we discuss assumptions and diagnostics.

6.1.1 Linear Model for a Continuous Predictor (Linear Regression Model)

Consider a set of n observations on which we have recorded the value of a continuous random response

Box 6.1 ⓡ ⓔ Worked Example of Single-Predictor Linear Regression: CWD in Lakes

Christensen et al. (1996) studied the relationships between CWD and shoreline vegetation and lake development in a sample of 16 lakes in North America. They defined CWD as debris >5 cm in diameter, and recorded – for several plots on each lake – the basal area $(m^2 \, km^{-1})$ of CWD in the nearshore water and the density $(no.km^{-1})$ of riparian trees along the shore. They also recorded the density of cabins along the shoreline.

CWD Basal Area vs. Riparian Tree Density
A scatterplot of CWD basal area against riparian tree density, with a LOESS smoother fitted, showed no evidence of a nonlinear relationship (Figure 6.1). Boxplots were slightly skewed, but the residuals from fitting the linear regression model were evenly spread, and there were no obvious outliers. One lake (Tenderfoot) had a higher Cook's D_i than the others, mainly because its high riparian density gave it a slightly higher leverage value. Omitting this lake from the analysis did not alter the conclusions, so it was retained. The variables were not transformed.

An OLS fit of a linear regression model for CWD basal area against riparian tree density gave us the following table.

	Coefficient	SE	Standardized coefficient	t	P
Intercept	−77.099	30.61	0	−2.52	0.025
Slope	0.116	0.02	0.797	4.93	<0.001
Correlation coefficient $(r) = -0.797, r^2 = 0.634$					
Source	df	MS	F	P	
Regression	1	32,055	24.30	<0.001	
Residual	14	1,319			

There is strong evidence for a positive regression slope (β_1) with CWD basal area increasing by $0.12 \, m^2 \, km^{-1}$ (95% CI: 0.07–0.17) for every extra riparian tree per kilometer. There is also strong evidence that β_0 is different from zero. Although the intercept is usually of little biological interest, CWD with no riparian trees might raise questions about the origin of this material. The r^2 value indicates that we can explain about 63% of the total variation in CWD basal area by the linear regression with riparian tree density.

We can predict CWD basal area for a new lake with 1,500 trees km^{-1} in the riparian zone. Plugging 1,500 into our fitted regression model,

CWD basal area $= -77.099 + 0.116 \times 1,500,$

the predicted CWD is $96.18 \, m^2 \, km^{-1}$. The SE of this predicted value is 10.84, resulting in a 95% CI for true mean (predicted) CWD basal area of 14.9–177.5.

CWD Basal Area Against Cabin Density
A scatterplot of CWD basal area against cabin density, with a LOESS smoother, showed some evidence of a nonlinear relationship (Figure 6.2). The boxplot of cabin density was highly skewed, with several zero values. The residuals from fitting the linear regression model with untransformed data suggested increasing spread of residuals with an unusual value (Arrowhead Lake) with a low (negative) predicted value and a much higher Cook's D_i than the others (Figure 6.2). Following Christensen et al., we transformed cabin density to log (cabin density + 1) and refitted the linear model. The scatterplot of CWD basal area against log cabin density suggested a much better linear relationship (Figure 6.2). The boxplot of log cabin density was less skewed, but the residuals from fitting the linear regression model still showed increasing spread with increasing predicted values. Lake Arrowhead was no longer influential, but Lake Bergner was an outlier with a moderate Cook's D_i. We also fitted a linear model when both variables were log-transformed. The scatterplot (see the online box) of log CWD basal area against log cabin density

Box 6.1 (cont.)

suggested a slightly less linear relationship with log CWD basal area now negatively skewed, although the residuals showed constant spread and no particularly influential observations.

We will present the analysis with just cabin density transformed as per Christensen et al. The results of the OLS fit of a linear regression model to CWD against log (cabin density + 1) were as follows.

	Coefficient	SE	Standardized coefficient	t	P
Intercept	121.969	13.969	0	8.73	<0.001
Slope	−93.301	18.296	−0.806	−5.10	<0.001
$r^2 = 0.650$					

Source	df	MS	F	P
Regression	1	32,840	26.00	<0.001
Residual	14	1,263		

There is strong evidence for a positive regression slope (β_1) with CWD decreasing by 93.3 m^2 km^{-1} (95% CI: −134.5 to −54.1) for an increase of one log cabin (+1) per kilometer. Again, the intercept is also different from zero, indicating the CWD for a lake without cabins. The r^2 value indicates that we can explain about 65% of the total variation in CWD by the linear regression.

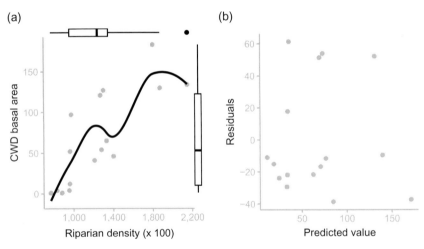

Figure 6.1 Regression diagnostics for relationship between CWD basal area against riparian vegetation density. Panel (a) shows a scatterplot (with LOESS smoother, smoothing parameter = 0.5) and marginal boxplots to show the distribution of each variable. Panel (b) shows the residual plot from a linear regression of basal area on riparian density.

variable (Y) and the value of a continuous fixed predictor (X). The linear regression model for one predictor was described in Section 4.2.1 and illustrated in Figure 4.1. Recall that for any particular value of X (x_i), there is a probability distribution of possible Y-values (and error terms) that are independently and identically normally distributed with zero mean and constant variance and

the true population regression line joins the means of these distributions of Y-values.

We can write the model to express the value of any individual observation, y_i:

$$y_i = \beta_0 + \beta_1 x_i + \varepsilon_i$$

or as the mean or expected value of y_i, $\mathrm{E}(y_i)$, for any x_i:

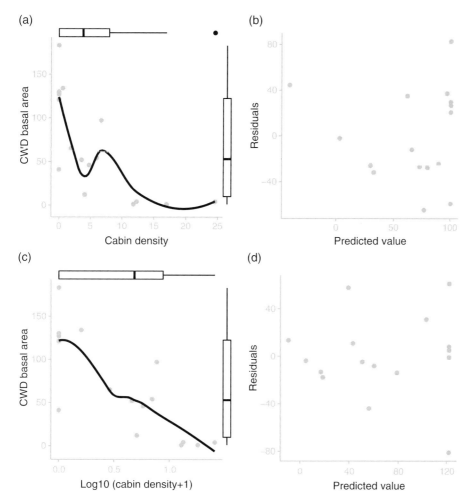

Figure 6.2 Regression diagnostics for the relationship between CWD basal area against cabin density. Panel (a) shows a scatterplot (with LOESS smoother, smoothing parameter = 0.5) and the (b) shows the residual plot from a linear regression of basal area on cabin density. Panels (c) and (d) show the same diagnostics following log transformation of the cabin density predictor.

$$\mathrm{E}(y_i) = \beta_0 + \beta_1 x_i.$$

For the aphid data from Shibao et al. (2004), we would fit:

$$(\text{soldier percentage})_i = \beta_0 + \beta_1 (\text{aphid density})_i + \varepsilon_i,$$

where $n = 16$ galls.

For the lake data from Christensen et al. (1996), we would fit:

$$(\mathrm{CWD})_i = \beta_0 + \beta_1 (\text{riparian tree density})_i + \varepsilon_i,$$

where $n = 16$ lakes.

In these models:

• y_i is the value of Y for the ith observation when the predictor $X = x_i$. For example, it is the proportion of soldiers for gall i, for which the aphid density is x_i, or the CWD for lake i, for which the riparian tree density is x_i.

• β_0 is a population parameter to be estimated: the intercept (mean of the probability distribution of Y when $X = 0$).

• β_1 is a population parameter to be estimated: the slope (the change in Y for a unit change in X).

• ε_i is random or unexplained error associated with observation i. Each ε_i measures the difference between each observed y_i and the mean of y_i. This mean is the value of y_i predicted by the population regression model, which we never know.

Box 6.2 ® € Worked Example of Single-Predictor Linear Regression: Soldier Production in Aphids

Shibao et al. (2004) studied whether the density of aphids in a colony affected the proportion of soldiers in that colony. They collected galls formed by the aphid *Tuberaphis styraci* from host trees in Japan. They recorded several characteristics of the galls (diameter, height, inner surface area, number of holes) and the aphids using that gall (number of aphids, proportion of soldier individuals, and aphid density – no. aphids/surface area). The variables of interest here are the percentage of soldiers and the aphid density (aphids mm^{-2}).

The scatterplot of percentage soldiers against aphid density is shown in Figure 6.3 and shows no evidence of a nonlinear relationship. The residual plot (not presented) showed an even spread, and the diagnostics demonstrated no large Cook's D_i values. The results of an OLS fit of a linear regression model of the percentage solders against aphid density is as follows.

	Coefficient	SE	Standardized coefficient	t	P
Intercept	27.78	3.55	0.00	7.83	<0.001
Slope	29.30	4.68	0.86	6.25	<0.001

Correlation coefficient $(r) = 0.858$, $r^2 = 0.736$

Source	df	MS	F	P
Regression	1	2,968.1	39.12	<0.001
Residual	14	75.9		

There is strong evidence for a positive regression slope (β_1) with the estimate for the percentage of soldiers in a gall increasing by 29.3 (95% CI: 19.3–39.3) for every unit increase in aphid density. Note that the F statistic is the square of the t statistic, allowing for rounding. The intercept is of little biological interest. The r^2 value indicates that we can explain about 74% of the total variation in the percentage of soldiers by the linear regression with aphid density.

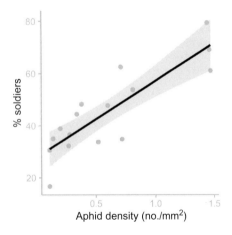

Figure 6.3 Scatterplot of soldier percentage vs. aphid density, with OLS regression fitted through data and 95% confidence band for regression.

The main aim of regression analysis is to estimate β_0 and β_1 based on our sample of n. The two commonest methods for estimating parameters of linear models are OLS and ML. If the ε_i are normally distributed, OLS and ML estimates of β_0 and β_1 are identical. Here we will focus on OLS, the most commonly used for normal-based linear models. To calculate SEs and CIs for these parameters, we also need to estimate σ_ε^2 (the variance of the ε_i, the error or residual variance).

We will look at the assumptions underlying OLS inference for single-predictor linear models in Section 6.4, but regression models have two other specific assumptions or conditions:

• We assume that a straight-line relationship is the most appropriate model. If not, we can use nonlinear models or nonparametric smoothing functions (Sections 5.1.2 and 13.9) or linear models including polynomial terms (Section 8.1.11).

• We also assume that the x_i are known constants (Kutner et al. 2005). There are two components to this, which biologists might find confusing.

– The predictor variable is "controlled" (Legendre & Legendre 2012) or fixed, that is, it is not a random variable with a probability distribution. This is sometimes termed a Model I regression. The fixed-X assumption is probably not met for many regressions in biology because X and Y are usually random variables recorded from a bivariate distribution. For example, Shibao et al. (2004) may have chosen galls to represent a range of aphid densities (some control over the predictor) or taken a random sample of galls and recorded the proportion of soldiers and aphid density from each gall. In the latter situation, any repeat of this study would likely use galls with different aphid densities. The inference and prediction methods we describe below still hold when X is a random variable if (1) we have clearly distinguished a response and predictor variable and our focus is on developing a model that predicts the response from the predictor; (2) the model specified at the start of this section and the assumptions described in Section 6.4 apply conditional on the x_i; and (3) the probability distribution of X is independent of the parameters of the model. Under these circumstances, X is considered fixed by being conditional on the x_i values rather than fixed by the sampling design (McCulloch et al. 2008). However, there are situations in biological research where X is random, and the distinction between response and predictor variables is less clear. The focus of our regression modeling is then the value of the estimate of the regression slope (β_1) – for example, modeling allometric (scaling) relationships (Warton et al. 2006) or comparing two measurement methods. More appropriate methods for such cases are presented in Section 6.1.7.

– Response and predictor variables are measured without error, so the recorded values of both variables for each sampling unit are the true values. When measurement error is present, the term errors-in-variables is sometimes used. Measurement error can be of

two sorts (Buonaccorsi 2010). First is sampling error, where the recorded value is an estimate based on subsampling each unit. For example, both CWD basal area and riparian tree density values for a lake (Christensen et al. 1996) are based on samples from around each lake rather than examining the entire lake shoreline. Sampling error is probably minimal in the Shibao et al. (2004) study, as the counts for each gall should be exact. The second is actual error in measuring due to inconsistencies in the measuring device, instrument, or researcher.

Measurement error in the predictor is the main problem and can cause a biased estimate of the regression slope. Unfortunately, it can only be assessed by having repeated observations on each sampling unit, an uncommon situation in biology. However, this may be necessary if we expect measurement error to be significant. OLS estimation and inference for our linear model essentially assumes that the errors in X are small relative to the range of values. If the focus is on testing for an association between Y and X and developing a predictive model to predict from new X-values with equivalent measurement error (which is likely), our regression inferences described below hold (Warton et al. 2006). Measurement errors will be considered further in Section 6.1.7, where our main interest is the true value of the regression slope.

6.1.2 Model Parameters

The OLS estimates of β_0 and β_1 are the values that produce a sample regression line $\hat{y}_i = b_0 + b_1 x_i$ that minimizes $\sum(y_i - \hat{y}_i)^2$, the sum of the squared deviations between each observed y_i and the value predicted by the sample regression line ($\hat{y}_i$). This is the sum of squared vertical distances between each observation and the fitted regression line (see Figure 4.2).

6.1.2.1 Regression Slope

β_1 is the population slope (e.g. the change in the proportion of soldiers for a one-unit change in aphid density or the change in CWD basal area for a one tree km^{-1} change in riparian tree density). We are most interested in this parameter as it measures how much the response variable changes as the predictor changes. If we can infer a causal relationship, it measures the effect of the predictor on the response.

b_1 is the estimate of β_1 from our sample. The OLS estimate of b_1 is the covariance between Y and X divided by the SS of X.

There is also a close mathematical relationship between linear regression and bivariate correlation and

Table 6.1 ANOVA table for simple linear regression of Y on X

Source of variation	SS	df	MS	Expected mean square
Regression	$\sum\limits_{i=1}^{n}(\hat{y}_i - \bar{y})^2$	1	$\dfrac{\sum\limits_{i=1}^{n}(\hat{y}_i - \bar{y})^2}{1}$	$\sigma_\varepsilon^2 + \beta_1^2 \sum\limits_{i=1}^{n}(x_i - \bar{x})^2$
Residual	$\sum\limits_{i=1}^{n}(y_i - \hat{y}_i)^2$	$n-2$	$\dfrac{\sum\limits_{i=1}^{n}(y_i - \hat{y}_i)^2}{n-2}$	σ_ε^2
Total	$\sum\limits_{i=1}^{n}(y_i - \bar{y}_i)^2$	$n-1$		

we can calculate b_1 from the sample correlation coefficient between Y and X as:

$$b_1 = r\,(s_Y/s_X),$$

where s_X and s_Y are the sample SDs of X and Y and r is the sample correlation coefficient between X and Y.

6.1.2.2 Intercept

β_0 is the population intercept (e.g. the mean proportion of soldiers when aphid density $= 0$ or the mean basal area of CWD for lakes with no riparian trees). It rarely has a useful biological interpretation as 0 is often outside the range of relevant X-values. The true regression model should logically go through the origin in many cases, but including an intercept is standard practice for linear models as no-intercept models are difficult to interpret (Section 6.1.6).

b_0 is the sample estimate of β_0. As the OLS regression line must pass through $\bar{y}$ and $\bar{x}$, the estimate (b_0) is derived by solving a simple rearrangement of the sample regression equation, substituting b_1, $\bar{y}$, and $\bar{x}$.

6.1.2.3 Predicted Values

Once we have estimated the model parameters, we have the sample regression equation:

$$\hat{y}_i = b_0 + b_1 x_i,$$

where $\hat{y}_i$ is the value of Y predicted by the fitted regression line for each x_i (e.g. the predicted soldier percentage for gall i or CWD basal area for lake i). We predict new Y-values from new X-values by substitution into the sample regression equation. In doing this, however, be wary of predicting from X-values outside your sample data range.

6.1.2.4 Error Terms and Their Variance

ε_i represents random or unexplained error associated with the ith observation – for example, each error term is measuring the difference between the soldier proportion at a gall and the true mean soldier proportion for galls

with that aphid density or the difference between the CWD basal area and the true mean CWD basal area for lakes with that riparian tree density. We can estimate the error terms for the observations in our sample by calculating the difference between each observed y_i and each predicted $\hat{y}_i$. This difference estimate is a residual (e_i):

$$e_i = y_i - \hat{y}_i.$$

The estimate of the error variance is the variance of these residuals, the residual mean square from the ANOVA of the fitted model (Section 6.1.5; Table 6.1). Residuals are central to the analysis of linear models. They provide the basis of the OLS estimate of σ_ε^2 and they are valuable diagnostic tools for checking assumptions and fit.

6.1.2.5 Standardized Coefficients

The value of the slope depends on the units of X and Y. For example, if CWD basal area was measured per 10 km rather than per kilometer, the slope would be 10 times greater. There may be some interest in comparing regression coefficients for different datasets and a standardized regression slope b_1^*, termed a beta coefficient, is more suited:

$$b_1^* = b_1\,(s_Y/s_X).$$

This is also the sample correlation coefficient. We can achieve the same result by standardizing X and Y (each to a mean of 0 and an SD of 1) and calculating the usual sample regression slope. The value of b_1^* estimates the slope of the regression model independent of X and Y units. For example, Christensen et al. (1996) also recorded CWD density for each lake. The estimated slopes for regression models of basal area and density against riparian tree density were 0.116 and 0.652, respectively, suggesting a much steeper relationship for basal area. However, the standardized slopes were 0.797 and 0.874, indicating that when the units of measurement were considered, the strength of the relationships of riparian tree density with CWD basal area and density were similar. The linear regression model for standardized variables does not include an intercept because its estimate would always be zero.

Standardized regression coefficients are sometimes used in multiple linear regression models to evaluate the relative importance of the predictors. However, there are more appropriate measures of importance (Section 9.1).

6.1.3 Inference for Parameters
6.1.3.1 *Standard Errors and Confidence Intervals*
Now we have point estimates for σ_{ε}^2 and β_1, we can look at the sampling distribution and SE of b_1 and CIs for β_1. The central limit theorem applies to b_1, so its sampling distribution is normal with an expected value (mean) of β_1. The SE of b_1 is the standard deviation of its sampling distribution. Confidence intervals for β_1 are calculated as usual when we know the SE of a statistic and use the t distribution. The 95% CI for β_1 is:

$$b_1 \pm t_{0.05, n-2} s_{b_1}.$$

Note that we use $n - 2$ df for the t statistic. The interpretation of CIs for regression slopes is as described for means in Section 2.4.5. To illustrate using a 95% CI, under repeated sampling we would expect 95% of these intervals to contain the fixed, but unknown, true slope of our linear regression model. The SE for b_0 and CI for β_0 can be calculated similarly.

We can also determine confidence bands for the regression line, which are CIs on the mean value of Y at each x_i (Kleinbaum et al. 2013; Kutner et al. 2005). To illustrate with the data relating the proportion of soldiers to aphid density from Shibao et al. (2004), Figure 6.5 in Box 6.4 shows the confidence bands that would include the true population regression line 95% of the time under repeated sampling of galls and their aphids. Note that the bands are wider further away from $\bar{x}$, indicating we are less confident about our estimate of the true regression line at the extremes of the range of observations.

6.1.3.2 *Statistical Hypotheses*
The null hypothesis (H_0) commonly evaluated in linear regression analysis is that $\beta_1 = 0$ – that is, there is no linear relationship between Y and X. For example, soldier proportion is independent of aphid density and CWD is unrelated to riparian tree density. Null hypotheses about single parameters in OLS fitted linear models can usually be evaluated using a single parameter t test, as described in Section 2.5.1. For a simple linear regression model, we calculate a t statistic from our data:

$$t = (b_1 - \theta)/s_{b_1}.$$

Here, θ is the value of β_1 specified in the H_0. We compare the calculated value to a t distribution with $(n - 2)$ df to determine the probability of obtaining a t value equal to or greater than our observed t from repeated sampling.

While $\theta = 0$ is the most common situation, some other value may be relevant. A common example is allometric relationships between different body components and processes (Warton et al. 2006). However, when the focus of the analysis is specifically on the value of the regression slope and the predictor variable is not fixed, methods for what is sometimes termed Model II regression (Section 6.1.7) are preferred.

We can also evaluate the H_0 that $\beta_0 = 0$ in the same way, though it is not usually of much biological interest unless we are considering excluding an intercept from our final model, something rarely recommended (Section 6.1.6).

In both cases, we can also evaluate the null hypotheses that the parameters equal zero by comparing the fit of models with and without the relevant term and using an F test. This model comparison approach has broader applicability across the full range of linear models.

6.1.4 Inference for Predicted Values
The predicted Y-values have a normal sampling distribution, and the SE of the new predicted Y is:

$$s_{\hat{y}} = \sqrt{\mathrm{MS}_{\mathrm{Residual}} \left[1 + \frac{1}{n} + \frac{(x_p - \bar{x})^2}{\sum_{i=1}^{n} (x_i - \bar{x})^2} \right]},$$

where x_p is the new value of X from which we are predicting. The other terms have been used in previous calculations. Confidence intervals (more correctly called prediction intervals) for a predicted Y can be calculated as usual using this SE and the t distribution with df $= n - 2$ and converted to prediction bands as with confidence bands in Section 6.1.3. The prediction intervals for individual predicted Y-values from a given X would be wider than the CI for the predicted mean for the same X. The SE is based on the $\mathrm{MS}_{\mathrm{Residual}}$ (variance of the model error terms and hence of Y) and the variation associated with using predicted Y to estimate the mean of Y (Kleinbaum et al. 2013).

6.1.5 Model Comparison and the Analysis of Variance
Measuring the fit of a linear model to sample data can be done in two broad ways, based on how we estimate the

model parameters. If ML is used, fit is determined by the size of the likelihood or log-likelihood, and models can be compared using likelihood-ratio statistics (Section 4.4.2). Ordinary least squares models use explained and unexplained sums-of-squares (SS). Whatever approach is used, we can compare the fit of models with and without specific parameters to assess how much those parameters (associated with specific predictors) contribute to model fit.

Partitioning the variation in a response variable into amounts explained and unexplained by the linear combination of predictors was introduced in Section 4.5.1. This partitioning is usually presented as an ANOVA (Table 6.1). For a single-predictor regression model, the total variation in Y is expressed as a sum of squared deviations of each observation from the sample mean (SS_{Total}) with df $= n - 1$ (one df is lost to estimate the mean). Sometimes the total variation in Y is expressed as an "uncorrected" total sum-of-squares $- SS_{Total\ uncorrected}$ (see Kutner et al. 2005). This is simply $\sum y_i^2$ and can be "corrected" by subtracting $n\bar{y}^2$ (termed "correcting for the mean") to convert $SS_{Total\ uncorrected}$ into the SS_{Total} we have used. The uncorrected total SS is occasionally used when regression models are forced through the origin (Section 6.1.6) and in nonlinear regression.

The SS_{Total} can be partitioned into two, usually additive, components:

• $SS_{Regression}$ is the variation explained by the linear regression with X. It is measured as the sum of squared differences between each $\hat{y}_i$ and $\bar{y}$ (Figure 4.2). This is a measure of how well the estimated regression model predicts $\bar{y}_i$ for each x_i. The $df_{Regression} = 1$, as we are estimating a single parameter.

• $SS_{Residual}$ is the variation in Y not explained by the linear relationship with the predictor, also called the error or unexplained variation. It is calculated as the difference between each observed Y and the value predicted by our estimated regression, $\hat{y}_i$ (Figure 4.2). This is a measure of how far the Y-values are from the fitted regression line. The $df_{Residual} = n - 2$, because we have estimated two parameters (β_0 and β_1) to determine the $\hat{y}_i$.

The SS and df are additive (Table 6.1):

$$SS_{Regression} + SS_{Residual} = SS_{Total},$$

$$df_{Regression} + df_{Residual} = df_{Total}.$$

The SS in Table 6.1 represent the fit $\left(SS_{Regression}\right)$ and lack-of-fit $\left(SS_{Residual}\right)$ for the regression model with a slope and an intercept, termed the full model as it contains all parameters of interest. We can compare the fit of this model to one that omits the regression slope (i.e. sets it to

zero), termed a reduced model, to evaluate the importance of including the predictor variable:

$$y_i = \beta_0 + \varepsilon_i.$$

The predicted Y-value for each x_i from this model is the intercept, which equals $\bar{y}$. Therefore, the $SS_{Residual}$ from this reduced model is the sum of squared differences between the observed Y-values and $\bar{y}$, which is SS_{Total} from our ANOVA for the full model (Table 6.1).

The difference between the unexplained variation of the full model ($SS_{Residual}$) and the unexplained variation from the reduced model (SS_{Total}) is termed $SS_{Difference}$. It equals the $SS_{Regression}$ from the full model. It measures how much more variation in Y is explained by the full model than by the reduced model. Describing the $SS_{Regression}$ as the variation explained by the regression model is really explaining how much more variation in Y the full model explains over the reduced model.

The partitioning of the SS provides a descriptive measure of model fit called r^2 (also termed R^2 or the coefficient of determination), which measures the proportion of the total variation in Y explained by its linear relationship with X. When we fit the full model, it is usually calculated as:

$$r^2 = \frac{SS_{Regression}}{SS_{Total}} = 1 - \frac{SS_{Residual}}{SS_{Total}}.$$

Anderson-Sprecher (1994) argued that r^2 is better explained in terms of the comparison between the full model and a reduced (no slope parameter) model:

$$r^2 = 1 - \frac{SS_{Residual(Full)}}{SS_{Residual(Reduced)}}.$$

These two versions are identical for models with an intercept, but the latter emphasizes that r^2 measures how much the fit is improved by the full model compared to the reduced model. We can also relate explained variance back to the bivariate correlation model because r^2 is the square of the correlation coefficient r. Values of r^2 range between 0 (no relationship between Y and X) and 1 (all points fall on the fitted regression line). Therefore, r^2 is not an absolute measure of how well a linear model fits the data, only a measure of how much a model with a slope parameter fits better than one without.

Great care should be taken if using r^2 for comparing the fit of different models. It is inappropriate for comparing models with different parameters (Section 9.2.1) and can be problematic for comparing models based on different transformations of Y (Scott & Wild 1991).

Although the SS is a useful measure of variation, it depends on sample size. SS_{Total} increases as more observations are included. In contrast to the SS, the variance (mean square, or MS) does not depend on the number of components because it is an average of the squared deviations. The next step in the ANOVA is to convert the SS to MS by dividing them by their df (Table 6.1). The $SS_{Difference}$ from the fit of full and reduced models can be converted to the $MS_{Difference}$ by dividing by the difference in the $df_{Residual}$ from the two models (which equals 1 in this case). The $MS_{Difference}$ from the fit of the two models will equal the $MS_{Regression}$, another measure of model fit.

Mean square values are not additive: $MS_{Regression} + MS_{Residual} \neq MS_{Total}$ and MS_{Total} does not play a role in analyses of variance.

$MS_{Regression}$ and $MS_{Residual}$ are sample variances, and statisticians have determined the expected values of these MS – that is, the average of all possible values of these MS or the population values these MS estimate (Table 6.1):

• $MS_{Residual}$ estimates σ_ε^2, the common variance of the ε_i. The implicit assumption here (Sections 4.6.1 and 6.4) is that the variance of ε_i (and therefore of y_i) is the same for all x_i (homogeneity of variance), and thus can be summarized by a single variance. If this assumption is not met, $MS_{Residual}$ does not estimate a common variance σ_ε^2, and OLS-based inference may be unreliable.

• $MS_{Regression}$ also estimates σ_ε^2 plus additional variation determined by the strength of the absolute relationship between Y and X (i.e. β_1^2 multiplied by SS_X).

Based on the expected values of these MS, we can develop a test for the H_0 that $\beta_1 = 0$. Under H_0, it is apparent from Table 6.1 that $MS_{Regression}$ and $MS_{Residual}$ both estimate σ_ε^2, because $\beta_1^2 \sum (x_i - \bar{x})^2$ becomes zero. Therefore, the ratio of $MS_{Regression}$ to $MS_{Residual}$ should be 1. If $\beta_1 \neq 0$, the expected value of $MS_{Regression}$ is larger than that of $MS_{Residual}$ and their ratio should be >1.

The ratio of two variances or MS follows the F distribution (Section 2.5.5). We can use the appropriate probability distribution of F (defined by numerator and denominator df) to determine the probability from repeated sampling of obtaining our sample F or one more extreme under H_0. The F test essentially compares the fit (to the data) of a model with a slope term to one that does not. The F from the ANOVA equals the square of the t statistic described in Section 6.1.3. Hence, the two approaches are identical for evaluating the H_0 of zero slope in models with a single predictor.

6.1.6 Regression Through the Origin

Often we know that Y must equal zero when $X = 0$. For example, if riparian trees are the only source of CWD,

CWD basal area must be zero if a lake has no riparian trees, and the weight of an organism must be zero when its length is zero. It might be tempting to force our regression line through the origin by fitting a linear model without an intercept term:

$$y_i = \beta_1 x_i + \varepsilon_i.$$

There are several difficulties when interpreting the results of fitting this model. First, our minimum observed x_i often does not extend to zero, so forcing our regression line through the origin involves extrapolating the fitted regression line outside our data range and assuming the relationship is linear outside this range. If Y must logically be zero when X is zero, yet our estimated intercept differs from zero, the relationship between Y and X must be nonlinear, at least for small values of X. Like Hocking (2013), we recommend that it is better to have a model that fits the observed data well than one that goes through the origin but provides a worse fit to the observed data.

Second, residuals from the no-intercept model no longer sum to zero, and the usual partition of SS_{Total} into $SS_{Regression}$ and $SS_{Residual}$ doesn't work. The $SS_{Residual}$ can even be greater than SS_{Total} (Kutner et al. 2005), and the value of r^2 for a no-intercept model may not be comparable to r^2 from the full model. The residuals are still comparable, but the $MS_{Residual}$ is probably better for comparing the fit of models with and without an intercept (Chatterjee & Hadi 2012).

6.1.7 Regression with X Random

The linear regression model we have been using so far is sometimes called Model I, because X is a fixed variable. As we have discussed (Section 6.1.1), many applications of linear regression in biology are unlikely to involve fixed X-values; indeed, the (x_i, y_i) pairs are commonly a sample from a bivariate distribution of two random variables. We can use the usual OLS regression model when Y and X are random if we can clearly distinguish a response variable from a predictor and our main aims are measuring the associations, evaluating hypotheses about the model parameters (especially the regression slope), and predicting Y from X. However, OLS will produce a biased estimate (usually an underestimate) of the slope when X is random, especially if X is measured with error.

There are situations in biology where we wish to estimate the slope of a line of best fit between two random variables, without treating either as a response or predictor. For example, in allometry, the interest is how morphological and physiological variables scale against

Box 6.3 ® Worked Example of Single Random Predictor Linear Regression: Brain and Body Weight in Mammals

For the data from Allison and Cicchetti (1976), the two variables of interest (brain weight and body weight) will be treated as random, with the focus on estimating the regression slope of the linear relationship between two random variables. The original variables were log-transformed, with both variables having strongly skewed distributions due to a few large-bodied (and large-brained) species.

Statistic	MA	Ranged MA	OLS
b_1	0.776	0.785	0.752
95% CI	0.718–0.836	0.727–0.847	0.695–0.809
b_0	0.913	0.908	0.927
95% CI	0.878–0.946	0.872–0.941	0.844–1.011

The correlation coefficient was very high (0.96), as is often the case with morphometric studies, so we would not expect much difference in the regression slope estimates. The estimated regression slope from the RMA and the ranged MA models were larger than the OLS estimate, as OLS treats the predictor as fixed and underestimates the slope between two random variables. Not surprisingly, the estimates of the intercept also differed. There was overwhelming evidence that the regression slope differed from zero, with P-values from the OLS t test and a ranged MA randomization test both being <0.001 and the 95% CIs being relatively narrow and excluding zero.

each other. Allison and Cicchetti (1976) collated brain and body mass data for 62 mammal species. The interest was in the slope of the relationship between the log-transformed variables to compare to some theoretical value, rather than predicting one variable from the other. The analyses of these data are presented in Box 6.3.

Fitting a linear regression model when Y and X are random has sometimes been termed Model II regression (Legendre & Legendre 2012; Sokal & Rohlf 2012), although this name is not informative.

There is error variability associated with Y (σ_ε^2) and X (σ_δ^2) and the OLS estimate of β_1 is biased toward zero (Box 6.3). We will still use Y and X for consistency, although we are not necessarily identifying a response and a predictor. The extent of the bias depends on the ratio of these error variances (Legendre & Legendre 2012):

$$\lambda = \sigma_\varepsilon^2 / \sigma_\delta^2.$$

If X is fixed, $\sigma_\delta^2 = 0$ and the usual OLS estimate of β_1 is unbiased. The greater the error variability in X relative to Y, the greater the downward bias in the OLS estimate of β_1. Remember that the usual OLS regression line is fitted by minimizing the sum of squared vertical distances from each observation to the fitted line (Figure 6.4). Here, $\sigma_\delta^2 = 0$ (fixed X) and $\lambda = \infty$. The choice of method for estimating a linear regression model when Y and X are random depends on our best guess of λ, which will come

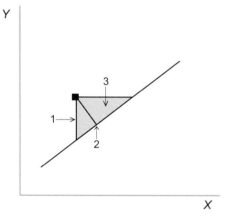

Figure 6.4 Distances or areas minimized by OLS (1), MA (2), and RMA (shaded area 3) linear regressions of Y against X.

from our knowledge of the two variables, the scales on which they are measured, and their sample variances.

Major axis (MA) line of best fit is estimated by minimizing the sum of squared perpendicular distances from each observation to the fitted line (Figure 6.4). It is also the first axis from a principal components analysis (PCA: Section 15.1) on the covariance matrix and is sometimes called the "fitted axis," with the "residual axis" at right angles to it (Warton et al. 2006). For MA, the variances associated with the random error in X and Y are assumed to be equal, so $\lambda = 1$. The estimate of the MA slope is:

$$b_{1(\text{MA})} = \frac{1}{2s_{XY}} \left(s_Y^2 - s_X^2 + \sqrt{\left(s_Y^2 - s_X^2\right)^2 + 4s_{XY}} \right).$$

It can also be calculated using the estimate of the slope of the OLS regression and the correlation coefficient:

$$b_{1(\text{MA})} = \frac{d \pm \sqrt{d^2 + 4}}{2}.$$

If r is positive, use the positive square root and vice versa. Here,

$$d = \frac{b_{1(\text{OLS})}^2 - r^2}{r^2 b_{1(\text{OLS})}}.$$

Standard or standardized major axis (SMA), also called the reduced major axis (RMA), is the line determined from the MA on standardized Y and X, back-transformed to the original scales by multiplying it by s_Y/s_X:

$$b_{1(\text{SMA})} = \sqrt{s_Y^2/s_X^2} = \text{sign}(s_{XY})s_Y/s_X.$$

This line is the first principal component based on the correlation matrix (see Section 15.1) and is sometimes termed the geometric regression, being the geometric mean of the OLS estimate of the slope of Y on X and the reciprocal of the OLS estimate of the slope of X on Y. Interestingly, the SMA slope can also be calculated from the OLS estimate of the slope of Y on X divided by the correlation coefficient between Y and X: $b_{1(\text{SMA})} = b_{1(\text{OLS})}/r_{XY}$.

If the correlation coefficient is close to 1, the SMA slope will be similar to the OLS regression slope. The closer the correlation coefficient is to 0, the more the OLS slope will underestimate the true slope when both variables are random. The SMA can also be fitted by minimizing the sum of areas of the triangles formed by vertical and horizontal lines from each observation to the fitted line (Figure 6.4). The SMA assumes that variances associated with the random error in X and Y are proportional to σ_Y^2 and σ_X^2, respectively, so $\lambda = \sigma_Y^2/\sigma_X^2$.

We introduced the difficult issue of measurement error, especially in the predictor variable, in Section 6.1. Measurement error is a particular problem if the focus of the analysis is estimating the true slope of the line of best fit using MA and SMA methods. Warton et al. (2006) describe a method-of-moments regression for when measurement error is not trivial and can be quantified. The challenge is that measurement error can only be quantified by repeated y_i and x_i measurements for each observation, which is uncommon for biological research.

Warton et al. (2006) summarized methods for evaluating hypotheses that the true slope equals a specific value. In allometric research, the hypothesis of interest

is often isometry, so the null hypothesis of zero slope (no relationship), as is common with OLS regression, is less interesting. Warton et al. recommended using the usual OLS F test, which also applies to the MA and SMA methods and is an exact test. Legendre and Legendre (2012) preferred permutation tests for the MA slope, although they pointed out this doesn't work for the SMA slope because of the inherent standardization of both variables. Warton et al. (2006) described bootstrap tests based on residuals for both MA and SMA slopes, but preferred the F test. Note there is no test that the SMA slope equals zero (Legendre & Legendre 2012). Examining evidence for a nonzero correlation coefficient first, to show a relationship between Y and X, is usually recommended (Legendre & Legendre 2012; McArdle 1988). Warton et al. (2006) provided formulae for CIs of the MA and SMA slopes based on the F test. Other methods for CIs include using the t distribution and the SE of the slope (Sokal & Rohlf 2012).

Legendre and Legendre (2012) and Warton et al. (2006) made recommendations on the most appropriate methods when both variables are random, the latter focusing on allometric relationships in biology. The broad conclusions are:

• When there is a clear distinction between predictor and response variables, and the aim is to evaluate whether there is a linear relationship between the two variables and predict Y from X, OLS (or ML) is recommended to fit the standard regression model.

• Evaluation of the null hypothesis that the slope equals zero is identical for the OLS regressions of Y on X and X on Y. Both are identical to the test that the correlation coefficient $\rho = 0$. The sample correlation coefficient is simply the geometric mean of these two regression slopes (Rodgers & Nicewander 1988):

$$r_{XY} = \pm\sqrt{(b_{YX}b_{XY})}.$$

• MA requires that the two variables are measured on the same scale (i.e. in the same units). This is not an issue if the variables are log-transformed, which is commonly the case in research areas like allometry, and the power of Y and X is arbitrary.

• SMA estimates the slope more efficiently than MA and is less biased (McArdle 1988). Inference has generally been regarded as more difficult with SMA due to the inherent standardization, although Warton et al. (2006) provided methods that overcome most of these challenges. Interestingly, Jackson (2003) pointed out that the SMA line seems to most observers a more intuitive and better "line of best fit" than the OLS line since it lies halfway between the OLS line for Y on X and the OLS line for X on Y.

- Measurement error is a particular difficulty for these methods because the focus is on estimating the true slope (independent of error in Y or X). If there are repeated measurements on each observation, then Warton et al. (2006) describe how to determine measurement error and incorporate it into slope estimation using method-of-moments regression.

6.2 Linear Model for a Single Categorical Predictor (Factor)

The other common single-predictor linear model used by biologists is a continuous response variable and a categorical predictor (often termed a factor). There are usually multiple independent units within each factor group. This chapter will focus on fixed predictors (factors), where all groups of interest have been included in the study, and we consider random predictors in Chapter 10. This model might apply to experimental or observational studies.

6.2.1 Experimental vs. Observational Studies

The main difference between these study types is whether randomization of independent units occurs by assigning the independent units randomly to the groups or by choosing independent units randomly from a pool within each factor group. The determination of factor groups may also differ.

6.2.1.1 Completely Randomized (Experimental) Designs

Here the investigator determines the groups of the factor and experimental units are randomly assigned to the different groups in a completely randomized (CR) design (Section 3.2.3). These CR designs provide strong inference about causal effects of the predictor on the response. For example, Giri et al. (2016) tested whether jointly altering the amounts of three micronutrients and three coenzymes in the diet of farmed Atlantic salmon changed the levels of desirable omega-3 fatty acids in fish flesh. They grew groups of salmon in large tanks, with each tank receiving one of several micronutrient/coenzyme groups. Here, the interest is not just whether the diets have different effects, but whether there are trends associated with the level of enrichment. The full data analyses are given in Box 6.4.

6.2.1.2 Observational (Nonexperimental) Designs

Here the factor groups are naturally occurring categories and sampling units are randomly selected from the populations of possible units within each group. It is more difficult to infer a causal relationship between the predictor and the response from such designs even though the situation may be more natural, without the artifacts that manipulative experiments sometimes introduce.

For example, Medley and Clements (1998) studied the response of diatom communities to heavy metals, especially zinc, in Rocky Mountains streams in the United States. One of their analyses was to classify the 34 stations into four groups based on zinc levels and test whether diatom species diversity differed between groups. The full analyses of these data are presented in Box 6.5.

Fitting linear models to data from these designs has two major purposes:
1. Describe the linear relationship between Y and X – that is, estimate the parameters of the linear model and evaluate hypotheses about those parameters, focusing on the differences between population group means.
2. Determine how much of the variation (uncertainty) in Y can be explained by the predictor groups and how much remains unexplained.

When the predictor is categorical (and our focus is on differences between group means), the linear models are often termed analyses of variance (ANOVAs). This terminology is somewhat confusing because ANOVA is a key part of most linear models, not just those with categorical predictors. We will treat the analyses of these datasets as linear model analyses, of which ANOVA is a part, just like the regression models described earlier in this chapter.

6.2.2 Linear Model for a Categorical Predictor

We'll start with a simple design with p factor groups and n observations in each group. In contrast to a regression model, we can specify the model in three ways that we introduced in Section 4.7.1.2. Statistical details are in Box 6.6, but each model specifies y_{ij}, the value of the response variable for observation j in group i.

6.2.2.1 Linear Effects Model

From Giri et al. (2016), $p = 4$ diets and $n = 3$ tanks:

$$(\text{n-3 LC-PUFA})_{ij} = (\text{overall mean}) + (\text{effect of diet})_i + \varepsilon_{ij}.$$

From Medley and Clements (1998), $p = 4$ zinc levels and $n = 8$ or 9 stations:

$$(\text{diatom diversity})_{ij} = (\text{overall mean}) + (\text{effect of zinc level})_i + \varepsilon_{ij}.$$

The effects represent the differences between each group mean and the overall mean. The effects model is common,

Box 6.4 Ⓡ Ⓔ Worked Example of Single-Factor Design with Ordered Treatment Groups: Fatty Acid Production in Salmon

Giri et al. (2016) were interested in the capacity of Atlantic salmon (*Salmo salar*), grown in an aquaculture situation, to deliver high levels of omega-3 long-chain polyunsaturated fatty acids (PUFA) in their tissues. Specifically, they tested whether several micronutrients (iron, zinc, magnesium) and coenzymes (riboflavin, biotin, and niacin) could increase the conversion from short- to long-chain PUFA. They used four treatments, a diet lacking these micronutrients and coenzymes (T-0), a diet with normal levels (T-100), and two levels of fortification, where the enzymes and micronutrients were 300% and 600% greater than normal (T-300 and T-600).

The experimental units were 1,000-L tanks, each having 24 fish, with three tanks for each of the four diets. After 84 days, fish were euthanized and their tissues analyzed. Fish (the observation units) were analyzed individually, and the data were then averaged to produce a single value for each large tank (the experimental unit).

The first analysis compares total omega-3 long-chain PUFAs (n-3 LC-PUFA, as mg/g) in fillets of fish (response variable across four types of diet; categorical predictor variable), with diet as a fixed factor. With only three tanks per treatment, we dispensed with boxplots, but the residual plots showed no unusual patterns and constant variance. The parameter estimates with SEs based on $MS_{Residual}$ from the ANOVA (and see Figure 6.5) were as follows.

Parameter	OLS estimate ($\pm$ SE)	
μ_i	Group mean n-3 LC-PUFA (all $n = 3$):	
T-0	9.06 ± 0.31	
T-100	10.59 ± 0.31	
T-300	10.70 ± 0.31	
T-600	10.47 ± 0.31	
μ	Overall mean: 10.20 ± 0.15	
$\alpha_i = \mu_i - \mu$	T-0:	$9.055 - 10.20 = -1.150$
	T-100:	$10.59 - 10.20 = 0.39$
	T-300:	$10.70 - 10.20 = 0.49$
	T-600:	$10.47 - 10.20 = 0.27$
ε_{ij}	Obs. 1: $9.67 - 9.06 = 0.61$	
	Obs. 2: $8.60 - 9.06 = -0.45$	
	Obs. 3: $8.90 - 9.06 = -0.15$	
	Obs. 4: $10.54 - 10.59 = -0.05$	
	etc.	

The results from an ANOVA from fitting a linear model with diet as the predictor variable were as follows.

Source	SS	df	MS	F	P	η^2
Diet	5.355	3	1.785	6.33	0.017	0.704
Residual	2.256	8	0.282			
Total	7.611	11				

There is evidence for an effect of diet on total omega-3 long-chain PUFAs (and the test of the null hypothesis of no difference between diets is "statistically significant"). In this case, we could do an unplanned comparison to further examine differences between diets. However, the diets form a quantitative sequence (T-0, T-100, T-300, T-600), from a diet depleted in coenzymes and micronutrients to increasingly fortified diets. A more useful examination of the data would be to ask whether n-3 LC-PUFA levels are related to the level of these dietary components. In the absence of further information, the simplest question would be whether there is a linear increase in the target fatty acids as we move from the most depleted to the most fortified diet. A linear contrast or planned comparison (see Section 6.2.7.1) will do this.

Box 6.4 (cont.)

Source	SS	df	MS	F	P
Contrast (linear trend)	2.850	1	1.730	6.13	0.038
Residual	2.256	8	0.282		

This contrast measures how much variation among treatments ($SS_{Treatment}$) is associated with a linear trend, and the F statistic is determined using the residual variation from the main analysis. There is evidence for a linear trend in total omega-3 long-chain PUFAs with increasing micronutrients and enzymes (see Figure 6.5). If our biological understanding had suggested that a quadratic relationship was more appropriate, we could create that contrast, and the F would be 8.851, $P = 0.018$.

While we have used these trends as examples, remember that the pattern of interest *must* be specified before examining the data or starting the analysis.

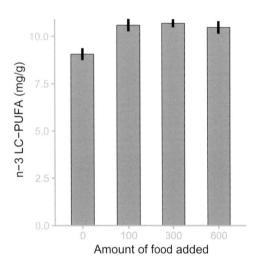

Figure 6.5 Bar graph showing mean concentrations of n-3 LC-PUFAs in tissues of Atlantic salmon fed diets with four levels of a group of micronutrients and coenzymes, ranging from depleted (T-0) to normal (T-100), up to highly enriched (T-600). Error bars are SEs from three replicate tanks per treatment, and the value for each tank is the mean of 24 fish.

perhaps because it includes a term measuring the effect of our predictor variable on the response, analogous to the regression slope in linear regression models.

6.2.2.2 *Means Model*

From Giri et al. (2016):

$$(\text{n-3 LC-PUFA})_{ij} = (\text{mean of diet group})_i + \varepsilon_{ij}.$$

From Medley and Clements (1998):

$$(\text{diatom diversity})_{ij} = (\text{mean of zinc level group})_i + \varepsilon_{ij}.$$

The means model is simpler than the effects model but doesn't provide a straightforward measure of group effects.

6.2.2.3 *Regression (Dummy Variable) Model*

Using reference coding (Box 6.6) for the Medley and Clements (1998) example, specifying background zinc level to be the reference group, the model is (Box 6.6):

$$(\text{diatom diversity})_{ij} = (\text{mean of background group}) +$$
$$\beta_1(\text{difference between low and background})_{ij} +$$
$$\beta_2(\text{difference between medium and background})_{ij} +$$
$$\beta_3(\text{difference between high and background}) + \varepsilon_{ij}.$$

The regression (dummy variable) model provides contrasts between groups that may be biologically relevant.

6.2.3 Model Parameters

In practice, we usually code the effects model into linear models software, although we are interested in parameters from all three models; these are standard in most software. Illustrated using the data from Giri et al. (2016), these include (Box 6.6):

• the overall mean (μ) (e.g. mean fatty acid level across all fish tanks from the four diet groups, estimated by the overall sample mean ($\bar{y}$));

Box 6.5 ⓡ ⓔ Worked Example of Single-Factor Design: Diatom Communities in Metal-Affected Streams

Medley and Clements (1998) sampled 4–7 stations on each of six streams in a region of the Rocky Mountains of Colorado, United States, known to be polluted by heavy metals. They recorded zinc concentration and species diversity of the diatom community, and several other ecological responses.

They classified each station as background ($<20\,\mu g\,L^{-1}$, 8 stations), low (21–$50\,\mu g\,L^{-1}$, 8 stations), medium 51–$200\,\mu g\,L^{-1}$, 9 stations), and high ($>200\,\mu g\,L^{-1}$, 9 stations) Zn, then compared these zinc groups.

We will fit a model to compare mean diatom species diversity (response) across the four zinc groups (fixed categorical predictor). Boxplots of species diversity against zinc group (Figure 6.6) showed no obvious skewness; two sites with low species diversity were highlighted in the background and medium zinc groups as possible outliers. The residual plot from this model (Figure 6.6) revealed no unusually large residuals or unequal spread of residuals.

First, the parameter estimates with SEs based on $MS_{Residual}$ from the ANOVA (see Table 6.2).

Parameter	OLS estimate ($\pm$ SE)	
μ_i	Group mean diversity:	
Background	1.80 ± 0.165 ($n = 8$)	
Low	2.03 ± 0.165 ($n = 8$)	
Medium	1.72 ± 0.155 ($n = 9$)	
High	1.28 ± 0.155 ($n = 9$)	
μ	Overall mean diversity: 1.71*	
$\alpha_i = \mu_i - \mu$	Background:	$1.80 - 1.71 = 0.09$
	Low:	$2.03 - 1.71 = 0.33$
	Medium:	$1.72 - 1.71 = 0.01$
	High:	$1.28 - 1.71 = -0.43$
ε_{ij}	Obs. 1: $2.27 - 1.80 = 0.473$	
	Obs. 2: $2.20 - 1.80 = 0.403$	
	etc.	

* This is the unweighted mean of the group means, which will differ from the simple mean of all 34 observations because of unequal sample sizes.

The ANOVA results from fitting the model were as follows.

Source	SS	df	MS	F	P	η^2
Zinc	2.567	3	0.856	3.939	0.018	0.283
Residual	6.516	30	0.217			
Total	9.083	33				

There is evidence for an effect of zinc group on mean diatom diversity, and Figure 6.6 shows a drop in diatom diversity at high zinc levels. We can also create dummy variables and fit a (multiple) linear regression model for these data. We will use reference group coding, with the background zinc being the logical reference group. Most software does this automatically, but the coding for the dummy variables is as follows.

Zinc	Zinc1	Zinc2	Zinc3
Background (reference)	0	0	0
Low	1	0	0
Medium	0	1	0
High	0	0	1
Interpretation	Low vs. Background	Medium vs. Background	High vs. Background

Box 6.5 (cont.)

The resulting coefficients, with tests of the null hypotheses, are as follows.

	Estimate	SE	t	P	Interpretation (H_0)
Intercept (Background)	1.798	0.165	10.91	<0.001	Background mean = 0
Low (Zinc 1)	0.235	0.233	1.01	0.321	Low – Background = 0
Medium (Zinc 2)	−0.080	0.226	−0.35	0.727	Medium – Background = 0
High (Zinc 3)	−0.520	0.226	−2.30	0.029	High – Background = 0

Only high zinc levels result in an effect on diatom diversity compared to background levels.

If we had no comparisons of particular interest, we could use an unplanned pairwise comparison of group means, such as Tukey's test or a stepwise method like the REGW (Ryan–Einot–Gabriel–Welsch) procedure. We have presented mean differences with Tukey (**bold**) and the REGW (*italics*) adjusted P-values for each pairwise comparison in parentheses.

	Background	Low	Medium
Low	0.235 (**0.746**/*0.560*)		
Medium	0.080 (**0.985**/*0.985*)	0.315 (**0.515**/*0.319*)	
High	0.520 (**0.122**/*0.060*)	0.755 (**0.012**/*0.005*)	0.440 (**0.209**/*0.144*)

The only pairwise comparison suggesting a difference is between low and high zinc sites.

We could also analyze these data with more robust methods, especially if we had been concerned about underlying nonnormality or outliers. The Kruskal–Wallis nonparametric test based on rank sums tests the H_0 that there is no difference in the location of the distributions of diatom diversity between zinc groups, irrespective of the shape of these distributions.

Zinc group	Rank sum
Background	160.0
Low	183.0
Medium	166.5
High	85.5

The Kruskal–Wallis H statistic = 8.737 with $P = 0.033$ (based on a χ^2 distribution with df = 3), evidence for a difference in the location of the distributions (i.e. medians) between zinc groups.

Another robust approach is a randomization (or permutation) test, where we reallocate observations to the four groups at random many times to generate a distribution of a suitable test statistic. We used the percentage of total SS attributable to zinc groups as the statistic and 5,000 randomizations. The percentage of SS_{Total} accounted for by SS_{Groups} was 28.3%, and the P-value was 0.014, again supporting the conclusion there is an effect of zinc on diatom diversity. It is also possible to use other randomization procedures, such as the PERMANOVA (permutational analysis of variance) described in Chapter 16.

Finally, if we were more interested in the pattern of variances between the zinc groups, we could use Levene's test, based on medians:

Levene-median: $F_{3,30} = 0.020, P = 0.996$.

There was no evidence for a difference in population variances between the four zinc groups.

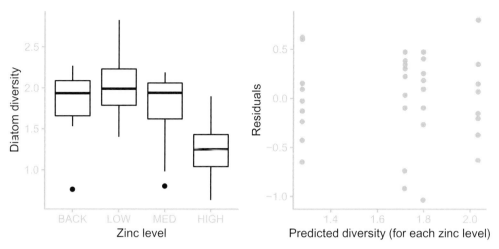

Figure 6.6 Boxplots of diatom diversity against zinc level group and residual plot from fit of a single-factor ANOVA model relating diatom diversity to zinc level group.

- the mean for each factor group (μ_i) (e.g. the mean fatty acid level across all fish tanks for diet group i, estimated by the sample mean for group i ($\bar{y}_i$));
- the effect of group i (α_i) (e.g. the effect of diet group i on fatty acid levels measured as the difference between the mean fatty acid level for diet group i and the overall mean fatty acid level, estimated by the difference between the overall sample mean and the sample mean for group i ($\bar{y} - \bar{y}_i$)); and
- the coefficients (βs) from the regression version of the model representing either contrasts between the means for each group and the overall mean (the α_i effects) or contrasts between the means for each group and the reference group mean.

To calculate SEs and CIs for the relevant parameters, we also need to estimate σ_ε^2 (the common variance of ε_{ij}).

6.2.3.1 *Predicted Values*

Once we have estimates of μ and α_i, we can substitute these into our effects model to derive the predicted Y-values ($\hat{y}_{ij}$; Box 6.6), which are simply the relevant group means ($\bar{y}_i$). As with other linear models, we are estimating the mean of the response variable for each value of X (i.e. for each group). However, in contrast to our regression model, prediction of the response for new values of the predictor variable is less relevant for the effects model because our fixed categorical predictor represents all the factor groups of interest.

6.2.3.2 *Error Terms and Their Variance*

ε_{ij} is random or unexplained error associated with observation j from group i, the difference between the actual value of the response variable and its true mean for that group.

For example, this measures the error associated with each observation of fatty acid level for fish from any possible tank within the four diets. We can estimate these error terms using sample residuals ($e_{ij} = y_{ij} - \hat{y}_{ij} = y_{ij} - \bar{y}_i$), which are important for checking the assumptions underlying the fit of the model (Section 6.4). The variance of the model error terms is estimated by the variance of these residuals, the $MS_{Residual}$ from the ANOVA (Section 6.2.5; Table 6.2).

For interval estimation and tests of hypotheses about model parameters using OLS, we must make certain assumptions (Gauss–Markov assumptions; see Section 4.6.1) about the error terms (ε_{ij}). We will examine them in more detail in Section 6.4.

6.2.4 Inference for Parameters
6.2.4.1 *Standard Errors and Confidence Intervals*

The SEs for groups means and differences between means are the SDs of the relevant sampling distributions, assumed to be normal based on the central limit theorem. There are two estimates of the SE we can use. The usual one in linear models software is based on the $MS_{Residual}$, assuming the within-group population variances of the error terms (and response variable) are the same. Alternatively, we could estimate the SE for each mean or mean difference based solely on the relevant observations in the specific group(s). Confidence intervals are usually based on the t distribution, as for regression models.

6.2.4.2 *Hypothesis Tests*

Evaluating hypotheses with a categorical predictor and more than two groups is a bit more complicated than for a regression model, where we had only a single slope parameter to test. Now we might wish to

Box 6.6 The Single-Factor Linear Model, Overparameterization, and Estimable Functions

Consider a dataset of p groups or treatments and n units in each group (Figure 6.7). For now, we assume equal sample sizes, but they can vary across groups (see Section 6.2.6). We can use three related linear models to describe this design – a linear effects model, a cell means model, and a dummy variable approach.

The linear effects model is $y_{ij} = \mu + \alpha_i + \varepsilon_{ij}$. This equation expresses the value of observation j from group i, as an overall population mean (μ), a deviation from this mean measuring the effect of group i (α_i), and random or unexplained error resulting in observation j differing from its group mean (ε_{ij}). μ is the equivalent of the intercept in the regression model and provides a reference point for measuring the effect of each group. α_i is conceptually equivalent to the slope in the simple linear regression model. It indicates the statistical effect of the predictor on the response. ε_{ij} is random or unexplained error associated with observation j from group i.

This model is structurally similar to the simple linear regression model described in Section 6.1.1. Like the regression model, it has two components: the model $(\mu + \alpha_i)$ and the error (ε_{ij}).

The linear effects model is what statisticians call "overparameterized" (Searle 1993) because the number of group means (p) is less than the number of parameters to be estimated $(\mu, \alpha_1, \ldots, \alpha_p)$. Not all parameters in the effects model can be estimated by OLS unless we impose some constraints because there is no unique solution to the set of normal equations (Searle 1993). The usual constraint, sometimes called a sum-to-zero constraint (Yandell 1997), a $\sum$-restriction (Searle 1993), or a side condition (Maxwell et al. 2018), is that the sum of the group effects equals zero. We can also set one parameter, either μ or one of the α_i, to zero (set-to-zero constraint; Yandell 1997), although this approach is mostly useful when one group is clearly a control or reference group (see dummy variable coding below).

The **cell means model** simply replaces $\mu + \alpha_i$ with μ_i and uses group means (μ_i) instead of group effects for the model component:

$$y_{ij} = \mu_i + \varepsilon_{ij}.$$

The cell means model is no longer overparameterized because the number of parameters in the model component is the same as the number of group means. This model can be useful for more complex designs with missing cells where not all main effects and interactions are estimable (Section 7.2.1).

For the **dummy variables approach**, we convert the factor groups to dummy variables, each taking one of two values, and fit a linear regression with multiple predictors. We need one fewer dummy variable than groups.

The regression or dummy variable model is:

$$y_{ij} = \beta_0 + \beta_1(\text{dummy}_1)_{ij} + \beta_2(\text{dummy}_2)_{ij} \ldots + \beta_{p-1}\left(\text{dummy}_{p-1}\right)_{ij} + \varepsilon_{ij}.$$

In this model, $\beta_0, \beta_1, \beta_2$, etc. are regression coefficients. The predictor variables are now dummy variables. Fitting this type of model is sometimes called dummy coding.

Dummy variables can be coded in two ways, both of which avoid overparameterization:

• **Reference coding** (sometimes called dummy coding) uses 0s and 1s and designates one group as the reference. It could be a group that naturally provides a basis to which to compare the other groups, such as an experimental control. Each predictor (and its model coefficient) represents the difference between the population mean of one other group and the reference. The intercept is the population mean of the reference group.
• **Effects coding** uses −1s and +1s. Each predictor (and its model coefficient) represents the difference between the population mean of each group and the overall population mean, that is the α_i. The intercept is the overall population mean (unweighted mean of the group means if sample sizes are unequal).

The advantage of these dummy variable models is that the parameters represent specific contrasts between the groups that might be of biological interest, especially with reference coding when there is a natural reference or control group.

The basic results from estimation and hypothesis testing will be the same as when fitting the effects or means models, but estimates of group effects will often be coded to compare with a reference category, so your output will only have $p - 1$ effects. You should always check which group is the reference when fitting a model of this type;

Box 6.6 (cont.)

software often defaults to the first one when sorted alphabetically. The intercept (β_0) is the mean of the reference group (reference coding) or the overall mean (effects coding).

Regardless of the approach, these models rely on assumptions about the error terms (ε_{ij}), which generally also apply to the response variable. These assumptions are examined in detail in Section 6.4.

As with the linear regression model, model parameters can be estimated with OLS or ML, with the former being more common for normally distributed error terms:

- μ is estimated by $\bar{y}$, which is $\left(\sum \bar{y}_i\right)/p$;
- μ_i is estimated by $\bar{y}_i$, which for each i is $\left(\sum y_{ij}\right)/n_i$;
- α_i is estimated by $\bar{y}_i - \bar{y}$; and
- β_i is estimated by $\bar{y}_i - \bar{y}_R$ for group i, where R is the reference group.

The predicted or fitted values of the response variable from our model are:

$$\hat{y}_{ij} = \bar{y} + (\bar{y}_i - \bar{y}) = \bar{y}_i.$$

So, any predicted Y-value is simply predicted by the sample mean for that factor group.

The residuals, $y_{ij} - \bar{y}_i$, can estimate the linear model's error terms (ε_{ij}). Note that the sum of the residuals within each factor group equals zero. The OLS estimate of σ_ε^2 is the sample variance of these residuals and is the residual (or error) MS (see Table 6.2) from the ANOVA and is used to calculate SEs for parameter estimates.

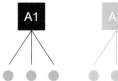

Figure 6.7 General data layout for a single-factor design where factor A has $p = 2$ groups and there are $n = 3$ replicates.

assess whether the effect of all groups equals zero $\left(H_0\!: \alpha_1 = \ldots = \alpha_i = \ldots \alpha_p = 0\right)$, which is equivalent to a test that all group means are equal $\left(H_0\!: \mu_1 = \ldots = \mu_i = \ldots \mu_p = \mu\right)$. This test is done using an F statistic from our ANOVA. With only two groups, the F test of these equivalent null hypotheses equates to the t test in Section 2.5.5. As we now have multiple groups, we might also wish to test for differences between specific groups or combinations of groups – these usually simplify to a comparison of two group (or group combinations) means and can be done using t or F statistics – see Section 6.2.7.

6.2.5 Model Comparison and the Analysis of Variance

As introduced earlier, the ANOVA partitions the total variation in the response variable into its components or sources. The total variation in Y is expressed as an SS. The SS_{Total} for Y is the sum of the squared differences between each y_{ij} and the overall mean $\bar{y}$. The df is the total number of observations across all groups minus 1. SS_{Total} can be partitioned into two additive components (Table 6.2):

- SS_{Groups} is the variation explained by the predictor (factor) and is calculated as the difference between each $\bar{y}_i$ and the overall mean $\bar{y}$ – that is, it measures how different the group means are and how much of the total variation in Y is explained by the difference between factor groups. The df associated with the variation between group means is $p - 1$.
- $SS_{Residual}$ is variation in Y not explained by the predictor (factor), also called the error or unexplained variation, and represents the variability within each group, calculated as the difference between each y_{ij} and relevant group mean $\bar{y}_i$. It measures how different the observations are within each group, summed across groups. The df associated with the $SS_{Residual}$ is the number of observations in each group minus 1, summed across groups. These SS and df are additive:

$$SS_{Groups} + SS_{Residual} = SS_{Total},$$
$$df_{Groups} + df_{Residual} = df_{Total}.$$

Equivalently to regression models, the SS in Table 6.2 represent the fit $\left(SS_{Groups}\right)$ and lack-of-fit $\left(SS_{Residual}\right)$ for our effects model. This is the full model,

Table 6.2 ANOVA table for single-factor linear model with categorical predictor showing partitioning of variation

Source of variation	SS	df	MS	Expected mean square
Between groups	$\sum_{i=1}^{p} n_i(\bar{y}_i - \bar{y})^2$	$p - 1$	$\dfrac{\sum_{i=1}^{p} n_i(\bar{y}_i - \bar{y})^2}{p - 1}$	$\sigma_\varepsilon^2 + \sum_{i=1}^{p} n_i \frac{\alpha_i^2}{p-1}$
Residual	$\sum_{i=1}^{p}\sum_{j=1}^{n}\left(y_{ij} - \bar{y}_i\right)^2$	$\sum_{i=1}^{p} n_i - p$	$\dfrac{\sum_{i=1}^{p}\sum_{j=1}^{n}\left(\bar{y}_{ij} - \bar{y}_i\right)^2}{\sum_{i=1}^{p} n_i - p}$	σ_ε^2
Total	$\sum_{i=1}^{p}\sum_{j=1}^{n}\left(y_{ij} - \bar{y}\right)^2$	$\sum_{i=1}^{p} n_i - 1$		

with all the parameters of interest. We can evaluate the importance of the predictor by comparing this model's fit to the reduced model that omits its effect (i.e. sets it to zero):

$$y_{ij} = \mu + \varepsilon_{ij}.$$

The predicted Y for each x_i (group) from this model is simply the overall sample mean ($\bar{y}$), as it was for the intercept-only regression model. The SS_{Residual} from this reduced model is the SS_{Total} from our ANOVA for the full model (Table 6.2). The difference between the unexplained variation of the full model (SS_{Residual}) and the unexplained variation from the reduced model (SS_{Total}) is termed $SS_{\text{Difference}}$ and equals the SS_{Groups} from the full model. It measures how much more variation in Y is explained by the full model than by the reduced model. As with regression models, the partitioning of the SS provides the r^2 value ($SS_{\text{Groups}}/SS_{\text{Total}}$) which is the proportion of the variation in Y explained by the differences between group means (the effects of our factor groups).

We convert the SS into MS (or variances) by dividing them by their df. The expected values of these mean squares (EMS) – that is, the means of the probability distributions of these sample variances or what population values these MS actually estimate – are presented in Table 6.2:

- The MS_{Groups} also estimates σ_ε^2 plus a component representing the squared effects of the p groups (we square the effects because positive and negative effects are of interest). Note these EMS are subject to the important constraint that the sum of the group effects equals zero ($\sum \alpha_i = 0$). Without this constraint, we cannot get unbiased estimators of individual treatment effects (Box 6.6).

- The MS_{Residual} estimates σ_ε^2, the common (pooled) population variance of the error terms, and hence of the Y-values, within groups.

Based on these expected values, we can develop a test for the two equivalent H_0s of interest described in Section 6.2.4.2: no difference between population group means or all group effects equal zero. We use the EMS to determine an appropriate F for testing these hypotheses. If the H_0 for a fixed factor is true, all $\alpha_i = 0$, so MS_{Groups} and MS_{Residual} both estimate σ_ε^2 and their ratio should be 1. If the H_0 is false, then some α_i will differ from zero, and we expect $MS_{\text{Groups}} > MS_{\text{Residual}}$ and their $F > 1$. Assuming homogeneity of within-group variances, our F statistic will follow an F distribution with df relevant to the two MS. We can determine the probability from repeated sampling of obtaining our sample F or one more extreme if the H_0 is true. The F test basically compares the fit to the data of a model that includes the effect of the predictor (factor) to the fit of a model that does not.

It should also be clear now why the assumption of equal within-group variances is so important. If variances differ, MS_{Residual} does not estimate a single population variance and we cannot construct a reliable F for testing H_0.

6.2.6 Unequal Sample Sizes (Unbalanced Designs)

Unequal sample sizes between groups cause no computational difficulties for fitting the linear model as we have described. However, they can cause other problems. First, the group means will be estimated with varying precision, making interpretation more difficult (Underwood 1997). Note that sample size is only one contributor to the precision of an estimate and some have suggested designing experiments with different sample sizes, with more observations from groups that are more variable or for which we are more interested in their mean (Mead et al. 2012). Second, the ANOVA F test is much less robust to violations of assumptions, particularly homogeneity of variances, if sample sizes differ (Section 6.4). The worst case is when larger variances are associated with smaller sample sizes. This is a particularly important reason to design experiments and sampling programs with equal sample sizes where possible. Third, estimation of group effects, particularly variance components for random effect models (Chapter 10), is more difficult.

One solution to unequal sample sizes in single-predictor designs is deleting observations until all groups have the same. We regard this practice as extreme; linear models can deal with unequal sample sizes and biological studies often suffer from small sample sizes. Deleting observations will exacerbate the situation, particularly if one group has a much lower sample size than others. If there is no evidence that the assumption of homogeneity of variance is seriously compromised and the difference in sample sizes is not large, we recommend simply fitting the linear model as above.

6.2.7 Specific Comparisons of Group Means

The F test described in Section 6.2.5 tests an overall null hypothesis of no group effects and therefore no differences between population group means. However, we are usually interested in more detail about the differences between groups, and this can take two forms. In most cases our research question involves particular (prespecified) patterns of differences between the means of different groups (or combinations of groups). We want to make specific or planned comparisons. In other cases, we may simply want to explore where differences might lie if there is evidence against the overall null hypothesis using unplanned comparisons. Comparisons of group means are illustrated in Boxes 6.4, 6.5, and 6.7.

6.2.7.1 *Planned Comparisons or Contrasts*

These are interesting and logical comparisons (often termed contrasts) of groups or combinations of groups, including fitting polynomial (e.g. linear, quadratic, etc.) trends. Each contrast commonly uses a single df. They should always be planned as part of the analysis strategy before the data are examined – the choice of contrasts is not determined from inspection of the data. Planned contrasts ideally should be orthogonal, as P-values associated with independent contrasts are more straightforward to interpret. This limits the number of contrasts to $p - 1$, although we recommend proceeding with nonorthogonal

contrasts if they answer important biological questions and the number of contrasts is not excessive. We find it convenient to distinguish between contrasts about differences and those about trends.

Contrasts about Differences

There are two types of contrasts about differences between group means:

- Model-defined contrasts are represented by the coefficients for each dummy variable in the regression version of the model (Section 6.2.2.3). They will be contrasts between each group and a reference group (reference coding) or between all but one group and the overall mean (effects coding). These contrasts are often default output from linear model routines in statistical software.
- User-defined contrasts are where the comparisons of interest are not represented by the regression coefficients. They can evaluate differences between any two group means or can be more complex, evaluating the difference between two combinations of groups. Each contrast uses one df from the $\text{df}_{\text{Groups}}$.

These planned contrasts are sometimes sequential. For example, Marshall and Steinberg (2014) examined the response of newly metamorphosed juvenile sea urchins to host seaweed (three groups: the adult host *Ecklonia radiata*, the metamorphosis host *Amphiroa anceps*, and control with no plants). They included two contrasts: *Ecklonia* vs. control, and then if that contrast was not "statistically significant," *Amphiroa* vs. *Ecklonia* and control pooled. Keough and Raimondi (1995) examined how different microbial communities on hard surfaces ("biofilms") affected the settlement of serpulid worms, again in a marine environment. They incorporated three specific planned contrasts to identify and remove potential experimental artifacts and test their main hypothesis (Box 6.7).

To specify a contrast, we need to define a linear combination of the p means that represents the comparison of interest:

$$c_1 \bar{y}_1 + \cdots + c_i \bar{y}_i + \cdots + c_p \bar{y}_p.$$

The c_i are contrast coefficients and sum to zero, ensuring a valid contrast, and $\bar{y}_i$ are group means. The absolute values of the coefficients are not relevant if they sum to zero. We use integers for simplicity. To omit a group from the contrast, we give it a coefficient of zero.

We can also define orthogonality in terms of coefficients. Two contrasts, A and B, are independent (orthogonal) if the sum of the products of their coefficients equals

Box 6.7 ⓡ ⓔ Worked Example of Single-Factor Design with Planned Comparisons: Serpulid Polychaete Recruitment and Biofilms

Keough and Raimondi (1995) tested how serpulid (polychaete worms) larvae responded to biofilms that develop on hard substrata in shallow marine waters. They compared biofilms developed in the laboratory to those in the field and to unfilmed surfaces. They needed an additional treatment to control for a potential artifact because surfaces exposed in the field were screened to prevent the confounding effect of settlement by invertebrates. The four treatments were: sterile substrata, biofilms in the field + net, biofilms developed in the lab, and lab biofilms + net. After one week in the field, they counted the newly settled worms on each experimental unit.

We have not shown the initial data screening stages. Still, the response variable was log-transformed to improve slightly skewed distributions and as part of a decision to report data for all species (many of which were strongly skewed) on a log-transformed scale for consistency. The H_0 was no difference between treatments in the mean log-transformed number of serpulid recruits per replicate. The residual plot from the single-factor model with log-transformed numbers of serpulid recruits revealed a single outlier, but similar data spread between groups, suggesting that the assumptions were met. The similarity of data ranges is probably a more reliable guide to the reliability of the ANOVA than the formal identification of outliers from boxplots when there are only seven observations per group.

The ANOVA was as follows.

Source	SS	df	MS	F	P
Biofilms	0.246	3	0.082	6.02	0.003
Residual	0.327	24	0.014		
Total	0.573	27			

This analysis suggests some overall difference between treatments in the log numbers of serpulid recruits. However, in this example, we are more interested in the planned contrasts between specific treatments.

The mean log number of serpulid recruits for each of the four biofilm treatments from Keough and Raimondi (1995) were (Figure 6.8):

Field (F): 2.11
Netted lab (NL): 2.18
Sterile lab (SL): 1.93
Un-netted lab (UL): 2.13

Planned comparisons were done, each testing a hypothesis about the biofilms. The contrasts were done in sequence, with each comparison depending on the result of previous ones. As single df contrasts, we could make the tests using a t or an F test; here we present the t statistics.

First, Keough and Raimondi (1995) tested whether a fine-mesh net over a surface affected recruitment, by comparing the netted and un-netted laboratory treatments. The H_0 is:

$\mu_{NL} = \mu_{UL}$ or $\mu_{NL} - \mu_{UL} = 0$.

We use the latter expression to define the linear contrast equation:

$(0)\bar{y}_F + (+1)\bar{y}_{NL} + (0)\bar{y}_{SL} + (-1)\bar{y}_{UL}$.

This contrast specifically represents the hypothesis and we use coefficients of zero to omit groups not part of the H_0. The table below indicates that we would not detect a difference.

Second, the laboratory and field films were compared. Because the two laboratory-developed films did not differ, we can pool them, so the H_0 is:

Box 6.7 (cont.)

$(\mu_{NL} + \mu_{UL})/2 = \mu_F$ or $(\mu_{NL} + \mu_{UL})/2 - \mu_F = 0$.

The linear contrast equation is:

$(+1)\bar{y}_F + (-0.5)\bar{y}_{NL} + (0)\bar{y}_{SL} + (-0.5)\bar{y}_{UL}$ or
$(+2)\bar{y}_F + (-1)\bar{y}_{NL} + (0)\bar{y}_{SL} + (-1)\bar{y}_{UL}$

Note that the coefficients for the two lab treatments produce the average of those two groups, which is contrasted to the field treatment. The table below indicates no detected difference.

Finally, we compare the whole set of substrata with biofilms to the single, unfilmed treatment. The H_0 is:

$(\mu_F + \mu_{NL} + \mu_{UL})/3 = \mu_{SL}$ or $(\mu_F + \mu_{NL} + \mu_{UL})/3 - \mu_{SL} = 0$.

The linear contrast equation is:

$(+1)\bar{y}_F + (-0.33)\bar{y}_{NL} + (-0.33)\bar{y}_{SL} + (-0.33)\bar{y}_{UL}$ or
$(+3)\bar{y}_F + (+1)\bar{y}_{NL} + (-1)\bar{y}_{SL} + (-1)\bar{y}_{UL}$

The coefficients for the three lab treatments represent the average of those three groups and are contrasted to the field treatment. There is a clear difference.

Contrast	t	P
NL vs. UL	0.795	0.435
F vs. average (NL & UL)	−0.808	0.427
SL vs. average (F & NL & UL)	−4.094	<0.001

Note that if the coefficients sum to zero and represent the contrast of interest, the size of the coefficients is irrelevant – for example, in the first example above we could have used 1, −1, 0, 0 or 0.5, −0.5, 0, 0 or 100, −100, 0, 0; the results would be identical.

Note also we can check that comparisons are orthogonal. For example, for the first two comparisons we can use the formal test of orthogonality: $\sum c_{i1} c_{i2} = (0)(1) + (1)(0.5) + (0)(0) + (-1)(-0.5) = 0 - 0.5 + 0 + 0.5 = 0$.

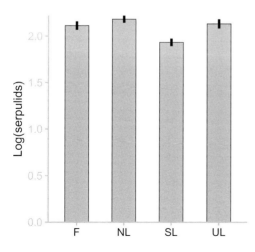

Figure 6.8 Mean log numbers of serpulid polychaetes for four surfaces differing in the kinds of microbial films. Error bars are SEs.

zero ($\sum c_{iA} c_{iB} = 0$). It is sometimes not intuitively obvious whether two comparisons are orthogonal, and we check by doing this calculation.

The SE for a contrast is given by

$$\sqrt{\mathrm{MS_{Residual}} \sum_{i=1}^{p} \frac{c_i^2}{n_i}}.$$

The CIs can be calculated using the t distribution based on $\mathrm{df_{Residual}}$, as usual.

The formal test of each contrast can be done in two functionally equivalent ways. First, the $\mathrm{SS_{Groups}}$ can be partitioned into the contribution due to each contrast. Each $\mathrm{SS_{Contrast}}$ will have one df and, therefore, $\mathrm{SS_{Contrast}} = \mathrm{MS_{Contrast}}$. The null hypothesis associated with each contrast is evaluated with an F statistic ($\mathrm{MS_{Contrast}}/\mathrm{MS_{Residual}}$). Second, a t test can also be used for contrasts with the modification that the contrast SE is used in the denominator.

Contrasts about Trends

If the groups can be ordered, tests for trends across the group means may be more informative than tests about specific differences. These trends are examined using simple polynomial functions through the group means. A linear polynomial represents a straight-line relationship through the group means, a quadratic represents a U-shaped relationship, and the cubic represents a more complex pattern with two "changes of direction" (Figure 6.9). For example, in the data analysis from Giri et al. (2016), we evaluated linear and quadratic trends in mean omega-3 LC-PUFAs across the four diet treatments with increasing micronutrients and coenzymes (Box 6.4).

The logic of fitting orthogonal polynomials is like that for the difference contrasts. We use contrast coefficients to define the polynomial of interest, with one df, and evaluate it with an F or t. When sample sizes are equal, the SS for a linear contrast will be the same as the $\mathrm{SS_{Regression}}$ from fitting a linear regression model. The

$\mathrm{SS_{Residual}}$, and therefore the test of linearity, will differ in the two cases because they partition the $\mathrm{SS_{Total}}$ differently. $\mathrm{SS_{Residual}}$ from the effects model will be smaller but with fewer df, as only one df is used for the regression but $(p - 1)$ is used for the groups. The difference in the two $\mathrm{SS_{Residual}}$ (from the regression model and the effects model) is termed "lack-of-fit" (Kutner et al. 2005), representing the variation not explained by a linear fit but possibly explained by other (e.g. quadratic) components.

The SS from a quadratic contrast will be the same as the $\mathrm{SS_{Extra}}$ from fitting a quadratic regression model over a linear regression model to the original observations (Section 8.1.11). The quadratic polynomial is testing whether there is a quadratic relationship between the response variable and the factor over and above a linear relationship, and so on. Sometimes, the remaining SS after $\mathrm{SS_{Linear}}$ is extracted from $\mathrm{SS_{Groups}}$ is used to test for departures from linearity (Kirk 2012).

When sample sizes are unequal or the groups unevenly spaced, contrast coefficients can still be determined by solving simultaneous equations (Kirk 2012). Most statistical software will provide these coefficients. Alternatively, we could simply fit a hierarchical series of polynomial regression models, testing the linear model over the intercept-only model, the quadratic model over the linear model, etc. (Section 8.1.11). These two approaches do not produce the same SS when sample sizes are different, although the difference is usually not large (Maxwell et al. 2018). We prefer using the contrast coefficients and treating the test for a linear trend as a planned contrast between group means.

The rules for contrast coefficients still apply. The coefficients for each polynomial should sum to zero and their absolute values are irrelevant. Successive polynomial contrasts (linear, quadratic, etc.) are independent (orthogonal) if the number of polynomials doesn't exceed the $\mathrm{df_{Groups}}$.

6.2.7.2 Unplanned Pairwise Comparisons

As the name suggests, unplanned pairwise comparisons compare all possible pairs of group means in a *post hoc*

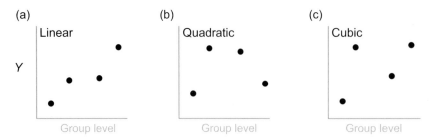

(a) Linear   (b) Quadratic   (c) Cubic

Y

Group level   Group level   Group level

Figure 6.9 Diagrammatic representation of polynomial trends across four equally spaced quantitative groups.

exploratory fashion to discover which groups are different. They are traditionally recommended only if a test from the original linear model revealed a "statistically significant" effect of the factor – that is, evidence against the null hypothesis. These multiple comparisons are clearly not orthogonal because there are many more than $p - 1$ pairs. They can also involve or encourage data snooping (searching for statistically significant results), or picking winners (Day & Quinn 1989), from a large collection of tests. Because there are so many nonindependent comparisons, many have argued that we should protect against the increased risk of at least one Type I error. There are two broad approaches to controlling the family-wise (across the collection of contrasts) probability of at least one Type I error when comparing all possible pairs of group means. First, we could use the methods described in Section 2.5.4 to adjust the statistical significance levels for each pairwise comparison, including control of false discovery rates.

Second, we can use unplanned pairwise multiple comparison (UPMC) tests designed specifically for pairwise comparisons of group means. They calculate tail probabilities against the H_0 of no difference for each comparison, depending on how far apart two groups in each comparison are in the ordered ranking of group means. They are often based on the studentized range statistic (Q). There are many unplanned multiple comparison tests available, and they are of two types:

• Simultaneous tests use the value of the test statistic based on the total number of groups in the analysis, despite how many means are between any two being compared. These tests also permit simultaneous adjusted CIs on pairwise differences between means. The best known is Tukey's HSD (honestly significant difference) test, known to keep the family-wise Type I error rate at or below the specified level. The cost is that each individual comparison has a significance level lower than the family-wise one, and the power to detect real differences is reduced. We illustrate Tukey's test in Box 6.5. There are other simultaneous comparison methods. Fisher's least significant difference (LSD) method is based on pairwise t tests and applied only if the original ANOVA F test is "statistically significant" (hence "protected"). It does not control family-wise Type I error rate well. Scheffé's method is very conservative and not very efficient for comparing all pairs of means.
• Stepwise tests use different values of the test statistic for comparing means closer together and are generally more powerful than simultaneous tests. However, their control of the family-wise Type I error rate is not always strict, and simultaneous CIs are not available. One of the most powerful

stepwise multiple comparison procedures that controls the family-wise error rate well is the REGW procedure (Box 6.5). Other stepwise tests sometimes used in biological research (e.g. Student–Neuman–Keuls [SNK] test, Duncan's multiple range test) don't provide adequate Type I error control under some circumstances (Day & Quinn 1989).

Both test types can handle unequal sample sizes, using minor modifications (e.g. the Tukey–Kramer test). Dunnett's test is a t test modified specifically for comparing each group to a control group. Under this scenario, there are fewer comparisons than when comparing all pairs of group means, so it can be more powerful than other multiple comparison tests.

Pairwise multiple comparison procedures based on ranks of the observations are available (Day & Quinn 1989). Some tests are robust to unequal variances, including Dunnett's T3, Dunnett's C, and Games–Howell tests (Day & Quinn 1989; Kirk 2012). They are best used with robust linear model methods described in Section 6.6.

6.3 Predictor Effects

Estimating model parameters and evaluating null hypotheses about them is one part of our linear model analysis. Another aspect is to estimate the effect size, the statistical effect of the predictor on the response. Effect sizes provide an important standardized measure that can be used in meta-analyses and in sampling and experimental design through power analysis (Section 3.3).

6.3.1 Continuous Predictor (Regression) Models

In a simple linear regression model, we can measure the predictor effect on the response variable in several ways. Nakagawa and Cuthill (2007) recommended that the unstandardized regression slope (with CIs) be presented, although the r^2 value and the standardized regression coefficient can be useful for meta-analyses. The simple correlation coefficient can also be used (Harrison 2011), but this would seem more appropriate for true random X regression (Section 6.1.7) or correlation studies, rather than models that predict a response from a predictor.

6.3.2 Categorical Predictor Models

When the predictor variable is categorical, measuring effect size is often of more interest. One type of measure is the proportion of total variation in the response variable explained by the differences between groups (effect of the predictor). This is sometimes termed PEV (proportion of explained variance; Petraitis 1998) or association (Ialongo

Table 6.3 Measures of ES for single-predictor linear models

Examples are from Christensen et al. (1996) for the "effect" of riparian tree density on CWD basal area, and Medley and Clements (1998) for the "effect" of zinc level group on mean diatom diversity and for the specific difference between background zinc and high zinc groups.

Measure	Formula	Example
Continuous predictor		
Regression slope (b_1)	$r(s_Y/s_X)$	0.023
Correlation coefficient (r)	$\dfrac{\sum_{i=1}^{n}[(x_i - \bar{x})(y_i - \bar{y})]}{\sqrt{s_X^2 s_Y^2}}$	0.634
(also standardized regression slope)	$b_1(s_Y/s)$	0.797
Categorical predictor		
Eta-squared (η^2)	SS_{Groups}/SS_{Total}	0.284
Omega-squared (ω^2)	$\dfrac{SS_{Groups} - (p-1)MS_{Residual}}{SS_{Groups} + MS_{Residual}}$	0.206
Cohen's effect size (f)	$\sqrt{\dfrac{\frac{p-1}{p}(MS_{Groups} - MS_{Residual})}{\sum_{i=1}^{p} n_i}}{MS_{Residual}}}$	0.509
Cohen's d	$\dfrac{|\bar{y}_1 - \bar{y}_2|}{\sqrt{(s_1^2 + s_2^2)/2}}$	0.237 (with CF = 0.225)
Hedge's g	$\dfrac{|\bar{y}_1 - \bar{y}_2|}{\sqrt{\dfrac{(n_1 - 1)s_1^2 + (n_2 - 1)s_2^2}{n_1 + n_2 - 2}}}$	1.074 (with CF = 1.019)
Bias correction factor for d or g	$\left[1 - \dfrac{3}{4(n_1 + n_2 - 2) - 1}\right]$	0.949

2016) and measures suitable for single-factor analyses are in Table 6.3. The three most common are:

- η^2 (eta-squared), which is simply the r^2 from the ANOVA;
- ω^2 (omega-squared) introduced by Hays (1994), which adjusts η^2 by incorporating the error variance and number of groups; and
- Cohen's f^2, which is based on η^2. Note that $\omega^2 = f^2/(1 + f^2)$ (Petraitis 1998).

Omega-squared is more conservative than η^2. Cohen's f^2 is sometimes used as an effect size for *a priori* power calculations (Sections 3.3 and 6.7). All these measures are more commonly used in disciplines like psychology rather than biology. It is important to recognize that the variance explained by a fixed factor is quite different to a true variance for a random effect measured by a variance component (Chapter 10).

A second type of effect size measure is a difference (or *d*) statistic (Ialongo 2016; Nakagawa & Cuthill 2007) that is more common in biological research. These statistics focus on the difference between two group means, so are not an overall measure of the predictor's effect on the response. The two most common *d* statistics are (Table 6.3):

- Cohen's d, which is a standardized measure of the difference between two group means; and
- Hedge's g, which is similar to Cohen's d, but allows unequal sample sizes and weights the within-group variances by these sample sizes.

Both statistics are standardized by the pooled standard deviation of the two groups. Nakagawa and Cuthill (2007) provide equations for calculating asymptotic SEs for the effect size (ES) statistics that can be converted to CIs. They also suggest a bias correction for small sample sizes. When the two groups come from a dataset with more than two groups, such as in our two examples from Section 6.2, the pooled within-group SD $(\sqrt{MS_{Residual}})$ would be appropriate, just as for specific comparisons of two group means in Section 6.2.7.

6.4 Assumptions

We introduced the Gauss–Markov assumptions underlying OLS estimation in Section 4.6. We assume that error terms from our linear model are independently and identically distributed (i.i.d.) with zero mean and constant variance for different predictor values. For CIs and hypothesis tests, we also assume the error terms are normally distributed, although this assumption is less important. Regression models also often assume that X is a fixed variable measured without error (see Section 6.1.7) and the relationship between Y (transformed if necessary) and X is best described by a straight line. Evaluating these assumptions as part of fitting a model is important, to ensure reliability of our parameter estimates and because deviations from these assumptions may prompt more suitable, alternative analyses.

6.4.1 Zero Conditional Mean

The assumption that the model error terms are independent of the predictor variable and have zero mean for each x_i is often ignored. A plot of residuals against X can evaluate it – no pattern would suggest the assumption is reasonable.

6.4.2 Independence

Independence of errors is most often violated when we have a time series or repeated measurements from the same sampling or experimental units. These situations require more sophisticated statistical models, which we address in later chapters, particularly Chapters 10–12.

6.4.3 Variance Homogeneity

The two common causes of variance heterogeneity in single-predictor linear models are:
1. The underlying distribution of the errors, and therefore usually the response variable, is one where the variance is functionally related (usually positively) to the mean.
2. There are unusual observations (outliers) that result in the variance being greater at some predictor variable values; see Section 5.2 for strategies to deal with outliers. Plots of residuals against predicted values should show a consistent spread of the residuals if this assumption is met. The alternative is usually an increasing spread of residuals as predicted values increase, suggesting an underlying skewed distribution. If the distribution matches one from the exponential family (e.g. Poisson), a GLM is appropriate; otherwise, transforming the response variable might make the errors homogeneous.

6.4.4 Normality

Normality of model error terms for each value of X is an oft-stated assumption for reliable CIs and hypothesis tests using OLS. Some authors recommend against checking this assumption (e.g. Gelman & Hill 2006) because OLS is very robust to nonnormality. We feel that the underlying distribution of error terms and the response variable is often useful biological information even if lack of normality does not invalidate the proposed analysis. Unequal variances often result from skewed distributions, so identifying the distribution of the error terms and response variable can often suggest appropriate remedial steps for variance heterogeneity, such as transformations (Section 5.4). For regression models with single Y-values for each x_i, we can only check the overall distribution of the model errors (ignoring X) using boxplots and probability plots. If we have multiple observations within each group of a categorical predictor, then we can examine the distributions of residuals for each group.

Nonnormal distributions can also sometimes make us think about the biological processes that produced the distribution.

6.5 Model Diagnostics

Model diagnostics allow us to check the appropriateness of our linear model. It is important to determine how much each observation can influence the outcome of our model fitting. We don't want our conclusions to be determined by one or two unusual values. Influence in linear models can come from at least two sources that are best explained by a regression model with a continuous predictor variable. Think of a regression line as a seesaw, balanced on the mean of X. An observation can influence, or tip, the regression line more easily the further it is from the mean (i.e. the closer to the end of the range of X-values, measured by leverage) or if it is far from the fitted regression line (i.e. has a large residual, analogous to a heavy person on the seesaw) – see Figure 6.10. Diagnostics, especially graphical methods, can also evaluate whether our data meet the underlying assumptions for OLS estimation and hypothesis testing.

6.5.1 Residuals

Plotting sample residuals to assess assumptions was introduced in Section 5.1.3. The pattern in the residuals is our best estimate of patterns in the model error terms. Besides checking homogeneity of variances, independence, and normality, residuals are used to evaluate model fit and detect influential observations. One problem with sample residuals is that their variance is unlikely to be constant for different x_i. We can use them for detecting variance heterogeneity, but they are less useful for detecting outliers, so they are usually adjusted to make them more

Table 6.4 Types of residuals for single-predictor linear models. For regression models, h_i is the leverage for observation i

	Continuous predictor (regression models)	Categorical predictor
Residual	$e_i = y_i - \hat{y}_i$	$e_{ij} = y_{ij} - \bar{y}_i$
Standardized residual	$\dfrac{e_i}{\sqrt{MS_{Residual}}}$	$\dfrac{e_i}{\sqrt{MS_{Residual}}}$
Studentized residual	$\dfrac{e_i}{\sqrt{MS_{Residual}(1 - h_i)}}$	$\dfrac{y_{ij} - \bar{y}_i}{\sqrt{\dfrac{MS_{Residual}(n_i - 1)}{n_i}}}$
Studentized deleted residual	$e_i\sqrt{\dfrac{n-1}{SS_{Residual}(1 - h_i) - e_i^2}}$	$e_{ij}\left[\dfrac{n - p - 1}{SS_{Residual}(1 - 1/n_i) - e_{ij}^2}\right]$

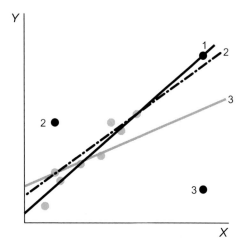

Figure 6.10 Residuals, leverage, and influence. The solid regression line is fitted through the observations with gray symbols. Observation 1 is an outlier for both Y and X (large leverage) but not from the fitted model and is not influential. Observation 2 is not an outlier for either Y or X but is an outlier from the fitted model (large residual). Regression line 2 includes this observation and its slope is only slightly less than the original regression line, so observation 2 is not particularly influential (small Cook's D_i). Observation 3 is not an outlier for Y, but it does have large leverage and it is an outlier from the fitted model (large residual). Regression line 3 includes this observation and its slope is markedly different from the original regression line, so observation 3 is very influential (large Cook's D_i, combining leverage and residual).

directly comparable across the observations in our dataset (Table 6.4). Standardized residuals use the $\sqrt{MS_{Residual}}$ as an approximate SE for the residuals. Unfortunately, this SE doesn't solve the problem of nonconstant variances of the residuals, so a more sophisticated modification is needed. Studentized residuals incorporate leverage (h_i; defined below). They do have constant variance, so different studentized residuals can be compared validly.

Large (studentized) residuals for a particular observation indicate that it is an outlier from the fitted model compared to the other observations. In terms of diagnostic plots (Section 6.5.4), the patterns for standardized and studentized residuals are generally similar. Studentized residuals are sometimes compared to a t distribution with df $= n - 1$ to determine the probability of getting a specific studentized residual, or one more extreme (but see Schützenmeister et al. 2012). Another form of the studentized residual is where the predicted value for observation i is calculated from the fitted model omitting this observation; this is also termed a studentized deleted residual, PRESS residual, or jackknife residual, and can sometimes detect outliers that might be missed by usual residuals (Kleinbaum et al. 2013; Kutner et al. 2005).

6.5.2 Leverage

Leverage measures how extreme an observation is for the X variable, so an observation with high leverage is an outlier in the X-space (Figure 6.10). Leverage basically measures how much each x_i influences $\hat{y}_i$ (Kutner et al. 2005) and X-values further from their mean influence the predicted Y-values more than those close to $\bar{x}$. Leverage is often given the symbol h_i because the values for each observation come from the hat matrix (**H**), a matrix that relates the y_i to the $\hat{y}_i$.

Leverage values normally range between $1/n$ and 1, and a useful criterion is that any observation with a leverage $> 2p/n$ (where p is the number of parameters in the model including the intercept) should be checked (Hoaglin & Welsch 1978). Some statistical packages may use other criteria for warning about observations with high leverage. The main use of leverage values is when incorporated in measures of influence.

When effects or means models are fitted with categorical predictors, leverage for each observation is difficult to interpret because it is essentially determined by the group sample size (leverage $= 1/n_i$).

6.5.3 Influence Measures

With a continuous predictor, Cook's distance statistic, D_i, measures each observation's influence on the estimates of the regression parameters. It incorporates leverage and the residual for each observation and basically measures the influence of each observation on the estimated regression slope (Figure 6.10). A large D_i indicates that removing that observation would considerably change the regression parameters' estimates. Cook's D_i can be used in two ways. First, we can scan the D_is of all observations informally and note any values much larger than the rest. An approximate guideline is that any observation with a $D_i > 1$ is particularly influential (Weisberg 2013). Second, we can compare D_i to an $F_{1,n}$ distribution. DFFITS$_i$ is an alternative influence measure that calculates the change in the predicted value for observation i when it is excluded from model fitting. It is a studentized measure, divided by the SE of its predicted value.

We use the same strategies for dealing with influential observations in linear models as we do for outliers more generally (Section 5.2). First, we check that the observation is valid (i.e. not a data entry error or some other mistake) and then we fit the model with and without the observation to evaluate its effect on the model parameters. If the conclusions and biological interpretation change, and the observation is not clearly an error, the only option is to report both analyses.

When effects or means models are fitted to data with categorical predictors, Cook's measure is not particularly helpful, being based on leverage.

6.5.4 Diagnostic Plots

The diagnostic measures described above are best used as part of a suite of graphical methods for evaluating model fit. We will emphasize three particularly useful graphs for single-predictor linear models we introduced in Section 5.1: scatterplots, boxplots, and residual plots.

6.5.4.1 *Scatterplots*

A scatterplot of Y against X should always be the first step in the analysis for regression models. Scatterplots can indicate unequal variances, nonlinearity, and outlying observations. They can also be used with smoothing functions (Section 5.1.2) to explore the relationship between Y and X without being constrained by a specific linear model. The effect of potential transformations can be easily observed by transforming scatterplot axes (see Box 6.8).

6.5.4.2 *Boxplots*

Boxplots are valuable along the two axes of a scatterplot to display the distribution of response and predictor variables and when the predictor variable is categorical. We are particularly interested in skewed distributions, unequal spread (indicating unequal variances) of the boxplots across groups and unusual values. However, unusual values are not necessarily influential. As with scatterplots, the transformations can be evaluated easily by transforming the response variable axis. While boxplots commonly plot the response variable, we can also use residuals. They may be more appropriate because assumptions apply to model error terms. The sample size needs to be at least 10 for the plots to be easily interpretable – boxplots with small sample sizes can produce confusing patterns.

6.5.4.3 *Residual Plots*

The most informative way of examining residuals is to plot them against x_i or $\hat{y}_i$. Residual plots commonly use predicted Y-values because, in more complex models with multiple predictor variables, there is still a single plot in contrast to plotting residuals against each predictor separately. These plots can tell us whether the model's assumptions are met and indicate observations that do not match the model well. The ideal pattern in the residual plot is a scatter of points with no obvious pattern.

If the distribution of Y-values is positively skewed (e.g. lognormal, Poisson) we would expect larger $\hat{y}_i$ to be associated with larger residuals. A wedge-shaped pattern, with a broader spread of residuals for larger $\hat{y}_i$, as shown for our worked example (Box 6.8), indicates increasing variance with increasing $\hat{y}_i$. This pattern is often associated with nonnormality and violating the assumption of homogeneity of variance. A curved pattern in the residuals can detect nonlinearity, and outliers from the fitted model also stand out as having large residuals (and large Cook's distances if influential). Plots of studentized residuals can be particularly informative about unequal variances and outliers by including simultaneous tolerance intervals and bands (Schützenmeister et al. 2012).

For categorical predictors, residual plots will show groups of residuals, but the interpretation is the same. We want there to be little pattern with similar spread of residuals within each group and no strong outliers.

6.5.5 Transformations

When continuous variables have specific skewed distributions, such as lognormal or Poisson, transforming them to a different scale will often render their distributions closer to normal (Section 5.4). Transformations can

Box 6.8 ® ☻ Worked Example of Linear Regression Diagnostics and Transformations: Diversity in Mussel Clumps

Peake and Quinn (1993) investigated the relationship between the number of species of macroinvertebrates, the total abundance of macroinvertebrates, and area of clumps of mussels on a rocky shore in southern Australia. The variables of interest are clump area (dm^2), number of species, and number of individuals.

Number of Species Against Clump Area
A scatterplot of number of species against clump area, with LOESS smoothing, suggested a nonlinear relationship (Figure 5.8). Although only clump area was positively skewed, Peake and Quinn (1993) transformed both variables because of the species–area relationships for other seasons in their study and the convention in species–area studies to transform both variables.

The scatterplot of log number of species against log clump area (Figure 6.11) linearized the relationship effectively except for one of the small clumps. The residual plot also showed no evidence of nonlinearity, but that same clump had a larger residual and was relatively influential (Cook's $D_i = 1.02$). Re-examination of the raw data indicated no problems with this observation and omitting it did not alter the conclusions from the analysis (b_1 changed from 0.386 to 0.339, r^2 from 0.819 to 0.850, all tests still $P < 0.001$), so it was not excluded from the analysis. In fact, just transforming clump area produced the best linearizing of the relationship with no unusually large residuals or Cook's D_i statistics, but both variables were transformed for the reasons outlined above.

The results of the OLS fit of a linear regression model to log number of species and log clump area were as follows.

	Coefficient	SE	Standardized coefficient	T	P
Intercept	−0.270	0.138	0	−1.98	0.060
Slope	0.386	0.038	0.905	10.22	<0.001
Correlation coefficient (r) = 0.905, r^2 = 0.819					
Source	df	MS	F	P	
Regression	1	1.027	104.35	<0.001	
Residual	23	0.010			

The t test and the ANOVA F test indicate that $\beta_1 \neq 0$. We would also conclude that β_0 is different from zero, indicating that the relationship between species number and clump area must be different for small clump sizes since the model must theoretically go through the origin. The r^2 value indicates that we can explain about 82% of the total variation in log number of species by the linear regression with log clump area.

Number of Individuals Against Clump Area
A scatterplot of individuals against clump area suggested a nonlinear relationship. The residual plot from a linear regression model showed a clear pattern of increasing spread of residuals against increasing predicted number of individuals (Figure 6.12). The boxplots indicated that both variables were positively skewed, so we transformed them to logs.

The scatterplot and residual plots for log-transformed data were much improved (Figure 6.13), with symmetrical boxplots for both variables and no pattern in the residuals).

The results of the OLS fit of a linear regression model to log number of individuals and log clump area were as follows.

	Coefficient	SE	Standardized coefficient	t	P
Intercept	−0.576	0.259	0	−2.24	0.036
Slope	0.835	0.071	0.927	11.82	<0.001

Box 6.8 (cont.)

(cont.)

	Coefficient	SE	Standardized coefficient	t	P
Correlation coefficient (r) = 0.927, r^2 = 0.859					
Source	df	MS	F	P	
Regression	1	4.809	139.62	<0.001	
Residual	23	0.034			

There is strong evidence that β_1 differs from zero. The r^2 value indicates that we can explain about 86% of the total variation in log number of individuals by the linear regression with log clump area.

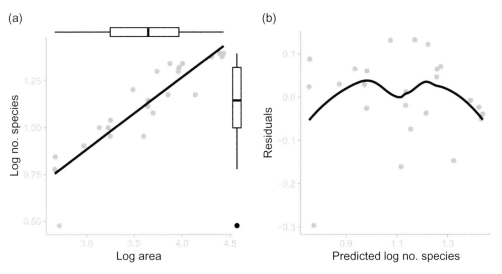

Figure 6.11 Scatterplot (with linear regression line fitted) of log number of species against log clump area and residual plot (with LOESS smoother, smoothing parameter = 0.5) from linear regression of log number of species against log clump area.

improve the fit of a linear model to observed data and help meet other OLS assumptions such as homogeneity of variance (compare Figures 6.12 and 6.13). In linear regression models, they can also improve the linearity of their relationship (Figures 5.8 and 6.11).

There are two things to watch for when considering transformations. First, our model interpretation and any predictions will be in terms of transformed Y and/or X – for example, predicting log number of species from log clump area (Box 6.8). However, predictions can be back-transformed to the original measurement scale if required (Section 5.4). Second, if the response variable's distribution is from the exponential family, such as Poisson, GLMs (Chapter 13) may be more effective than

transformation. Warton et al. (2016) also looked at over-dispersed distributions not from the exponential family, such as negative binomial and Poisson lognormal. They argued that for count data that included zeroes, the choice between GLMs and linear models on transformed data should be based on how well the data fit the model, rather than preconceived ideas. They also recommended that resampling methods can be used with GLMs to provide stricter Type I error control if needed.

6.6 Robust Linear Models

Ordinary least squares estimation may sometimes be inappropriate because assumptions aren't met or there are

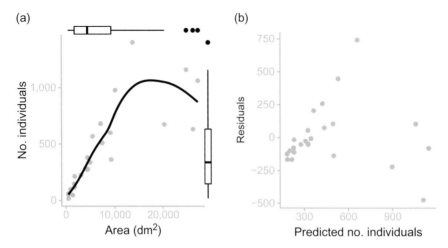

Figure 6.12 Regression diagnostics for the relationship between number of individuals against clump area. Panel (a) is a scatterplot (with LOESS smoother, smoothing parameter = 0.5) and (b) shows residuals from linear regression of number of individuals against clump area.

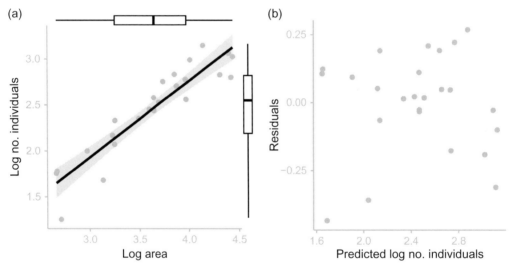

Figure 6.13 Scatterplot (with linear regression line and 95% confidence band fitted) of log number of individuals against log clump area, with accompanying residual plot from linear regression of log number of individuals against log clump area.

influential observations. While GLMs based on ML can apply for some distributions, and transformations, especially to logs, have broad applicability in biology, we sometimes need to consider robust statistical methods. They allow estimation and testing of model parameters but are less sensitive to deviations of the underlying distribution of error terms from that specified, and to extreme observations.

6.6.1 Rank-Based ("Nonparametric") Methods

The term nonparametric can take two meanings in linear models. Biologists commonly think of nonparametric

procedures as those that do not require distributional assumptions of error terms or the response variable. Another meaning is fitting models that do not rely on a specific model structure (Section 13.9).

6.6.1.1 *Continuous Predictor (Regression)*

The simplest nonparametric regression analysis is based on the $[n(n-1)]/2$ OLS slopes of the regression lines for each pair of X-values (the slope for (x_1, y_1) and $(x_2, y_2), (x_2, y_2)$ and $(x_3, y_3), (x_1, y_1)$ and (x_3, y_3), etc.). The Theil (or Theil–Sen) estimator of $\beta_1 (b_1)$ is the median of these slopes and the nonparametric estimator of β_0 (b_0) is the median of all the $y_i - b_1 x_i$ differences (Hollander et al. 2013; Sokal & Rohlf 2012). Thiel's test for the H_0 that $\beta_1 = 0$ is based on the sum of the differences between all pairs of y_i, converted to $+1$ if the difference is positive and -1 if negative. The test is essentially the equivalent of Kendall's rank correlation test (for a description, see Hollander et al. 2013).

6.6.1.2 *Categorical Predictor*

There are two broad rank-based tests for comparing groups in a single-factor design. First is the Kruskal–Wallis test, which is a rank-randomization test and an extension of the Mann–Whitney–Wilcoxon test described in Section 2.5.6.3 for two groups. It evaluates the H_0 there is no difference in the location of the distributions between groups, and is based on ranking the pooled data, determining the rank sums within each group and calculating the H statistic, which follows a χ^2 distribution with df $= p - 1$ (Hollander et al. 2013; Sokal & Rohlf 2012). Although the Kruskal–Wallis test is nonparametric in the sense it does not assume that the underlying distribution is normal, it does assume that the shapes of the distributions are the same in the different groups. The H_0 tested is a difference in location. This implies that variances should be similar (Hollander et al. 2013). Therefore, the Kruskal–Wallis test is not recommended for testing when variances are unequal. However, it is useful for dealing with outliers. The Kruskal–Wallis test is sometimes described as a "nonparametric ANOVA," but this is a little misleading. There is no partitioning of variance and the H_0 is no longer about means unless the distributions are symmetric. Day and Quinn (1989) and Hollander et al. (2013) describe relevant multiple comparison procedures controlling Type I error rates.

In the rank transform (RT) method, we transform the data to ranks and then simply fit our linear model using OLS on the ranked data and test the H_0 with an F test. For a single-factor design, an RT F test will produce the same result as the Kruskal–Wallis test if ties are handled equivalently, but it is a more general procedure and can potentially be used for complex models with interactions (Section 7.1.14). The RT approach also does not deal with unequal variances. If the variances are unequal on the raw scale, the ranks may also have unequal variances. We would also need to conduct the usual model-checking diagnostics on the ranked data.

6.6.2 Generalized (Weighted) Least Squares

If the variance of y_i varies for each x_i, we can weight each observation by the reciprocal of an estimate of its variance: $w_i = 1/s_i^2$. We then fit our linear regression model using generalized least squares, which minimizes $\sum w_i (y_i - \hat{y}_i)^2$. This is the principle of weighted least squares (Kutner et al. 2005). It matches ML estimation of the linear model with the same weighting (Gelman & Hill 2006). When we have multiple independent Y-values for each x_i, such as with a categorical predictor, estimating s_i^2 and w_i is straightforward. Two approaches can be used for regression models with only a single Y-value at each x_i. The first is to group nearby observations and calculate s_i^2 (Kutner et al. 2005), although there are no clear guidelines for how many observations to include in each group. The second is to use the absolute value of each residual ($|e_i|$) from the OLS regression as an estimate of σ_i. Kutner et al. (2005) suggested that the predicted values from an OLS regression of $|e_i|$ against x_i or $\hat{y}_i$ could be used to calculate the weights for each observation, where w_i is the inverse of the square of this predicted value. These weights can be used in statistical software with a generalized or weighted least squares option or, equivalently, OLS regression used once y_i and x_i in each pair has been multiplied (i.e. weighted) by w_i.

6.6.3 Other Robust Methods

We introduced robust methods for estimating population means in Section 2.4.2.1. These methods are specifically designed to be more robust than OLS against outliers but have also been adapted to deal with heterogeneous variances, particularly when comparing the means (or medians) of different groups of a categorical predictor.

6.6.3.1 *Categorical Predictors: Handling Heterogeneous Variances*

Several procedures have been developed for testing equality of population group means (and specific comparisons; see Day & Quinn 1989) when variances are very different (Wilcox 2022). One of the earliest was Welch's test, which uses adjusted df to protect against increased Type I errors under variance heterogeneity (Day & Quinn 1989). Wilcox (2022) described an extension of the Yuen–Welch test that tests the H_0 of no difference between trimmed means, including a measure of ES. The test statistic is compared to an F distribution and Wilcox argued this test maintained good Type I error control. The challenge is choosing an appropriate quantile of the distribution to be trimmed; 0.2 (20%) is often recommended. A bootstrap version is better with small sample sizes and low levels of trimming (<0.2). Other tests for comparing group means under heterogeneous variances include the Brown–Forsythe test and James second-order method (Wilcox 2022).

Introduced in Section 2.4.2.1 for estimating the mean of a population, M-estimators can be adapted to linear models. In a linear regression model, M-estimators minimize the sum of some function of the residuals e_i. Recall that OLS minimizes $\sum e_i^2$; two alternatives are:

- Least absolute deviations (LAD, sometimes termed least absolute residuals – see Berk 1990) minimize the residuals' absolute value, $\sum |e_i|$. By not squaring the residuals, extreme observations have less influence on the fitted model.

- M-estimators. The residuals (e_i) are scaled by a robust measure of their standard deviation, like MAD (mean absolute deviation), and weighted depending on how far they are from zero (Berk 1990) using what is termed an objective function. Two common functions, and associated weights, are the Huber and Tukey bisquare functions (Kutner et al. 2005). Both downweight larger residuals in a pattern determined by the tuning constant (e.g. Huber weights decline when $|e_i|$ exceeds the tuning constant). As this is essentially a weighted (generalized) least squares problem, the parameters are estimated using IRLS (iteratively reweighted least squares; see Box 4.3) and the tuning constant for each method (e.g. the Huber constant is 1.345) is chosen to ensure the procedure is at least 95% as efficient as OLS when applied to normally distributed data. These methods can also be used for regression models and those focused on differences between group means, often with bootstrap resampling (Wilcox 2022) because the sampling distributions of the estimated coefficients may not be normal, unless sample sizes are large.

6.6.4 Resampling and Permutation Methods

The bootstrap methods for estimating parameters and calculating CIs in Section 2.4.7.1 also apply to linear models. For linear regressions, two approaches are available (Kutner et al. 2005). Random X resampling (also called case resampling), which will apply for many regression models in biology, resamples the observations $(x_i, y_i$ pairs) with replacement many times, and fits the linear model each time to get the regression slope. If the predictor is fixed, then fixed-X resampling is more appropriate. Here, the residuals from the OLS model are resampled with replacement, with each set of residuals fitted against the original X-values to generate the bootstrapped regression slopes. In both methods, the final bootstrap estimate is the mean, and the SE is the SD, of the bootstrapped coefficients. Confidence intervals can be determined as described in Section 2.4.7.1. We assume that the distribution of regression coefficients is normal and use the appropriate quantiles (e.g. 0.025, 0.975 for 95% CI) of the standard normal (z) distribution or derive

the intervals directly from the distribution of bootstrapped coefficients (termed the reflective method by Kutner et al. 2005). A test of the H_0 that the regression slope equals zero can be done by seeing if zero falls inside or outside the interval.

Randomization (or permutation) tests of parameters (Section 2.5.6.2) can also be applied to linear models. Manly and Navarro Alberto (2022) describe a method to test the H_0 that $\beta_1 = 0$ in a linear regression model. The observed value of b_1 is compared to the distribution of b_1 found by pairing the y_i and x_i values at random many times and calculating b_1 each time, with the P-value being the percentage of values of b_1 from this distribution equal to or larger than the observed value of b_1. Anderson (2001) focused on fixed-X regression models and recommended just reordering (permuting) Y many times with X unchanged and using r^2 as the test statistic.

We can also use a randomization test to test the H_0 of no difference between groups for a categorical predictor (Anderson 2001; Edgington & Onghena 2007; Manly & Navarro Alberto 2022). The procedure randomly allocates observations (or even residuals for models with multiple predictors) to groups (keeping the same sample sizes) many times to produce the distribution of a test statistic – for example, F or SS_{Groups} or MS_{Groups} – under the H_0 of no group effects. If this H_0 is true, we expect all randomized allocations of observations to groups are equally likely. We simply compare our observed statistic to the randomized distribution of the statistic to determine the probability of getting our observed statistic, or one more extreme, by chance. Randomization of observations and residuals produces similar results. Note that a randomization test to compare group means may not be robust against unequal variances; the H_0 can be rejected because of different variances with no differences between the means (Manly & Navarro Alberto 2022).

Some of the permutation approaches for fitting linear models to multivariate data described in Chapter 16 can also be used when there is a single predictor variable.

6.7 Power of Single-Predictor Linear Models

Power analysis (Section 3.3) is commonly used to determine the appropriate sample size(s) when designing a study. This approach is based on using decision criteria (the Neyman–Pearson protocol) for statistical hypothesis testing, as we are specifying the alternate hypothesis as an ES of interest. We are essentially asking how many observations we need to be confident (at a specified level, i.e. power) that we will detect an effect (e.g. regression slope,

Box 6.9 Variation in Formal Implementations of Power Analysis for Linear Models

There are two possible sources for confusion when calculating power for a simple linear model. First, ESs can be expressed differently. With a categorical predictor, the effect is sometimes described as the pattern of means, or with fixed effects, the α_i values. Other authors, such as Cohen (1988), combine the variation among means and an estimate of σ_ε^2 to produce a standardized effect.

For example, for a two-sample t test, Cohen's ES parameter is $d = (\mu_1 - \mu_2)/\sigma_\varepsilon$ and for a single-factor ANOVA test his parameter f is given by

$$f = \sqrt{\frac{\sum\limits_{i=1}^{p} \alpha_i^2/p}{\sigma_\varepsilon^2}},$$

which can then be estimated from the α_i values specified by the alternative hypothesis, and an estimate of residual variance.

Second, the noncentrality parameter is most often expressed as λ. However, Searle (1993; Searle & Gruber 2016) defines the noncentrality parameter as $\lambda/2$, and sometimes noncentrality is defined as $\sqrt{\lambda}/(p-1)$ or as $\varphi = \sqrt{(\lambda/p)}$.

Suppose power is to be calculated using tabulated values. There, most authors provide power values tabulated against φ (e.g., Winer et al. 1991), although Cohen (1988) provides very extensive tables of power against f and n. Note that $f = \varphi/\sqrt{n}$. This reflects a difference in approach, with using φ representing a standardization using the SE of the mean, and f a standardization using σ_ε.

These different formulations are mathematically equivalent in terms of the final result of the calculations, but it is confusing initially to encounter different definitions of ostensibly the same parameter. You must check the formulation used by a particular author or software package. A good check is to use a standard example from one of the major texts and run it through the new calculations. When the same answer is obtained, begin your calculations.

difference between group means) of a certain size if it exists, given the residual variance. To do these analyses, we need three key pieces of information: the ES (usually standardized – see Section 6.3) we wish to detect if it occurs; an estimate of the residual variance; and the desired level of confidence (power). Implementations of power are not uniform across software packages, and some care is needed when specifying parameters (Box 6.9)

6.7.1 Regression Models

Since H_0s about the regression slope and the correlation coefficient can be tested with t tests, power calculations are relatively straightforward based on noncentral t distributions (Kutner et al. 2005; see also Section 3.3). At the design stage, we wish to know how many observations we need to detect a regression slope of a certain size if it exists. An example of power analysis for a linear regression model is provided in Box 6.10.

6.7.2 Categorical Predictor Models

We wish to know how many observations we need to detect a difference of a certain size between group means

for categorical predictors. The overall test of the H_0 of equal population group means is evaluated with an F test, so power calculations rely on noncentral F distributions. The exact shape of this distribution depends on df_{Groups}, df_{Residual} and on how different the true population means are under the desired ES. This difference is summarized by the noncentrality parameter (λ), or $\phi = \sqrt{(\lambda/p)}$ where p is the number of groups. It is defined as:

$$\lambda = \frac{\sum\limits_{i=1}^{p} \alpha_i^2}{\sigma_\varepsilon^2/n} = \frac{n\sum\limits_{i=1}^{p} \alpha_i^2}{\sigma_\varepsilon^2}.$$

The noncentrality parameter λ incorporates the ES [group effects (α_i) squared] and the within-group variance σ_ε^2.

An example of power calculations is included in Box 6.11. These power calculations are straightforward for two groups, but become more complicated with more groups. When there are only two groups, the ES is a difference between the two means. However, when we have more than two groups, the alternative hypothesis could, for example, have the groups equally spaced or

We will use Aranda et al.'s (2014) data on the bryophytes of islands in the Macaronesian archipelagos (Azores, Madeira, Canary Islands, and Cape Verde) in the North Atlantic. We will use bryophyte species richness and the predictor variable annual precipitation to explore the power test for simple linear regression. First, we need basic summary statistics for our power analysis.

Mean bryophyte species richness was 216. The range of annual precipitation is 113.1–1492.9.

A Priori Power Analysis

Suppose we are designing a study to detect a dependence of bryophyte richness on rainfall and have nominated a 50% change in species richness across the range of precipitation sites as an important change. Under this scenario, richness would rise by 108 species over a rainfall range of 1,379.8 mm. This change corresponds to an H_1 slope of $108/1379.8 = 0.0783$. This is our effect size.

We need an estimate of variance, and with regression analysis the variation of interest is the residual variance. At the planning stage of a study, we do not have this information, which might come from pilot data or other studies. In this case, we can use the variation in Y-values, s_y, as an estimate. For this dataset, the SD of bryophyte richness is 132.4. We also need the SD of the predictor variable – in this case, we'll use our existing data, and $s_x = 429.4$. Depending on the software being used, we may enter these two values independently or combine them before entering data (s_y/s_x – in this case 0.308).

At the planning stage, we only need to nominate a target α (we'll be boring and use 0.05) and consider whether our test will be one- or two-tailed. In this case, bryophyte richness should rise as we go from arid islands to wetter ones, so we will use a one-tailed test.

Using this information, and the software package G*Power (one-tailed bivariate regression, power = 0.08), we get a required sample size of >90 – we need to visit many islands!

two the same and one different. These different patterns of means will lead to different λ values and, hence, power. The difficulty of specifying H_A becomes greater with more groups unless we have a very specific H_A that details a particular arrangement of our groups (e.g. a linear trend across groups).

If the number of groups is not too large, one option is to calculate the power for several group arrangements. For example, we can calculate the power characteristics for four arrangements of groups for a given difference in means (between the largest and smallest), such as groups equally spaced, one group different from all others (which are equal), and so on. Examples of such power curves are plotted in Figure 6.14, where for a given ES you can easily see the range of power values. There is little difference for very large or very small differences between the largest and smallest group means. For planning experiments, it may be enough to know the range of power values for a given effect and decide around one specific arrangement of groups, considering where that arrangement fits on the power spectrum of alternatives.

6.8 Key Points

- Linear models with a continuous response (Y) and a continuous predictor (X) variable are termed regression models. Models with a continuous Y and a fixed categorical X (factor) are sometimes termed "analysis of variance" (ANOVA) models. OLS is commonly used to fit both types of models and estimate their parameters.
- The focus of regression models is estimating the change in the response for a unit change in the predictor, measured by the regression slope. Null hypotheses about regression slopes can be evaluated using t or F statistics from an ANOVA comparing the fit of models with and without the slope.
- A measure of the strength of the fit of the model is r^2, the proportion of the total SS due to the linear regression on X.
- OLS regression strictly assumes that the X variable is fixed, whereas it is common in biology for both Y and X to be random. For most biological research, OLS is still suitable when X is random, but if the focus is specifically

Box 6.11 ® ℮ Worked Example of Power Analysis for Single-Factor Design: Serpulid Recruitment onto Surfaces with Different Biofilms

One of the other response variables in the study of recruitment by Keough and Raimondi (1995), the number of bryozoans in the genus *Bugula*, showed no differences between the filming treatments, so power becomes an issue. For *Bugula*, the ANOVA was as follows.

Source	df	SS	MS	F	P
Biofilms	3	22.8	7.59	0.338	0.798
Residuals	24	538.3	22.43		
Total	27	561.1			

The mean for the unfilmed (SL) treatment was 6.86. We can use this information to look at the power of the test for this genus.

Suppose we define an ecologically important ES as doubling settlement over the value for unfilmed surfaces, our ES = 6.86. First, let's look at the overall power of the test. Suppose that we wanted to detect any effect of biofilms, in which case the SL treatment would have a value of 6.8, and the other three would be 13.72. The grand mean would be 12.01, giving estimates of α_i of −5.15, 1.72, 1.72, and 1.72. For these values, $\sum \alpha^2 = 35.29$ and, from the table above, our estimate of σ_ε^2 is 22.43.

Using the formula at the start of Section 6.7.2, $\lambda = (7 \times 35.29)/22.43 = 11.01$ and Cohen's $f = 0.627$ (using the formula in Box 6.10). Substituting one value into any software packages that calculates power, using $df_{Groups} = 3$ and $df_{Residual} = 24$ we get power of 0.73. Remember, power is the probability of statistically detecting this ES if it occurred. This experiment had a good, but perhaps less than desirable, chance of detecting a doubling of settlement of this form.

To see how our specification of H_A affects power, let's look at the power for a pattern that is one of the hardest to detect using an overall F test, a gradual trend across treatments from largest to smallest mean. Using the example here, the four means would be 6.86, 9.14, 11.44, and 13.72. Then, the power is 0.58. If a steady change across treatments is our main interest, there's a >40% chance we'd miss a doubling in settlement with this experiment. With most biological experiments being costly (in time or money), that's a substantial risk.

Rather than demonstrating that power was low for one response variable, we could use this information to plan a new study. We can use the observed variation among experimental units for bryozoans to determine the number of replicates we'd need to detect a doubling in settlement for these organisms. Using the simple (and easy to detect) pattern of one mean different from the other three, we would only need to increase our sample size from seven to eight to be 80% confident. If it's important to detect a trend across treatments, we'd need $n = 11$.

on the value of the slope relating Y to random X without interest in prediction (e.g. allometric studies) then MA or SMA methods are appropriate.

• Models with a categorical predictor focus on estimating differences or contrasts between group means, although an ANOVA F statistic of the null hypothesis of no group differences is commonly presented. These contrasts can derive from the dummy variables used to fit the model as a regression (e.g. comparing each group to a reference group) but other user-defined contrasts can be specified (e.g. comparing specific groups).

Null hypotheses about contrasts can be based on t or F statistics from an ANOVA, as with regression models.

• OLS estimation has key assumptions about the model error terms for CIs and hypothesis tests of parameters: they are normally distributed, have similar variance for different X, and are independent. In most cases, these assumptions apply equally to Y. Transformations of Y (e.g. to logs) are often useful if the transformed response makes biological sense.

• These assumptions are best checked using graphical methods, especially scatterplots, boxplots, and plots of

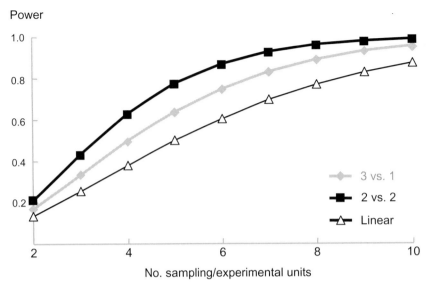

Figure 6.14 Power vs. sample size curves for hypothetical comparison of four groups, with the target difference between largest and smallest means of 10, with within-cell standard deviation of 6.2. The curves show power for three patterns – one treatment different from the others, which are identical; two "large" and two "small" means; four means equally spaced from largest to smallest.

residuals against predicted values where the aim is consistent spread of residuals with no outliers. Diagnostics for influential observations in regression models, such as Cook's D, are also valuable.

Further Reading

Single-predictor models: Most applied statistical analysis and linear models books will cover single-predictor models. For example, Sokal and Rohlf (2012) and Kutner et al. (2005) have several chapters on single-predictor linear models, with Gelman and Hill (2006) and Gelman et al. (2020) emphasizing regression analyses.

Random predictor (Model II) regression: Legendre and Legendre (2012) provide an excellent overview of the issues when fitting regression models with random predictors.

Robust, permutation and nonparametric linear models: Wilcox (2022) is a detailed text on many of these approaches, with Manly and Navarro Alberto (2022) and Edgington (2007) covering randomization methods.

7 Linear Models for Crossed (Factorial) Designs

A common experimental or sampling design in biology incorporates two (or more) categorical predictors (factors) that are fully crossed – that is, all combinations of the factors are included in the design and every group of each factor occurs combined with every group of the other factors. Such designs are also termed factorial. They allow us to measure two sorts of factor effects:

• the main effect of each factor independent of (pooling over) the other factors; and

• the interaction between factors, measuring how the effects of one factor depend on the groups of other factors. The absence of an interaction means that the joint effect of the factors is predictable by adding their individual effects together. An interaction indicates a synergistic or antagonistic effect.

In this chapter, we focus on models for designs that cross categorical predictors that are fixed factors. Models for designs that include crossed random factors will be considered in Chapter 10, while models including interactions involving continuous predictors will be described in Chapter 8.

We will start with complete factorials, in which all possible combinations of the factors are used.

7.1 Two-Factor Fully Crossed (Factorial) Designs

There are three research scenarios in biology where a design that crosses two fixed factors might be applicable.

7.1.1 Completely Randomized (Experimental) Designs

In these designs, researchers control the groups of each factor, and the experimental units can be allocated randomly to the factor combinations. For example, Linton et al. (2009) examined the effect of an insecticide on red land crabs on Christmas Island. They had two fixed factors in their design. There were two bait types (with and without insecticide), and baits were provided to crabs at three dosage levels, reflecting different ways of delivering insecticide. The response variable was the ovarian mass of each crab. Both factors in this study were experimentally manipulated. The analysis of these data is given in Box 7.1.

7.1.2 Observational (Nonexperimental) Designs

Researchers may also rely on natural groupings in the data. Here, the observation units fall into these groupings. While they may be drawn randomly from a group of possible such units at each factor combination, we cannot control the distribution of these units among the different combinations. As an example of this design, Breitwieser et al. (2016) studied the effect of chemical contamination on scallops on the Atlantic coast of France. Their design had two fixed factors. There were four sites (one unpolluted and three polluted), and each site was sampled at two seasons with scallops collected at each combination of site and season, and several biomarkers recorded. This design was observational; neither factor was experimentally manipulated. The analysis of these data is presented in Box 7.2.

A second example comes from Tartu et al. (2016), who studied mercury exposure in Arctic kittiwakes. Their sampling design had two fixed factors, both with two groups: sex (males, females) and breeding stage (incubating, chick-rearing) with an unequal number of birds in each of the four combinations. The concentration of Hg from a blood sample from each bird was the response variable. This is also an unbalanced design, with different sample sizes in each factor combination. The analysis of these data is shown in Box 7.3.

Box 7.1 Ⓡ Ⓔ Worked Example of Two-Factor Crossed Design: Effects of Insecticides on Land Crabs

Linton et al. (2009) studied the effects of the insecticide pyriproxyfen on ovarian development in an endemic Christmas Island land crab, *Gecarcoidea natalis*. The insecticide was proposed to control numbers of introduced crazy ants, and it is an endocrine disruptor. The experiment tested whether the insecticide might pose risks to the crabs, which have a hormone similar to the one targeted in insects. The design consisted of feeding crabs a mixture of leaf litter and a bait. Half of the baits contained the insecticide, and the other half were controls ("bait type" factor). Baits were supplied at three rates, with two groups corresponding to those used in field applications ($2\,kg\,ha^{-1}$ and $4\,kg\,ha^{-1}$), with the third rate being *ad libitum* feeding ("bait dosage" factor). The experimental units were large plastic tubs, each containing a single female crab, and there were seven crabs for each factor combination. The response variable was the dry mass of the ovaries of each crab. A two-factor linear model including the fixed main effects of bait type and bait dosage and their interaction was fitted to these data.

The effects of interest were:

- the interaction between bait type and bait dosage on ovarian mass;
- the main effect of bait type on ovarian mass pooling dosages; and
- the main effect of bait dosage on ovarian dry mass pooling bait types.

There were no outliers and the residual plot (Figure 7.1) did not suggest major problems with OLS assumptions, although there was some suggestion of a relationship between mean and variance. Like Linton et al. (2009), we analyzed the untransformed data.

Source	df	MS	F	P	η^2
Bait type	1	2.61	7.36	0.010	0.152
Bait dosage	2	0.34	0.95	0.397	0.039
Type × dosage	2	0.58	1.63	0.209	0.067
Residual	36	0.35			

There was no evidence of a strong interaction. There was evidence against the null of no bait type effect ("statistically significant"). Figure 7.1 shows roughly 50% higher ovary mass for baits with insecticide, although the pattern showed some inconsistency with dosage. The amount of bait supplied had little overall effect.

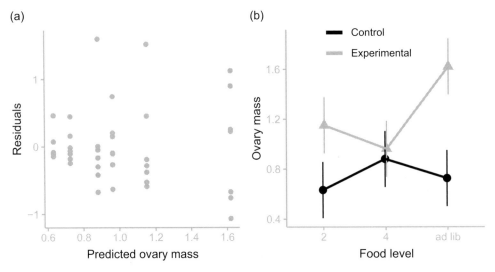

Figure 7.1 Residual plot and interaction plot for ovarian development in land crabs. Panel (a) shows residuals plotted against predicted values. Each vertical line of points represents one combination of factor groups. Panel (b) shows mean ovarian mass ($g\,ovary^{-1}\,kg^{-1}$ dry body mass), with standard errors, for three food regimes and control and insecticide-laced baits.

Box 7.2 Ⓡ Ⓔ Worked Example of Two-Factor Crossed Design: Biomarkers in Scallops

Breitwieser et al. (2016) surveyed variegated scallops, *Mimachlamys varia*, living at several sites along the French Atlantic coast. Four sites were chosen (factor "site"); three were chosen because they were potentially contaminated, and the fourth was considered relatively clean (and considered a reference site). The authors sampled scallops at two times of the year (March and September), chosen to correspond to before and at the end of the scallop's reproductive season. These two times were groups of the factor "season" (although strictly these groups represent two different sampling times that may or may not reflect seasonal differences). Breitwieser and colleagues measured several variables to assess the condition of scallops, and here we analyze their data using a biomarker, malondialdehyde (MDA). Malondialdehyde is a stress marker and was measured (μM/mg fresh tissue) in 10 scallops (the experimental units) from each combination of site and season. A two-factor linear model including the fixed main effects of site and season and their interaction was fitted to these data.

The effects of interest were:

• the interaction between site (including the contrast between contaminated and uncontaminated) and season on scallop MDA levels;
• the main effect of site (including the contrast between contaminated and uncontaminated) on scallop MDA levels, pooling seasons; and
• the main effect of season on scallop MDA levels, pooling sites.

Initial examination of residuals (Figure 7.2) suggested a relationship between variance and mean, mostly due to a few somewhat large MDA values at two sites. A log transformation of MDA slightly reduced the influence of these unusual values, but the residual plot did not change markedly, nor did the conclusions about the interaction effects, so we analyzed the data on the original scale.

Source	df	MS	F	P	η^2
Season	1	14018.3	9.31	0.003	0.093
Site	3	4648.5	3.09	0.032	0.094
Season × site	3	4377.0	2.91	0.040	0.088
Sites in March			1.22	0.308	
Sites in September			4.77	0.004	
Residual	72	1,505.3			

There was some evidence for an interaction between site and season (and the *P*-value for the interaction was "statistically significant"). The interaction plot (Figure 7.2) suggests a straightforward interpretation, with all sites similar in the March sampling, when MDA values were high, whereas in September one site, Port-Neuf, had elevated values, while the other three were similar and lower than in March. Breitwieser et al. (2016) attributed these site differences to concentrations of cadmium. While the interaction could be interpreted easily from the interaction plot in this case, we could use more formal analysis. Examination of site effects at each season separately (using $MS_{Residual}$ from the original model for *F*s) indicates no site effects in March but strong effects in September.

The four sites represented three contaminated sites and one uncontaminated site (Loix-en-Ré). We might be interested in including this contrast in the above analysis, particularly to interpret the interaction effect. The interaction between season and the difference in mean MDA between Loix-en-Ré and the average of the three contaminated sites was weak ($t = -0.053$, $P = 0.957$), indicating that the season by site interaction was not due to seasonal change in the difference between the contaminated sites and the reference site. Additionally, the effect of this contrast, pooling the two times, was also weak ($t = 0.787$, $P = 0.434$).

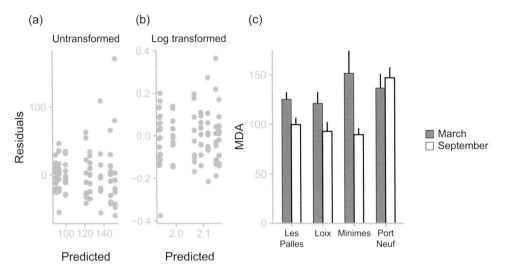

Figure 7.2 Residual plots and interaction plot from analyses of scallop MDA activity (μM/mg fresh tissue weight). Panels (a) and (b) show residuals using untransformed and log-transformed response variables, and panel (c) shows means and standard errors, plotted as log(MDA) to reflect the analyses.

7.1.3 Designs That Combine Completely Randomized Factors with Nonrandomized (Observational) Factors

Some designs combine an experimental factor with an observational factor. For example, Hostetler et al. (2016) examined the effects of methamphetamine on hypothalamic neuropeptides in the brains of laboratory-housed prairie voles. Individual voles were the experimental units, and there were two fixed factors. Treatment, representing access to methamphetamine or water, was an experimental factor, crossed with sex (male vs. female), an observational factor, with five or six voles in each of the four combinations. This example is not analyzed, as the model and analysis strategy are the same as the previous examples. However, it illustrates how experimental and observational factors can be incorporated into a single crossed design.

7.1.4 The Factorial Linear Effects Model

A two-factor crossed design is illustrated in Figure 7.4 with a factor relationship diagram.

Consider a dataset consisting of two factors. Factor A has p groups ($i = 1$ to p), factor B has q groups ($j = 1$ to q) crossed with each group of A, and there are n_i sampling or experimental units ($k = 1$ to n_i) within each combination of A and B. The combinations of the two factors are called cells, and there are pq cells. Every group of factor B is crossed with every group of factor A and vice-versa. It doesn't matter which factor is labeled A or B, as fully crossed designs are symmetrical, although the order of terms in linear model statements can sometimes be important. For the moment, we'll consider the situation with the same number of units (n) in each combination of A and B. Unequal sample sizes will be discussed in Section 7.1.7.

From Linton et al. (2009), $p = 3$ dosages (factor A), $q = 2$ bait types (factor B) and $n = 7$ crabs within each cell. From Breitwieser et al. (2016), $p = 4$ sites (factor A), $q = 2$ seasons (factor B), and $n = 10$ scallops per cell.

We distinguish between two types of means in multifactor crossed designs (Table 7.1).

• Marginal means are the means for the groups of one factor pooling over the groups of the second factor, so the marginal mean A_1 is the mean for the first group of A, pooling over the groups of B. For example, the marginal mean for Loix-en-Ré from Breitwieser et al. (2016) is the mean MDA level from all possible scallops from that site, regardless of season. The marginal mean for each group of A is μ_i, and the marginal mean for each group of B is μ_j. Marginal means can be calculated in two ways:

1. Unweighted, where they are averages of the relevant cell means (sometimes termed least squares means). Unweighted means are likely to be of interest if any differences in sample sizes do not reflect underlying differences in populations, such as when we lose a few observations in a designed experiment due to random processes (e.g. equipment malfunction).

2. Weighted, where the cell means are weighted by the cell sample sizes before averaging. Analyzing

Box 7.3 Ⓡ Ⓔ Worked Example of Two-Factor Unbalanced Crossed Design: Mercury Concentrations in Kittiwakes

Tartu et al. (2016) collected data on blood mercury concentrations in male and female (factor: sex, with two groups) Arctic black-legged kittiwakes (*Rissa tridactyla*) that were collected during the incubation period and while chicks were being reared (factor: breeding stage, two groups). During incubation, they collected 48 females and 44 males, with 17 and 21, respectively, during chick-rearing. The authors analyzed these data with a GLM based on a normal distribution and identity link. In contrast, keeping with the theme for this chapter, we fitted a two-factor OLS linear model including the fixed main effects of site and season and their interaction. With these unequal sample sizes, we have the options of Type I, II, or III SS.

The residuals from the fit of the full model are presented in Figure 7.3. There is a strong pattern with increasing spread of residuals with increasing cell means. A log transformation could be considered for these data, especially given the unbalanced natures of the dataset, which may make the OLS model fit and inference more sensitive to unequal variances if the interpretation of the interaction term makes sense on a log scale. However, we will analyze the untransformed data as the original authors did.

The results of the analysis based on different types of SS are presented here.

Source	df	Type I SS	F	Type II SS	F	Type III SS	F
Sex	1	6.36	44.68	7.26	51.02	4.77	33.49
Stage	1	6.98	49.02	6.98	49.02	6.75	47.40
Sex × stage	1	0.39	2.77	0.39	2.77	0.39	2.77
Residual	126	0.14		0.14		0.14	

1. For sex × stage, $P = 0.099$. Fs for main effects all $P < 0.001$.
2. If stage enters the model first, then Type I SS were: stage: 6.08; sex: 7.26.

The residual and the interaction effect were unaffected by the unbalanced sample sizes. The interaction effect was minor relative to the main effects (Figure 7.3) and not "statistically significant." The SS and subsequent Fs for the main effects differ between the three types of SS, although the changes are not dramatic, despite sample sizes per cell ranging from 17 to 48. Both main effects were strong, with males having higher blood mercury levels than females, and overall levels were higher during incubation than during chick-rearing.

Note that the same model fitted to log-transformed mercury concentrations indicated the minor interaction effect was completely removed ($P = 0.44$), the main effects were still strong, and conclusions were again unaffected by the different types of SS.

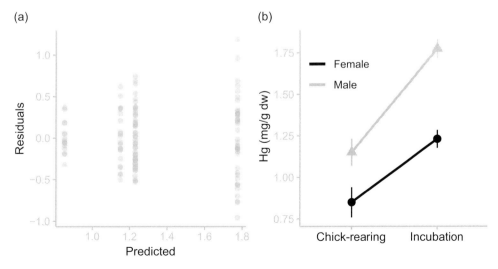

(a)

(b)

Figure 7.3 (a) Residual plot and interaction plot from analyses of mercury content of male vs. female kittiwakes at different breeding stages. (b) Means and standard errors.

Table 7.1 OLS estimates and standard errors for cell and marginal means from fitting a two-factor crossed linear model (Kutner et al. 2005)

	Estimate	Standard error
Overall mean	$\bar{y} = \dfrac{\sum_{i=1}^{p}\sum_{j=1}^{q}\bar{y}_{ij}}{pq}$	$\sqrt{\dfrac{\text{MS}_{\text{Residual}}}{(pq)^2}\sum_{i=1}^{p}\sum_{j=1}^{q}\dfrac{1}{n_{ij}}}$
Marginal means Unweighted	$\bar{y}_i = \dfrac{\sum_{j=1}^{q}\bar{y}_{ij}}{q} \quad \bar{y}_j = \dfrac{\sum_{i=1}^{p}\bar{y}_{ij}}{p}$	$\sqrt{\dfrac{\text{MS}_{\text{Residual}}}{q^2}\sum_{j=1}^{q}\dfrac{1}{n_{ij}}} \text{ and } \sqrt{\dfrac{\text{MS}_{\text{Residual}}}{p^2}\sum_{i=1}^{p}\dfrac{1}{n_{ij}}}$
Weighted	$\bar{y}_i = \dfrac{\sum_{j=1}^{q}n_{ij}\bar{y}_{ij}}{n_i} \quad \bar{y}_j = \dfrac{\sum_{i=1}^{p}n_{ij}\bar{y}_{ij}}{n_j}$	
Cell means	$\bar{y}_{ij} = \dfrac{\sum_{i=1}^{p}\sum_{j=1}^{q}\sum_{k=1}^{n_{ij}}y_{ijk}}{n_i}$	$\sqrt{\dfrac{\text{MS}_{\text{Residual}}}{n_{ij}}}$

Standard errors (SE) of marginal means use harmonic mean of sample sizes across relevant groups of the second factor.

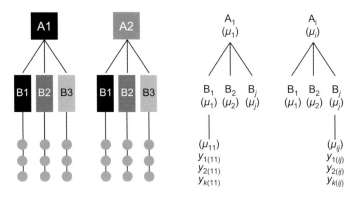

Figure 7.4 Factor relationship diagram for two-factor crossed design, with p levels of factor A ($i = 1$ to p), q levels of factor B ($j = 1$ to q), where the levels of B are the same and crossed with each level of A, and n replicate units within each combination (cell) of A and B ($k = 1$ to n).

weighted means might be appropriate when the differences in sample size represent real differences in relative population sizes of the different factor combinations. If the design is balanced, unweighted and weighted marginal means are the same.

• Cell means are the means of the observations within each combination of A and B. For example, the mean MDA level from all scallops within each site and season combination. The cell means for each combination of A and B are μ_{ij}.

The linear effects model for a factorial design with two fixed factors extends the model used for single-factor designs in Chapter 6. The two-factor effects model is:

$$y_{ijk} = \mu + \alpha_i + \beta_j + (\alpha\beta)_{ij} + \varepsilon_{ijk}.$$

From Linton et al. (2009):

(mass of ovaries)$_{ijk}$ = (overall mean ovary mass)
+(effect of dosage)$_i$ +(effect of bait type)$_j$
+(effect of interaction between dosage and bait type)$_{ij}$
+ε_{ijk}.

From Breitwieser et al. (2016):

(MDA activity)$_{ijk}$ =(overall mean MDA activity)
+ (effect of site)$_i$ +(effect of season)$_j$
+ (effect of interaction between site and season)$_{ij}$
+ ε_{ijk}.

These models include parameters for the two main effects and the interaction between A and B. Note that we use β for the effect of B; don't confuse this with the regression coefficients βs from the regression version of the model (Box 7.4). We will describe fitting this model with OLS, based on the Gauss–Markov assumptions described in Section 7.1.13, but ML will produce very similar estimates when error terms are normal. The statistical details of the two-factor linear effects model, including estimation of its parameters, are provided in Box 7.4. We could also fit a means model or a regression model based on dummy variables, also described in Box 7.4. The means model is useful when we have designs with missing cells. The regression model produces a potentially useful set of default contrasts depending on how we code the dummy variables.

7.1.5 Model Parameters

As with single-factor models, we usually code the effects model into our software, but will often be interested in marginal and cell means and parameters from the regression model based on dummy variables. Illustrated using the data from Breitwieser et al. (2016), these include (Box 7.4):

- The overall mean (μ), such as the mean MDA activity in all scallops across eight site and season combinations. It is estimated ($\bar{y}$) by the unweighted mean of cell means, or equivalently the unweighted mean of the factor A marginal means or the factor B marginal means.
- The marginal mean for each factor group $\left(\mu_i \text{ or } \mu_j\right)$ pooling over the second factor, such as the mean MDA activity across all scallops for site i, pooling season, or for season j pooling site. These are estimated by the sample marginal means $\left(\bar{y}_i \text{ or } \bar{y}_j\right)$.
  - The main effects for each factor group $\left(\alpha_i \text{ or } \beta_j\right)$, such as the effect of site i or season j on MDA activity across all scallops measured as the difference between the mean MDA activity for site i and the overall mean or the difference between the mean MDA activity for season j and the overall mean. These effects are estimated by the difference between the overall sample mean and the sample mean for A group i or B group j ($\bar{y} - \bar{y}_i$ or $\bar{y} - \bar{y}_j$).
- The cell means $\left(\mu_{ij}\right)$, such as the mean MDA activity in each combination of groups from factors A and B, estimated by the cell sample means $\left(\bar{y}_{ij}\right)$.
  - The effect of the interaction between factors A and B $\left(\alpha\beta_{ij} = \mu_{ij} - \mu_i - \mu_j + \mu\right)$, such as how much the effect of site on MDA activity of scallops varied between seasons and vice-versa. The interaction effects are estimated by $\bar{y}_{ij} - \bar{y}_i - \bar{y}_j + \bar{y}$. If the

differences between groups of one factor are consistent across the groups of the second factor, $(\alpha\beta)_{ij} = 0$. An interaction can make main effects difficult to interpret. We cannot easily assess the effect of factor A or B independent of the other. This also implies that the tests of main effects are conditional on the outcome of evaluating the interaction.

- The coefficients (βs) from the regression version of the model representing either contrasts between the means for each group and the overall mean (the α_i effects) or contrasts between the means for each group and the reference group mean.

To calculate SEs and CIs for the relevant parameters, we also need to estimate σ_ε^2 (the common variance of ε_{ij}).

7.1.5.1 Predicted Values

Once we have estimates of our model parameters, we can substitute these into our effects model to derive the predicted Y-values ($\hat{y}_{ijk}$; Box 7.4), which are simply the relevant cell means $\left(\bar{y}_{ij}\right)$.

7.1.5.2 Error Terms and Their Variance

ε_{ijk} is the random error associated with MDA activity of a scallop at each site and season. We can estimate the error terms by calculating sample residuals $\left(e_{ijk}\right)$ which are the differences between each observed y_{ijk} and each predicted $\hat{y}_{ijk}$:

$$e_{ijk} = y_{ijk} - \hat{y}_{ijk} = y_{ijk} - \bar{y}_{ij}.$$

We will use these residuals to evaluate the assumptions and fit of our model. The OLS estimate of the variance of these error terms $\left(\sigma_\varepsilon^2\right)$ is the $MS_{Residual}$ term from the ANOVA from the fit of the full model (Table 7.3).

7.1.6 Inference for Parameters
7.1.6.1 Standard Errors and Confidence Intervals

The focus with categorical predictors is on estimates of cell and marginal means. Details of these OLS estimates and their SEs are in Table 7.1. SEs are based on the $MS_{Residual}$ from the ANOVA. Confidence intervals for parameters are based on the t distribution as usual.

Hypothesis Tests

The three key hypotheses of interest are that the two main effects, and the interaction effect, equal zero. These null hypotheses are evaluated using F tests and associated P-values as part of the ANOVA, comparing full and reduced models. There may also be hypotheses related to subsets of factor groups, which can be tested as contrasts using F or t tests. As with the linear models

Box 7.4 The Factorial Linear Model and Its Parameters for Categorical Predictors

The linear effects models for a factorial design with two fixed factors is an extension of the model used for single-factor designs in Chapter 6. The effects model is:

$$y_{ijk} = \mu + \alpha_i + \beta_j + (\alpha\beta)_{ij} + \varepsilon_{ijk},$$

where:

y_{ijk} is the replicate observation k from the combination of the group i of factor A and group j of factor B (i.e. cell ij);
μ is the overall population mean of the response variable;
α_i is the effect of group i of factor A, pooling the groups of factor B. This is the main effect of factor A, defined as the difference between each A marginal mean and the overall mean $(\mu_i - \mu)$;
β_j is effect of the jth group of factor B, pooling the groups of factor A, which is the difference between each B marginal mean and the overall mean. This is the main effect of factor B, the difference between each B marginal mean and the overall mean $\left(\mu_j - \mu\right)$;
$(\alpha\beta)_{ij}$ is the effect of the interaction of the ith group of A and the jth group of B and is defined as $\left(\mu_{ij} - \mu_i - \mu_j + \mu\right)$. The interaction effect can be described in two ways. First, the interaction is the difference between the actual value of μ_{ij} and the value predicted if the effects of factors A and B were additive, that is if the two factors were affecting the response variable independently. We can estimate the interaction effect by comparing the full model above with a reduced model that omits the interaction term. Second, the interaction measures whether the effect of factor A on the response variable depends on the group of factor B and vice-versa. The two factors are not affecting the response independently of each other – their effects are multiplicative rather than additive.
ε_{ijk} is random or unexplained error associated with replicate k from the combination of the group i of factor A and group j of factor B. These error terms are assumed to be normally distributed at each combination of factor groups, with a mean of zero and a variance of σ_ε^2.

This fixed effects model is overparameterized because the number of means (combinations of factors plus overall mean) is less than the number of model parameters to be estimated $(\mu, \alpha_1, \alpha_2, \ldots, \beta_1, \beta_2, \ldots, (\alpha\beta)_{11}, (\alpha\beta)_{12}, \ldots)$. Overcoming this problem so we can estimate model parameters requires a series of "sum-to-zero" constraints:

$$\sum_{i=1}^{p} \alpha_i = 0, \quad \sum_{j=1}^{q} \beta_j = 0, \quad \sum_{i=1}^{p} \alpha\beta_{ij} = 0, \quad \sum_{j=1}^{q} \alpha\beta_{ij} = 0.$$

These constraints appear formidable but simply imply that the sum of the effects of factor A, pooling B, and the sum of the effects of factor B, pooling A, are both zero. Additionally, the sums of the interaction effects for each group A and for each group B are also zero. These constraints are necessary for fitting effects models, although their importance has been debated; see Hocking (2013), Searle (1993), and Yandell (1997).

An alternative to imposing constraints on the effects model is to fit a much simpler means model:

$$y_{ijk} = \mu_{ij} + \varepsilon_{ijk},$$

where μ_{ij} is the mean of cell ij and ε_{ijk} is random or unexplained variation. The means model basically treats the analysis as a single-factor model comparing all cells, and tests specific hypotheses about interactions and main effects. The means model estimates A and B means by averaging the cell means across rows or columns (Searle 1993), so it has certain advantages for unbalanced designs by ignoring the sample sizes completely. Means models are mainly useful for missing cells designs (see Section 7.2.1).

Finally, we could fit a multiple regression model based on dummy variables that represent particular contrasts for main effects and the interaction. We will illustrate this model using the Breitwieser et al. (2016) example studying biomarkers in scallops from four sites, with the one unpolluted site (Loix-en-Ré; LeR) set as the reference, and across two seasons, with March set as the reference:

Box 7.4 (cont.)

$$y_{ijk} = \beta_0 + \beta_1(\text{PN vs LeR}) + \beta_2(\text{LM vs LeR}) + \beta_3(\text{LP vs LeR}) + \beta_4(\text{Season : March vs September})$$
$$+ \beta_5(\text{PN vs LeR} \times \text{Season}) + \beta_6(\text{LM vs LeR} \times \text{Season}) + \beta_7(\text{LP vs LeR} \times \text{Season}) + \varepsilon_{ijk}$$

In this model, based on reference coding, β_0 is the mean of the reference cell (LeR:March), each main effect coefficient represents a contrast between that factor's reference group and the other groups, and the interaction coefficients represent the interactions between the main effect contrasts. For example, β_2 is measuring how much the contrast of PN vs. LeR varies across the two seasons. One challenge of reference coding in multifactor models is the need to identify a reference group for all factors. Alternatively, we could use effects coding where the coefficients represent differences from the overall factor mean.

Estimating the parameters of the different versions of the factorial linear model follows the methods outlined for a single-factor model in Chapter 6, with the added complication of estimating interaction effects.

Cell means $\left(\mu_{ij}\right)$ for each combination of A and B are estimated from the sample mean of the observations in each cell, based on the sample size of the particular cell if sample sizes are unequal.

The factor group (marginal) mean for each group of A pooling groups of B is simply the mean of the sample means for each cell at group i of factor A, averaged across the groups of B (see Table 7.2). An analogous calculation can be done for factor B means. These are unweighted means and ignore any difference in sample sizes between cells.

An alternative approach is to calculate a weighted marginal mean, which averages the observations for each group of A taking into account different n_{ij} within each cell. If we have a fully balanced design (all n_{ij} equal), the unweighted and weighted estimates of factor group means will obviously be the same. If we have unequal numbers of observations per cell (some n_{ij} different), the estimates will be different. In unbalanced crossed designs, only Type III SS are based on unweighted marginal means. Standard errors for these means are based on the $\text{MS}_{\text{Residual}}$ from the ANOVA (Table 7.3).

The estimates of $\alpha_i(\mu_i - \mu)$ and $\beta_j\left(\mu_j - \mu\right)$ are the differences between the mean of each A group or each B group and the overall mean, $\bar{y}_i - \bar{y}$ and $y_j - \bar{y}$, respectively. Interaction effects measure how much the effect of one factor depends on the group of the other factor and vice-versa. If there was no interaction between the two factors, we would expect the cell means to be represented by the sum of the overall mean and the main effects:

$$\mu_{ij} = \mu + \alpha_i + \beta_j.$$

Therefore, the effect of the interaction between the ith group of A and jth group of B $(\alpha\beta)_{ij}$ can be defined as the difference between the ijth cell mean and its value we would expect if there was no interaction:

$$\alpha\beta_{ij} = \mu_{ij} - \mu_i - \mu_j + \mu,$$

which is estimated by:

$$\bar{y}_{ij} - \bar{y}_i - \bar{y}_j + \bar{y}.$$

This represents those effects not due to the overall mean and the main effects.

In practice, biologists rarely calculate the estimated factor or interaction effects, instead focusing on contrasts of marginal or cell means.

described in previous chapters, P-values are just one piece of evidence for evaluating model effects.

7.1.7 Model Comparison and Analysis of Variance

7.1.7.1 *Balanced Designs*

The ANOVA partitions the total variation in the response variable into its components or sources. For a two-factor factorial model, there are four sources of variation (Table 7.3). The SS_A measures the sum of squared differences between each A marginal mean and the overall mean; the SS_B measures the sum of squared differences between each B marginal mean and the overall mean; the SS_{AB} measures the sum of squared differences for the interaction contrast involving cell means, marginal means, and the overall mean; finally, the $\text{SS}_{\text{Residual}}$

Table 7.2 Illustration of marginal and cell means for a two-factor factorial design

Data from Breitwieser et al. (2016), where factor A is site, factor B is season, and the response variable is MDA concentration per scallop for each of 10 scallops per cell.

	B_1	B_2	B_j	Marginal means A
A_1	μ_{11}	μ_{12}	μ_{1j}	$\mu_{i=1}$
A_2	μ_{21}	μ_{22}	μ_{2j}	$\mu_{i=2}$
A_3	μ_{31}	μ_{32}	μ_{3j}	$\mu_{i=3}$
A_4	μ_{41}	μ_{42}	μ_{4j}	$\mu_{i=4}$
A_i				μ_i
Marginal means B	$\mu_{j=1}$	$\mu_{j=2}$	μ_j	Grand mean μ

Factor $A(A_i)$	Site	Factor B (B_j) season B_1 March	B_2 September	Factor A marginal means
A_1	Loix-en-Ré	$\bar{y}_{11} = 121.367$	$\bar{y}_{12} = 92.905$	$\bar{y}_{i=1} = 107.136$
A_2	Les Palles	$\bar{y}_{21} = 125.425$	$\bar{y}_{22} = 99.754$	$\bar{y}_{i=2} = 125.590$
A_3	Minimes	$\bar{y}_{31} = 151.668$	$\bar{y}_{32} = 89.585$	$\bar{y}_{i=3} = 120.627$
A_4	Port-Neuf	$\bar{y}_{41} = 136.698$	$\bar{y}_{42} = 147.015$	$\bar{y}_{i=4} = 141.857$
Factor B marginal means		$\bar{y}_{j=1} = 133.790$	$\bar{y}_{j=2} = 107.315$	$\bar{y} = 120.552$

measures the difference between each observation and its relevant cell mean, summed across all cells. The ANOVA provides a method to evaluate the main effects and the interaction using Fs, similar to that described for a single-factor design in Section 6.2.5.

With balanced sample sizes across cells, the SS for the two-factor factorial model represent an additive partitioning of the total SS in the response variable so each SS supplies independent information (Table 7.3):

$$SS_{Total} = SS_A + SS_B + SS_{AB} + SS_{Residual}.$$

The df are calculated as usual (the number of components making up the source of variation minus 1), with $df_{AB} = df_A \times df_B$. While equations for these SS (assuming equal numbers of observations in each cell) are shown in Table 7.3, the SS and df are derived more generally by comparing the fit or lack of fit of relevant full and reduced models. The difference in lack-of-fit ($SS_{Residual}$) between the full model and the reduced model that omits the effect of interest is calculated. Using Breitwieser et al. (2016) as an example, the SS_{AB} is the difference in the $SS_{Residual}$ from the fit of the full model:

$(\text{MDA activity})_{ijk} = (\text{overall mean MDA activity}) +$
$(\text{effect of site})_i + (\text{effect of season})_j +$
$(\text{effect of interaction between site and season})_{ij} + \varepsilon_{ijk}$

to the fit of the reduced model:

$$y_{ijk} = \mu + \alpha_i + \beta_j + \varepsilon_{ijk}$$

$(\text{MDA activity})_{ijk} = (\text{overall mean MDA activity})$
$+ (\text{effect of site})_i$
$+ (\text{effect of season})_j + \varepsilon_{ijk}.$

The reduced model omits the interaction effect – that is, sets $(\alpha\beta)_{ij} = 0$.

In complex models with more predictors there are other full and reduced model options. They produce the same SS for balanced designs but increasingly divergent outcomes with increasing imbalance (see Section 7.1.7.2).

Mean squares are determined by dividing each SS by its df. The expected values of these MS are provided in Table 7.3. The $MS_{Residual}$ estimates σ_ε^2 (the variation in the error terms in each cell, pooled across all cells). Mean squares for the main effects and the interaction estimate the residual variance plus a measure of the relevant effects.

The MS can be used to construct F statistics to evaluate each effect in the model:

- If the effect of factor A is zero (i.e. $\alpha_1 = \alpha_2 = \cdots = \alpha_i = 0$) or equivalently $\mu_1 = \mu_2 = \cdots = \mu_i = \cdots = \mu_p$, MS_A and $MS_{Residual}$ have the same expected value so these two MS are used in an F to evaluate the effect of A. From Breitwieser et al. (2016), this is evaluating whether the marginal mean MDA activity is the same for different sites pooling seasons – that is, site does not affect MDA activity.
- If the effect of factor B is zero, MS_B and $MS_{Residual}$ have the same expected value, so they are used in an F to

Table 7.3 ANOVA table for two-factor factorial model (both factors fixed) assuming equal n ($n > 1$) in each cell

Source	SS	df	MS	EMS	F
A	$nq \sum_{i=1}^{p} (\bar{y}_i - \bar{y})^2$	$p - 1$	$\dfrac{SS_A}{p - 1}$	$\sigma_\varepsilon^2 + nq \dfrac{\sum_{i=1}^{p} \alpha_i^2}{p - 1}$	$\dfrac{MS_A}{MS_{Residual}}$
B	$np \sum_{j=1}^{q} (\bar{y}_j - \bar{y})^2$	$q - 1$	$\dfrac{SS_B}{q - 1}$	$\sigma_\varepsilon^2 + np \dfrac{\sum_{j=1}^{q} \beta_j^2}{q-1}$	$\dfrac{MS_B}{MS_{Residual}}$
AB	$n \sum_{i=1}^{p} \sum_{j=1}^{q} (\bar{y}_{ij} - \bar{y}_i - \bar{y}_j + \bar{y})^2$	$(p - 1)(q - 1)$	$\dfrac{SS_{AB}}{(p - 1)(q - 1)}$	$\sigma_\varepsilon^2 + n \dfrac{\sum_{i=1}^{p} \sum_{j=1}^{q} (\alpha\beta)_{ij}^2}{(p - 1)(q - 1)}$	$\dfrac{MS_{AB}}{MS_{Residual}}$
Residual	$\sum_{i=1}^{p} \sum_{j=1}^{q} \sum_{k=1}^{n} (y_{ijk} - \bar{y}_{ij})^2$	$pq(n - 1)$	$\dfrac{SS_{Residual}}{pq(n - 1)}$	σ_ε^2	
Total	$\sum_{i=1}^{p} \sum_{j=1}^{q} \sum_{k=1}^{n} (y_{ijk} - \bar{y})^2$	$pqn - 1$			

evaluate the H_0. From Breitwieser et al. (2016), this is evaluating whether the marginal mean MDA activity is the same for both seasons pooling site – that is, season does not affect MDA activity.

- If there is no interaction effect (i.e. $(\alpha\beta)_{ij} = 0$) MS_{AB} and $MS_{Residual}$ have the same expected value, so they are used to evaluate the interaction. This is testing there are no effects besides the overall mean (intercept) and the main effects. From Breitwieser et al. (2016), the effect of site on mean MDA activity is the same for both seasons, and equivalently, the effect of season on MDA activity is the same for all sites.

As we will see in Section 7.1.8, it is sensible to evaluate the interaction effect first and only assess the main effects if there is no strong evidence for an interaction.

7.1.7.2 Unbalanced Designs

If sample sizes are unequal, there is no simple partitioning of the SS. Each SS shares information with the other SS – that is, the effects are nonorthogonal. There are multiple options for full and reduced models for each effect, with the different SS outcomes termed, somewhat unhelpfully, Type I, Type II, and Type III (Table 7.4). While these different strategies result in identical SS for balanced designs, this is not the case for main effects with unequal sample sizes. The main effects are based on marginal means that can be unweighted or weighted by the sample sizes in each cell (Table 7.1).

SS_Residual and SS_AB are unaffected by the unequal cell sizes. Thus, we compare the fit of the additive (no interaction) model and the multiplicative (with interaction) model and the difference in fit of these two models (difference in their $SS_{Residual}$) is the SS_{AB}. This effect is consistent regardless of the strategies in Table 7.4 we use, for both balanced and unbalanced designs, because the interaction effect is not measured using marginal means. The three types of SS differ in evaluating main effects, related to how the interaction term is treated (Hector et al. 2010; Shaw & Mitchell-Olds 1993).

Type I SS are determined from the improvement in fit gained by adding each term to the model in a hierarchical sequence; they are sometimes termed sequential SS because the models are built sequentially (Hector et al. 2010). The order of terms in the model is important. The SS due to factor B will be different if it enters the model after or before factor A. Type I SS are also based on marginal means weighted by sample sizes and hence evaluate effects weighted by sample sizes.

The remaining two SS are termed adjusted SS because the SS for any term in the model are adjusted for the other terms. Type II SS are assessed by comparing a model with the term of interest to a model lacking that term but including all additional terms at the same or lower level, called the "higher-order terms omitted" approach. Each main effect is evaluated using a full model that omits the interaction term. This strategy has merit, as we would not usually examine main effects if the interaction was present, so omitting the interaction to examine the main effects makes sense. Type II SS do not depend on the order of terms in the model but still evaluate effects based on marginal means weighted by cell sample sizes.

Table 7.4 Strategies for model comparison in a two-factor factorial ANOVA model

	Type I[12]	Type II[2]	Type III
$A \times B$ interaction	$y_{ijk} = \mu + \alpha_i + \beta_j + (\alpha\beta)_{ij} + \varepsilon_{ijk}$ $y_{ijk} = \mu + \alpha_i + \beta_j + \varepsilon_{ijk}$	$y_{ijk} = \mu + \alpha_i + \beta_j + (\alpha\beta)_{ij} + \varepsilon_{ijk}$ $y_{ijk} = \mu + \alpha_i + \beta_j + \varepsilon_{ijk}$	$y_{ijk} = \mu + \alpha_i + \beta_j + (\alpha\beta)_{ij} + \varepsilon_{ijk}$ $y_{ijk} = \mu + \alpha_i + \beta_j + \varepsilon_{ijk}$
A main effect	$y_{ijk} = \mu + \alpha_i + \varepsilon_{ijk}$ $y_{ijk} = \mu + \varepsilon_{ijk}$	$y_{ijk} = \mu + \alpha_i + \beta_j + \varepsilon_{ijk}$ $y_{ijk} = \mu + \beta_j + \varepsilon_{ijk}$	$y_{ijk} = \mu + \alpha_i + \beta_j + (\alpha\beta)_{ij} + \varepsilon_{ijk}$ $y_{ijk} = \mu + \beta_j + \varepsilon_{ijk}$
B main effect	$y_{ijk} = \mu + \alpha_i + \beta_j + \varepsilon_{ijk}$ $y_{ijk} = \mu + \alpha_i + \varepsilon_{ijk}$	$y_{ijk} = \mu + \alpha_i + \beta_j + \varepsilon_{ijk}$ $y_{ijk} = \mu + \alpha_i + \varepsilon_{ijk}$	$y_{ijk} = \mu + \alpha_i + \beta_j + (\alpha\beta)_{ij} + \varepsilon_{ijk}$ $y_{ijk} = \mu + \alpha_i + \varepsilon_{ijk}$

[1] Note that Type I SS depend on the order of factors A and B – we have described factor A entered first then factor B.

[2] Calculating Type I and II SS for main effects assumes there is no interaction.

Type III SS are often the default in commercial statistical software and use the same full model for evaluating the two main effects and the interaction. This is sometimes called the "higher level terms included" approach (Hector et al. 2010) because higher-level interaction terms are included. In contrast to Type I and II SS, Type III SS are based on unweighted marginal means and therefore are not influenced by different sample sizes in each cell.

We illustrate fitting a linear model to an unbalanced design using the example of Tartu et al. (2016), analyzed in Box 7.3, where cell sizes were 17–48. While the main effects SS differed between the approaches for the Tartu et al. (2016) data, the conclusions from the analysis did not.

Some comments on these different strategies for determining SS in factorial models:
- All three strategies produce the same SS if the design is balanced because the main effects and the interaction SS represent independent components of the total variation.
- For unbalanced designs, the residual and interaction SS are the same for all three strategies.
- The difference between the strategies for main effects SS is related to the imbalance in cell sample sizes – minor differences in sample size won't affect the SS much.
- Type III SS are the most common (as they are the default for many commercial statistical programs) and use unweighted marginal means. However, they also include the interaction in the full model for testing main effects (so if an interaction is present, main effects can be difficult to interpret). The reduced models include an interaction without including one of the main effects associated with that interaction.
- Type I and II SS for main effects assume no interaction effect and it is omitted from the full and reduced models.
- Type I and II SS are based on weighted marginal means, so are influenced by the cell sample sizes.

- Type I SS for main effects depend on the order in which main effects enter the model. The SS for the main effect of the factor entered first ignores the effect of the other factor (Maxwell et al. 2018). Type II SS for main effects do not depend on the order of model terms and consider the other factor when evaluating main effects.

If the focus is on comparisons of marginal means and researchers do not want their analysis influenced by differences in sample size, then like many authors, we recommend Type III SS. Searle (1993) also pointed out that Type III SS were the equivalent of his preferred SS developed using the cell means model. If the analysis is a model-fitting exercise with the aim to find the "best" model or the researcher would prefer to use weighted means, Type II SS may be more appropriate. Type II SS are derived from a model selection approach (the interaction effect is omitted if found not to be strong), so careful assessment of the interaction effect is important. We also support Hector et al.'s (2010) recommendation that researchers should think carefully about which model comparisons make the most biological sense, rather than aiming for the "correct" ANOVA table.

Whichever strategy is chosen, the SS are converted to MS by dividing by the df. F statistics for each effect are developed as described above, with the $\text{MS}_{\text{Residual}}$ being the denominator for the Fs. However, the effects being evaluated relate to weighted means for Type I and II SS, in contrast to unweighted means for Type III SS.

7.1.8 More on Main Effects and Interactions

If there are interactions, interpreting main effects is more difficult. The main effect of a factor (comparison of marginal means) pools over the groups of the other factor, which is not appropriate if the effects of the two factors are not independent. Figure 7.5 shows a range of interactions between two factors; note that interactions can be moderate ($B_2 > B_1$ for all groups of A, but the size of the

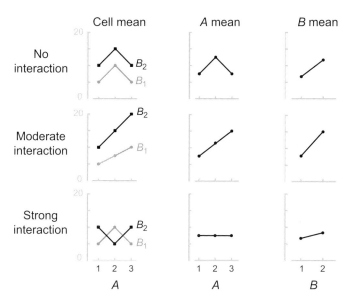

Figure 7.5 Interaction plots of the cell means of two factors, A (A_1, A_2, A_3) and B (B_1, B_2) and plots of marginal main effects means for each factor separately for interactions of three different strengths.

difference varies) or severe ($B_2 > B_1$ for A_1 and A_3, but this difference is reversed for A_2). Underwood (1997) provided clear examples of how large interactions can cause misleading interpretation of main effects in an ecological context. He showed how a strong interaction could cause statistically weak main effects when the effects of a factor were strong but inconsistent across the other factor.

Evaluation of the interaction (e.g. based on evidence from appropriate graphs, the P-value for the interaction effect, etc.) should be done first. If the interaction is absent or weak, assessment of main effects can proceed. If the interaction is strong, main effects may be difficult to interpret. Interaction effects nearly always offer important biological insights. For example, Linton and colleagues knowing whether the effect of insecticide depends on food supply to crabs might influence how the insecticide would be applied in the field.

Sometimes, however, researchers might have little interest in the interaction and want to focus on main effects. Under these circumstances, the key question is whether the interaction effect is strong enough to preclude careful analysis of main effects (Kutner et al. 2005). It becomes a matter of using all the available evidence (P-values, interaction plots, effect sizes) to judge how much an interaction constrains interpretation of main effects.

7.1.9 Interactions and Transformations
Transformations of the response variable deserve special mention for factorial linear models. Transformations of variables with skewed distributions can greatly improve

normality and homogeneity of within-cell variances of the model error terms (Sections 5.4 and 6.5). However, as we introduced in Section 5.4.3, they can also affect the interpretation of interaction terms, as the interaction may result from the scale on which the response variable is measured.

To illustrate, in the middle panels of Figure 7.5, the effects of factors A and B are multiplicative, with the differences between the two B groups changing for different A groups. On a log scale, this interaction effect would probably disappear and an additive model without an interaction term would fit the transformed data well. In the lower panels, the effects of B are reversed for each A group. A log transformation would not change the nature of this interaction term much. This phenomenon was also demonstrated in the example from Breitwieser et al. (2016), where the site $\times$ season interaction was apparent for both untransformed and log-transformed MDA activity (Box 7.2; Figure 7.2).

Log transformations can make multiplicative effects on the raw scale additive on the transformed scale, and other transformations can also alter interaction strengths. Whether to transform data before fitting linear models with interactions then also depends on whether the biological interaction of interest is best represented on the transformed scale.

7.1.10 Specific Comparisons of Marginal Means
If there are no strong interactions, interpreting main effects is relatively straightforward and can incorporate

specific contrasts of marginal means. These contrasts were described in Section 6.2.7, and include model-defined contrasts, user-defined contrasts, and unplanned comparisons.

In the scallop example in Box 7.2, an obvious planned contrast would be comparing the MDA activity at the unpolluted site to the average of the three polluted sites, although an interaction means a contrast on main effects could be misinterpreted.

7.1.11 Interpreting Interactions

The interaction between two factors is often of considerable biological interest and nearly always deserves further analysis if present. Follow-up analyses on interactions may be based on visual examination of cell means or more formal tests of simple main effects or contrasts across the interaction term.

7.1.11.1 *Graphs*

Plotting cell means with the response variable on the vertical axis, the groups of one factor on the horizontal axis, and the second factor represented by multiple bars or lines (see Figures 7.1 and 7.2) is sometimes called an interaction plot. With a line graph, an interaction is indicated by deviation of the lines from parallel. We illustrate the effects of interactions on interpretation of main effects in Figure 7.5; Underwood (1997) provided a similar example and detailed explanation.

The interaction plot for mean MDA activity of scallops across sites and seasons (Breitwieser et al. 2016) clearly shows that the interaction between the two factors is because mean MDA activity shows little difference between seasons at Port-Neuf. In contrast, MDA activity was greater in March for the other three sites (Figure 7.2).

7.1.11.2 *Simple Main Effects*

Simple main effects measure the effect of factor A at each B group separately (or vice-versa) and can help interpret interactions. For example, in the Breitwieser et al. (2016) study, the sites represented a comparison of polluted and unpolluted locations, so evaluating the simple main effects of site at each season separately makes sense (Box 7.2).

Simple main effects are logically planned into the design and analysis in anticipation of a possible interaction between the two factors. They are basically single-factor linear models fitted at each group of the other factor, but they also represent a set of specific contrasts based on partitioning the SS_A, SS_B, and SS_{AB}. With equal cell sizes, the simple main effects for factor A at each B group partition the SS and df for A and AB;

simple main effects for B at each A group partition the SS and df for B and AB. Often, we might only wish to examine simple main effects for one factor, as in the two examples above. This also ensures that the SS for the simple main effects don't exceed the sum of the SS_A, SS_B, and SS_{AB}. If we calculate both sets of simple main effects, the SS of all these effects will exceed the sum of the SS_A, SS_B, and SS_{AB}, because these simple effects are not orthogonal, and some of SS_{AB} is being used for more than one effect. Fs for simple main effects should use the original $MS_{Residual}$ as the denominator, because we have already decided that is our best estimate of the residual variance from our linear model.

7.1.11.3 *Treatment–Contrast and Contrast–Contrast Interactions*

An extension of simple main effects is to include contrasts for one or both factors and examine the interaction between the contrast and the second factor or between a contrast in one factor and a contrast in the second factor. Treatment–contrast interactions partition the interaction term by examining contrasts (e.g. group 1 versus group 2) and trends (e.g. linear, quadratic) in one factor against the groups of the second factor. Using the Breitwieser et al. (2016) study, we might examine whether the contrast in MDA activity between the one unpolluted and the three polluted sites interacts with season (Box 7.2).

Contrast–contrast interactions are a particular case of treatment–contrast interactions and examine the interaction between contrasts or trends in one factor and contrast or trends in the second factor.

7.1.12 Predictor Effects

The measures of effect size based on the proportion of variation in the response variable (PEV measures) explained by the predictor(s) described in Section 6.3.2 for single-factor models also apply for multifactor models. However, complications arise in determining the "total" variation from which the proportion attributable to each effect should be calculated. Some authors recommend using partial eta-squared or partial omega-squared (Maxwell et al. 2018; Tabachnick & Fidell 2019 – see equations below from Olejnik and Algina (2003):

Partial eta-squared:

$$\eta_P^2 = \frac{SS_{Effect}}{SS_{Effect} + SS_{Residual}},$$

Partial omega-squared:

$$\omega_P^2 = \frac{\text{SS}_{\text{Effect}} - \text{df}_{\text{Effect}}\text{MS}_{\text{Residual}}}{\text{SS}_{\text{Effect}} + (N - \text{df}_{\text{Effect}})MS_{\text{Residual}}}.$$

Both measure the variation due to an effect as a proportion of that variation plus the within-cells (residual) variation, ignoring the variation contributed by the other effects in the model. Olejnik and Algina (2003) described generalized versions of these two statistics applicable across a range of designs and are based on distinguishing manipulated (i.e. experimental) factors from measured factors (see also Lakens 2013).

The different measures of effect size described in Section 6.3.2 for single-factor models are also relevant here for main effects and more useful than PEV measures for power calculations and sample size determination (Section 7.3).

7.1.13 Assumptions

Fortunately, the Gauss–Markov assumptions underlying inference for factorial linear models based on OLS are basically the same as we have discussed for single-factor models (Section 6.4). A key assumption is independence of the error terms from the model. Problems with the assumption of independence might arise if we have clustered sampling or experimental designs, including nested, split-unit, and repeated measures structures; analyses of such designs need to incorporate random effects for the clusters and use mixed models (Chapters 10–12). Another important assumption is homogeneity of within-cell variances. Normality of error terms is a less restrictive condition, especially with equal cell sizes, and mainly causes problems if an underlying skewed distribution results in unequal variances. We can check the homogeneity of variance and normality assumptions using the same techniques (boxplots, mean vs. variance plots, and residuals vs. cell mean plots) already described in Chapters 4–6. The focus should be on residuals as our best estimate of model error terms, but understanding the distribution of the response variable is also informative and usually will match the distribution of the residuals.

Options for dealing with violations of the OLS assumptions are the same as those outlined in Section 6.4 onwards. If the underlying distribution of the response variable, and the mean–variance relationship, matches one of the exponential distributions (e.g. Poisson), then GLMs would be appropriate. This decision should be made at the start of the analysis and not as part of a hunt for statistical "significance." Generalized least squares may also be a suitable alternative to OLS for modeling data with heterogeneous variances and some forms of nonindependence (Section 6.6.2). Transformations (log

or power transformations) are common in biological research, but interpretation of interaction terms can change. Finally, some of the robust methods described in the next section may be useful, but they become harder to apply when the design and model becomes more complicated by more predictors.

7.1.14 Robust Factorial ANOVAs

Robust methods can be useful for linear models with nonnormal distributions, although OLS inference is pretty reliable across a range of distribution types. The main strength of robust methods is for analyzing data with outliers and the focus generally is on statistical hypothesis testing. Three broad approaches described in Sections 2.5.6 and 6.6 are rank-based methods, M-estimators, and randomization tests.

The rank transform (RT) method, whereby the data are converted to ranks and the usual linear model fitted to the ranked data, was introduced for single-predictor linear models in Section 6.6.1. Although this method may be useful for evaluating main effects, it is inappropriate for interactions because of the nonlinear nature of RT data. Wilcox (2022) described trimmed mean and M-estimator (usually with bootstrap resampling) methods, introduced for single-factor models in Section 6.6.3, for analyzing two-factor crossed models, and other rank-based methods.

Randomization (permutation) methods can also be used for evaluating effects in factorial linear models, although exact tests based on F distributions for interaction effects are impossible (Anderson & Ter Braak 2003; Edgington & Onghena 2007). Approximate tests for interactions can involve permuting residuals (randomization residuals equation; accounts for variability due to main effects) or raw data. The permutation of residuals may be more powerful, especially with unbalanced data and lognormal distributions (Anderson & Ter Braak 2003; Manly & Navarro Alberto 2022). Manly and Navarro Alberto (2022) reviewed different permutation test approaches. In recent years, randomization procedures have been developed that allow fitting of complex linear models. These techniques were developed for multivariate data (Chapter 16) and are reviewed by Anderson (2017). She points out that the permutation approach can also be applied to univariate data and allow fitting of models to many datasets with awkward data properties.

7.2 Complex Factorial Designs

Extending linear models to three or more factors where all cells have more than one observation is relatively straightforward, except for interpreting complex inter-actions. For example, Long and Porturas (2014) examined the effect of multiple stressors on the performance of a saltmarsh plant important for ecological restoration. They established plots in a saltmarsh at two sites, and at each site plots were allocated randomly to combinations of two salinities and two scale insect treatments. Site, salinity, and scale were fixed factors, and the response variable was the number of days to senescence of the focus plant in each plot. Their three-factor model included a single three-factor interaction (e.g. is the interaction between salinity and scale consistent for the two sites?), three two-factor interactions (e.g. is the difference between salinities the same for the two scale treatments, pooling sites?) and three main effects (e.g. is there a difference between salinities, pooling scale treatment, and site?). Note that the three-factor interaction is symmetrical: "is the interaction between salinity and scale treatment con-sistent for the two sites?" and "is the interaction between salinity and site consistent across the scale treatments?" etc. The full analysis of these data is given in Box 7.5.

We fit the three-factor model in the same way as a two-factor model by comparing the fit of appropriate full and reduced models. The effects are examined sequen-tially as for two-factor models: examine the three-factor interaction first and proceed to the two-factor interactions only if the higher-order interaction is not important, then examine main effects only if they are not involved in important two-factor interactions. With complex linear models with multiple predictors and interactions, the degrees of freedom associated with higher-order inter-actions can be large and even minor interaction effects might be detected. As we have emphasized often, a range of evidence, particularly effect sizes, should be used when interpreting any linear model effect. This is especially so with interaction terms.

The strategies for exploring complex interactions follow those outlined in Section 7.1.8. The equivalent of simple main effects are simple interaction effects, where the AB interaction is examined at each group of C or the AC interaction is examined at each group of B, etc. If the simple interaction effects are strong, they could then be followed by evaluating simple main effects. One diffi-culty with complex models is that the number of effects of interest, especially when further evaluating complex interactions, increases quickly. The best strategy is to plan a subset of simple interactions or main effects, with con-trasts if needed, that make the most sense, rather than examining all possible combinations.

In the saltmarsh example there was a three-factor interaction, so the authors sensibly examined the salinity

× scale interaction at each site separately. This showed an interaction for the North site, but not South site (Box 7.5; Figure 7.6). A next step might be to explore the salinity × scale interaction at the North site.

These designs can extend to more than three factors. For example, Shaw and Shackleton (2010) examined the effects of insect supplementation (remove all dead insects, leave all dead insects, add a maggot, control), water treatment (removal, control), site (two sites 200 m apart), and rosette size (large, small) on growth and seed set of a biennial herb, the teasel (*Dipacus fullonum*) near London. All four factors were considered fixed. The analysis of these data is presented in Box 7.6.

As we move into more complex multi-predictor designs and their linear models, especially those that include continuous predictors, the strategy of fitting a full model and evaluating all the main effects and interactions as a single analysis might not be the most appropriate. With fully crossed designs, or nested designs (see Chapter 10), comprising up to three or four factors, the approach we have described here is informative. However, we might be more interested in a model selection process, particularly if prediction is the major goal of the modeling, and criteria for selecting the "best" model (or set of models) will be discussed in Chapter 9.

7.2.1 Missing Cells
We described analyzing unbalanced factorial designs in Section 7.1.7. An extreme form of unequal sample sizes is where there are no observations for one or more cells in a multifactor design (Table 7.5). Such designs are difficult to analyze with linear models because not all marginal and cell means can be estimated, and therefore not all main effects and interactions can be evaluated. Generally, inference about model parameters that involve the miss-ing cell(s) is impossible (Milliken & Johnson 2009).

There is no single correct analysis for missing cells designs, and different approaches test different hypotheses, all of which might be of interest. The recommended approach is to fit a means model (Box 7.4) and examine subsets of interactions and, if relevant, main effects, based on contrasts between cell means (Milliken & Johnson 2009). The residual from the means model is used for all subsequent tests when the factors are fixed. To test the interaction effects, we need to determine which interaction contrasts are estimable – that is, do not depend on the missing cells. For the 2×3 ($A \times B$) factorial design in Table 7.5 with the A_1B_3 cell missing, there is only one interaction contrast that is estimable $(\mu_{11} - \mu_{21} - \mu_{12} + \mu_{22})$, the effect of A at B_1 versus B_2. If there are more than two groups of both factors, and

Box 7.5 ℝ 𝔼 Worked Example of Three-Factor Crossed Design: Restoration of Cordgrass

Long and Porturas (2014) investigated the impact of herbivorous scale insects (*Haliaspis spartinae*) and salinity on senescence of the saltmarsh cordgrass *Spartina foliosa*. Their focus was the potential for salinity stress to modify the herbivory coming from scale insects, and they experimentally removed scale insects or left them intact (factor: scale) on plots with salinity at ambient levels or elevated (factor: salinity) They changed salinity by adding sea salt. The experiment was repeated at two sites chosen to be very different in overall elevation (factor: site, a fixed effect in this context) within a marsh in southern California. These three factors form a three-way factorial design, and from each experimental plot (of which there were between six and eight at the two sites), the time to senescence of a single focal stem of the cordgrass was measured.

Inspection of residuals (Figure 7.6) suggested no issues with the data, so we proceeded with the analysis. The ANOVA indicated evidence for an interaction between all three factors ($F_{1,51} = 5.78, P = 0.020; MS_{Residual} = 896.9$).

In the context of this study, salinity and herbivory are the central interests, so we proceeded by looking at how the combined effects of salinity and scale change with site. We did this by fitting a salinity by scale model (Type III SS) for each site separately, and also with an interaction plot (Figure 7.6), which shows the complexity of the interaction.

Source	df	MS	F	P	η^2
North					
Salinity	1	6,174	6.88	0.011	0.100
Scale	1	3,590	4.00	0.051	0.058
Scale × salinity	1	6,320	7.05	0.011	0.102
South					
Salinity	1	196	0.22	0.642	0.004
Scale	1	12	0.01	0.907	<0.001
Scale × salinity	1	392	0.44	0.512	0.008
Residual	51	896.9			

At the North site the presence of scale reversed the effect of salinity (interaction effect) – in the absence of scale, plants senesced more quickly when salinity was enhanced, but when scale were present leaves senesced earlier at ambient salinity than when salinity was enhanced. At the South site, there was no interaction between scale insects and salinity, nor any main effects.

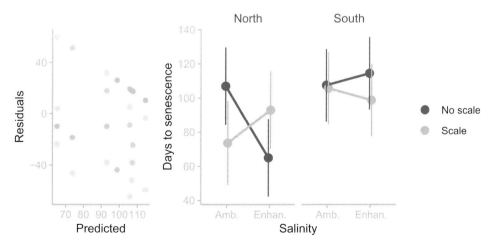

Figure 7.6 Residual plot and interaction plots showing combined effects of scale insects, salinity, and location on senescence times. The analysis showed an interaction between all three factors, and because the two process of greatest interest were those of scale insects and salinity, we have visualized the three-factor interaction as a pair of interaction plots, one for each site, with each plot showing how salinity and scale insects combined to affect senescence.

depending on the pattern of missing cells, there may be more than one estimable interaction contrast, and sums of nonestimable contrasts might also be estimable (Searle 1993). We could also use Type III SS to compare the fit of the full effects model to the fit of the reduced model omitting the interaction. This F is testing whether all the estimable interaction contrasts are zero. Since there is only one estimable interaction contrast, this test is the same as obtained from the contrast as part of the cell means model. When there is more than one estimable contrast in the interaction term, interpreting the ANOVA from fitting an effects model is more difficult, and Milliken and Johnson (2009) advise against using effects models for missing cells designs.

What about main effects if the interaction is not important? The recommended approach is to determine a set of contrasts of marginal means (for the part of the

dataset without missing cells) or cell means that test sensible hypotheses based on the available data. This is where the cell means model is very important. Using the design in Table 7.5 as an example, we can contrast B_1 vs. B_2 using marginal means ($H_0: \mu_{B1} = \mu_{B2}$) because all cells have observations. We can also contrast B_1 vs. B_3 and B_2 vs. B_3, but only using cell means for A_2.

Note these analyses do not represent orthogonal partitioning of the SS_{Total}, because these designs are extreme examples of imbalance, and we have pointed out there is no simple partitioning of the SS in unbalanced designs. We should also mention Type IV SS, another example of the SS numbering system first developed by the SAS statistical software package. When there are no missing cells, Type IV SS are the same as Type III SS. When there are missing cells, Type IV SS are calculated for all the

---

Box 7.6 Ⓡ Ⓔ Worked Example of Four-Factor Crossed Design: Are Teasel Carnivorous?

Shaw and Shackleton (2010) studied the effects of four factors (insect supplementation, water treatment, site, and rosette size) on growth and seed set of a possibly carnivorous biennial herb, the teasel (*Dipacus fullonum*) near London. The design was fully balanced with three plants per factor combination. Preliminary analyses of both response variables (using residual plots) suggested a need for transformation. In both cases, small predicted values were associated with small spreads of residuals. For seed set, variance of the residuals increased steadily with predicted values (Figure 7.7), whereas residuals for growth showed a pattern of strongly increasing then decreasing spread of the residuals. In both cases, log transformation of the response variable removed the issues with the distribution of residuals.

The ANOVA results from the model fitting are shown below.

Source	df	Seed set			Growth		
		MS	F	P	MS	F	P
Site	1	8.597	56.64	<0.001	10.30	278.75	<0.001
Size	1	2.563	16.89	<0.001	2.819	76.30	<0.001
Insect	2	0.502	3.31	0.045	0.047	1.28	0.287
Water	1	0.002	0.02	0.900	0.028	0.77	0.385
Site × size	1	0.013	0.08	0.774	0.022	0.59	0.445
Site × insect	2	0.378	2.49	0.094	0.073	1.96	0.151
Site × water	1	0.003	0.02	0.895	0.009	0.24	0.629
Size × insect	2	0.127	0.84	0.440	0.041	1.11	0.338
Size × water	1	0.667	4.40	0.041	0.252	6.81	0.012
Insect × water	2	0.027	0.18	0.836	0.035	0.94	0.399
Site × size × insect	2	0.101	0.66	0.519	0.072	1.96	0.152
Site × size × water	1	0.152	1.00	0.322	0.001	0.02	0.903
Site × insect × water	2	0.074	0.49	0.616	0.023	0.61	0.547
Size × insect × water	2	0.508	3.35	0.044	0.036	0.98	0.382
Site × size × insect × water	2	0.039	0.26	0.773	0.009	0.26	0.775
Error	48	0.152			0.037		

Box 7.6 (cont.)

As most higher-order interaction effects were minor, Shaw and Shackleton (2010) only presented results for main effects and two-factor interactions, while we provide the full ANOVA table. For the log-transformed seed set, there was some evidence for a three-factor interaction involving size, water treatment, and insect supplementation, and a very strong main effect of site (Figure 7.8), which was not involved in any interactions. Further analysis might examine the water treatment and supplementation interaction for each size class separately and pooling sites. For log-transformed growth, there was some evidence of an interaction between size and water treatment (Figure 7.8). There was also a very strong main effect of site, again not involved in any interactions. Further analysis might focus on interpreting the size class by water treatment interaction, pooling insect supplementation, and sites.

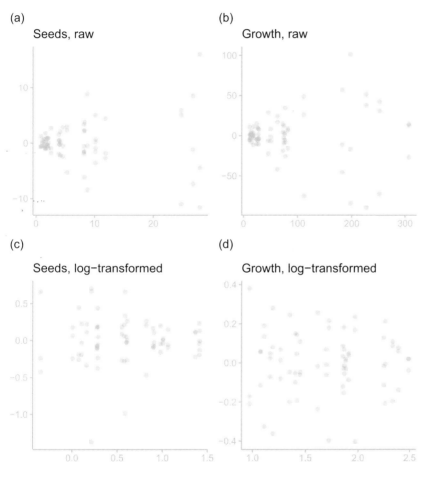

Figure 7.7 Residual plots from the Shaw and Shackleton (2010) example. Residual plots show residuals against predicted values for an experiment with 4 factors and 24 total factor combinations. The lower panels show residuals after log transformation of the response variable. For simplicity, we have omitted scales on axes, because it is the pattern of residuals that is of greatest interest.

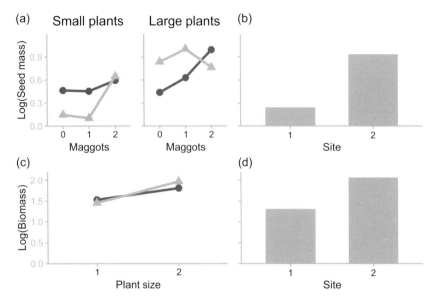

Figure 7.8 Interaction plots from the Shaw and Shackleton (2010) example. Panels (a) and (b) show the results for seed production, and illustrate the three-factor interaction between size, water treatment, and insects, along with the main effect for site. Panels (c) and (d) show the total plant biomass (growth), and illustrate the two-factor interaction between size and water treatment, plus the independent main effect of site. For water treatment, black denotes control and gray the water removal. Note that error bars are not shown, because the effects are pooled across other factors, so simply calculated SE would not be representative.

estimable contrasts although SAS selects a subset of them (see Milliken & Johnson 2009 for details). Searle (1993) and Yandell (1997) have argued strongly that the default Type IV SS may not be useful for many missing cells designs. It is a more sensible strategy to think carefully about what subset of hypotheses are of most interest from all those that can be tested when there are missing cells.

Another approach to missing cells designs is to analyze subsets of the dataset with observations in all cells. In the example in Table 7.5, we could delete all data from B_3 and fit a two-factor model to the remaining 2×2 data. SS_{AB} for this subset analysis is the same as the Type III SS from the full dataset because the only estimable interaction contrast is the one from this subset. The first contrast of B in the cell means analysis is also the main effect of B from an analysis of the subset of the data, omitting B_3. The other subset with observations in all cells uses A_2 data only for all B groups, although that becomes a single-factor analysis. When both factors have more than two groups, and there is at least one missing cell, there may be more than one subset suitable for a factorial model. But if there is a complex pattern of missing cells (disconnected data, *sensu* Searle 1993), the subsets might be small compared to the full dataset – that is, most rows and columns may need to be deleted to form an analyzable subset of the data.

The analysis of missing cells factorial designs will involve analyzing balanced subsets of the data, especially for interactions, and sensible contrasts of cell means for examining components of main effects. We have only discussed fixed factor models here. The tests of relevant contrasts in mixed models with missing cells are messy because of the difficulty of calculating an appropriate error term. Automated approaches are not typically useful, and you need to think about sensible subsets to compare.

7.2.2 Fractional Factorial Designs

Sometimes we might wish to explore (or "screen") the effects of several factors, but there are so many combinations of factor groups that the experiment is logistically impossible because it would require too many experimental units. Fractional factorial designs are often used in these situations, especially when we have many factors with two groups each. These are called 2^p designs, where p is the number of two-level factors. If we had four factors ($p = 4$), there would be 16 cells in a fully factorial design, and each unit increase in sample size would require 16 additional experimental units. A fractional factorial design uses a subset of factor combinations. Their main issue is that two or more factor effects are aliased, where one effect

Table 7.5 Example of a balanced 2×3 factorial design with one missing cell

(a) Design layout

	B_1	B_2	B_3
A_1	μ_{11}	μ_{12}	
A_2	μ_{21}	μ_{22}	μ_{23}

(b) Estimable contrasts with a cell means model

Source	df
All cells	4
A:	
A_1 vs. A_2 for B_1 and B_2 $(\mu_{11} + \mu_{12} = \mu_{21} + \mu_{22})$	1
B:	
B_1 vs. B_2 marginal means $(\mu_{11} + \mu_{12} = \mu_{21} + \mu_{22})$	1
B_1 vs. B_3 for $A_2 (\mu_{21} = \mu_{23})$	1
B_2 vs. B_3 for $A_2 (\mu_{22} = \mu_{23})$	1
$A \times B$:	
A_1 vs. A_2 at B_2 vs. B_3 $(\mu_{11} - \mu_{21} - \mu_{12} + \mu_{22})$	1
Residual	5 cells $\times\ n - 1$

includes effects of other factors or interactions (Gunst & Mason 2009). This is also sometimes called confounding of effects (see also Section 3.2). For example, a common fractional design for a 2^p configuration is a half-fraction design, where only half the possible factor combinations are used in the design. For a 2^3 experiment this would mean each main effect would be confounded with a two-factor interaction and the three-factor interaction cannot be estimated (Kuehl 2000). Whether this confounding of effects is a problem depends on researchers' prior knowledge or expectations about the importance of interaction effects.

There are two ways to reduce the required number of units. First, fractional factorial designs are commonly unreplicated, so there is only one replicate unit within each cell used. There is no estimate of the within-cells variation, so some higher-order interaction terms must be the residual for F tests when we fit our linear model. Second, the logical basis of these designs is the assumption that most of the important effects will be main effects or relatively simple (e.g. two-factor) interactions, and complex interactions will be unimportant. Fully factorial experiments with many factors allocate most of their resources, measured by df, to higher-order interactions (Gunst & Mason 2009), so designing an experiment where higher-order interactions cannot be estimated can be a more efficient allocation of resources if we suspect those higher-order interactions are not important. The experiment is conducted using a subset of cells that allows estimation of main effects and simple interactions but confounds them with higher-order interactions assumed to be trivial.

The combination of factor groups to be used is tricky to determine but, fortunately, most statistical software now includes experimental design modules that generate fractional factorial design structures. This software often includes algorithms such as Plackett–Burman and Taguchi designs, which set up fractional factorial designs in ways that try to minimize confounding of main effects and simple interactions.

It is difficult to recommend these designs for routine use in biological research. We know that interactions between factors are often biologically interesting, and it is usually difficult to decide *a priori* which interactions are less likely. We also know that within-cells variation can be large with biological systems, so unreplicated studies are inherently risky. Possibly such designs have a role in tightly controlled laboratory experiments where the response variable(s) can be measured with high precision and there is only small variability between experimental units, and experience suggests that higher-order interactions are not important. However, the main application of these designs will continue to be in industrial settings where additivity between factor combinations is a realistic expectation. Good references include Cochran and Cox (1957), Kuehl (2000), and Kutner et al. (2005).

7.3 Power and Sample Size in Factorial Designs

For factorial designs, power calculations are simplest for designs in which all factors are fixed. Power for tests of main effects can be done using the principles described in the earlier chapter. We effectively treat each main effect as a single-factor design but use $MS_{Residual}$ and $df_{Residual}$ from the full ANOVA in the power calculations. Power tests for interaction terms are more difficult, mainly because it is harder to specify an appropriate form for the effect size. Just as different patterns of means lead to different noncentrality parameters in single-factor designs, interactions can take many forms, particularly when there are more than two factors or where factors have many groups. Each pattern has its own noncentrality parameter. Calculating the noncentrality parameter (and hence, power) is easy, but specifying exactly which pattern of means would be expected under some alternative hypothesis is far more difficult. Despite the difficulty specifying effects, the fixed effect factorial models have the advantage that power for all effects is increased by increasing the number of replicates in each treatment combination, and any such steps taken to increase the power of a test on main effects will also improve power of tests of interactions. Interaction tests often have more degrees of freedom than corresponding main effects, so power may be more of a problem for tests of main effects.

7.4 Key Points

- Studies with two fixed categorical predictors or factors arranged in a crossed or factorial design (i.e. all combinations of both factors are represented) include two types of means: (1) marginal means, the means of each factor group pooling over the second factor and (2) cell means, the means of each combination of factor groups.
- Linear models with a continuous response (Y) and two fixed crossed factors are commonly fitted using OLS and include:
  - main effects, the effect of each factor pooling over the groups of the second factor, measured as differences in marginal means; and
  - interaction effects, the change in the effect of each factor depending on the groups of the second factor, measured as a contrast involving marginal and cell means.

- The focus of the analysis is estimating the main and interaction effects and deriving an ANOVA with F statistics for each main effect and interaction.
- The ANOVA is straightforward if the sample sizes are the same for each cell in the design. With unbalanced designs, there are three different SS – Types I, II, and III – which give different values for the main effects. Type III SS are commonest. Be sure to check which SS is the default in your statistical software.
- Main effects can be difficult to interpret if there is a strong interaction; evaluate the interaction effects first before deciding whether to look at main effects. Contrasts among marginal means, either based on dummy variables or user-defined contrasts, are nearly always relevant for important main effects.
- Interactions are nearly always biologically interesting. Important interactions can be explored by plotting cell means or using simple main effects for one factor separately for each group of the second factor.
- OLS estimation has the usual key assumptions about the model error terms for CIs and hypothesis tests of parameters: they are normally distributed, have similar variance for different X, and are independent. Transformations of Y can be useful but remember that transformations can make multiplicative effects additive and change the interpretation of the interaction term.
- Extensions to three or more crossed fixed factors are common in biological research. Again, evaluate the most complex interaction first before moving onto simpler interactions and then main effects. For these more complex models, simplifying the final model by omitting unimportant interaction terms is a common strategy among biologists.

Further Reading

Most introductory and linear models textbooks (e.g. Kutner et al. 2005, Sokal & Rohlf 2012) cover factorial designs in detail, and Underwood (1997) has an excellent, traditional, section on interactions and their interpretation. Milliken and Johnson (2009) emphasize factorial designs and their analyses with messy data (e.g. unbalanced, missing cells etc.).

8 Multiple Regression Models

In Chapter 6, we examined linear regression models with a single continuous predictor variable. In this chapter, we will consider more complex linear models, including regression models with multiple continuous predictors and with continuous and categorical predictors.

8.1 Linear Model for Multiple Continuous Predictors: Multiple Regression

A common extension of a simple linear regression model is where we have recorded more than one continuous predictor variable on each sampling or experimental unit. When the predictors are all continuous, the models are called multiple linear regressions. We will illustrate them using two examples.

Cricket Jump Distance

Ercit et al. (2014) tested whether carrying eggs influenced the jump distance of female crickets and made them less able to escape predators. They measured the jump distance (response variable) of each cricket when exposed to a simulated predator attack, along with three predictors: number of eggs being carried, the dry mass of one hind leg, and pronotum length as a proxy for body size. The analyses of these data are presented in Box 8.1.

Bird Abundance in Forest Patches

Understanding which aspects of habitat and human activity affect the biodiversity of remnant habitat patches is an important conservation biology aim. Loyn (1987) wanted to identify habitat characteristics related to the abundance and diversity of forest birds. He selected forest patches in southeastern Australia and recorded the number of species and abundance of forest birds in each patch as two response variables. The predictors for each patch included

area, the year the patch was isolated by land clearing, the distance to the nearest other patch, the distance to the nearest larger patch, an index of stock grazing history, and altitude. These data are analyzed in Box 8.2.

8.1.1 The Multiple Linear Regression Model

Consider a set of n observations on which we have recorded the values of a continuous response (Y) and p predictor variables X_1 to X_p. As with the simple linear regression model, the predictors are considered fixed (i.e. not a random variable) and there is a population of possible Y-values at each combination of predictor variable values. In biological research, the predictors are often random and we will discuss the implications of this in Section 8.1.12.3. The multiple linear regression model, including estimation of its parameters, is detailed in Box 8.3.

For the cricket jump distance example from Box 8.1, $p = 3$ and the linear model would be (omitting the subscripts):

$$\text{jump distance} = \beta_0 + \beta_1(\text{egg number}) \\ + \beta_2(\text{pronotum length}) + \beta_3(\text{leg mass}) \\ + \varepsilon.$$

For the bird abundance example from Box 8.2, $p = 5$ and the linear model would be:

$$\text{bird abundance} = \beta_0 + \beta_1(\text{patch area}) \\ + \beta_2(\text{distance to nearest patch}) + \beta_3(\text{grazing}) \\ + \beta_4(\text{altitude}) + \beta_5(\text{year isolated}) + \varepsilon.$$

These are additive models where all the explained variation in Y is due to the additive (independent) effects of the response variables. They do not allow for interactions (multiplicative effects) between the predictors, although these are possible (even likely) and will be discussed in Section 8.1.10.

Box 8.1 Ⓡ Ⓔ Worked Example of Multiple-Predictor Linear Regression: Cricket Jump Distance

Ercit et al. (2014) focused on whether egg number influenced jump distance (cm) of female tree crickets (*Oecanthus nigricornis*), considering differences in crickets' body size (pronotum length, mm) and leg strength (leg mass, mg). They initially fitted a model that included all three predictors plus interaction terms. However, the sample size ($n = 29$) is relatively small for a model to have seven parameters plus the intercept, so we will focus just on the main effects of the three predictors.

A scatterplot matrix (Figure 8.1) revealed no strong nonlinearities in the relationships between distance jumped and the predictors, nor any obvious outliers. The boxplot for distance jumped was relatively symmetrical. There was, however, an unsurprising positive relationship between pronotum length and leg mass, suggesting that collinearity may be an issue with this model. Their correlation was 0.575, with no strong correlations between them and egg number (0.185, 0.153). As the pronotum–leg mass correlation didn't exceed our collinearity cut-off of 0.7, and the variance inflation factors were small (1.04–1.52), we, like Ercit et al. (2014), fitted the model with all three predictors. We will consider the possible collinearity further below.

The model fitted was:

$$(\text{distance})_i = \beta_0 + \beta_1(\text{egg number}) + \beta_2(\text{pronotum length}) + \beta_3(\text{leg mass}) + \varepsilon.$$

The residual plot from fitting this model showed an even spread of residuals, and the influence diagnostics included no concerningly large Cook's D values (the largest being 0.35).

The parameter estimates from this linear model (with 95% CI for the regression coefficients) were as follows.

	Coeff.	95% CI	SE	Std. coeff.	t	P
Intercept	−19.06	−47.85 to 9.74	13.98	0	−1.36	0.185
Egg number	−0.11	−0.16 to −0.07	0.02	−0.70	−5.14	<0.001
Pronotum	18.12	3.83 to 32.41	6.94	0.43	2.61	0.015
Leg mass	0.38	−7.17 to 7.92	3.66	0.02	0.10	0.918

There is a negative effect of egg number on jump distance independent of pronotum length and leg mass. There is also evidence that crickets with larger pronotums, independent of egg number and leg mass, jump further. Leg mass (independent of the other predictors) has little effect on jump distance. The partial regression plots (Figure 8.2) illustrate these relationships. The SEs for pronotum length and especially leg mass are high relative to the parameter estimates, reflecting their collinearity. The ANOVA from the fit of the full model produced an F of 10.62 with df = 3, 25 and $P < 0.001$.

Because of the collinearity, we also fitted two simpler models that included only one of the predictors involved.

For the model including egg number and pronotum length and the model including egg number and leg mass, the results were as below, respectively.

	Coeff.	95% CI	SE	Std. coeff.	t	P
Intercept	−19.58	−45.89 to 6.74	12.80	0	−1.53	0.138
Egg number	−0.11	−0.16 to −0.07	0.02	−0.69	−5.25	<0.001
Pronotum	18.52	6.97 to 30.08	5.62	0.44	3.30	0.003

	Coeff.	95% CI	SE	Std. coeff.	t	P
Intercept	16.37	8.62 to 24.11	3.77	0	4.34	<0.001
Egg number	−0.11	−0.16 to −0.06	0.02	−0.65	−4.40	<0.001
Leg mass	5.77	−1.11 to 12.65	3.35	0.26	1.72	0.097

Box 8.1 (cont.)

Our conclusions don't change. The effect of egg number, independent of the other two predictors, is still present in both models. The effect of leg mass in the second model has a larger standardized coefficient and a smaller *P*-value than in the original analysis, so our interpretation for that predictor does change slightly once we remove the other correlated predictor.

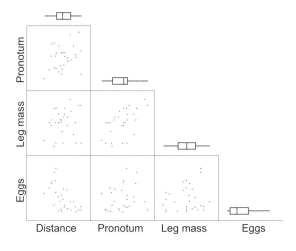

Figure 8.1 Scatterplot matrix (with boxplots) for data from Ercit et al. (2014).

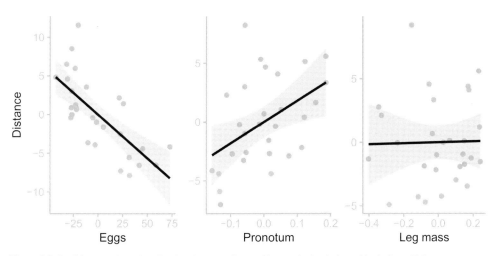

Figure 8.2 Partial regression plots for the three predictors. The vertical axis is residuals from OLS regression of jump distance against all predictors except the one labeled, and the horizontal axis is residuals from OLS regression of the labeled predictor against the remaining predictors. The fitted regression slope and confidence bands are also shown.

Estimation of parameters of multiple regression models with normally distributed error terms is usually by OLS (Box 8.3). To calculate SEs and CIs for these parameters, we also need to estimate σ_ε^2 (the common variance of ε_i and therefore of y_i).

8.1.2 Model Parameters

8.1.2.1 *Intercept and Partial Regression Slopes*

As with simple regression models, the intercept has little biological meaning unless the predictors have been centered. However, it is nearly always included, as

Box 8.2 Ⓡ Ⓔ Worked Example of Multiple-Predictor Linear Regression: Bird Abundance in Remnant Forest Patches

Loyn (1987) selected 56 forest patches in southeastern Victoria, Australia, and related the abundance of forest birds in each patch to six predictor variables: patch area (ha), distance to nearest patch (km), distance to nearest large patch (km), stock grazing intensity (1–5, indicating light to heavy), altitude (m), and years since isolation (years). The aim was to develop a predictive model relating bird abundance to the predictors and identify those most useful for bird abundance.

We'll treat stock grazing intensity as a continuous variable for now, but it could also be considered categorical. The nearest patch was also the closest large patch for nearly half the patches, so only distance to nearest patch was included in the analysis. Patch area had a skewed distribution primarily driven by two patches with unusually large areas (1,771 and 973 ha), with the next largest being 144 ha. These outliers resulted in a nonlinear relationship between abundance and patch area (Figure 8.3). There was also an outlier in distance to the nearest patch, with one patch more than twice as far (1,427 km) from the nearest patch as the next furthest (597 km). Both predictors were log-transformed to reduce these outliers' influence, also producing linear relationships.

A correlation matrix indicated some moderate correlations between predictors, especially between log area and grazing (larger patches have less grazing pressure) and year isolated and grazing (patches isolated earlier had greater grazing pressure), but none exceeded 0.7.

	Log area	Log distance	Grazing	Altitude	Year isolated
Log area	1.00				
Log distance	0.30	1.00			
Grazing	−0.56	−0.14	1.00		
Altitude	0.28	−0.22	−0.41	1.00	
Year isolated	0.28	−0.02	−0.64	0.23	1.00

The variance inflation factors (VIFs) were small (1.36–2.52), so all predictors were retained. An additive multiple linear regression model was fitted (omitting subscripts):

$$\text{bird abundance} = \beta_0 + \beta_1 \log \text{ area} + \beta_2 \log \text{ distance} + \beta_4 \text{ grazing} + \beta_5 \text{ altitude} + \beta_6 \text{ year isolated} + \varepsilon.$$

The residual plot from this model looked fine, with no evidence of increasing spread as predicted values increase. The largest Cook's D was 0.23, suggesting no observations were particularly influential.

The parameter estimates from this model were as follows.

	Coeff.	95% CI	SE	Std. coeff.	t	P
Intercept	−131.85	−309.86 to 46.19	88.64	0	−1.49	0.143
Log area	7.30	4.62 to 9.98	1.34	0.55	5.46	<0.001
Log distance	−1.30	−5.96 to 3.35	2.32	−0.05	−0.56	0.577
Grazing	−1.68	−3.53 to 0.17	0.92	−0.23	−1.82	0.075
Altitude	0.02	−0.02 to 0.07	0.02	0.09	0.94	0.353
Year isolated	0.08	−0.01 to 0.16	0.04	−0.18	1.74	0.087

There was a strong positive partial regression slope for bird abundance against log area (larger patches have more birds) and weaker evidence for a negative partial regression slope for grazing intensity (patches with more intense grazing have fewer birds) and a positive partial regression slope for year isolated (more birds in patches isolated more recently) – see Figure 8.4. There was little effect of log distance to nearest patch or altitude on bird abundance. The overall r^2 was 0.685, indicating this linear combination of predictors can explain about 69% of the variation in bird abundance. The ANOVA from the full model fit produced an F of 21.68 with df = 5, 50 and $P < 0.001$.

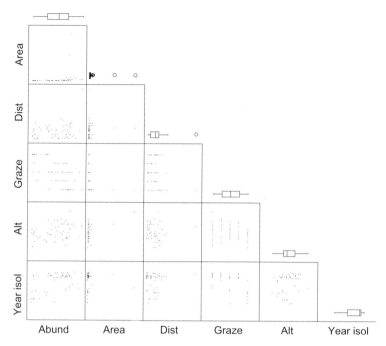

Figure 8.3 Scatterplot matrix (with boxplots) for untransformed data from Loyn (1987).

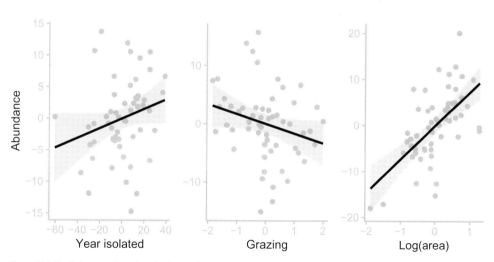

Figure 8.4 Partial regression plots for three of the predictors from a linear model relating bird abundance in forest patches to log patch area, log distance to nearest patch, grazing intensity, altitude, and years since isolation for the 56 patches surveyed by Loyn (1987). The vertical axis is residuals from OLS regression of bird abundance against all predictors except the one labeled, and the horizontal axis is residuals from OLS regression of the labeled predictor against the remaining predictors. The fitted regression slope and confidence bands are also shown.

Box 8.3 The Multiple Linear Regression Model and Its Parameters

The multiple linear regression model we usually fit to the data is:

$$y_i = \beta_0 + \beta_1 x_{i1} + \cdots + \beta_j x_{ij} + \cdots + \beta_p x_{ip} + \varepsilon_i,$$

where

y_i is the Y-value for observation i when the predictor variable $X_1 = x_{i1}, X_2 = x_{i2}, X_j = x_{ij}$, etc.;

β_0 is the population intercept, the true mean value of Y when $X_1 = 0, X_2 = 0$, etc.;

β_1 is the partial population regression slope for Y on X_1 holding $X_2, \ldots, X_p$ constant. It measures the change in Y per unit change in X_1 holding the value the other $p - 1$ X-variables constant;

β_j is the partial population regression slope for Y on X_j holding the other predictors constant;

ε_i is random or unexplained error associated with observation i. Each ε_i measures the difference between each observed y_i and the mean of y_i; the latter is the value of y_i predicted by the population regression model for a particular combination of X-values. We assume that for a particular set of X-values, these error terms are normally distributed and their mean is zero. We also assume that their variance, σ_ε^2, is the same for all X combinations. We also assume these ε_i terms are independent of, and therefore uncorrelated with, each other.

Fitting the model and obtaining parameters extends the methods for simple linear regression, although the computations are complex. We need to estimate the parameters ($\beta_0, \beta_1, \ldots, \beta_p$ and σ_ε^2) of the multiple linear regression model based on our random sample of n observations. Once we have those estimates, we can determine the sample regression line:

$$\hat{y}_i = b_0 + b_1 x_{i1} + \cdots + b_j x_{ij} + \cdots + b_p x_{ip},$$

where:

$\hat{y}_i$ is the value of y_i predicted by the fitted regression line;

b_0 is the sample estimate of β_0, the Y-intercept; and

$b_1, \ldots, b_p$ are the sample estimates of $\beta_1, \ldots, \beta_p$, the partial regression slopes.

We can estimate these parameters using OLS or ML and the estimates are the same if the data are normal. As with simple regression, we will focus on OLS estimation. The actual calculations for the OLS estimates of the model parameters involve solving a set of simultaneous normal equations, one for each parameter in the model, and are best represented with matrix algebra (Box 8.4).

The OLS estimates of $\beta_0, \beta_1, \beta_2$, etc., are the values that produce a sample regression line that minimizes the sum of the squared deviations between each observed y_i and the value predicted for each x_{ij}. Each deviation $(y_i - \hat{y}_i)$ is a residual from the fitted regression plane and represents the vertical distance between the regression plane and the Y-value for each observation. The OLS estimate of σ_ε^2 is the sample variance of these residuals and is the residual (or error) mean square (MS) from the ANOVA.

forcing a regression model through the origin with multiple predictors (especially when relationships between Y and the predictors can be positive and negative) makes little sense.

The slope parameters $\left(\beta_1, \beta_2, \ldots, \beta_j, \ldots, \beta_p\right)$ are termed partial regression slopes (coefficients) and they are commonly described as measuring the change in Y per unit change in a particular X, holding the other $p - 1$ X-variables constant. Another way of describing partial regression coefficients is that they are the regression

coefficients for Y on that component of a particular X that is independent of the other $p - 1$ X-variables (Box 8.5).

8.1.2.2 Predicted Values

Once we have estimated the model parameters, we have our fitted regression line for predicting new values of Y (Box 8.3):

$$\hat{y}_i = b_0 + b_1 x_{i1} + \cdots + b_j x_{ij} + \cdots + b_p x_{ip}.$$

Box 8.4 Matrix Algebra Approach to OLS Estimation of Multiple Linear Regression Models and Determination of Leverage Values

Consider an additive linear model with one response variable (Y) and p predictor variables ($X_1, \ldots, X_p$) and a sample of n observations. The linear model will have $p + 1$ parameters, a slope term for each X-variable, and an intercept. Let $\mathbf{Y}$ be a vector of observed Y-values with n rows, $\hat{\mathbf{Y}}$ be a vector of predicted Y-values with n rows, and $\mathbf{X}$ be an $n \times (p + 1)$ matrix of the values of the X-variables (one X-variable per column) plus a column for the intercept. The linear model can be written as:

$$\mathbf{Y} = \boldsymbol{\beta}\mathbf{X} + \boldsymbol{\varepsilon},$$

where $\boldsymbol{\beta}$ is a vector of model parameters $\left(\beta_0, \beta_1, \ldots, \beta_p\right)$ with $p + 1$ rows and $\boldsymbol{\varepsilon}$ is a vector of error terms with n rows. The OLS estimate of $\boldsymbol{\beta}$ can be found by solving the normal equations:

$$\mathbf{X}'\mathbf{X}\mathbf{b} = \mathbf{X}'\mathbf{Y}.$$

The OLS estimate of $\boldsymbol{\beta}$ then is:

$$\mathbf{b} = (\mathbf{X}'\mathbf{X})^{-1}(\mathbf{X}'\mathbf{Y}),$$

where $\mathbf{b}$ is a vector of sample partial regression coefficients $\left(b_0, b_1, \ldots, b_p\right)$ with $p + 1$ rows. Note that $(\mathbf{X}'\mathbf{X})^{-1}$ is the inverse of $(\mathbf{X}'\mathbf{X})$ and is critical to the solution of the normal equations and hence the OLS estimates of the parameters. The calculation of this inverse is very sensitive to rounding errors, especially when there are many parameters, and correlations (linear dependencies – see Rawlings et al. 1998) among the X-variables (i.e. collinearity). Correlations exaggerate the rounding errors problem and make parameter estimates unstable and their variances large (see Box 8.7).

The matrix containing the variances of, and the covariances between, the sample partial regression coefficients $\left(b_0, b_1, \ldots, b_p\right)$ is:

$$\mathbf{s}_b^2 = \mathrm{MS}_{\mathrm{Residual}}(\mathbf{X}'\mathbf{X})^{-1}.$$

Using the variances of the sample partial regression coefficients, we can calculate SEs for each partial regression coefficient.

We can also create a matrix $\mathbf{H}$ whereby:

$$\mathbf{H} = \mathbf{X}(\mathbf{X}'\mathbf{X})^{-1}\mathbf{X}'.$$

$\mathbf{H}$ is an $n \times n$ matrix, usually termed the hat matrix, whose n diagonal elements are leverage values (h_{ii}) for each observation (Kutner et al. 2005). These leverage values measure how far an observation is from the means of the X-variables. We can then relate $\mathbf{Y}$ to $\hat{\mathbf{Y}}$ by:

$$\hat{\mathbf{Y}} = \mathbf{H}\mathbf{Y}.$$

The hat matrix transforms observed $\mathbf{Y}$ into predicted $\mathbf{Y}$.

As with simple regression models, be cautious about predicting from X-values outside the range of the sample data.

8.1.2.3 *Error Terms and Their Variance*

The ε_i represent random or unexplained error associated with the ith observation – for example, the difference between the observed jump distance for cricket i and the true mean jump distance for all crickets with that combination of egg number, pronotum length, and leg mass.

We can estimate the error terms for our sample observations calculating the difference between each observed y_i and predicted $\hat{y}_i$ – that is, the residuals (e_i). As with all linear models, these residuals are important for checking the assumptions underlying our model. Our main interest is the variance of these error terms, and the OLS estimate of this variance is the MS residual term from the ANOVA for the fit of our model (see Table 8.1). This $\mathrm{MS}_{\mathrm{Residual}}$ is also used to calculate SEs for parameter estimates.

Box 8.5 ® Worked Example of Partial Regression Coefficients: Cricket Jump Distances

Each partial regression coefficient represents the slope of the linear relationship between a response variable and the component of each predictor variable independent of the remaining predictor variables in the model. We will use the example of cricket jump distances from Box 8.1 to illustrate how partial regression slopes are determined.

Recall the model fitted to these data:

$$\text{distance} = \beta_0 + \beta_1 \,(\text{egg number}) + \beta_2 \,(\text{pronotum length}) + \beta_3 \,(\text{leg mass}) + \varepsilon.$$

To determine b_1, the estimate of the partial regression coefficient for egg number:

1. Fit a model with egg number as the response to the remaining predictors (pronotum length and leg mass):
   Model 1: egg number $= \beta_0 + \beta_A \,(\text{pronotum length}) + \beta_B \,(\text{leg mass}) + \varepsilon$.
2. Calculate the residuals from this model. They represent the component of egg number independent of the other predictors:
   −22.64, 58.84, −27.06, 31.76, 25.83, −3.03, −26.44, 32.10, 5.52, 41.89, 23.06, −11.74, −1.53, 45.33, −20.09, 22.59, −31.86, −10.20, −42,90, −26.54, −21.92, 72.87, −26.60, −5.31, −21.83, 21.57, −27.48, −23.24, −30.96.
3. Now fit a model relating jump distance to these residuals:
   Model 2: distance $= \beta_0 + \beta_1 \,(\text{residuals from Model 1}) + \varepsilon$.
4. The estimated regression coefficient from this model (−0.11) is the partial regression coefficient for egg number, independent of the other predictors, from the original multiple regression model with all three predictors (see Box 8.1).

8.1.2.4 *Standardized Partial Regression Slopes*

As with simple regression models, we can calculate standardized regression slopes by regressing the standardized response variable against the standardized predictors, or alternatively, calculate for predictor j:

$$b_j^* = b_j \frac{s_{X_j}}{s_Y}.$$

These slopes indicate the change in the response variable for a unit change in the predictor when both are measured in SD units. They are the same as the correlation coefficients relating the response variable to each predictor, as long as the predictors are not correlated. The regression model based on standardized variables doesn't include an intercept, because its OLS (and ML) estimate will always be zero. Gelman and Hill (2006) recommended standardizing by dividing by two SDs rather than one, as this makes the standardized coefficients for continuous predictors more compatible with those for binary predictors, such as the dummy variables used to code categorical predictors.

Standardized coefficients from a single model are, in principle, comparable regardless of the scales on which the predictors are measured. However, while each regression coefficient is conditional on the other predictors in the model, each predictor is standardized using its own SD, which ignores the other predictors in the model and any correlation among the predictors (collinearity).

Therefore, coefficients standardized in this way may not be useful measures of relative importance of each predictor (see Section 9.1).

Hypothesis tests on individual standardized coefficients will be identical to those on unstandardized coefficients. Regression models using standardized (or simply centered) predictors are important for detecting and treating multicollinearity and interpreting interactions between predictors (Sections 8.1.9–8.1.10).

8.1.3 Inference for Parameters
8.1.3.1 *Standard Errors and Confidence Intervals*

Standard errors for regression model parameters (s_{b_j}) are the estimated SDs of their sampling distributions (Box 8.4). Confidence intervals are calculated as usual using a t distribution with $n - p - 1$ df. Note that the CIs for model parameters (slopes and intercept) depend on the number of observations and predictors. Therefore, for a given SE, our confidence in our estimates of model parameters is reduced when we include more predictors. Standard errors and prediction intervals can also be determined for new Y-values (see Kutner et al. 2005).

8.1.3.2 *Statistical Hypotheses*

The null hypotheses commonly evaluated with multiple regression models are that each partial regression coefficient (slope) equals zero – that is, there is no linear relationship between Y and X_j, holding the other predictors constant. For example, the most biologically interesting null hypothesis with the cricket jump distance example from Ercit et al. (2014) is that there is no linear relationship between jump distance and number of eggs carried, holding pronotum length and leg mass constant – that is, between jump distance and that component of egg number independent of pronotum length and leg mass. These p null hypotheses are evaluated with partial t tests, each with $n - p - 1$ df:

$$t = \frac{b_j}{s_{b_j}},$$

where s_{b_j} is the SE of b_j.

We can also evaluate the null hypothesis that the intercept $\beta_0 = 0$ using a t statistic calculated by dividing the estimated intercept by its SE, but this hypothesis is rarely of any biological interest.

As for simple regressions, we need not be constrained to testing whether coefficients are zero; it is straightforward to modify the previous equation to include a nonzero hypothesis.

These hypotheses can also be evaluated with F tests by comparing the fit of the full regression model to the fit of a reduced model that omits the parameter of interest. For each regression coefficient, $F = t^2$ and the P-value will be the same.

8.1.4 Model Comparison and Analysis of Variance

Like the simple linear regression models described in Chapter 6, we can fit multiple regression models with OLS, using ANOVA to partition the total variation in Y (SS_{Total}) into two additive components (Table 8.1). The first is the variation in Y explained by its linear relationship with $X_1, \ldots, X_p$, termed $SS_{Regression}$. The second is the variation in Y not explained by this linear relationship, termed $SS_{Residual}$ and measured as the difference between each observed y and its value predicted by the regression

model ($\hat{y}_i$). These SS in Table 8.1 are identical to those in Table 6.1 for simple regression models. In fact, the partitioning of the SS_{Total} for the simple linear regression model is just a special case of the multiple regression model where $p = 1$, although the calculations for multiple regression models are more complex.

The proportion of the total variation in Y explained by the regression model is the multiple r^2:

$$r^2 = \frac{SS_{Regression}}{SS_{Total}} = 1 - \frac{SS_{Residual}}{SS_{Total}}$$
$$= 1 - \frac{Full\ SS_{Residual}}{Reduced\ SS_{Residual}}$$

Here, the reduced model is one with just an intercept and no predictors (i.e. all $\beta_i = 0$). Interpretation of r^2 in multiple linear regression must be done carefully. As in simple regression, r^2 is not directly comparable between models based on different transformations. It is also inappropriate for comparing models with different numbers or combinations of predictors. As more predictors are added to a model, r^2 cannot decrease, so models with more predictors will always appear to fit the data better. Comparing the fit of models with different numbers of predictors needs alternative measures, which we describe in the next chapter.

The $SS_{Regression}$ and $SS_{Residual}$ can be converted into variances (MS) by dividing by the appropriate degrees of freedom. The expected values of these two MS are again just an extension of those we described for simple regression (Table 8.2). The expected value for $MS_{Residual}$ is σ_ε^2, the variance of the error terms (ε_i), which are assumed to be constant across each combination of $x_{i1}, \ldots, x_{ij}$, etc. The expected value for $MS_{Regression}$ is more complex (Kutner et al. 2005), but importantly it includes the square of each regression slope plus σ_ε^2. This leads naturally to the F statistic $MS_{Regression}/MS_{Residual}$ for testing the null hypothesis that all partial regression slopes equal zero. For example, the H_0 for the model from Ercit et al. (2014) is that there is no relationship for female tree crickets between jump distance and egg number, pronotum length, or leg mass, independent of the other predictors in the model.

Table 8.1 ANOVA table for a multiple linear regression model with an intercept, p predictors, and n observations

Source of variation	SS	df	MS
Regression	$\sum_{i=1}^{n} (\hat{y}_i - \bar{y})^2$	p	$\dfrac{\sum_{i=1}^{n} (\hat{y}_i - \bar{y})^2}{p}$
Residual	$\sum_{i=1}^{n} (y_i - \hat{y}_i)^2$	$n - p - 1$	$\dfrac{\sum_{i=1}^{n} (y_i - \hat{y}_i)^2}{n - p - 1}$
Total	$\sum_{i=1}^{n} (y_i - \bar{y})^2$	$n - 1$	

Table 8.2 Expected values of MS from ANOVA for a multiple linear regression model with two predictors

Mean square	Expected value
$MS_{Regression}$	$\sigma_\varepsilon^2 + \dfrac{\beta_1^2 \sum\limits_{i=1}^{n}(x_{i1} - \bar{x}_1)^2 + \beta_2^2 \sum\limits_{i=1}^{n}(x_{i2} - \bar{x}_2)^2 + 2\beta_1\beta_2 \sum\limits_{i=1}^{n}(x_{i1} - \bar{x}_1)(x_{i2} - \bar{x}_2)}{2}$
$MS_{Residual}$	σ_ε^2

Box 8.6 ℝ Worked Example of Model Comparison Approach to Multiple-Predictor Linear Regression: Cricket Jump Distances

We will illustrate model comparisons in multiple regression models using the example of cricket jump distance.

The H_0 that all partial regression slopes $= 0$ is evaluated by comparing the fit (or lack of fit) of these two models:

jump distance $= \beta_0 + \beta_1$ (egg number) $+ \beta_2$ (pronotum length) $+ \beta_3$ (leg mass) $+ \varepsilon_i$

ANOVA : $SS_{Residual} = 323.72$, $df_{Residual} = 25$

jump distance $= \beta_0 + \varepsilon_i$

ANOVA : $SS_{Residual} = 736.42$, $df_{Residual} = 28$

Change in $SS_{Residual}$, which we'll call SS_{Diff}, is 412.71, with 3 df; so $MS_{Diff} = 137.57$. $MS_{Residual}$ for the first (full) model is 323.72/25 = 12.95. We can use these numbers to construct $F = MS_{Diff}/MS_{Res} = 137.57/12.95 = 10.62$ $(P < 0.001)$. This F is the same as that produced from the ANOVA on the full model.

To evaluate a H_0 for any individual partial regression slope $\left(\beta_j = 0\right)$, we compare the fit of the full model to one that omits the predictor of interest. For example, for egg number, we would compare

jump distance $= \beta_0 + \beta_1$ (egg number) $+ \beta_2$ (pronotum length) $+ \beta_3$ (leg mass) $+ \varepsilon$

ANOVA : $SS_{Residual} = 323.72$, $df_{Residual} = 25$

jump distance $= \beta_0 + \beta_2$ (pronotum length) $+ \beta_3$ (leg mass) $+ \varepsilon$

ANOVA : $SS_{Residual} = 666.21$, $df_{Residual} = 26$

Difference in $SS_{Residual} = 342.59$ with 1 df; $MS_{Residual} = 342.59$. $F = 342.59/12.95 = 26.45$ $(P < 0.001)$. This F is the square of the t statistic (5.14) for this predictor in Box 8.1.

More generally, we can use the model comparison approach to partition the variance in Y attributable to each predictor or combination of predictors (Box 8.6). The null hypothesis described above can be evaluated by comparing the fit of the full model including all predictors to a model with no predictors (all regression coefficients $= 0$) and just an intercept. We calculate the difference in either explained (SS_{Extra}) or unexplained (SS_{Drop}) variation, convert it to a MS (by dividing by the difference in df) and calculate an F statistic using the $MS_{Residual}$ from the fit of the full model. The difference in residual (unexplained) variation between these two models provides the $SS_{Regression}$ we see in the ANOVA from fitting the full model.

We can also test the null hypotheses that each $\beta_j = 0$ in the same way (Box 8.6) by comparing the fit of the full model to a reduced model without the term of interest. We can calculate either the change in SS explained by including β_j in the model:

$$SS_{Diff} = \text{Full } SS_{Regression} - \text{Reduced } SS_{Regression},$$

or the change in residual SS when β_j is omitted from the model:

$$SS_{Diff} = \text{Reduced } SS_{Residual} - \text{Full } SS_{Residual}.$$

We can then use an F test, now termed a partial F test, to test whether a single partial regression slope $= 0$:

$$F_{1,n-p} = \frac{MS_{Diff}}{\text{Full } MS_{Residual}}.$$

These partial F tests are equivalent to the partial t tests described in Section 8.1.3. One advantage of the F tests is that they can be used for any subset of regression coefficients, not just a single coefficient.

The methods described in this section are for evaluating each predictor, commonly by evaluating the null hypothesis that it equals zero. However, these tests do not provide much information about the relative importance of each predictor and are also not the preferred approach for building or selecting the "best" predictive models. We describe methods suitable for these tasks in Chapter 9.

8.1.5 Assumptions of Multiple Linear Regression Models

As with simple linear models, interval estimation and hypothesis tests of the parameters of the multiple linear regression model based on OLS rely on several assumptions about the model error terms at each combination of $x_{i1}, x_{i2}, \ldots, x_{ip}$. We assume that the error terms, and therefore usually the Y-values:

- are normally distributed;
- have constant variance; and
- are independent of each other.

We check these assumptions as for simple linear models. Boxplots and probability plots of the residuals can be used to check for normality, plots of residuals against $\hat{y}_i$ can detect heterogeneity of variance (Section 8.1.7), and plots of residuals against each X_j can detect autocorrelation, especially if X_j is a time sequence.

Linear regression models also assume that each X is a fixed variable, which is unlikely in biological research. Both of our examples illustrate this point: Ercit et al. (2014) probably did not choose specific combinations of egg number, pronotum length, and leg mass for their sample crickets, but they probably chose crickets with a range of egg numbers, while reducing variation in the other two variables (used to adjust for overall body size and strength). Loyn (1987) did not choose forest patches with specifically chosen values of area, distance to the nearest patch, stock grazing intensity, year patch was isolated by clearing, or altitude, although he almost certainly chose patches to span the range of values for each variable. As we argued for simple linear regression models, when we clearly distinguish a response variable from predictor variables and we are developing predictive models, the usual OLS approach to fitting regression models is appropriate. Techniques for estimating slopes when all variables are considered truly random are introduced in Section 8.1.12.3.

An added assumption that affects multiple linear regression models (and some of the related techniques

we will consider in Chapter 9) is that the predictors cannot be strongly correlated with each other. The situation of correlated predictors is called (multi)collinearity and it is so important that we discuss its impacts and solutions separately in Section 8.1.9.

Finally, the number of observations must exceed the number of predictors, or the matrix calculations (Box 8.4) will fail. Green (1991) proposed specific minimum ratios of observations to predictors, $n = p + 104$ for testing individual predictors, and these guidelines have become recommendations in some texts (e.g. Tabachnick & Fidell 2019). These numbers may be unrealistic for much biological research. Babyak (2004) and Kutner et al. (2005) were more lenient, the former recommending a minimum $n:p$ ratio of 10–15 in psychosomatic research and the latter (nondiscipline-specific) 6–10. We can only suggest that researchers try to maximize the numbers of observations and if trade-offs in terms of time and cost are possible, reducing the numbers of predictors to allow more observations is nearly always preferable to reducing the number of observations.

8.1.6 Model Diagnostics

Diagnostic checks of the assumptions underlying statistical inference, checking the model fit, and finding potential outliers and influential observations are particularly important when there are multiple predictors. We are usually dealing with large datasets, and scanning the raw data or simple bivariate scatterplots – which might have worked for simple regressions – will rarely be sufficient for a multiple regression. Fortunately, other diagnostic checks for simple regression models apply equally well for multiple regressions.

8.1.6.1 *Leverage*

Leverage measures how extreme each observation is from the means of the p predictors (the multivariate centroid; see Chapter 14), so in contrast to simple regressions, leverage for multiple regression models considers all the predictors in the model. Leverage values $> 2p/n$ should be cause for concern, although such values would usually be detected as influential by Cook's D_i.

8.1.6.2 *Residuals*

Residuals in multiple regression models are interpreted in the same way as for simple regression. Regardless of how many predictors we have, there is only a single residual value for each observation. These residuals can be standardized or studentized (see Table 6.4) and large residuals indicate outliers that could be influential.

8.1.6.3 *Influence*

Measures of how influential each observation is on the fitted model include Cook's D_i and DFITS$_i$, and these are as relevant for multiple regression as they were for simple regression. Observations with $D_i > 1$ are usually considered influential and should be checked carefully.

8.1.7 Diagnostic Graphics

Graphical techniques are often the most informative checks of assumptions and for the presence of outliers and influential values in linear models.

8.1.7.1 *Scatterplots*

Bivariate scatterplots between the predictors are important for detecting multicollinearity, and scatterplots between Y and each X_j, particularly in conjunction with smoothing functions, will indicate any obvious nonlinearities. Scatterplot matrices (SPLOMs; see Chapter 5) are the easiest way of displaying these bivariate relationships. However, scatterplots between Y and X_1, Y and X_2, etc., ignore the other predictors in the model and therefore do not represent the relationship we are modeling – that is, the relationship between Y and X_j holding all other Xs constant.

A scatterplot that does show this relationship for each predictor variable is the added variable, or partial regression, plot, which is a scatterplot between two sets of residuals. The residuals for the vertical axis of the plot come from the OLS regression of Y against all p predictors except X_j. The residuals for the horizontal axis of the plot come from the OLS regression of X_j against all p predictors except X_j. This scatterplot shows the relationship between Y and X_j, holding the other X-variables constant, and will also show outliers that might influence the partial regression slope for X_j. If we fit an OLS regression of this Y residual against the X_j residual, its slope is the partial regression slope for X_j from the full regression model of Y on all p predictors.

These partial regression plots for our two worked examples are in Figures 8.2 and 8.4. For example, the partial regression plot for log patch area (Figure 8.4) has the residuals from a model relating bird abundance to all predictors except log patch area on the vertical axis and the residuals from a model relating log patch area to the other predictors on the horizontal axis. Note the strong positive relationship for log patch area and the weak negative relationships for grazing and year isolated. There was little pattern in the plots for the other three predictors.

8.1.7.2 *Residual Plots*

There are several options for plotting residuals. A plot of residuals against $\hat{y}_i$, as recommended for simple regression models, is the most common and can detect heterogeneity of variance (wedge-shaped pattern) and outliers. Plots of residuals against each X_j can detect outliers specific to that X_j, nonlinearity between Y and that X_j, and can also detect autocorrelation if X_j is a time sequence. Finally, residuals can even be plotted against predictors, or interactions between predictors, that were not included in the model, to assess whether these predictors or their interactions might be important, even if they were initially excluded from the model (Kutner et al. 2005).

8.1.8 Transformations

Our general comments on transformations for simple linear models are just as relevant for multiple regression models. Transformations (commonly to logs) of the response variable can remedy nonnormality and heterogeneity of variance of error terms. Transformations of one or more of the predictors might be necessary to deal with nonlinearity and influential observations due to high leverage. For example, two of the predictors in Loyn's bird study were log-transformed to deal with observations with high leverage (Box 8.2). Transformations can also reduce the influence of interactions between predictors on the response variable – that is, make an additive model a more appropriate fit than a multiplicative one.

As always, the transformed variable must have a sensible biological interpretation and transforming variables to a logarithmic scale is most often used in biological research because lognormal distributions are common for measurement variables that can't be negative but have no theoretical upper limit. They are also often used when there are biological reasons to expect a nonlinear relationship. If the response variable has a distribution from the exponential family, GLMs (Chapters 4 and 13) are often more appropriate than transforming and fitting an OLS linear model.

8.1.9 Collinearity

Correlations between predictors can have important detrimental effects on the estimated regression coefficients. It's often hard to avoid collinearity with real biological data, where predictors are often correlated with each other to some extent. For example, we might expect that pronotum length of crickets would be correlated with their leg mass (Box 8.1) and we might expect more intense grazing the longer the forest patch has been isolated and lighter grazing for bigger patches since domestic stock cannot easily access larger forest fragments (Box 8.2). This contrasts with the multiple categorical predictors in Chapter 7, which were from experimental or sampling designs where factors were crossed *and* orthogonal, so collinearity among the predictors (factors) was not an issue.

Ordinary least squares estimation of parameters of multiple linear regression models requires calculations that involve matrix inversion (Box 8.4). Collinearity causes computational problems because it makes the determinant of the matrix of X-variables close to zero. Values in the inverted matrix are then very sensitive to small differences in the numbers in the original data matrix (Tabachnick & Fidell 2019) – that is, the inverted matrix is unstable. This has three effects on our model and statistical inference (Dormann et al. 2013; Freckleton 2011; Kutner et al. 2005; Legendre & Legendre 2012):

• Estimates of the partial regression slopes are unstable and small changes in the data or adding or deleting one of the predictors can change the estimated regression coefficients considerably, even changing their sign.

• SEs of the estimated slopes, and therefore CIs for the model parameters, are inflated when some predictors are correlated (e.g. Box 8.1). There may be evidence from the full model ANOVA that at least one partial regression slope differs from zero, but it may not be apparent when we evaluate the individual partial regression slopes. This reflects lack of power for individual tests on partial regression slopes because of their inflated SEs.

• It can be difficult to reliably select a suitable model with fewer predictors – that is, model selection can be compromised (see Section 9.2). Model validation can also be unreliable if the pattern of collinearity differs between the training and validation subsets of the data (see Section 9.2.5).

If we are not extrapolating beyond the range of our predictor variables and we are making predictions from data with a similar pattern of collinearity as the data to which we fitted our model, collinearity doesn't necessarily prevent us from generating a regression model that fits the data well and has good predictive power (Kutner et al. 2005). It does, however, mean that we may not be confident in our parameter estimates. A different sample from the same population, even using the same values of the predictors, might produce very different parameter estimates.

Box 8.7 ℝ Illustration of Multicollinearity among Predictors

Here we illustrate the effects of collinearity in a multiple regression model with one response variable (Y) and two predictor variables (X_1, X_2). Two artificial datasets were generated for the three variables from normal distributions. In the first dataset, X_1 and X_2 are relatively uncorrelated ($r = 0.26$). A multiple linear regression model, including an intercept, was fitted to these data.

	Coefficient	SE	t	P
Intercept	3.07	0.94	3.27	0.004
X_1	0.49	0.14	3.57	0.002
X_2	−0.03	0.17	−0.18	0.857

The VIF for X_1 and X_2 was 1.08, so there was no sign of collinearity between the two predictors. There was strong evidence for a positive partial regression slope for X_1, but little evidence against a zero slope for X_2.

For the second dataset, the values of X_2 were rearranged between observations (but the values, their mean, and SD were the same) so they are highly correlated with X_1 ($r = 0.95$), which along with Y is unchanged. Again, a multiple linear regression model, including an intercept, was fitted.

	Coefficient	SE	t	P
Intercept	3.22	0.83	3.88	0.001
X_1	0.71	0.41	1.71	0.105
X_2	−0.29	0.51	−0.57	0.578

The VIF for X_1 and X_2 was 9.76 (close to the recommended cut-off of 10), confirming the collinearity between the two predictors. The SE for the partial regression slope of Y against X_1 is much bigger than for the first dataset and the evidence against the H_0 this partial regression slope equals zero is much weaker (and no longer "statistically significant"), despite the values of Y and X_1 being identical to the first dataset.

8.1.9.1 *Detecting Collinearity*

Collinearity can be detected in several ways (e.g., Chatterjee & Hadi 2012; Kutner et al. 2005; Legendre & Legendre 2012) and we illustrate some of them in Boxes 8.1 and 8.2.

First, examine scatterplots between the predictors and look for obvious, strong, relationships. A scatterplot matrix is very helpful and, if the response variable is included, can also indicate nonlinear relationships between the response variable and the predictors (e.g. Figures 8.1 and 8.3).

Second, look at the correlations between the predictors. While there are rarely consistent cut-offs for this examination, a recommendation that correlation coefficients should be <0.7 has gained some traction (see summary in Dormann et al. 2013). Collinearity can be more complex than just pairwise correlations between predictors (Kutner et al. 2005) – that is, a predictor can be strongly correlated with two other predictors together but not with either one individually, so bivariate correlations are usually supplemented with other diagnostic tools.

Third, the VIF measures the change in the variance of each partial regression coefficient from a situation where the predictors are not correlated. The VIF for X_j is simply $1 - r^2$ from the OLS regression of X_j against the remaining $p - 1$ predictors. A similar measure is tolerance, which is the inverse of the VIF. Again, simple cut-offs are tricky to recommend. An approximate guide is to worry about VIF values ≥ 10 (Hocking 2013) or a mean VIF (across all predictors) much larger than 1 (Kutner et al. 2005), although Vatcheva et al. (2016) showed that even individual VIFs around 5 could be a problem.

Fourth, we can extract the principal components from the correlation matrix among the predictors (see Chapter 15). Principal components with eigenvalues (i.e. explained variances) near zero indicate collinearity among the original predictors. Three statistics are commonly used to assess collinearity in this context. First, the condition index is the square root of the largest eigenvalue divided by each eigenvalue $\left(\sqrt{(\lambda_{max}/\lambda)}\right)$. There will be a condition index for each principal component and values >30 indicate collinearities that require attention (Chatterjee & Hadi 2012). The second is the condition number, which is simply the largest condition index $\left(\sqrt{(\lambda_{max}/\lambda_{min})}\right)$. Third, Hocking (2013) proposed just using λ_{min} and suggested that values <0.5 indicated collinearity problems.

It is worth noting that examining eigenvalues from the correlation matrix of the predictor variables implicitly standardizes the predictors to zero mean and unit variance, so they are on the same scale. In fact, most collinearity diagnostics give different results for unstandardized and standardized predictors and two of the solutions to collinearity described below are based on standardized predictor variables.

Our recommendation is to use a combination of pairwise correlations and VIFs to diagnose collinearity concerns. Biological insight is also important, and it makes sense not to include predictors that are essentially measuring the same biological entity or underlying process.

8.1.9.2 *Dealing with Collinearity*

Numerous solutions to collinearity have been proposed. All result in estimated partial regression slopes that are likely to be more precise (smaller SEs) but are no longer unbiased. The first approach is the simplest: omit predictors if they are highly correlated with other predictors that remain in the model. Multiple predictors that are really measuring similar biological entities (e.g. a set of morphological measurements that are highly correlated) represent redundant information and little can be gained by including them all in a model. Unfortunately, omitting variables may bias estimates of parameters for those variables that are correlated with the omitted variable(s) but remain in the model. Estimated partial regression slopes can change considerably when some collinear predictors are omitted or added. Nonetheless, retaining only one of a number of highly correlated predictors that contain biologically and statistically redundant information is a sensible first step to dealing with collinearity.

The second approach is based on a PCA of the predictor variables and is termed principal components regression (PCR). The (up to p) principal components are extracted from the correlation matrix of the predictor variables. Each one is a linear combination of the predictors, and they are uncorrelated. Y is regressed against these principal components, rather than the individual predictors. Components that contribute little to the total variance among the X-variables and are unrelated to Y are usually deleted and the regression model of Y against the remaining components refitted, so the number of variables is $\ll p$. The regression coefficients for Y on the principal components are not always informative, however, because the components are combinations of the original predictors. Therefore, we back-calculate the partial regression slopes on the original standardized variables from the partial regression slopes on the reduced number of principal components. The back-calculated regression slopes are standardized because the PCA is usually based on a correlation matrix of X-variables, so we don't have to worry about an intercept term. Because PCR requires an understanding of PCA, we will describe it in more detail in Section 15.1.8.

Deciding which components to omit is critical for PCR. Simply deleting those with small eigenvalues (little relative contribution to the total variation in the X-variables) can be very misleading (Chatterjee & Hadi 2012; Jackson 2003). The strength of the relationship of each component with Y must also be considered.

Another method for dealing with collinearity is partial least squares (PLS), which also combines multiple regression with PCA. In contrast to PCR, where the components are derived in order of explained variance among the predictors, PLS derives components that explain the covariance between the predictor *and* the response. Therefore, a PLS regression has the advantage over PCA regression by considering the relationship with the response variable as part of deriving the components (sometimes termed latent vectors). Partial least squares is also used when the number of predictors is large compared to, or even exceeds, the number of observations, and can also handle a multivariate response structure. An excellent and detailed description of PLS can be found in Abdi (2003).

The third approach is ridge regression, another biased regression estimation technique that is somewhat controversial. A small biasing constant is added to the normal equations that are solved to estimate the standardized regression coefficients (Chatterjee & Hadi 2012; Kutner et al. 2005). This constant biases the estimated regression coefficients but reduces their variability and hence their SEs. The choice of the constant is critical. The smaller its value, the less bias in the estimated regression slopes (when the constant is zero, we have an OLS regression). The larger its value, the less collinearity. A range of values (e.g. starting with 0.001) is usually tried and a diagnostic graphic, the ridge trace, is used to determine the constant that is the best compromise between variation in the estimated regression slopes and VIFs. Kutner et al. (2005) provided a clear worked example.

Careful thought about the choice of predictors can reduce collinearity problems before any analysis. Do not include clearly redundant variables that are basically measuring similar entities or processes. If the remaining predictors are correlated to an extent that might affect our confidence in the estimates of the partial regression slopes, then we prefer PCR over ridge regression, because PCA is also a useful check for collinearity, and is often done anyway.

8.1.10 Interactions in Multiple Regression

So far, we have described additive models. They don't consider situations where the relationship between the response and each predictor might depend on the values of the other predictor(s). In many biological situations, however, we might anticipate interactions between the predictors so that their effects on Y are multiplicative, just as we described for categorical predictors (factors) in Chapter 7. Multiple linear regression models including an interaction are sometimes referred to as moderated models, where the statistical effect of one predictor on the response is moderated by the second predictor.

For example, Paruelo and Lauenroth (1996) analyzed the geographic distribution and the effects of climate variables on the relative abundance of C3 grasses (compared to other plant functional types). They used data from central North America, and we focused on latitude and longitude as the two predictors. The analyses are presented in Box 8.8.

The additive multiple linear regression model is:

$$\text{C3 grasses} = \beta_0 + \beta_1(\text{latitude}) + \beta_2(\text{longitude}) + \varepsilon.$$

Remember that the interpretation of the partial regression coefficients is the change in the abundance of C3 grasses for a one-unit change in the predictor (e.g. latitude), holding the other predictor (e.g. longitude) constant. This assumes that the partial regression slope of C3 on latitude is independent of longitude and vice-versa.

But what if we allow the relationship between C3 plants and latitude to vary for different longitudes? Then we need to include an interaction between our predictors, and our model becomes multiplicative:

$$\text{C3 grasses} = \beta_0 + \beta_1(\text{latitude}) + \beta_2(\text{longitude})$$
$$+ \beta_3(\text{latitude} \times \text{longitude}) + \varepsilon.$$

The new parameter, β_3, represents the interactive effect. It measures the dependence of the partial regression slope of C3 against latitude on the value of longitude and the dependence of the partial slope against longitude on latitude.

The interaction term is assessed by comparing the fit of the full model with a reduced model that omits the interaction term, using a partial t or equivalent F statistic. The fit of this model shows an interaction between latitude and longitude (Box 8.8), where the relationship between C3 plant abundance and latitude varies with longitude. When the predictors interact, it is difficult to interpret their separate partial regression slopes.

There are two main challenges with including interaction terms. The first is that lower-order terms will usually be highly correlated with their interactions, e.g. X_1 and X_2 will be correlated with their interaction X_1X_2. This results in all the problems associated with collinearity. A commonly recommended solution to this

Box 8.8 ⓡ Worked Example of Multiple-Predictor Linear Regression with Interactions: Geography, Climate, and Abundance of C3 Plants

Paruelo and Lauenroth (1996) analyzed the geographic distribution and the effects of climate variables on the relative abundance of several plant functional types, particularly C3 grasses. The grasses were chosen because they utilize atmospheric C differently in photosynthesis and are expected to respond differently to elevated CO_2. There were 73 sites across North America. There were six potential predictors, but due to collinearity Paruelo and Lauenroth (1996) separated the predictors into two groups for analysis. One group included latitude and longitude, and that is the analysis we will focus on: the relationship between the relative abundance of C3 plants and latitude and longitude, or the geographic pattern in C3 relative abundance across North America. We fitted a multiplicative model including an interaction term that measured how the relationship between C3 plants and latitude could vary with longitude and vice-versa:

$$(C3) = \beta_0 + \beta_1 (\text{latitude}) + \beta_2 (\text{longitude}) + \beta_3 (\text{latitude} \times \text{longitude}) + \varepsilon.$$

The results from fitting this model were as follows.

	Coeff.	95% CI	SE	Std. coeff.	t	P
Intercept	6.75	0.89 to 12.62	2.94	0	2.30	0.025
Latitude	−0.16	−0.31 to −0.01	0.07	−3.29	−2.19	0.032
Longitude	−0.08	−0.13 to −0.02	0.03	−1.86	−2.66	0.001
Interaction	0.00	0.00 to 0.00	<0.005	4.52	2.64	0.010

The residual plot showed a larger spread for intermediate predicted values but no obvious outliers, and the largest Cook's D was 0.128, so we can trust the model fit. There was evidence of an interaction between latitude and longitude; the effects of latitude on relative abundance of C3 plants varies with longitude. We looked at simple slopes for the relationship between C3 plant abundance and latitude at three values for longitude: mean longitude, mean − 1 SD, and mean + 1 SD.

	Estimate Coeff.	SE	t	P
Mean longitude − 1 SD (100°)	0.03	0.01	5.16	<0.001
Mean longitude (106°)	0.04	0.01	8.20	<0.001
Mean longitude + 1 SD (113°)	0.05	0.01	6.48	<0.001

The steepness of the partial regression slope for C3 plant abundance against latitude increases with longitude (i.e. as one moves east to west across North America). This interaction is also evident in the three-dimensional plot with a spline smoothing surface showing the predicted values (Figure 8.5), although a bit more difficult to interpret.

We wouldn't usually examine the individual ("main") effects of latitude and longitude because (1) there is an interaction, and (2) there is very strong collinearity between the predictors and their interactions (VIFs range between 66.8 and 400.9).

If we were interested in the effects of latitude and longitude, we could refit the full model based on centered predictors.

	Coeff.	95% CI	SE	Std. coeff.	t	P
Intercept	0.265	0.22 to 0.31	0.00	0	11.91	<0.001
Latitude	0.038	0.03 to 0.05	<0.001	0.77	8.20	<0.001
Longitude	0.000	−0.01 to 0.01	<0.001	0.00	0.01	0.994
Interaction	0.002	0.00 to 0.00	<0.001	0.25	2.64	0.010

Box 8.8 (cont.)

Note that the estimated regression coefficient and its SE for the interaction term are identical to those from the uncentered model, as are the t statistic and P-value for the test of the H_0 that $\beta_3 = 0$. The VIFs are between 1.021 and 1.220, suggesting that collinearity is no longer a problem (but see Section 8.1.10). While the main effects of latitude and longitude might be difficult to interpret because of their interaction, the standardized coefficients and P-values suggest a much stronger pattern in C3 plant abundance with latitude than longitude.

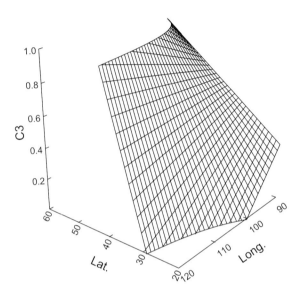

Figure 8.5 Smoothing spline (see Section 13.9) response surface relating relative abundance of C3 plants to latitude and longitude for 73 sites in North America (Paruelo & Lauenroth 1996). The surface shows predicted values from an OLS fit to a linear model incorporating latitude, longitude, and their interaction.

problem is to center the predictors, so the interaction is then the product of the centered values (Aiken & West 1991; Kutner et al. 2005), although there are differing views (Dalal & Zickar 2012; Echambadi & Hess 2007). If X_1 and X_2 are centered, neither will be strongly correlated with their interaction. Predictors can also be standardized (subtract the mean and divide by the SD), which has an identical effect in reducing collinearity.

Whether centering predictors is necessary with interactions depends on the analysis strategy. Consider a model with two predictors and one two-way interaction. Centering doesn't affect the interaction parameter; its estimate and SE and the associated t statistic and P-value are the same as with uncentered predictors. If this interaction is important, then usually the focus is just on interpreting it (see below) so collinearity between the individual predictors and their interaction is much less relevant.

If the two-way interaction is determined to be unimportant, there are two approaches. First, we might use a

model-building/selection strategy where unimportant interaction terms are omitted from the model. Centering doesn't make any difference here as centering the predictors does not change the estimates of the partial regression slopes for individual predictors, nor their hypothesis tests, when there is no interaction term in the model. Alternatively, analogous to the approach often used by biologists with multiple categorical predictors (Chapter 7), we might keep the interaction in the model but still try to interpret the partial regression slopes associated with the individual predictors. Under these circumstances, collinearity may make interpretation of the partial regression slopes difficult and centering predictors is recommended (Aguinis & Gottfredson 2010; Aiken & West 1991). Centering may also make the partial regression slopes for the individual predictors more interpretable when there is an interaction term in the model. Without centering, the partial regression slopes represent the relationship between the response and each predictor when the other predictor = 0. After centering, the partial regression slopes represent the relationship between the response and each predictor when the other predictor is at its mean value.

Models with more than two predictors can, of course, have three-way or higher-order interactions. Higher-order interactions are examined first, and centering is only needed if a model including all terms is retained and lower-order effects are evaluated.

The second challenge is that with even moderate numbers of predictors, including all two-way and higher-order interaction terms will result in very complex models with many parameters to be estimated. Unless the sample size is very large, we will exhaust the available df and not be able to fit the model. There are two ways we might decide which interactions to include, especially if our sample size does not allow us to include them all. First, we can use our biological knowledge to predict likely interactions and only incorporate this subset. Second, we can plot the residuals from an additive model against the possible interaction terms (new variables formed by simply multiplying the predictors) to see if any of these interactions are related to variation in the response variable.

8.1.10.1 *Probing Interactions*

A useful approach to probing interaction terms in multiple regression models is to examine simple slopes, the slopes of the relationships between the response and one predictor at specified values of the second predictor, analogous to simple main effects for categorical predictors we described in Section 7.1.11. In the model with two predictors and their interaction, the change in the (mean) response for a one-unit change in X_1, holding X_2 constant, is:

$$\beta_1 + \beta_3 X_2,$$

and the change in the (mean) response for a one-unit change in X_2, holding X_1 constant, is

$$\beta_2 + \beta_3 X_1.$$

These represent changes in the response for a change in one predictor conditional on specific values of the second predictor.

Let's focus on the relationship between the response and X_1. The OLS estimate of the first of these equations is $(b_1 + b_3 x_{i2})$, the simple slope of the regression of Y on X_1 for any particular value of X_2 (indicated as x_{i2}). We can then choose values of X_2 and calculate the estimated simple slope and use it for inference about the population simple slope for the specific values of X_2 chosen. Cohen et al. (2003) suggested using three different values of

X_2: $\bar{x}_2, \bar{x}_2 + s, \bar{x}_2 - s$, where s is the sample standard deviation of X_2, although there may be more appropriate values to choose based on biological knowledge.

Statistical inference for a simple regression slope is based on the SE:

$$\sqrt{s_{11}^2 + 2x_2 s_{13}^2 + x_2^2 s_{33}^2},$$

where s_{11}^2 and s_{33}^2 are the variances of b_1 and b_3 respectively, s_{13}^2 is the covariance between b_1 and b_3, and x_2 is the value of X_2 chosen. The variance and covariances are obtained from a covariance matrix of the regression coefficients. The usual t statistic can then be calculated as the estimated simple slope divided by its SE. Some statistical software will have routines for simple slopes in multiple regression models, but they can be calculated using any linear models package, as outlined in Box 8.9.

8.1.11 Regression Models with Polynomial Terms

Curvilinear models generally fall into the class of non-linear regression modeling because they are best fitted by models that are nonlinear in the parameters (e.g. power functions). One type of curvilinear model, the polynomial regression, can be fitted by OLS (i.e. it is still a linear model) and is widely used in biology. Polynomial models have some predictors also included as quadratic, cubic or higher terms. We introduced polynomials in the context of fitting trends through group means in Box 6.4; our focus here will be on polynomials of continuous predictors.

As an example, Caley and Schluter (1997) used existing databases and distribution maps to evaluate whether regional species richness could be used to predict local species richness across a range of taxa (butterflies, amphibians, fish, birds, etc.). Local species richness was the response and regional species richness was the predictor. The data analysis is in Box 8.10.

Let's consider a model with one predictor variable (X_1). A second-order polynomial model is:

$$y_i = \beta_0 + \beta_1 x_{i1} + \beta_2 x_{i1}^2 + \varepsilon_i$$

$$\text{local species richness} = \beta_0 + \beta_1(\text{regional species richness}) + \beta_2(\text{regional species richness})^2 + \varepsilon,$$

where β_1 is the linear coefficient and β_2 is the quadratic coefficient for the predictor.

The question of most interest here is whether a second-order polynomial fits better than the simpler first-order model. We use the model comparison approach

Box 8.9 Calculating Simple Slopes Using Linear Models Packages

Simple slope tests can be done easily with most statistical software. For example, we use these steps to calculate the simple slope of Y on X_1 for a specific value of X_2, such as $\bar{x}_2 + s$.

1. Create a new variable (called the conditional value of X_2, say CVX_2), which is x_{i2} minus the specific value chosen.
2. Fit a multiple linear regression model for Y on X_1, CVX_2, X_1 by CVX_2.
3. The partial slope of Y on X_1 from this model is the simple slope of Y on X_1 for the specific value of X_2 chosen.
4. The statistical software can then generate the SE and t test.

This procedure can be followed for any conditional value. Note we have calculated simple slopes for Y on X_1 at different values of X_2. Conversely, we could have easily calculated simple slopes for Y on X_2 at different values of X_1.

If we have three predictor variables, we can have three two-way interactions and one three-way interaction:

$$y_i = \beta_0 + \beta_1 x_{i1} + \beta_2 x_{i2} + \beta_3 x_{i3} + \beta_4 x_{i1} x_{i2} + \beta_5 x_{i1} x_{i3} + \beta_6 x_{i2} x_{i3} + \beta_7 x_{i1} x_{i2} x_{i3} + \varepsilon_i.$$

In this model, β_7 is the regression slope for the three-way interaction between X_1, X_2, and X_3 and measures the dependence of the regression slope of Y on X_1 on the values of different combinations of both X_2 and X_3. Equivalently, the interaction is the dependence of the regression slope of Y on X_2 on values of different combinations of X_1 and X_3 and the regression slope of Y on X_3 on values of different combinations of X_1 and X_2. If we focus on the first interpretation, we can determine simple regression equations for Y on X_1 at different combinations of X_2 and X_3 using sample estimates:

$$\hat{y}_i = (b_1 + b_4 x_{i2} + b_5 x_{i3} + b_7 x_{i2} x_{i3}) x_{i1} + (b_2 x_{i2} + b_3 x_{i3} + b_6 x_{i2} x_{i3} + b_0).$$

Now we have $(b_1 + b_4 x_{i2} + b_5 x_{i3} + b_7 x_{i2} x_{i3})$ as the simple slope for Y on X_1 for specific values of X_2 and X_3 together. Following the logic we used for models with two predictors, we can substitute values for X_2 and X_3 into this equation for the simple slope. Aiken and West (1991) suggested using $\bar{x}_2$ and $\bar{x}_3$ and the four combinations of $\bar{x}_2 \pm s_{x_2}$ and $\bar{x}_3 \pm s_{x_3}$.

1. Create two new variables (called the conditional values of X_2 and X_3, say CVX_2 and CVX_3), which are x_{i2} and x_{i3}, minus the specific values chosen.
2. For each combination of specific values of X_2 and X_3, fit a multiple linear regression model for Y on X_1, CVX_2, CVX_3, X_1 by CVX_2, X_1 by CVX_3, CVX_2 by CVX_3, and X_1 by CVX_2 by CVX_3.
3. The partial slope of Y on X_1 from this model is the simple slope of Y on X_1 for the chosen specific values of X_2 and X_3.

Simple slopes for Y on X_2 or X_3 can be calculated by reordering the model's predictor variables.

to compare the fit of the full model with that of a reduced model lacking the quadratic term. We can then develop the usual partial t or F statistic and associated P-value to assess the quadratic term.

In our example from Caley and Schluter (1997), the model with the quadratic term was a better fit than the simpler linear model.

Polynomial regressions can be extended to third-order (cubic) models:

$$y_i = \beta_0 + \beta_1 x_{i1} + \beta_2 x_{i1}^2 + \beta_3 x_{i1}^3 + \varepsilon_i.$$

Polynomial models can also include higher-order (quartic, quintic, etc.) terms and more predictors. Our experience

in biology is that polynomials above cubic are rarely used and providing biological interpretations for terms beyond cubic is often difficult.

There are two additional considerations when fitting polynomial regressions. First, x_{i1}^2 is just an interaction term (i.e. x_{i1} by x_{i1}). Polynomial terms in these models will always be correlated with lower-order terms, so collinearity can be a problem. We offer the same advice here as we did in Section 8.1.10 when fitting models with interaction terms. Centering the predictors is often recommended to reduce collinearity, although it is really only important if the full quadratic (or higher) model is fitted and we wish to make inferences about the coefficients for the lower-order terms.

Box 8.10 ⓡ Worked Example of Single-Predictor Polynomial Regression: Predicting Local Species Richness from Regional Richness

Caley and Schluter (1997) asked whether regional (500×500 km cells) species richness could be used to predict local species richness across a range of taxa (butterflies, amphibians, fish, birds, etc.) and continents. We will focus on their dataset for North America, where "local" was 10% of the region. There were eight sampling units (i.e. taxa: gymnosperms, angiosperms, butterflies, amphibians, reptiles, fish, birds, and mammals), and local species richness was the response and regional species richness the predictor. Although there was some evidence that local and regional species richness were skewed, we will analyze untransformed variables like the original authors. Caley and Schluter (1997) forced their models through the origin, but that can make interpretation more difficult, so we will include an intercept. First, we will fit a second-order polynomial:

$$(\text{local species richness}) = \beta_0 + \beta_1(\text{regional species richness}) + \beta_2(\text{regional species richness})^2 + \varepsilon.$$

	Coefficient	95% CI	SE	t	P
β_0	8.124	-9.22 to 25.47	6.75	1.20	0.283
β_1	0.248	-0.19 to 0.69	0.17	1.46	0.203
β_2	0.003	0.001 to 0.005	<0.001	3.50	0.017

There is evidence that the regression coefficient associated with the quadratic term differs from zero, and including a quadratic term improves the model's fit. We can approach this from a model comparison perspective by comparing the fit of the full model to one that omits the quadratic term. The $SS_{Residual}$ from the full model is 377 (5 df), and the $SS_{Residual}$ from the reduced model is 1,299 (6 df). Therefore, SS_{Diff} is 923 with one df and F [(regional species richness)² | regional species richness] $= 12.2$ with $P = 0.017$; this F statistic is the square of the t statistic, and the P-values are identical. It is apparent from Figure 8.6 that, despite the small sample size, the second-order polynomial model provides a better visual fit than a simple linear model.

We cannot easily interpret the linear coefficient in this model because it is highly correlated with the quadratic term (VIF $= 15.20$). To interpret all terms in this full model, we need to use centered predictors to reduce collinearity.

	Coefficient	95% CI	SE	t	P
β_0	59.16	0.46–72.17	5.06	11.69	<0.001
β_1	0.80	0.69–0.92	0.04	18.20	<0.001
β_2	0.003	0.000–0.005	<0.001	3.50	0.017

Note that the quadratic term is unaffected, but now the standard error for the linear term is much smaller because collinearity is reduced (VIF $= 1.02$). There is strong evidence that the coefficient associated with the linear term differs from zero. It is apparent from Figure 8.6 that the linear fit to the data is also reasonable, but adding a quadratic term does improve the fit.

Second, be even more careful about extrapolation beyond the sample data range. For example, imagine predicting the local species richness for regional richness >240 on Figure 8.6. The fitted model (Box 8.10) would predict that local richness exceeded regional richness, which is impossible.

8.1.12 Other Issues in Multiple Linear Regression

8.1.12.1 *Regression Through the Origin*

Forcing a regression model through the origin by omitting an intercept is rarely a sensible strategy. This is even more true for multiple regression models because it implies that

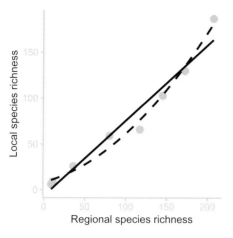

Figure 8.6 Scatterplot of local species richness against regional species richness for North America for a range of taxa (Caley & Schluter 1997) showing linear (solid line) and second-order polynomial (quadratic; dashed line) regressions.

$Y = 0$ when all $X_j = 0$. Even when this is so (e.g. allometric relationships), forcing our model through the origin will nearly always involve extrapolating beyond the range of our observed values for the predictors. Measures of fit for no-intercept models are difficult to interpret.

8.1.12.2 Weighted (Generalized) Least Squares

Weighting each observation by a value related to the variance in y_i (and possibly the covariance between observations) is one way of dealing with heterogeneity of variance and/or correlated observations, although determining the appropriate weights is not straightforward. Other ways of dealing with variance heterogeneity include transformations of the response to a different scale and GLMs.

8.1.12.3 X Random (Model II Regression)

The extension of methods described in Section 6.1.7 to situations when some predictors are random was reviewed by McArdle (1988). To calculate the RMA equivalent estimates for each β_j, first produce a correlation matrix among all the variables (Y and all p X-variables). Then run a PCA (see Chapter 15) on this correlation matrix and extract the eigenvector for the last component with the smallest eigenvalue (explained variance). The estimate of the regression slope for each predictor variable $\left(X_j\right)$ is:

$$b_j = \alpha_j/\alpha_Y,$$

where b_j is the regression slope for X_j, α_j is the coefficient for X_j and α_Y is the coefficient for Y from the eigenvector for the principal component with the smallest eigenvalue. McArdle (1988) refers to this method as the standard minor axis (SMA) and it simply becomes the reduced major axis (RMA) method when $p = 1$. These are standardized

regression slopes, because they are based on a correlation matrix, so the regression model does not include an intercept.

The choice between OLS and SMA is not as straightforward as that between OLS and RMA for simple one-predictor regression and, as Warton et al. (2006) pointed out, the standardization underlying SMA makes inference complicated.

8.1.12.4 Robust Regression

When there are problems with normality or especially if there are outliers that we cannot deal with via deletion or transformation, OLS may be unreliable. One approach is to use robust fitting methods that are less sensitive to outliers, as described for simple regression models in Section 6.6. Hollander et al. (2013) described the Jaeckel–Hettmansperger–McKean test statistic for evaluating the H_0 that a subset of regression coefficients equal zero. Wilcox (2022) described extensions of the Theil–Sen estimator for a simple regression slope that might be appropriate for multiple regression models. He also described other approaches based on M-estimators.

The randomization test of the H_0 that $\beta_1 = 0$ in simple linear regression can also be extended to multiple regression. We compare the observed partial regression slopes to a distribution of partial regression slopes determined by randomly allocating the y_i to observations but not altering the x_{i1}, x_{i2}, etc., for each observation. Other randomization methods can be used, including using the residuals, although the different methods appear to give similar results (Anderson & Legendre 1999; Manly & Navarro Alberto 2022).

8.1.12.5 Missing Data

Biological data comprising multiple variables can have missing values for some observations. In datasets suited to multiple regression modeling, we may be missing values for some variables for some sample units. It is important to distinguish missing values (no data) from zero values (data recorded but the value was zero). If missing values for the response variable reflect a biological process – for example, some organisms died during an experiment and therefore growth rate could not be measured – then analyzing the pattern of missing values in relation to the predictors may be informative. More commonly, we have missing values for our predictors due to random events such as equipment failure, incorrect data entry, or data being subsequently lost. In these circumstances, some linear models software will omit the entire sampling or experimental unit from analysis, even if data are only missing for one of the variables. Alternatives to deletion when missing data occur, including imputing replacement values, were discussed in Section 5.6.

8.1.12.6 *Power of Tests*

The tests of whether individual partial regression coefficients equal zero are based on t statistics and therefore the determination of power of these tests is the same as for any simple t test that a single population parameter equals zero. Our comments on power calculations for simple regression analyses (Section 6.7.1) apply similarly for multiple regression. As for other, more complex models, specifying an effect size can be difficult.

8.1.13 Categorical Predictors in Multiple Regression Models

The models so far in this chapter have included multiple continuous predictors. In Chapter 7, we looked at models with multiple categorical predictors. Many study designs have a combination of continuous and categorical predictors.

When we have mostly continuous predictors and one or two categorical predictors, we would like to include all predictors in a single model and evaluate it as a multiple regression model. We will illustrate this approach by returning to the bird example from Box 8.2. We originally treated grazing as a continuous variable with integer values from 1 (no grazing) to 5 (intense grazing). It is strictly an ordered categorical variable so we could reanalyze these data by fitting a model with grazing as a categorical predictor with five categories.

Categorical predictors are easily incorporated into general linear (regression) models by converting them to indicator or dummy variables. For Loyn's (1987) grazing history variable, there are five categories: zero, low, medium, high, and intense grazing. In Box 8.11 we show how to code them as dummy variables. For a predictor

with c categories, we need $c - 1$ dummy variables. We opted for reference coding in this case, with zero grazing as the reference.

For convenience, we simplified our model by only including log patch area and the dummy variables for grazing:

$$(\text{bird abundance}) = \beta_0 + \beta_1(\log \text{ area}) + \beta_2 \text{grazing}_1$$
$$+ \beta_3 \text{grazing}_2 + \beta_4 \text{grazing}_3 + \beta_5 \text{grazing}_4 + \varepsilon.$$

The analysis based on this model is presented in Box 8.11. Estimation of parameters is based on OLS as usual and partial t or F statistics. This model can be envisaged as separate parallel linear regression models between bird abundance and log area for each level of grazing. The partial regression slope for each dummy variable measures the difference in bird abundance between that category of grazing and the reference category (zero grazing) for any given log area. This model fits the same slope for the relationship between bird abundance and log patch area within each grazing category, so each partial regression slope measures the difference in intercepts between each grazing category and the reference category. Remember to check what coding your software uses, and if it's reference coding, which category is the reference.

The usual OLS assumptions apply. Collinearity is of particular interest, and we assume that continuous predictors are not strongly correlated with categorical ones. This assumption can be evaluated by treating the continuous predictor variable as the response and fitting a linear model against the categorical predictor. Essentially, we are assuming that the continuous predictors are unaffected by the categorical predictors.

Interaction terms between the dummy variables and the continuous variable could also be included. They evaluate how much the slopes of the regressions between bird abundance and log patch area vary between the levels of grazing. In the next section, we will look at how we might model the relationship (intercept and slope) between a response and a continuous predictor across different groups of a categorical predictor and adjust the mean responses in each category based on the relationship between the response and the continuous predictor.

8.2 Analysis of Covariance

When designing studies with categorical predictors (factors), biologists often measure additional continuous predictors (sometimes called covariates) on each sampling or experimental unit. The focus of this type of

Box 8.11 ⓡ Worked Example of Multiple-Predictor Linear Regression with Continuous and Categorical Predictors: Bird Abundance in Remnant Forest Patches

We simplified the model from Box 8.2 by just including grazing and log patch area, consistent with model selection (Box 9.1).

The results from fitting this model were as follows.

	Coeff.	95% CI	SE	Std. coeff.	t	P
Intercept	21.60	15.40 to 27.80	3.09	0	6.99	<0.001
Log(area)	6.89	4.30 to 9.48	1.29	0.52	5.34	<0.001
Grazing	−2.85	−4.28 to −1.42	0.71	−0.39	−4.01	<0.001

There was evidence for a negative effect of grazing pressure on bird abundance and the positive effect of patch area, holding the other predictor constant.

If we treat grazing more correctly as a categorical predictor, we need to code the grazing categories using dummy variables. Setting zero grazing as the reference category, the four dummy variables have the following values.

Grazing intensity	$Grazing_1$	$Grazing_2$	$Grazing_3$	$Grazing_4$
Zero (reference category)	0	0	0	0
Low	1	0	0	0
Medium	0	1	0	0
High	0	0	1	0
Intense	0	0	0	1

We then fitted the following model based on these dummy variables:

$$(bird\,abundance) = \beta_0 + \beta_1 (\log area) + \beta_2 grazing_1 + \beta_3 grazing_2 + \beta_4 grazing_3 + \beta_5 grazing_4 + \varepsilon,$$

which gave us the following results.

	Coeff.	95% CI	SE	Stand. Coeff.	t	P
Intercept	15.72	10.16 to 21.27	2.77	0	5.68	<0.001
Log(area)	7.25	4.73 to 9.77	1.26	0.55	5.77	<0.001
$Grazing_1$	0.38	−5.47 to 6.23	2.91	0.01	0.13	0.896
$Grazing_2$	−0.19	−5.31 to 4.93	2.55	−0.01	−0.07	0.941
$Grazing_3$	−1.59	−7.57 to 4.39	2.98	−0.05	−0.54	0.595
$Grazing_4$	−11.89	−17.78 to −6.01	2.93	−0.47	−4.06	<0.001

Interestingly, the strong negative effect of grazing we observed in the first analysis is driven by the difference in bird abundance between zero grazing and the most intense grazing. The other grazing intensities are not obviously different from zero grazing. We could have included grazing as a categorical predictor in our original analysis in Box 8.2, but this would have added an extra three parameters (replacing grazing as a continuous predictor with four dummy variables) to our model that already had six terms and the ratio of sample size to the number of parameters to be estimated would have become marginal.

design is on the differences in the mean of the response variable across the factor groups, taking into account the relationship between the response and the covariates. The covariates are often used to reduce the unexplained variation from the fit of the model to get more precise estimates of the factor effects. The factor effects are measured as differences between the mean responses for each group adjusted for the linear response–covariate relationship.

For example, in Box 6.6 we analyzed the sampling study of Medley and Clements (1998), who recorded the diversity freshwater diatoms at multiple locations (sampling units) along streams that fell into four distinct levels of zinc contamination (background, low, medium, and high). We fitted a simple linear model relating diatom diversity to zinc group. Imagine if the researchers had also measured a covariate for each location, such as maximum depth of the stream channel. Now the research question might be whether there is an effect of zinc group on diatom diversity taking into account the relationship between diatom diversity and water depth – that is, some of the unexplained variation from our original linear model might now be explained by the relationship between diatom diversity and depth.

The analysis of this type of design is sometimes called an ANCOVA, where the effect of the covariate on the response variable is removed from the unexplained variability by regression analysis. However, the term ANCOVA is not that helpful, because it is not a distinct analysis; it is just a linear model with categorical and continuous predictors, focusing on the factor effects adjusted for any covariates. This differs from traditional regression models, where all the predictors are of interest. Nonetheless, the term ANCOVA is so entrenched in the literature that we will stick with it. These models can also be used to compare the intercepts and/or slopes of two or more regression lines.

We will illustrate these analyses with two examples of essentially single-factor designs where a covariate was also recorded for each sampling/experimental unit. Partridge and Farquhar (1981) examined the effect of number and type of mating partners on longevity of male fruit flies. There was a single factor (partner type) with five groups. In addition to longevity, they measured the thorax length of each male fly as a covariate. If size (reflected in thorax length) explains some of the variation in longevity, then our assessment of the effect of partner type on longevity will be more precise when adjusted for thorax length. The analysis is given in Box 8.12.

Constable (1993) studied the role of sutures in the shrinking of the test of the sea urchin *Heliocidaris erythrogramma*. The categorical predictor was food regime,

with three groups. The response variable was suture width and body volume was the covariate. The main question was whether suture width was affected by food regime, adjusting for the relationship between suture width and body volume. The analysis is presented in Box 8.13.

8.2.1 Linear Models for Simple Analyses of Covariance

Consider a dataset where X_1 (factor A using our terminology for categorical predictors) is a fixed categorical predictor variable with p groups, X_2 is a continuous predictor variable (covariate), and we have a continuous response variable Y, with Y and X_2 recorded for each experimental or sampling unit within each group.

If we frame the model in terms of the effects of the factor groups on the response, we have a linear effects model with an added regression component for the covariate:

$$y_{ij} = \mu + \alpha_i + \beta(x)_{ij} + \varepsilon_{ij}.$$

In the fruit fly example, X_1 (factor A) is partner type ($p = 5$), X_2 is thorax length, the response variable Y is longevity, and the model is:

$$longevity = \mu + (partner\ type) + \beta(thorax\ length) + \varepsilon.$$

This model isn't quite what we are after as μ doesn't represent the overall mean (Kutner et al. 2005) because of the presence of the covariate. The ANCOVA model centers the covariate:

$$y_{ij} = \mu + \alpha_i + \beta(x_{ij} - \bar{x}) + \varepsilon_{ij}$$

$$longevity = overall\ mean + partner\ type$$
$$+ \beta[(thorax\ length) - (mean\ thorax\ length)]$$
$$+ \varepsilon.$$

The details of this linear effects model, including estimation of its parameters and means, are in Box 8.14. Most linear models software just requires the predictors to be specified, as centering the covariate happens in the background. However, using a centered covariate may be necessary for getting the correct contrasts in some software.

As an aside, we could also use dummy variable coding, and the fruit fly model (with reference coding) is:

$$(longevity) = \beta_0 + \beta_1\ partner_1 + \beta_2\ partner_2$$
$$+ \beta_3\ partner_3 + \beta_4\ partner_4 + \beta_5(thorax\ length) + \varepsilon.$$

Each dummy variable partial regression slope measures difference in response between that category of partner type and the reference category for any given thorax

Box 8.12 ℝ 𝔼 Worked Example of Single-Factor ANCOVA Design: Sex and Fruit Fly Longevity

Partridge and Farquhar (1981) studied the effect of the number of mating partners on the longevity of fruit flies. There were five treatments: one virgin female per day, one newly inseminated (pregnant) female per day, eight virgin females per day, eight newly inseminated (pregnant) females per day, and a control group with no females. There were 25 males, kept individually in vials, in each group. The thorax length of each fly was recorded as a covariate. If thorax length (a measure of body size) explains some of the variation in longevity, then the evaluation of the effect of mating on longevity adjusted for thorax length will be more precise. The raw data were extracted by reading from figure 2 in the original paper (see also the description and discussion in Hanley & Shapiro 1994). Our general H_0 was no effect of partner treatment on longevity of male fruit flies, adjusting for thorax length.

Initial boxplots of longevity and thorax length showed no skewness or unequal variances, and there were linear relationships between longevity and thorax length for each treatment (Figure 8.7). There was no evidence that the treatments affected thorax length ($F_{4,120} = 1.26, P = 0.289$). A full model relating longevity to treatment, thorax length, and their interaction was fitted to evaluate whether the regression slopes within each treatment were reasonably parallel. The residual plot (Figure 8.7) showed some unequal spread of residuals. Still, neither the longevity nor thorax length variables had wide ranges that might suggest a log transformation would be appropriate, so both were analyzed on their original scales. There was little evidence against parallel slopes ($F_{4,115} = 0.093, P = 0.984$), as evident in Figure 8.7.

A reduced model omitting the interaction term, and thus assuming parallel slopes, was then fitted, giving the following ANCOVA using Type III SS.

Source	df	MS	F	P
Treatment	4	2,402.9	21.75	<0.001
Thorax	1	13,168.89	119.22	<0.001
Residual	119	110.5		

There was strong evidence for differences between adjusted treatment means.

Adjusted means for longevity (unadjusted ignoring the covariate in parentheses) with 95% CI were (see also Figure 8.7) as follows.

Treatment	Adjusted	95% CI
Control (no females)	61.5 (63.6)	57.3–65.7
1 virgin female	54.5 (56.8)	50.3–58.7
1 pregnant female	64.2 (64.8)	60.0–68.3
8 virgin females	41.6 (38.7)	37.4–45.8
8 pregnant females	65.4 (63.4)	61.3–69.6

There are several ways we could examine the differences in adjusted male longevity between treatment groups in more detail. We could use the contrasts based on reference coding with the control group set as the reference group.

	Coefficient	95% CI	SE	t	P
Control group	61.5	57.3 to 65.7	2.1	29.15	<0.001
1 virgin vs. control	−7.0	−12.9 to −1.1	3.0	−2.36	0.020
1 pregnant vs. control	2.7	−3.2 to 8.5	3.0	0.89	0.375
8 virgin vs. control	−20.0	−25.9 to −14.0	3.0	−6.64	<0.001
8 pregnant vs. control	3.9	−2.0 to 9.9	3.0	−2.36	0.192

Box 8.12 (cont.)

The only groups for which there is evidence for a difference from the control group in longevity are the treatments with virgin females where mating was possible. Providing pregnant females does not change mean longevity of males compared to the control group.

These default reference coding contrasts may not be what we want. For example, it might be of more interest to focus on two specific contrasts: Is male longevity affected by the provision of virgin females versus pregnant females and, for virgin flies where mating was possible, is male longevity affected by the number of females (and therefore mating opportunities) provided?

	Coefficient	95% CI	SE	t	P
Pregnant vs. virgin	16.8	12.6–20.9	2.1	7.98	<0.001
1 virgin vs. 8 virgin	6.5	3.5–9.4	1.5	4.30	<0.001

Longevity of male flies provided with one or eight pregnant females is greater than that of male flies provided with one or eight virgin females, suggesting that increased mating shortens longevity. It is also clear that longevity is greater if only one virgin female was provided compared to eight.

Finally, let's compare the results from fitting the ANCOVA model above with the covariate (based on a pooled within-groups regression) to an ANOVA model that omits the covariate.

Source	df	MS	F	P
Treatment	4	2,984.8	13.61	<0.001
Residual	120	219.3		

The $MS_{Residual}$ is twice as large as that from the ANCOVA, indicating that including the covariate reduces the unexplained variability in longevity by around 50%.

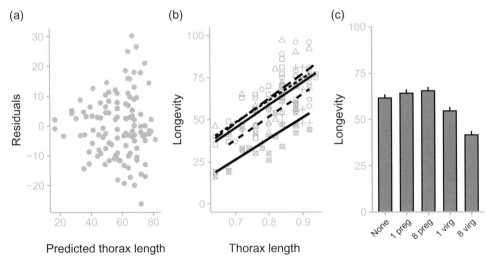

Figure 8.7 Data from Partridge and Farquhar (1981) showing residual plot, scatterplots with linear regression lines, and adjusted means (and SE) of longevity (days) modeled using thorax length (mm) for male fruit flies under each of the five partner treatment groups. To reduce clutter on this figure, we have omitted treatment group labels for the middle panel.

Box 8.13 🅡 🅔 Worked Example of Single-Factor ANCOVA Design: Shrinking in Sea Urchins

Constable (1993) studied the role of sutures in the shrinking of the test of the sea urchin *Heliocidaris erythrogramma* under different food regimes. The categorical predictor (factor) was food regime with three groups: high-food regime, low-food regime, and an initial (pre-feeding) sample. The response variable was the width (mm) of inter-radial sutures (ligaments connecting the skeletal plates) from each urchin, and initial body volume (ml) was the covariate. There were 24 urchins in each group. Constable (1993) transformed body volume to cube roots to linearize the relationship between suture width and body volume. While a scatterplot suggested the untransformed relationships were approximately linear (Figure 8.9), we will follow the original paper and apply the transformation. There was little evidence that the covariate was affected by food regime ($F_{2,69} = 1.363, P = 0.263$). We fitted a full model relating suture width to food regime, cube root of body volume, and their interaction. The residual plot showed an even spread with no concerning values. The regression slopes were heterogeneous (interaction $F_{2,66} = 4.701, P = 0.012$), with the slope for the low-food regime group shallower compared to the others (Figure 8.8).

Constable used the Wilcox (1987) modification of the Johnson–Neyman procedure to determine the covariate ranges over which there were statistically "significant" ($\alpha = 0.05$) differences among the food regime groups. The results were as follows.

Difference range for difference	Covariate (body volume)$^{1/3}$
Initial > low food	>2.95
High food > initial	>1.81
High food > low food	>2.07

This analysis indicates that the high-food group's suture widths were greater than the low-food regime group only in larger urchins: (body volume)$^{1/3}$ > 2.07. The comparisons of each food regime group to the initial group are also provided.

We could also use picked-points analysis on this interaction – that is, estimate the differences between food regime groups for specific covariate values. Here, we used the mean covariate value (2.67) and values ± SD (2.17, 3.16) and just focused on the high-food vs. low-food contrast.

Covariate ($^3\sqrt{\text{body volume}}$) value	Coefficient	SE	t	P
2.17	0.019	0.009	2.198	0.032
2.67	0.038	0.006	6.090	<0.001
3.16	0.056	0.010	5.630	<0.001

There was strong evidence that suture widths in the high-food regime group were greater than in the low-food regime group for the two larger values of cube root body volume (see also Figure 8.8); for the lowest covariate value, the evidence was less strong (but still statistically "significant").

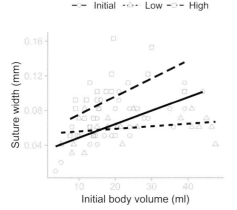

Figure 8.8 Scatterplot showing relationships between body volume and suture width of sea urchins for three food groups for the Constable (1993) example.

Box 8.14 The Linear Analysis of Covariance (ANCOVA) Model and Its Parameters

The usual form of the ANCOVA model is:

$$y_{ij} = \mu + \alpha_i + \beta(x_{ij} - \bar{x}) + \varepsilon_{ij}.$$

In this model we have the following:

y_{ij} is the value of the response variable for observation j in group i of factor A;
μ is the overall mean value of the response variable;
α_i is effect of ith group factor A, defined as the difference between the mean for group i and the overall mean $(\mu_i - \mu)$;
β is a combined regression coefficient representing the pooling of the regression slopes of Y on X within each group;
x_{ij} is the covariate value for replicate observation j from group i of factor A;
$\bar{x}$ is the mean covariate value;
ε_{ij} is random or unexplained error associated with observation j from the ith group of factor A, representing the component of the response variable not explained by the effects of the factor or the relationship with the covariate.

This model is overparameterized, so when factor A is fixed, the usual constraint $\sum \alpha_i = 0$ applies, and parameters in the effects model can be estimated using OLS, as in Box 6.6.

Note we have centered the X-values by subtracting the mean. If we don't, the model is:

$$y_{ij} = \mu + \alpha_i + \beta x_{ij} + \varepsilon_{ij},$$

and μ is now a population intercept (for $X = 0$) rather than an overall population mean of Y. It doesn't matter for the partitioning of variance and testing hypotheses which version we use, although the first is most common in the literature.

The focus in the linear model with a single categorical predictor is on estimating group means. In ANCOVA models, we wish to estimate group means adjusted for the covariate effects (i.e. adjusted means). These are the means of the adjusted values of the response variable (Figure 8.9). For group i, the adjusted mean represents the mean value of the response variable if the mean value of the covariate for that group equals the overall mean value for the covariate:

$$\mu_{i(\text{adj})} = \mu_1 - \beta(\bar{x}_i - \bar{x}).$$

This is estimated by:

$$\bar{y}_{i(\text{adj})} = \bar{y}_1 - b(\bar{x}_i - \bar{x}).$$

The SE of the adjusted mean is:

$$s_{\bar{y}_i(\text{adj})} = \sqrt{\frac{1}{n_i} + \frac{(\bar{x}_i - \bar{x})^2}{SS_{\text{Residual}(X)}}},$$

where MS_{Residual} is from the ANCOVA partitioning of variation (Table 8.3) and $SS_{\text{Residual}(X)}$ is from an ANOVA on the covariate.

We may also wish to estimate β, the pooled within-groups regression coefficient relating Y to X. This is the average of the individual regression slopes for each group, weighted by the variation (SS) in the covariate within each group. Fortunately, this pooled within-groups regression coefficient, and its SE, are standard output from most linear models software packages.

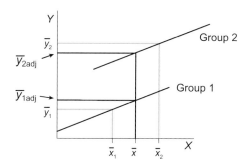

Figure 8.9 Diagrammatic representation of adjusted means in ANCOVA. The adjusted Y means are based on the overall X mean, not the X means for each group. Note that the difference between the adjusted Y means is smaller than the difference between the unadjusted Y means, although this does not always occur in ANCOVA adjustment.

length. Expressing it in this form shows how ANCOVA relates to multiple regression.

If there is no relationship between the response variable and the covariate (i.e. $\beta = 0$) the ANCOVA model simply becomes the single-factor effects model. If there are no effects of the factor groups, it becomes the simple linear regression model with the intercept β_0 replacing μ.

Although our model includes effects of a categorical predictor (α_i) on the response variable and the slope (β) of a regression line relating a continuous predictor to the response variable, the interpretation of the parameters is familiar. We measure the effects of factor A adjusting for the covariate (i.e. holding it constant). The ANCOVA model can therefore be considered as a linear model for Y against a categorical predictor on data adjusted by the regression slope of Y on the covariate X. Each adjusted observation (the value of an observation "corrected" for the effects of the covariate) can be expressed as:

$$y_{ij(\mathrm{adj})} = y_{ij} - \beta\left(x_{ij} - \bar{x}\right) = \mu + \alpha_i + \varepsilon_{ij}.$$

In this model, α_i is the effect of group i of factor A, adjusted for the effects of the covariate $\left(\mu_{i(\mathrm{adj})} - \mu_{(\mathrm{adj})}\right)$. Substituting the OLS estimate of the pooled within-groups regression slope (a weighted average of the individual within-group regression slopes), we obtain:

$$y_{ij(\mathrm{adj})} = y_{ij} - b\left(x_{ij} - \bar{x}\right).$$

Each adjusted value is the value of Y for an observation in any group adjusted (centered) to the mean value of the covariate. For example, the longevity of a male fruit fly is adjusted for the effects of the covariate X by subtracting a term that represents a shift, using the regression of longevity on thorax length volume, of the thorax length for that fly to the mean thorax length of all flies in the experiment (Figure 8.7):

longevity$_{\mathrm{adj}}$ = longevity
$$- b[(\text{thorax length}) - (\text{mean thorax length})].$$

As ANCOVA models combine linear effects and regression models, estimation of parameters with normally distributed error terms is by OLS. Details on estimating adjusted means and their SEs are provided in Box 8.14, and resulting CIs are based on the t distribution as usual. The pooled within-groups regression slope can also be estimated as a weighted (by the SS of the covariate within each group) average of the individual within-group regression slopes.

The primary H_0 of interest in a single-factor ANCOVA model with a single covariate is that there is no difference between adjusted population group means – that is, no difference in the population mean values of Y in each group when the covariate is adjusted to equal $\bar{x}$, using the estimate of pooled within-groups regression slope of Y on $X(\beta)$. This H_0 is also that all group effects, adjusted for the covariate, equal zero. It can be evaluated with an F statistic derived from the partitioning of the variation in adjusted Y-values. Methods for individual contrasts of adjusted means will be described in Section 8.2.7.

8.2.1.1 Predicted Values and Residuals

In practice, the ANCOVA model is a single categorical predictor model fitted to observations adjusted for the effects of the covariate. The predicted values from this model are based on the regression adjustment and the factor group:

$$\hat{y}_{ij} = \bar{y}_i - b\left(\bar{x}_i - \bar{x}_{ij}\right).$$

These predicted values are different for each observation within each group, in contrast to the model without a covariate where the predicted values within each group were the same (i.e. the group mean). The residuals from the fitted ANCOVA incorporate the effects of the continuous covariate and the categorical factor. As for all linear models, residuals are the basis of the OLS estimate of σ_ε^2 and they are valuable for checking model assumptions and fit.

8.2.2 Model Comparison and the Analysis of (Co)variance

The partitioning of the variation from fitting the single-factor ANCOVA model is essentially the same as for the

Table 8.3 ANOVA table for single-factor ANCOVA based on factor *A* being fixed and response variable adjusted for the effects of the covariate

Source of variation	df	Mean square	EMS	F
Factor *A* (adjusted)	$(p-1)$	$\dfrac{SS_{A(adjusted)}}{p-1}$	$\sigma_\varepsilon^2 + \dfrac{n\sum_{i=1}^{n}\alpha_i^2}{p-1}$	$\dfrac{MS_{A(adjusted)}}{MS_{Residual(adjusted)}}$
Residual (adjusted)	$p(n-1)-1$	$\dfrac{SS_{Residual(adjusted)}}{p(n-1)-1}$	σ_ε^2	
Total (adjusted)	$pn-2$			

single-factor ANOVA (Table 6.2), except the SS_{Total} based on adjusted *Y*-values is partitioned, and the variation attributable to the pooled regression slope of *Y* against the covariate is identified (Table 8.3). This $SS_{Total(adj)}$ is simply the SS_{Total} from an ANOVA on *Y* less the $SS_{Regression}$ from a linear regression of *Y* on *X*, the latter representing the adjustment to the *Y*-values based on the relationship between *Y* and *X*. This $SS_{Total(adj)}$ is partitioned into the variation due to the difference between adjusted group means $(SS_{A(adj)})$ and the variation not explained by the factor $(SS_{Residual(adj)})$. The df are calculated as usual except the $df_{Residual(adj)}$ is the total number of observations minus the number of groups minus 1 for the regression of *Y* on the covariate. The MS are the SS divided by the df as usual, and the expected values of these mean squares (EMS) are identical to those from a single-factor ANOVA, except that the analysis is based on *Y*-values adjusted for the covariate. Although this model is essentially a single-factor effects model, the inclusion of the covariate means that the order in which terms are entered into the model can affect the SS. Type II and III SS will be the same, but Type I SS may be different.

Evaluating the H_0 of no effects of the factor on adjusted mean response is a matter of comparing the full model to a reduced one without the factor effect:

$$y_{ij} = \mu + \beta\left(x_{ij} - \bar{x}\right) + \varepsilon_{ij}.$$

Here, β is the pooled within-groups regression slope of *Y* on *X*. The SS_{Total} from the ANCOVA (i.e. $SS_{Total(adjusted)}$) is simply the $SS_{Residual}$ from the full model and $SS_{Residual(adjusted)}$ is simply the $SS_{Residual}$ from the reduced model, analogous to the model-fitting procedure described in previous chapters. The usual *F* statistic can be used to compare the two models and to evaluate the H_0 of no factor effects and is standard output from linear models software.

8.2.3 Assumptions of ANCOVA Models

The assumptions for OLS estimation include those already described for linear models based on either continuous or categorical predictors. The error terms from our

ANCOVA model (based on adjusted *Y*-values) are assumed to be independently and identically distributed (i.i.d.) with zero mean and constant variance across the different factor groups. For CIs and hypothesis tests, we also assume the error terms are normally distributed. We check these assumptions using residuals. Because our ANCOVA model has a regression component, these residuals will be different for observations within each group as well as between groups. Plots of residuals against adjusted group means are the best check of the assumption of homogeneous variances. Transformations of *Y* will often help if the heterogeneous variances are due to skewed distributions of *Y* within each group and WLS and GLMs may also be applicable.

Because the ANCOVA model is a linear model with categorical and continuous predictors, there are additional relevant assumptions/conditions for reliable inference and interpretation:

• The relationship between *Y* and the covariate in each group should be linear. As always, scatterplots are a good way of checking this assumption and transformations of *Y* and sometimes the covariate may be necessary. Specific forms of nonlinearity between *Y* and the covariate may be dealt with by including a polynomial term as an extra covariate (see Section 8.1.11). For a second-order polynomial, the actual analysis will be based on centered covariate values for the linear term and squared then centered covariate values for the quadratic term (Huitema 2011). Collinearity between the linear and quadratic covariate terms is likely and our recommendations in Section 8.1.9 also apply here. Evaluating the homogeneity of within-groups regression slopes will also be more complex because of possible interactions between the factor and the linear covariate term and between the factor and the quadratic covariate term. These interactions will also exacerbate the collinearity issue.

• The covariate is assumed to be a fixed variable with no random error. This is the standard fixed *X* assumption of linear regression models. *X* is commonly a random variable in regression analysis and this usually results in underestimation of the true regression slope. If the assumptions about homogeneity of variance, range of covariate values, and parallel slopes hold, there is no reason to suspect that the underestimation of the true pooled within-groups regression coefficient between Y and the covariate will vary between groups. So any bias should not invalidate our interpretation of differences between adjusted means. Huitema (2011) pointed out that the issue of the covariate being random, and most likely measured with error, is more serious for nonrandomized observational study designs than ones where units are randomly allocated to experimental groups.

• The interpretation of adjusted means for each group, and contrasts between groups, is difficult if the covariate has a different distribution, especially the range of covariate values, across groups. This issue is basically one of no collinearity between the continuous and categorical predictors in the model – that is, the covariate is independent of the factor groups. In practice, biologists should avoid situations in which there is a range of covariate values present in one group but absent from others. Part of the problem is that the adjustment procedure can involve extrapolation of the regression between Y and the covariate beyond the range of covariate values in some groups. One important implication of this assumption is that adjusted means will be difficult to interpret if an ANCOVA model is used as a correction for different values of the covariate in each group of an experiment. A situation like this may also raise questions about the nature of the relationship (Box 8.15).

There is no hard and fast rule about what constitutes too little overlap of covariates between groups, but some is obviously required. Researchers sometimes use an ANOVA model to examine the H_0 of no difference in covariate means between factor groups, although we suggest that scatterplots of Y against the covariate for the different groups will highlight potential problems. Simply evaluating the "statistical significance" does not measure how different the covariates are between groups.

One approach to dealing with differences in covariate ranges between groups is to use what Huitema (2011) termed a *quasi*-ANCOVA model. Instead of including the covariate centered by subtracting the overall covariate mean as in the standard ANCOVA model, the covariate is centered by subtracting the mean covariate value for the group in which that observation occurs. This is essentially including the residuals from a model relating the covariate to factor groups (a simple effects model ANOVA) instead of the covariate values themselves in the ANCOVA model. While this method will result in estimates of

adjusted means that are the same as unadjusted means from an ANOVA on the response variable, fitting the ANCOVA model will still result in a smaller $MS_{Residual}$ if Y is related to the covariate than would a simple ANOVA model for Y.

• A fundamental condition underlying the application of the ANCOVA model and estimation of adjusted group means is that the within-group regression slopes relating Y to X are the same across groups, and this will be examined next.

8.2.4 Homogeneous Within-Group Regression Slopes

The interpretation of adjusted means relies on the slopes of the regressions of Y on the covariate being the same between factor groups (i.e. homogeneity of within-group regression slopes). The adjustment of the Y-values to produce adjusted group means and effects is based on a pooled within-groups regression coefficient. This pooled slope must be a reasonable representation of the individual slopes, which will only be true if the individual slopes are similar – that is, the individual regression lines are (close to) parallel. If the slopes of the within-group regression lines differ, assessing whether there are differences between groups depends on the covariate values used for the comparison, in contrast to the situation with parallel lines where the group differences are constant.

8.2.4.1 *Evaluating Within-Group Regression Slopes*

There are two equivalent ways to examine whether the within-group regression slopes are parallel. We can fit the ANCOVA effects model with a pooled within-groups regression slope and compare it to a model with a separate slope for each group (Maxwell et al. 2018). The second, more common, approach matches the way we would code the models in linear models software. Heterogeneous within-group regression slopes basically imply an interaction effect between the factor and the covariate. This interaction is interpreted like any other interaction in a linear model; it measures how much the effect of the factor on the response depends on the level of the covariate. To evaluate whether the within-group regression slopes are homogeneous, we compare the fit of a full model that includes the interaction between the factor and the covariate to one that omits the interaction.

We will illustrate with regression models for the fruit fly data. The full model with an interaction is:

Box 8.15 Ⓡ Ⓔ Overlapping Covariate Ranges Is a Statistical and a Biological Issue

We use a simple hypothetical dataset to illustrate some issues when covariate ranges do not match well. The dataset consists of two treatment groups, a single response variable (Y) and a single covariate (X).

Group 1		Group 2	
X	Y	X	Y
0.94	7.51	4.35	12.17
2.49	7.42	5.79	10.70
3.45	10.24	3.51	9.92
1.51	6.54	5.73	11.28
3.69	8.91	6.28	12.90
2.9	7.73	3.67	10.36
4.01	10.00	4.72	10.23
3.8	8.59	5.54	10.54
2.65	9.10	3.58	11.24
2.59	8.06	6.23	9.75

We can fit a simple ANCOVA model to these data, starting with checking for homogeneity of slopes, and find no concern there ($F_{1,16} = 2.8, P = 0.114$), so we could proceed to a simple model with a pooled slope. Fitting this model suggests an effect of treatments ($F_{1,16} = 4.847, P = 0.042$) with the adjusted mean of treatment 2 being 15–20% larger than treatment 1 (10.37 vs. 8.95). There is also a strong covariate effect ($F_{1,16} = 5.34, P = 0.034$ and see Figure 8.10), justifying its inclusion in the model.

If we look more closely at the data (which we would, of course, have done before fitting the models!), there are issues, the most serious of which is the disparity in covariate ranges (Figure 8.10). Because of this disparity, the adjusted means are calculated for a covariate value ($\bar{x}$) that is above most data points for one treatment and below most for the other treatment.

Following Huitema (2011), we fitted a model in which the covariate was the residual, calculated separately for each group ($x_{ij} - \bar{x}_i$). Figure 8.10 shows how effectively this technique removes the disparity in covariate ranges, and the overall conclusions about effects of treatments and covariates are unchanged. However, the treatment effect, reflected in the adjusted means, is slightly larger (10.91 and 8.41).

Using residuals has removed statistical concerns about covariate ranges, but there is a more fundamental question of whether we feel comfortable with a linear extrapolation beyond the range of each group. In doing so, we are presuming that the same biological relationship is appropriate across the whole covariate range. In many biological situations, this may not be the case. For example, a covariate is often used to remove effects of organism size when comparing response variables. It would not be unusual for effects of size not to be linear across the whole size range. In our hypothetical example, the LOESS smoother fitted through the whole dataset suggests that a nonlinear relationship may be more appropriate, a conclusion supported by the pattern of residuals. In this example, we should ask whether we are seeing treatment effects that are deviations from a common overall (linear) covariate relationship or if we are observing a nonlinear relationship in which the two treatment groups reflect different parts of this relationship.

There will not be a single answer to the problem of disparate covariate ranges. Still, it is important to consider the statistical issues *and* the underlying biological relationships, which should dictate the form of the model to be fitted.

$$(\text{longevity}) = \mu + (\text{partner type}) + (\text{thorax length})$$
$$+ (\text{partner type} \times \text{thorax length}) + \varepsilon.$$

The reduced model, assuming homogeneous regression slopes between groups, is:

$$(\text{longevity}) = \mu + (\text{partner type}) + (\text{thorax length}) + \varepsilon.$$

These two models are compared in Box 8.12.

One important problem that can occur, especially with data like morphometrics, is when we evaluate the

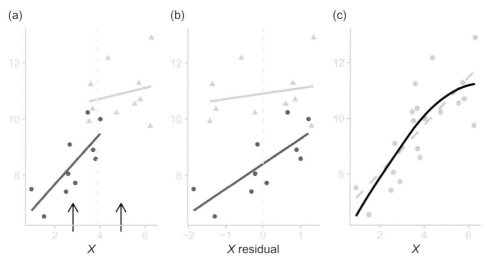

Figure 8.10 Hypothetical dataset showing issues when covariate ranges overlap little. Panel (a) shows the raw data, with linear regressions through each group. The dashed vertical line shows the overall covariate mean, which is the base for calculating adjusted means. The arrows show the mean covariate values for each group. Panel (b) shows data prepared for a quasi-ANCOVA, where the new covariate is the residual covariate value for data from the two groups. Panel (c) shows a LOESS smoother fitted to the entire dataset (smoothing parameter = 0.9), and the dashed line shows a least squares linear smoother.

homogeneity of slopes with a formal test for statistical significance. If the factor has many groups, there are many observations in each group, or the linear regression model of Y on X fits the data in each group very well (i.e. r^2 is very high), the interaction test may be very sensitive. It may reject H_0 even though a scatterplot suggests the regression lines are almost parallel. Any formal statistical test for parallel slopes should only be used and interpreted in conjunction with a sensible, often graphical, interpretation of differences in regression slopes.

8.2.4.2 Dealing with Heterogeneous Within-Group Regression Slopes

As with any linear models interaction, it is difficult to interpret lower-order (including main) effects if the interaction is important. While we might sometimes assess the main effects of the factor when the slopes are not parallel (Maxwell et al. 2018), the usual recommendation is to use other approaches to interpret the analysis. First, the slopes themselves might be the primary biological interest of the analysis and you can contrast slopes across treatment combinations. This is like using treatment–contrast interactions to examine the Y by covariate interaction (Section 7.1.11.3).

Second, if the factor (group) effects are the main interest, Huitema (2011) recommended a test called the Johnson–Neyman technique (see also Maxwell et al. 2018). This procedure compares groups in a pairwise fashion and identifies covariate ranges for which the group means for Y are different (at a pre-specified statistical significance level), and ranges where they aren't. It is essentially an unplanned comparison technique, which, with the Wilcox (1987) modification, adjusts P-values to account for the number of tests. Constable (1993) described the application of the Wilcox procedure to compare groups in his sea urchin example (see Box 8.13). The downside of this approach is that it is entirely based around tests of statistical significance; it is not designed to give measures of the size of the difference between groups.

Third, Huitema (2011) described the picked-points analysis, which requires the researcher to select several values of the covariate at which to compare factor groups. This is analogous to the simple slopes procedure we described for examining interactions in multiple regression models with continuous predictors (Section 8.1.10), except we are examining differences between factor groups for specific values of the continuous covariate. Huitema (2011) suggested choosing the mean of the covariate plus values one SD above and below this value. The picked-points analysis can be done with most software that allows simple slopes or simple main effects analyses, and is illustrated for the sea urchin example in Box 8.13.

8.2.5 Robust ANCOVA

As with other linear models involving categorical or continuous predictors, there may be situations where we wish to fit an ANCOVA model to our data but the underlying assumptions are not met. While we recommend solutions such as transformations or GLMs if the response variable distribution is binomial or Poisson, methods that have fewer underlying assumptions might sometimes be warranted.

The simplest robust approach is based on analyzing the ranks of the Y-values and the covariate (termed rank ANCOVA by Huitema 2011). The usual ANCOVA model is then fitted using the ranks. The adjusted means are now adjusted mean ranks (Huitema 2011). Evaluating homogeneity of regression slopes between groups is, however, problematic with ranked data. Ranking generally makes interactions difficult to interpret (Section 7.1.14), and Huitema (2011) demonstrated how rank transformation of Y and the covariate can change the interpretation of within-group regression slopes. Other methods for fitting robust ANCOVA models are described by Huitema (2011) and Wilcox (2022), including those based on robust linear model analyses (Wilcoxon estimation; see McKean 2004), smoothing functions (GAMS; see Section 13.9), and bootstrap methods (Section 6.6.4). If the focus is on the test of the H_0 and generating a P-value, randomization tests could also be used if we consider the ANCOVA model as a multiple regression and do multiple randomizations of experimental or sampling units to groups (as in a single-factor model), keeping the pairing between Y and the covariate.

8.2.6 Unequal Sample Sizes (Unbalanced Designs)

There are no specific difficulties associated with ANCOVAs with unequal sample sizes between groups beyond those we have already discussed in Chapters 6 and 7 for single and multifactor models. We should be more careful about checking assumptions with unequal sample sizes, and the choice of SS is important when deriving F statistics from a factorial ANCOVA. We have generally used Type III SS for our analyses in this chapter, although we recognize there is also a sensible argument for using Type II SS.

8.2.7 Specific Comparisons of Adjusted Means

The methods described in Section 6.2.7 for comparisons among factor group means apply to adjusted means with some modifications to account for the covariate.

8.2.7.1 Planned Comparisons

Contrasts among adjusted means can be done with the usual t statistic, modified to take the relationship between Y and the covariate into account:

$$t = \frac{c_1 \bar{y}_{1(\text{adj})} + c_2 \bar{y}_{2(\text{adj})} + \cdots}{\sqrt{\text{MS}_{\text{Residual}} \left[\dfrac{c_1^2}{n_1} + \dfrac{c_2^2}{n_2} + \cdots + \dfrac{(c_1 \bar{x}_1 + c_2 \bar{x}_2 + \cdots)^2}{\text{SS}_{\text{Residual}(X)}} \right]}}$$

This daunting equation is simply the usual t test for a contrast in a standard ANOVA, except it considers the covariate means and the covariate residual variation (Huitema 2011). The c_is are the contrast coefficients, $\text{MS}_{\text{Residual}}$ is from the ANCOVA, and $\text{SS}_{\text{Residual}(X)}$ is from an ANOVA on the covariate. There is an equivalent F approach.

8.2.7.2 Unplanned Comparisons

The Tukey–Kramer test described in Section 6.2.7 is appropriate for unplanned multiple comparisons of adjusted means. A modification to handle random covariates, the Bryant–Paulson–Tukey test, is unnecessary (Huitema 2011). Our earlier advice about unplanned multiple comparisons also applies here – avoid them if possible and try to plan a small number of sensible contrasts.

8.2.8 Factorial Designs

An extension of the simple ANCOVA model described so far is a factorial design that includes a covariate measured on each experimental or sampling unit. For example, Yavno and Fox (2013) looked at the effect of water velocity on morphological features of juvenile pumpkin-seed sunfish produced by adults from four populations in Canada. This is essentially a two-factor design with population and water velocity as the factors. The response variables were different morphological measurements, and the covariate was the centroid of nine points on the body as a measure of body size. The research question of interest is how the populations and water velocities affect fish morphology, adjusting for body size. The analysis of these data is presented in Box 8.16.

Consider a two-factor crossed design with a single covariate recorded along with the response variable for each observation. To avoid confusing the parameter for the regression slope with the label we've used for a second categorical predictor, we'll label this second

Box 8.16 Ⓡ Ⓔ Worked Example of Two-Factor Crossed ANCOVA Design: Morphology of Sunfish from Different Populations Living in Different Water Flows

Yavno and Fox (2013) collected introduced pumpkinseed sunfish from four sites in Canada, two flowing water sites and two still water sites, and returned them to a holding facility. Progeny from mating between individuals within each population were then allocated to artificial channels and subjected to still or flowing water for 80 days. Various morphological features for each fish were measured at the end of the experiment. Each morphological measurement was analyzed in a three-factor linear model, with two categorical predictors (collection site and water velocity) and a covariate, the centroid of nine points on the body. This centroid was a measure of body size. Each morphological variable and the centroid was log-transformed before analysis. For this example, we used body depth as the response variable (Figure 8.11).

Mean log (centroid) was somewhat affected by the interaction between sites and water velocities ($F_{3,248} = 3.34, P = 0.020$) and there were differences between sites ($F_{3,248} = 34.48, P < 0.001$) and between water velocities ($F_{1,248} = 15.80, P < 0.001$). The subsequent ANCOVA might require a different approach, such as a quasi-ANOVA based on the residuals of the covariate. However, the tests were very sensitive and the covariate ranges generally overlapped. Like Yavno and Fox (2013), we will proceed with the standard ANCOVA.

We used Type III for our analyses. The SS from models with categorical and continuous predictors are not orthogonal, and different types of SS might cause different results. Yavno and Fox (2013) did not specify which they used.

The fit of the full model, including all interactions, resulted in (we have omitted MS values) the following.

Source	df	F	P
Population	3	0.59	0.624
Water velocity	1	13.38	<0.001
Population × water velocity	3	0.73	0.535
Log centroid	1	3,198.31	<0.001
Population × log centroid	3	0.53	0.661
Water velocity × log centroid	1	13.29	<0.001
Population × water velocity × log centroid	3	0.79	0.499
Residual	240		

Only the interaction terms including the covariate are relevant from this ANOVA. There is little evidence that the relationships between log body depth and log centroid differed among groups in the population by water velocity interaction or among populations pooling water velocity. However, the slopes for the two water velocities were not parallel.

Using Huitema's (2011) recommendation of combining the different covariate by factor interactions into a single term, we get an F of 2.42 based on 7 df with $P = 0.021$, indicating evidence of heterogeneity of regression slopes across the two factors. Under these circumstances, we would not usually fit the model based on observations adjusted for the covariate because of the heterogeneous slopes.

predictor C with its population parameter γ. The ANCOVA model for this design (assuming parallel within-cell regression slopes) is:

$$y_{ijk} = \mu + \alpha_i + \gamma_j + \alpha\gamma_{ij} + \beta(x_{ijk} - \bar{x}) + \varepsilon_{ijk}$$

(log body depth) = overall mean + (population) + (water velocity) + (population × water velocity) + $\beta[(\text{log centroid}) - (\text{mean log centroid})]$ + error.

Essentially, the Y-values are adjusted using a within-cells regression slope pooled across all the A–C combinations:

$$y_{ijk(adj)} = y_{ijk} - b(x_{ijk} - \bar{x}).$$

This adjustment is based on the estimate (b) of the pooled within-cells regression slope. The analysis then uses these adjusted values in a two-factor crossed linear model. For most linear models software, the model fitted is the usual two-factor crossed ANOVA model including a covariate term.

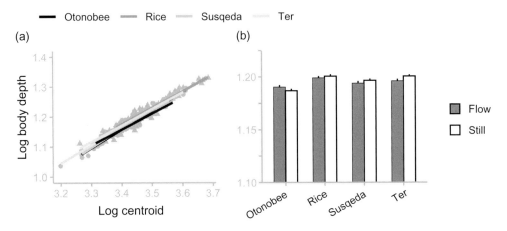

Figure 8.11 Sunfish depth from four sites and two water flow regimes. Panel (a) is a scatterplot with least squares regressions lines fitted through each site–flow combination, with the two flow regimes shown in different colors. Panel (b) shows adjusted means for the eight populations.

The effects of the two factors in crossed ANCOVA models are not orthogonal – that is, we have the same difficulty partitioning the $SS_{Total(adj)}$ as we do trying to partition the SS_{Total} in a crossed ANOVA design with unequal sample sizes (Section 7.1.7). Our recommendation for Type III SS in unbalanced factorial models also applies to factorial ANCOVAs, even when the sample sizes are equal, although there are arguments in favor of the other types of SS as well (see discussion for ANOVAs in Huitema 2011).

Since the adjusted Y-values are based on the pooled within-cells regression slope, we should evaluate whether the slopes are homogeneous before fitting the model. The full model using simplified terminology for the sunfish data, including all interactions, would be

(log body depth) = overall mean + (population) +
  (water velocity) + (population × water velocity) +
  (log centroid) + (population × log centroid) +
  (water velocity × log centroid) +
  (population × water velocity × log centroid) + ε.

Huitema (2011) recommended assessing the combined covariate interaction terms by comparing this model to the simpler model with no interaction terms. A possible alternative might be to recognize that in the more complex model, the evaluation of the interaction term of the two factors relies on homogeneity of regression slopes across all combinations, and the evaluation of each main effect relies on homogeneity of regression slopes across the groups of each factor, pooling the other factor. To evaluate each term in the model, we could examine the interaction between that term and the covariate separately and compare full and reduced models in a hierarchical

fashion, although interpretation can get complex when some main effects have heterogeneous slopes and some don't (Box 8.16).

8.2.9 Designs with Two or More Covariates

In designs with multiple covariates, the regression component of the ANCOVA becomes a multiple regression. The adjustment for a design with one factor and two covariates (X and Z) is:

$$\bar{y}_{i(adj)} = y_{ij} - b_{YX}(x_{ij} - \bar{x}) - b_{YZ}(z_{ij} - \bar{z}),$$

where b_{YX} is the estimate of the pooled within-groups regression slope relating Y to $X(\beta_{YX})$ and b_{YZ} is the estimate of the pooled within-groups regression slope relating Y to $Z(\beta_{YZ})$. The analysis is then done on these adjusted values, like when there is a single covariate. For most software, we simply include the multiple covariates when we fit the linear ANCOVA model.

There are some challenges when including multiple covariates:
• The risk of collinearity is exacerbated.
• Each covariate should have similar ranges across the factor groups, otherwise the ANCOVA adjustment requires extrapolation of either the Y on X or Y on Z regression lines.
• Checking homogeneity slopes is more difficult. Essentially the assumption is about parallelism of a series of planes (or higher-dimensional spaces with more than two covariates!), rather than simple lines. Huitema (2011) describes how to evaluate homogeneity of regression planes; an alternative might be to check the homogeneity of slopes for each covariate separately, by testing the interactions between the factor and each covariate.

Our experience is that biologists don't often have multiple covariates when the focus of the study is adjusting the response variable with ANCOVA to compare adjusted means across factor groups. When we start to have multiple continuous covariates in our study designs, it is more likely that a multiple regression model will be used with the question being about the relative importance of the different predictors or developing a predictive model.

8.3 Key Points

• Models with a continuous response and two or more continuous predictors are termed multiple regression models. Their focus is estimating the change in the response for a unit change in each predictor independent of the other predictors, termed a partial regression slope.
• Estimation is by OLS and tests of null hypotheses for each predictor are based on t or F statistics from comparing a full model to one that omits the predictor of interest.
• In addition to the usual assumptions of normality, homogeneity of variance and independence of model error terms, interpretation of partial regression slopes relies on the absence of strong correlations (collinearity) among the predictors.
• Scatterplot matrices, partial regression plots and residual plots (residuals against predicted values) are recommended ways of checking OLS assumptions. Diagnostic checks for influence include Cook's D statistic. Collinearity is checked by examining pairwise correlations among predictors, with correlations >0.7 causing concern, and the variance inflation factor, with VIFs >10 an issue.
• Interaction between continuous predictors can be incorporated into the models, although the number of parameters to be estimated might be a problem with small sample sizes, and interpretation of these interaction terms is more difficult than with categorical factors.
• These models can also include one or more categorical predictors, usually as dummy variables.
• A common design is one with one or more factors, as in the previous two chapters, but also with one or more continuous predictors (covariates). These models are sometimes termed analyses of covariance (ANCOVAs), and the focus is on the effect of the factor(s) adjusting for the linear relationship with the covariate(s).
• The interaction between a factor and a covariate in an ANCOVA measures how the slope of the relationship between the response and covariate varies among factor groups. Interpretation of the main effect of the factor adjusting for the covariate (comparing adjusted factor means) relies on these slopes being consistent across groups.

Further Reading

Multiple regression models: Most linear models (e.g. Kutner et al. 2005) and regression books cover multiple regression; Chatterjee and Hadi (2012) is very approachable, with good coverage of diagnostics and interactions.

Analysis of covariance: Good specialist texts on ANCOVA include Huitema (2011) and Milliken and Johnson (2001); both cover more complex designs.

9 Predictor Importance and Model Selection in Multiple Regression Models

The previous chapter described OLS fitting for linear models with multiple continuous predictors or a mix of continuous and categorical ones. In this chapter, we look at how to assess the relative importance of the different predictors in these models, recognizing that partial regression coefficients and their associated t statistics and P-values are not ideal for this purpose. We also consider how we can select a model with fewer predictors than the full model that still provides an optimum fit to our data and the ability to predict new observations. We'll also examine other modeling procedures, such as regression trees, as an alternative to traditional OLS multiple regression models.

We use two datasets to illustrate these methods:

- We introduced the dataset on the abundance of birds in remnant forest patches in Box 8.1. Further analyses of this dataset are presented in Box 9.1.
- Yasuhara et al. (2012) used existing databases to identify the environmental factors that best predicted diversity of Arctic shallow-water ostracods. The response variable was ostracod species richness. They used several environmental variables as predictors. This dataset has some strong collinearities among the predictors, and the analyses are in Box 9.2.

In this chapter, the observations are $i = 1$ to n, and the predictors are $j = 2$ to p. The full regression model includes all p predictors, and the possible models from this set of p predictors will be $k = 1$ to r.

9.1 Relative Predictor Importance

Evaluating the relative importance of the different predictors is not straightforward in complex models when collinearity among the predictors is present, as is often the case as the number of predictors rises. We will examine some of the commonly used methods, focusing on those measuring each predictor's contribution, by itself or with other predictors, to the explained variation in the response (Bi 2012; Grömping 2007). In their review of measures of relative importance, Johnson and LeBreton (2004) distinguished between single and multiple analysis methods. We will also use this distinction, but describe them as single and multiple model methods.

9.1.1 Single Model Methods

Single model methods are based on fitting a single regression model, usually the full model with all predictors. Three commonly used measures of relative importance are standardized regression slopes and the t statistics and P-values associated with them.

9.1.1.1 Standardized Partial Regression Slopes

We introduced standardized partial regression coefficients (slopes) in Section 8.1.2.4 and pointed out that Schielzeth (2010) argued they could be used as standardized effect sizes for each predictor. However, they are difficult to interpret unless the predictors are relatively uncorrelated, just like unstandardized coefficients. When collinearity is present, the standardized regression slopes do not naturally represent contributions to the overall explained variation in the response variable.

Bring (1994) proposed two improvements. First, he simplified the standardization by removing the standard deviation of the response, which doesn't affect the comparison of coefficients between the predictors. Second, he argued that standardization of partial regression slopes should use partial standard deviations for each predictor rather than ordinary standard deviations, so the size of the b_j^* relates to the reduction in r^2 when predictor X_j is omitted. This standardization is more appropriate when collinearity among the predictors exists. The partial standard deviation of predictor variable j (X_j) is:

Box 9.1 ® Worked Example of Relative Predictor Importance: Bird Abundance in Remnant Forest Patches

We calculated common measures of relative importance based on the fit of the following full model relating bird abundance to five predictors (grazing treated as continuous) for the dataset from Loyn (1987); see Box 8.2:

$$\text{(bird abundance)} = \beta_0 + \beta_1(\text{log patch area}) + \beta_2(\text{log distance to nearest patch}) + \beta_3(\text{grazing}) + \beta_4(\text{altitude}) + \beta_5(\text{year isolated}) + \varepsilon.$$

The r^2 for the fit of this model is 0.685 (see Box 8.2). The measures of relative importance (standardized regression coefficients and t statistics) were as follows.

Predictor	Standardized regression coefficient	Modified standardized coefficient[*]	t	LMG (95% bootstrap CI)	PMVD (95% bootstrap CI)
Log patch area	0.55	4.84	5.46	0.32 (0.22–0.46)	0.44 (0.20–0.62)
Log distance to nearest patch	−0.05	−0.50	−0.56	0.01 (0.00–0.07)	0.00 (0.00–0.06)
Grazing	−0.23	−1.61	−1.82	0.19 (0.12–0.28)	0.16 (0.00–0.44)
Altitude	0.09	0.83	0.94	0.05 (0.01–0.14)	0.01 (0.00–0.12)
Year isolated	0.18	1.54	1.74	0.10 (0.03–0.20)	0.06 (0.00–0.25)

[*] Based on partial standard deviation of each predictor (Section 9.1.1).

Note the close relationship between the modified standardized regression coefficients and the t values. The rank order of relative importance was consistent for all measures, with log patch area being the most important predictor. Grazing and year isolated were similar in importance when using the modified standardized coefficient (or t statistic), whereas both LMG (Lindeman, Merenda and Gold) and PMVD (proportional marginal variance decomposition) had grazing being approximately twice as important. The bootstrap CIs are reasonably broad and overlap for log patch area and grazing for both LMG and PMVD. Note that the sum of the LMG measures equals the r^2 from the fit of the full model.

We also used hierarchical partitioning based on r^2 as the goodness-of-fit measure (to match LMG).

Predictor	I	J	Total
Log patch area	0.32	0.22	0.55
Log distance to nearest patch	0.01	0.01	0.02
Grazing	0.19	0.27	0.47
Altitude	0.05	0.09	0.15
Year isolated	0.10	0.15	0.25

Note that the independent (I) contribution of each predictor matches the LMG measure. The closeness of the total contribution to r^2 of grazing and log patch area is driven mainly by grazing having a relatively high joint (J) contribution with other predictors.

We used the AIC_C (small sample adjustment of AIC) criterion to rank all the possible models (total possible models with five predictors was 32). The top five models are presented, with the unstandardized regression coefficients for each predictor, based on Δ_i values ≤ 2 having "substantial support" (Burnham & Anderson 2004).

Log patch area	Log distance to nearest patch	Grazing	Altitude	Year isolated	df	AIC_C	Delta (Δ_i)	Akaike weight
7.12		−1.90		0.08	5	372.3	0.00	0.21
7.08		−1.60	0.03	0.08	6	373.2	0.93	0.13
6.89		−2.85			4	373.3	0.99	0.13
7.49	−2.05	−1.94		0.07	6	373.9	1.55	0.10
8.15			0.04	0.13	5	374.1	1.75	0.09

Box 9.1 (cont.)

While the cut-off of Δ_i values ≤ 2 is somewhat arbitrary, the top five models all included log patch area, and grazing and year isolated were included in four of the models. The Akaike weight for the top model was not particularly high (0.211), and only the top 13 models had Akaike weights ≥ 0.001 (all of which included log patch area).

The sum of Akaike weights for each predictor across models that contained that predictor is as follows.

Predictor	Sum of weights
Log patch area	1.00
Grazing	0.80
Year isolated	0.69
Altitude	0.40
Log distance to nearest patch	0.30

Log patch area appeared in all models with weights greater than zero; hence its sum was 1. The sum of Akaike weights criterion ranked the predictors in the same order as did LMG, the consistency partly reflecting the lack of strong collinearity in this dataset.

Because of the model uncertainty (the number of models with similar AIC_C values and the top model having a relatively low weight), we used full (zero) model averaging to get a more reliable estimate of the partial regression coefficients. As a comparison, we provide unstandardized and standardized (by unconditional and by partial standard deviations; in italics) estimates. Revised SEs (Section 9.2.4) are presented, and approximate 95% CIs are based on $z = 1.96$ (bootstrapping could also have been used).

Parameter	Standardization	Coefficient	SE	95% CI
Intercept	None	−106.86	118.77	−339.64 to 25.93
	Both	0	0	0
Log patch area	None	7.39	1.42	4.62 to 10.16
	Unconditional SD	0.56	0.11	0.35 to 0.77
	Partial SD	5.16	1.13	2.94 to 7.37
Log dist. to nearest patch	None	−0.57	1.55	−6.51 to 2.73
	Unconditional SD	−0.02	0.06	−0.14 to 0.10
	Partial SD	−0.23	0.62	−1.43 to 0.98
Grazing	None	−1.73	1.24	−4.09 to −0.27
	Unconditional SD	−0.24	0.17	−0.57 to 0.10
	Partial SD	−1.91	1.44	−4.73 to 0.92
Altitude	None	0.01	0.02	−0.02 to 0.07
	Unconditional SD	0.04	0.08	−0.11 to 0.20
	Partial SD	0.45	0.81	−1.14 to 2.03
Year isolated	None	0.06	0.06	−0.00 to 0.19
	Unconditional SD	0.15	0.14	−0.12 to 0.43
	Partial SD	1.40	1.35	−1.25 to 4.05

The model-averaged standardized coefficients put the predictors in the same order of importance as the LMG measure above. The 95% CIs for the model-averaged coefficients, both unstandardized and standardized, only exclude zero for log patch area.

Box 9.2 ⓡ ⓔ Worked Example of Relative Predictor Importance: Ostracod Diversity

We fitted a model relating species richness of shallow-water ostracods to seven environmental predictors: water depth, bottom water temperature, salinity, productivity (particulate organic carbon flux to the ocean floor), productivity squared (because of commonly observed hump-shaped relationships between richness and productivity in marine systems), seasonal variation in productivity, and the annual number of ice-free days; $n = 129$. To be consistent with Yasuhara et al.'s original (2012) analysis, water depth and seasonal variation in productivity were both positively skewed and were log-transformed, although the same argument could apply to temperature. Additionally, all predictors were centered, although note that centering does not affect the recommended measures of relative importance. Scatterplots indicated some strong collinearities among the predictors, especially between log depth and salinity, temperature and number of ice-free days, log seasonal variation in productivity and number of ice-free days, and productivity and productivity squared (despite centering). The VIFs were also above 10 for productivity (27.66), productivity squared (20.77), seasonal variation in productivity (15.52), and number of ice-free days (20.76). Despite these collinearities, we will initially fit the model including all predictors (as Yasuhara and colleagues did) to illustrate measures of relative importance when collinearity is present:

$$(\text{richness}) = \beta_0 + \beta_1(\log \text{ depth}) + \beta_2(\text{temperature}) + \beta_3(\text{salinity}) +$$
$$\beta_4(\text{productivity}) + \beta_5(\text{productivity})^2 + \beta_6(\log \text{ productivity variation}) + .$$
$$\beta_7(\text{ice-free days}) + \varepsilon$$

The residual plot showed some pattern in the variances, but the largest Cook's D value was only 0.11, suggesting no particularly influential observations.

Predictor	Coefficient	95% CI	SE	t	P
Intercept	14.30	13.34 to 15.27	0.49	29.39	<0.001
Log depth	2.21	−1.84 to 6.24	2.04	1.08	0.282
Temperature	−1.43	−2.29 to −0.57	0.44	−3.28	0.001
Salinity	0.90	−0.16 to 1.96	0.54	1.69	0.094
Productivity	−0.06	−0.14 to 0.03	0.04	−1.32	0.191
Productivity squared	0.00	0.00 to 0.00	0.00	0.37	0.712
Log productivity variation	−35.63	−62.47 to −8.8	13.56	−2.63	0.010
Ice-free days	0.04	−0.05 to 0.14	0.05	0.87	0.389

The r^2 for the fit of this model was 0.319. The t statistics and P-values are difficult to interpret because of the collinearities present (note the relatively large SEs for some predictors), so we compared the usual regression output with other measures of relative importance of each predictor.

Predictor	Standardized regression coefficient	Modified standardized regression coefficient*	LMG (95% bootstrap CI)	PMVD (95% bootstrap CI)
Log depth	0.17	0.54	0.05 (0.02–0.11)	0.03 (0.00–0.17)
Temperature	−0.46	−1.64	0.08 (0.02–0.17)	0.09 (0.01–0.17)
Salinity	0.28	0.84	0.06 (0.03–0.13)	0.10 (0.00–0.20)
Productivity	−0.52	−0.66	0.03 (0.01–0.08)	0.05 (0.00–0.14)
Productivity²	0.13	0.19	0.03 (0.01–0.08)	0.00 (0.00–0.11)
Log prod. var.	−0.78	−1.32	0.04 (0.02–0.08)	0.05 (0.02–0.10)
Ice-free days	0.30	0.43	0.03 (0.02–0.05)	0.01 (0.00–0.06)

* Based on partial standard deviation of each predictor (Section 9.1.1).

Box 9.2 (cont.)

The two versions of standardized regression coefficients produce different orders of predictors, reflecting the effects of strong collinearities. The more robust (to collinearity) modified standardized regression coefficients, which suggest temperature and log seasonal variation in productivity are the most important, also differ from the LMG and PMVD measures, which include salinity in the top two predictors and downweight log seasonal variation in productivity.

We also used hierarchical partitioning based on r^2 as the goodness-of-fit measure.

Predictor	I	J	Total
Log depth	0.05	0.02	0.07
Temperature	0.08	−0.04	0.04
Salinity	0.06	0.04	0.10
Productivity	0.03	0.02	0.06
Productivity squared	0.03	0.02	0.05
Log productivity variation	0.04	−0.04	0.00
Ice-free days	0.03	−0.02	0.01

The independent contributions match LMG, but some of the joint contributions are negative, suggesting those predictors act as suppressors for other predictors in the model.

We used the AIC_C criterion to rank all the possible models (total possible models with seven predictors was 128). The top five models are presented, with the unstandardized regression coefficients for each predictor, based on Δ_i values ≤ 2 having "substantial support" (Burnham & Anderson 2004).

Log depth	Temp.	Salinity	Prod.	Prod²	Log (prod. var.)	Ice-free days	df	AIC_C	Delta (Δ_i)	Akaike weight
	−1.67	1.33	−0.04		−24.42		6	813.7	0.00	0.18
2.238	−1.70	0.83	−0.05		−25.02		7	814.6	0.91	0.11
4.883	−1.64		−0.05		−25.97		6	815.0	1.29	0.09
	−1.63	1.48		−0.00	−20.23		6	815.3	1.58	0.08
	−1.45	1.42	−0.04		−31.82	0.03	7	815.4	1.71	0.07

While the cut-off of Δ_i values ≤ 2 is somewhat arbitrary, the top five models all included temperature and log seasonal variation in productivity, and salinity and productivity were included in four of the models. The Akaike weight for the top model was low (0.18), and only the top 33 models had Akaike weights greater than or equal to 0.001 (all of which included temperature).

The sum of Akaike weights for each predictor across models that contained each predictor are as follows.

Predictor	Sum of weights
Temperature	1.00
Log productivity variation	0.96
Salinity	0.78
Productivity	0.74
Log depth	0.52
Productivity squared	0.43

Box 9.2 (cont.)

Temperature appeared in all models that had weights >0; hence its sum was 1. The sum of Akaike weights criterion ranked the other predictors somewhat differently from LMG, the inconsistency partly reflecting the strong collinearity in this dataset.

Because of the model uncertainty, we used full (zero) model averaging to get a more reliable estimate of the partial regression coefficients. Given the very different units of the predictors, we focused on standardized (by unconditional and by partial SDs) estimates. Revised SEs (Section 9.2.4) are presented, and approximate 95% CIs are based on $z = 1.96$ (bootstrapping could also have been used).

Parameter	Standardization	Coefficient	SE	95% CI
Intercept	Both	0	0	0
Log depth	Unconditional SD	0.14	0.18	−0.07 to 0.59
	Partial SD	0.65	0.93	−0.64 to 3.14
Temperature	Unconditional SD	−0.52	0.12	−0.76 to −0.28
	Partial SD	−2.39	0.70	−3.75 to −1.05
Salinity	Unconditional SD	0.29	0.20	0.08 to 0.67
	Partial SD	1.37	1.08	−0.01 to 3.52
Productivity	Unconditional SD	−0.32	0.28	−0.90 to 0.04
	Partial SD	−1.03	0.86	−2.76 to 0.00
Productivity squared	Unconditional SD	−0.06	0.22	−0.75 to 0.49
	Partial SD	−0.33	0.74	−2.65 to 1.15
Log productivity variation	Unconditional SD	−0.55	0.23	−0.98 to −0.16
	Partial SD	−1.74	0.70	−3.02 to −0.59
Ice-free days	Unconditional SD	0.05	0.24	−0.64 to 0.96
	Partial SD	0.05	0.48	−1.49 to 1.78

The 95% CIs for the model-averaged standardized coefficients exclude zero only for temperature and log seasonal variation in productivity, essentially matching the conclusions from the fit of the full model above.

This dataset had strong collinearities among some predictors. Yasuhara et al. (2012) reasonably argued that because their focus was on point estimation of coefficients, and they used model averaging to account for model uncertainty, they retained all predictors in their analysis. We will illustrate the effects of collinearity on measures of relative importance and model selection/averaging by re-analyzing these data with some collinear predictors removed. To maintain consistency with the first analysis, we used the same transformations and centering. First, we removed productivity squared as there was no evidence of a quadratic relationship between diversity and productivity. Log seasonal variation in productivity was highly correlated with the number of ice-free days (as Yasuhara et al. 2012 pointed out), so we only included the former in this revised analysis. The VIFs now ranged between 1.73 and 4.83 (so <5). Although salinity and log depth were still correlated (>0.8; deeper areas had greater bottom salinity), we analyzed this modified dataset. The residual plot was okay, and there were no particularly influential observations.

First, the fit of the full model is given.

Predictor	Estimate (95% CI)	SE	t	P
Intercept	14.30 (13.34 to 15.26)	0.48	29.54	<0.001
Log depth	2.24 (−1.69 to 6.17)	1.99	1.13	0.262
Temperature	−1.70 (−2.30 to −1.10)	0.30	−5.59	<0.001
Salinity	0.83 (−0.21 to 1.87)	0.52	1.59	0.115
Productivity	−0.05 (−0.07 to −0.02)	0.01	−3.24	0.002
Log productivity variation	−25.02 (−36.67 to −13.37)	5.89	−4.25	<0.001

Box 9.2 (cont.)

Not surprisingly, by reducing collinearity, some standard errors (and *P*-values) are lower than when all predictors were included.

Measures of relative importance (standardized regression coefficients and *t* statistics are absolute values) are as follows.

Predictor	Standardized regression coefficient	Modified standardized regression coefficient	LMG (95% bootstrap CI)	PMVD (95% bootstrap CI)
Log depth	0.17	0.56	0.04 (0.02–0.12)	0.03 (0.00–0.17)
Temperature	−0.55	−2.76	0.11 (0.03–0.20)	0.09 (0.04–0.17)
Salinity	0.26	0.78	0.07 (0.03–0.14)	0.10 (0.00–0.20)
Productivity	−0.41	−1.60	0.05 (0.01–0.13)	0.05 (0.01–0.15)
Log productivity var.	−0.55	−2.10	0.04 (0.02–0.09)	0.05 (0.02–0.09)

The CIs have changed somewhat from the model with all predictors, but temperature is still the most important predictor, whereas salinity is inconsistent across the measures.

We will skip the results of the model selection process, but the sum of Akaike weights for each predictor across models that contained each predictor are shown.

Predictor	Sum of weights
Temperature	1.00
Log productivity variation	1.00
Productivity	0.98
Salinity	0.76
Log depth	0.53

Temperature and log seasonal variation in productivity occurred in all models with Akaike weights above zero. Compared to the previous analysis, productivity has a larger sum.

Finally, model averaging is presented.

Parameter	Standardization	Coefficient	SE	95% CI
Intercept	Both	0	0	0
Log depth	Unconditional SD	0.14	0.18	−0.06 to 0.58
	Partial SD	0.69	0.96	−0.61 to 3.21
Temperature	Unconditional SD	−0.54	0.10	−0.73 to −0.34
	Partial SD	−2.71	0.50	−3.69 to −1.73
Salinity	Unconditional SD	0.27	0.20	0.07 to 0.65
	Partial SD	1.32	1.08	−0.03 to 3.49
Productivity	Unconditional SD	−0.40	0.14	−0.67 to −0.15
	Partial SD	−1.65	0.61	−2.79 to −0.58
Log productivity variation	Unconditional SD	−0.54	0.14	−0.80 to −0.28
	Partial SD	−2.08	0.52	−3.08 to −1.08

> **Box 9.2** (cont.)
>
> Again, some SEs are smaller, and the conditional standardized coefficients based on the partial SD have changed somewhat due to the altered collinearity structure. The pattern of averaged coefficients for the predictors is comparable to the model with all predictors included. In contrast to the previous analysis, the CI for productivity excludes zero in addition to temperature and log seasonal variation in productivity. In this example, the collinearity associated with the two predictors removed (productivity squared and number of ice-free days) did not greatly affect the measures of relative importance and model averaging. But different collinearity structures, especially if collinearity affects the stronger predictors, may have considerable impacts.

$$s_{X_j}^* = \frac{s_{Xj}}{\sqrt{\text{VIF}}} \sqrt{\frac{n-1}{n-p}}.$$

VIF is the variance inflation factor as defined in Section 8.1.9. The modified standardized coefficient is simply the unstandardized coefficient multiplied by this partial standard deviation. Cade (2015) also recommended using Bring's modification, especially in model averaging (Section 9.2.4). This modified standardized coefficient will be closely related to the t statistic associated with each predictor. Additionally, depending on the variation of each predictor and the collinearity among the predictors, these standardized coefficients won't be constrained between -1 and $+1$.

9.1.1.2 *Tests on Partial Regression Slopes*

Given how commonly biologists use statistical significance tests to evaluate the null hypotheses about each partial regression slope, not surprisingly, the t statistics and associated P-values from these tests might also be considered as measures of relative importance for each predictor. We have noted the correspondence between Bring's (1994) modification to standardized regression coefficients and the associated t statistics. However, these tests of null hypotheses about individual regression coefficients are very influenced by collinearity and are difficult to interpret when collinearity is severe. While P-values provide one line of evidence for evaluating the individual null hypotheses about each partial regression slope, they cannot easily be used as measures of relative importance when collinearity is present. As we have mentioned several times earlier, they cannot be used as importance measures in general.

9.1.2 Multiple Model Methods

In contrast to the single model methods described above, multiple model methods better assess the relative contribution of each predictor to the explained variation in the response.

9.1.2.1 *Change in Explained Variation*

The relative importance might be measured by the change in r^2 by adding that predictor to an existing model. For example, we could determine the difference in r^2 between a model with all predictors except the one of interest and the full model with all predictors. This is essentially the Type III SS associated with each predictor and, as a measure of relative importance, is termed "last" by Grömping (2006). The change in r^2 from adding a predictor to a model with all other predictors is:

$$r_{X_j}^2 = \frac{\text{SS}_{\text{Extra}}}{\text{Reduced SS}_{\text{Residual}}},$$

where SS_{Extra} is the increase in $\text{SS}_{\text{Regression}}$, or the decrease in $\text{SS}_{\text{Residual}}$, when X_j is added to the model, and Reduced $\text{SS}_{\text{Residual}}$ is unexplained SS from the model including all predictors except X_j. The ranking of predictors based on this change in r^2 will match the ranking based on t statistics associated with each predictor. These r^2 values do not represent a natural partitioning of the total r^2 for a model, so they are not recommended as measures of relative importance.

9.1.2.2 *LMG and Hierarchical Partitioning*

The LMG metric averages the contribution of each predictor over all possible models and predictor orderings, including that predictor. For example, consider a dataset with three predictors. For predictor X_1, LMG averages the improvement in r^2 (termed the squared semipartial correlation) gained by adding X_1 to models in which it doesn't occur, including different orderings of models with the other two predictors. So the LMG metric is based on Type I SS (see Section 7.1.7). More generally, for predictor X_j out of p predictors and r possible orderings (permutations) of the predictors (Bi 2012; Grömping 2006):

$$\text{LMG}(X_j) = \frac{1}{p!} \sum_{\substack{\text{permutation}}} \text{seq} r^2 (\{X_j\}|r),$$

where seqr^2 is the sequential change in r^2 when predictor X_j is added to a model. The sum of the LMGs for each predictor equals the r^2 from the fit of the full model. Calculation of the LMG metric has also been called dominance analysis (Bi 2012).

A related method is hierarchical partitioning, introduced into the biological literature by Mac Nally (1996). It extends the LMG metric in two ways. First, it allows other measures of goodness-of-fit besides r^2. For example, likelihoods can also be partitioned, so it can be used with GLMs. Second, it provides two components of a predictor's contribution to the goodness-of-fit measure, in this case, r^2. The independent contribution of each predictor matches the LMG metric, measuring the average increase in r^2 by adding the predictor individually to all possible relevant models (and orderings). The joint contribution of each predictor measures the change in r^2 when the predictor is added in partnership with other predictors to all relevant models and orderings. Grömping (2006) pointed out this joint contribution is equivalent to the difference between the LMG metric and r^2 explained by including each predictor in a model by itself. Hence, the sum of the average independent and average joint contribution is the r^2 for the model relating Y to each predictor by itself.

9.1.2.3 *Proportional Marginal Variance Decomposition*

The PMVD metric has been developed from LMG with the change that we use a weighted average of the change in r^2 from adding each predictor to all possible models and orderings. The weights are data-dependent and, to some extent, dependent on the order of the predictors in the model. Determining the weights is not particularly intuitive, but using a weighted average ensures that a predictor with a zero partial regression coefficient will also have a zero contribution to the r^2. In situations with moderate collinearity, this condition may not be met by LMG, although its importance is debatable (Grömping 2007).

9.1.3 Recommendations of Relative Importance

LMG and hierarchical partitioning, if both the independent and joint contributions to r^2 are relevant, are the recommended methods for assessing the relative importance of predictors in regression models with collinearity. Confidence intervals can be calculated using a bootstrapping procedure based on resampling the observations (random predictors) or the residuals (fixed predictors); see Grömping (2007). We show these in Boxes 9.1 and 9.2.

Other useful methods include those based on model selection criteria, especially information criteria, model averaging, and different modeling approaches such as regression trees and random forests. They will be considered in the rest of this chapter.

9.2 Model Selection

Datasets that biologists might analyze by fitting linear models, especially for observational research, often have many predictor variables that could be included in the model. It is common to want to find the smallest subset of predictors that provides the "best fit" to the observed data. There are two apparent reasons for this (Mac Nally 2000), related to the two primary purposes of fitting linear models: explanation and prediction. First, the "best" subset of predictors should include the most important ones in explaining the variation in the response variable, providing another approach to determining the relative importance of the predictors. Second, other things being equal, the precision of predictions from the model will be greater with fewer predictors in the model.

Before describing the different model selection approaches, there are some key points to emphasize:

•   The number of possible models can get large, even with moderate numbers of predictors. With p continuous predictors, there are 2^p possible models, excluding interactions – for example, 64 models with six predictors. With interactions and possible dummy variables from categorical predictors, that number rises sharply.

•   There will rarely be a single "best" subset of predictors for any dataset, particularly if there are many predictors and some collinearity. There will usually be quite a few models, with different numbers of predictors, that provide similar fits to the observed data. It is now common practice to combine the results of some or all of the alternative models, using methods like model averaging (Section 9.2.4), rather than just choosing a single model based on some criterion of fit.

•   Even if the analysis selects a best model or an average of a set of models, this does not mean that the model is any good in an explanatory or predictive sense (Mac Nally et al. 2017). It is important that the goodness-of-fit of the selected model(s) is evaluated. Ideally, cross-validation is used to assess the model's predictive power (see Section 9.2.5).

•   Regardless of our approach to model selection, biological knowledge plays an important role. The choice of predictors to include in the full set for model selection should be based, at least in part, on our knowledge of which predictors are likely to influence the response (e.g., Burnham et al. 2011; Garamszegi et al.

Table 9.1 Criteria for selecting "best" fitting OLS model in multiple linear regression

Formulae are for a specific model with p predictors included. Note that p excludes the intercept

Criterion	Formula
Adjusted r^2	$1 - \dfrac{\mathrm{MS_{Residual}}}{\mathrm{Full\ SS_{Total}}/(n-1)}$
Mallow's C_p	$\dfrac{\mathrm{SS_{Residual}}}{\mathrm{Full\ MS_{Residual}}} - n + 2(p+1)$
AIC	$n\left[\ln(\mathrm{SS_{Residual}}/n)\right] + 2(p+2)$
AIC$_C$	$\mathrm{AIC} + \dfrac{(2p+2)(p+3)}{n-(p+1)}$
Schwarz/Bayesian Information Criterion (BIC)	$n\left[\ln(\mathrm{SS_{Residual}}/n)\right] + (p+2)\ \ln(n)$

2009; Whittingham et al. 2006). Another component of the choice of predictors is ensuring that there are not strong collinearities among them, as strong collinearities can affect model selection methods.

9.2.1 Model Selection Criteria

Criteria for choosing between competing linear models from a set of possible predictors are usually designed to balance a measure of how well the model fits (or predicts) the observed data against the risk of "overfitting," where adding extra predictor variables may suggest a better fit even when they add very little explanatory or predictive power. For example, while r^2 is a valuable measure of OLS fit for a single model, it is not suitable for comparing models with different numbers of predictors because it increases as predictors are added, even if those predictors contribute nothing (Box 9.1). We will therefore focus on criteria that measure fit and penalize overfitting. We will use P to indicate the full set of predictors, p is the number of predictors in a specific model, n is the number of observations, and we will assume that an intercept is always fitted. If the models are all additive (i.e. no interactions), the number of parameters is $p+2$ (the number of predictors plus the intercept and residual variance).

9.2.1.1 Comparisons to the Full Model

Several measures are based on comparing the fit (or lack of fit, e.g. using $\mathrm{SS_{Residual}}$) of a specific model with p predictors to the full model with all P predictors (Table 9.1). We will highlight three of them:

• Adjusted r^2, a modified version of r^2 that considers the number of predictors by using the $\mathrm{MS_{Residual}}$ instead of the $\mathrm{SS_{Residual}}$ in its calculation.

• Mallow's C_p rewards large sample sizes and penalizes larger numbers of predictors. C_p should be small and as close to p as possible.

• The F statistic (and associated probability) from comparing a specific reduced model omitting one or more predictors to the full model including all P predictors, as we outlined in Section 8.1.5 for evaluating individual regression coefficients. This approach is focused on determining whether a particular predictor should remain or be omitted in the search for the best model (Heinze et al. 2018).

The first two are commonly default output from multiple regression routines in software packages, although both have been superseded in modern model selection by the information criteria outlined below. The F approach is more commonly applied as part of stepwise model selection, often using automated algorithms for deciding the predictors to include or omit (Section 9.2.2).

9.2.1.2 Information Criteria

The remaining measures are in the category of information criteria and emphasize the level of evidence for each possible model (Heinze et al. 2018), aiming for the most parsimonious model by explicitly penalizing unnecessarily complex models. Both are commonly formulated in terms of likelihood as the measure of fit, but can be adapted for use with OLS models (Table 9.1). For both, smaller values indicate better, more parsimonious models:

• AIC, introduced by Akaike (1978), is based on the Kullback–Leibler principle of statistical information (Burnham & Anderson 2004; Burnham et al. 2011). A modification of the AIC, AIC$_C$, is often recommended for when the sample size is small relative to the number of predictors in the full model: $n/(p+1) < 40$ (Symonds & Moussalli 2011).

• Bayesian information criterion (BIC) was developed by Schwarz (1978) and is sometimes termed the Schwarz information criterion (SIC). It is, to some extent, based on the "true" model being among the candidate models assessed. The BIC adjusts for sample size and number of predictors differently from the AIC and more harshly penalizes models with more predictors (Heinze et al. 2018).

The BIC has been criticized (Burnham & Anderson 2004; Burnham et al. 2011) and may be particularly problematic with small (but biologically realistic) sample sizes. Indeed, the AIC and its small sample size modification (AIC_C) are much more commonly used in biological research. Other information criteria can be useful for specific data types, such as overdispersion (Chapter 13).

Choosing an appropriate criterion for model fit is one part of model selection, but we also need a suitable method for searching possible models. We will describe two ways. The first is stepwise model selection, which biologists have commonly used, especially in a multiple regression context, but is now falling out of favor. The second usually involves comparing and ranking all possible models based on the information criteria (e.g. AIC) described above.

9.2.2 Traditional Stepwise Selection

One challenge with model selection before modern computer power was the number of possible models that needed to be searched and compared. Several automatic search algorithms were developed that searched a (usually small) subset of the possible models in a stepwise fashion, adding or removing predictors depending on the statistical significance of tests of null hypotheses that their regression coefficients equal zero.

Forward selection starts with a model with no predictors. It then adds the one (we'll call X_a) with the greatest F statistic (or t statistic or correlation coefficient) for the simple regression of Y against that predictor. If the H_0 this slope equals zero is rejected, a model with that variable is fitted. The next predictor (X_b) to be added is the one with the highest partial F statistic for X_b, given that X_a is already in the model $[F(X_b | X_a)]$. If the H_0 this partial slope equals zero is rejected, then the model with two predictors is refitted, and a third predictor added based on $F(X_c | X_a, X_b)$. The process continues until a predictor with a nonstatistically significant partial regression slope is reached, or all predictors are included.

Backward selection (elimination) is the opposite of forward selection. All predictors are initially included, and the one with the smallest and nonstatistically significant partial F statistic is dropped. The model is refitted, and the next predictor with the smallest and nonstatistically significant partial F statistic is dropped. The process continues until there are no more predictors with nonstatistically significant partial F statistics or no predictors left.

Stepwise selection is a forward selection procedure where, at each stage of refitting the model, predictors can also be dropped using backward selection. Predictors added early in the process can be omitted later and vice-versa.

For all three types of variable selection, the decision to add, drop, or retain variables in the model is based on a specified size of partial F statistics or P-values (i.e. statistical significance levels). These are sometimes termed F (or P)-to-enter and F-to-remove. The choice of threshold will greatly influence which variables are added to or removed from the model, especially in stepwise selection. Statistical significance levels different from 0.05 are sometimes recommended to avoid under- or overfitting (Kutner et al. 2005). However, as always, the choice of significance levels is arbitrary.

There are four main problems with using stepwise procedures, especially those based on P-values and statistical significance, for model selection (e.g. Burnham & Anderson 2002; Mac Nally 2000; Whittingham et al. 2006):

- Many P-values for null hypothesis tests of partial regression slopes are generated in variable selection procedures. These values are difficult to interpret due to the multiple testing problem (see, e.g. Section 2.5.4). In response, some authors have recommended that a type of Bonferroni adjustment to significance levels be considered (Kleinbaum et al. 2013). While model selection is not suited to the statistical null hypothesis testing framework, Richards (2015) argued that testing could be a useful approach to model selection if used carefully, especially within a likelihood framework.
- Different algorithms (backward vs. forward vs. stepwise) and the order in which predictors are specified can result in different final models. These stepwise procedures can also produce a final model with a high r^2, even if there is little relationship between the response and the predictor variables included in that model (Flack & Chang 1987).
- The estimates of regression coefficients will usually be inaccurate because a significance test for each predictor is based only on comparing two models, one with and one without that predictor, rather than comparing the full range of models with or without the predictor.
- Stepwise model selection methods focus on finding the single "best" model and don't consider model uncertainty. With many predictors, there is rarely one model clearly a much better fit to our data than all the others, and inference is better based on a suite of models that fit the data well.

9.2.3 All Subsets and Information Criteria

The modern approach to model selection is to compare all possible models (sometimes termed all subsets) using an appropriate information criterion, most commonly AIC or

AIC$_C$. While comparing all possible models has been criticized as a "poor strategy" (Burnham & Anderson 2002: 147) and described as a fishing expedition by Symonds and Moussalli (2011), we emphasize the point we made at the start of Section 9.2: predictors to be included in the full set should already have been through a screening process, so they are plausible explanations of the response and do not have strong collinearities with each other. Once the set is chosen, all the possible models can then be ranked from best to worst according to AIC or AIC$_C$.

Values of AIC or AIC$_C$ for specific models are not individually interpretable, so Burnham and Anderson (Burnham & Anderson 2002, 2004; Burnham et al. 2011) proposed calculating delta (Δ) for each model k:

$$\Delta_k = \text{AIC}_k - \text{AIC}_{\text{minimum}},$$

where AIC$_{\text{minimum}}$ is the lowest AIC value from the set of models (i.e. the best model). Δs can also be calculated using AIC$_C$ values. They have been interpreted as strength of evidence for or against each model compared to the one with the lowest AIC. An approximate guide is that models with $\Delta \leq 2$ have substantial support (i.e. comparable to the best model) and those with $\Delta > 10$ have no support, whereas interpretation of Δ values between 2 and 10 is a bit more problematic – see slightly different recommendations in Richards (2005), Burnham et al. (2011), and Symonds and Moussalli (2011). This approach is sometimes termed multimodel statistical inference, promoted as an alternative to traditional null hypothesis testing and statistical "significance" (e.g., Anderson et al. 2000; Garamszegi et al. 2009).

Another measure of evidence for a model is calculated by converting the Δs to model likelihoods (the relative likelihood of a model given the data) and then expressing the likelihood of a model (k) as a proportion of the sum of the likelihoods:

$$w_k = \frac{\exp(-\Delta_k/2)}{\sum\limits_{k=1}^{r} \exp(-\Delta_r/2)},$$

where the denominator is summed over all r possible models. These are Akaike weights and are now probabilities for each model, given the data and all possible models (Burnham et al. 2011). They can be interpreted as probabilities that any model is the best Kullback–Leibler model given the data (Burnham & Anderson 2004). Not surprisingly, the rank order of models will be the same if we use Δ or Akaike weights.

The Akaike weights can also be a potential measure of relative importance for each predictor by simply summing the weights for models containing that predictor (Burnham & Anderson 2002, 2004; Symonds & Moussalli 2011). The predictor with the largest sum of Akaike weights (SW) is considered to have the highest relative importance, and the remaining predictors can then be ranked in order of importance. Galipaud et al. (2014) criticized the use of SWs as a measure of predictor importance (see also Galipaud et al. 2017), arguing that model-averaged estimates of regression coefficients (see below) were preferable (see also Cade 2015). Giam and Olden (2016) responded by defending SWs and proposed other measures of predictor importance that combined the independent contributions of each predictor to r^2 (i.e. LMG or hierarchical partitioning) with Akaike weights.

9.2.4 Model Averaging

While one model will have the lowest AIC value, there are often quite a few models with different combinations of predictors with AIC values close to the best model, indicating considerable model uncertainty. Rather than relying on a single "best" model or the full model with all predictors for making predictions, it would be better to combine the estimates of regression coefficients from a range of models that represent different combinations of predictors using a method called model averaging. Usually, a subset of the top models is chosen for model averaging, based on either a specified Δ (e.g. 2 or 10) or 95% CI (choosing models until the Akaike weight reaches 0.95) – see Grueber et al. (2011). Once the top set is chosen, there are two approaches to model averaging (Burnham & Anderson 2002; Burnham et al. 2011; Lukacs et al. 2010; Symonds & Moussalli 2011), both based on calculating the weighted average of the estimates of regression coefficients from the top set.

Natural averaging is where each regression coefficient is averaged only over the models in which its predictor occurs:

$$\bar{\hat{\beta}} = \frac{\sum\limits_{k=1}^{r} w_k \hat{\beta}_k}{\sum\limits_{k=1}^{r} w_k},$$

where $\hat{\beta}_k$ is the estimated regression coefficient for a specific predictor, and w_k is the Akaike weight, from model k.

Full or zero model averaging is where each regression coefficient is averaged, weighted by Akaike weights (probabilities of each model), across all models in the top set:

$$\bar{\hat{\beta}} = \sum\limits_{k=1}^{r} w_k \hat{\beta}_k,$$

where the terms on the right-hand side are the same as in the previous equation. The coefficient is set to zero for any model that doesn't include that predictor. This equation simply lacks the denominator.

Choosing between these two approaches is not straightforward, and they can produce different results, especially for weaker predictors. Nakagawa and Freckleton (2011) preferred natural averaging because zero averaging shrinks the average toward zero for predictors not always included in models with high weights. Symonds and Moussalli (2011) argued for full or zero model averaging where there are numerous top models with similar AIC scores. Lukacs et al. (2010) also recommended this latter approach, arguing it would produce more reliable inference about weaker predictors in the averaged model. This is the approach described by Dormann et al. (2018) using predicted values rather than regression coefficients. We use full (zero) model averaging in our worked examples in Boxes 9.1, 9.2, and 9.3.

Grueber et al. (2011) argued that it is easier to interpret standardized (or at least centered) model-averaged coefficients than those averaged on their original scale, especially when interactions or polynomial terms are present. Certainly, comparing regression coefficients as effect size measures from the final averaged model requires standardized regression coefficients. Cade (2015) proposed using Bring's (1994) standardization based on the partial SD because this considers the collinearity among the predictors. However, Burnham and Anderson (2002), Lukacs et al. (2010), and Schielzeth (2010), among others, recommended using the SD unconditional on the other predictors.

Some final key points about model averaging:
• Model averaging can also use predicted values rather than estimates of regression coefficients (Burnham et al. 2011). Dormann et al. (2018) argued this approach is more reliable, as interpretation of partial regression coefficients depends on the model structure and may not be comparable across models with different combinations of predictors.
• Methods for calculating precision (SEs) and CIs for the model-averaged coefficients or predictions have been proposed (Burnham & Anderson 2004; Grueber et al. 2011; Lukacs et al. 2010; Symonds & Moussalli 2011), although their reliability has been questioned; bootstrapping might be a better approach (Burnham & Anderson 2004; Dormann et al. 2018). In our worked examples (Boxes 9.1–9.3), we used the revised SEs from Burnham and Anderson (2004) and, for convenience, calculated approximate CIs based on $z = 1.96$.
• Model averaging is not without its critics. For example, Richards et al. (2011) showed by simulations

that while model averaging will usually improve the precision of estimates of model parameters compared to simply using the model with the lowest AIC, it doesn't necessarily improve prediction accuracy. Cade (2015) also argued strongly that model-averaged coefficients are difficult to interpret when collinearity among the predictors is present, which is common. However, he pointed out that averaged standardized coefficients based on the partial SD partly address this concern (see Boxes 9.1 and 9.2).
• Model averaging can also be approached from a Bayesian perspective using the BIC to approximate the Bayes factors most useful for Bayesian model averaging (see Bolker 2008; Hoeting et al. 1999; Hooten & Hobbs 2015).

9.2.5 Model Validation

However a final model is chosen, we should check that it is specified appropriately and whether it is a good fit and useful for prediction (Mac Nally et al. 2017). The appropriate specification can be assessed using the diagnostic tools described in Section 8.1.7, especially plots of residuals. Goodness-of-fit for OLS-based linear models is usually measured by r^2, and there are modifications for other types of models that include random effects or are fitted as GLMs for nonnormal responses (see later chapters).

There are two ways to assess the predictive capacity of a model. The best is to collect a new sample of data and use the model to predict those new observations. Collecting new data is not always possible for model validation, especially when data collection is expensive or logistically difficult, such as larger-scale ecological field research. The other issue with testing the predictive capacity of a model with new data is whether the new data are comparable to the data on which the model was built. For example, the new data will represent a sample taken at a different time and possibly place, which may affect the reliability and interpretation of predictions.

An alternative is to split (usually randomly) the current dataset into two parts, a training or model-building set to estimate the model and a validation set for testing the model. This approach is often termed cross-validation. A significant constraint is that the original sample size needs to be large enough so each data split is adequate for its purpose. If the sample size is not large enough to split equally, the priority should be that the model-building set is large enough.

A key limitation of the standard cross-validation technique is that it relies on a single validation set and conclusions about the model's predictive ability are sensitive to which observations fall into the training set. Other

Table 9.2 Results from k-fold cross-validation of the full model fit for the Loyn (1987) data (see Boxes 8.2 and 9.1)

Recommended k-values of 5, 10, and 56 were used. The $\sqrt{MS_{Residual}}$ from the original full model was 6.326.

k-value	Training RMSE	Validation RMSE
5	5.839	7.077
10	5.936	6.721
56 (leave-one-out)	5.969	6.759

methods have been developed to overcome this problem by assessing the consistency in predictive ability with different components of the dataset:

• k-fold cross-validation, where the dataset is partitioned into k subsets, and over k repetitions each subset is allocated as the validation set, and the remainder are used for model-building. The value of k is commonly set to 5 or 10 (Kuhn & Johnson 2013).

• Leave-one-out cross-validation, where the validation set comprises a single observation, with the remaining observations used to build the model. The ability of the model to predict each observation, in turn, is assessed so all observations are predicted. This approach is k-fold cross-validation, where k is set to the number of observations. Whichever method is chosen, each observation in the new (or validation) dataset is predicted by the model, and a criterion used to assess how good that prediction is. A common criterion is the root mean squared (prediction) error for the validation dataset:

$$\text{RMSE} = \sqrt{\frac{\sum_{i=1}^{n}(y_i - \hat{y})^2}{n}},$$

where y_i is the value of Y for observation i in the new (or validation) dataset, $\hat{y}_i$ is the value of Y for observation i predicted by the model, and n is the number of observations in the new (or validation) dataset. For k-fold or leave-one-out cross-validation, the average RMSE across the repeated predictions is used. Lower values indicate the model is a better predictor, but RMSE is a relative measure. Kutner et al. (2005) suggest the model is a good predictor if the RMSE (they used the term mean square prediction error) for the validation set is comparable to the $\sqrt{MS_{Residual}}$ for the training set.

We used k-fold cross-validation for the fitted full model from the Loyn (1987) dataset (Table 9.2). We compared k-values of 5, 10, and n (representing leave-one-out cross-validation). The RMSE values for the validation sets were larger than those from the training sets and larger than the $MS_{Residual}$ from the full model fit for all

k-values. Still, the differences were not great, indicating that the model provides reasonably good predictions. Of course, the real test would be to predict data from new forest patches.

Kleinbaum et al. (2013) proposed calculating the cross-validation correlation (as an r^2) between the observations in the new (or validation) dataset and the predicted values and subtracting that from the r^2 determined between the observed and predicted values of the original dataset to get a measure of shrinkage. They suggest that shrinkage values of <0.10 indicate a reliable predictive model. Gelman and Hill (2006) also pointed out the value of visual assessments, such as plots of the new observed versus predicted values and residuals versus predicted values.

9.3 Regression Trees

An alternative to linear (regression) models for developing descriptive, especially predictive, models is regression tree analysis (Breiman et al. 1984; De'ath & Fabricius 2000; Kuhn & Johnson 2013). An "upside-down" tree is created where the root at the top contains all observations, which are divided into two branches at a node, then each branch is further split into two at subsequent nodes, and so on. A branch that terminates without further branching is called a leaf or terminal node. This method for building trees is termed recursive binary-partitioning. Regression trees are sometimes included in statistical software under the acronym CART (classification and regression trees). The main distinction between classification and regression trees is that the former is based on categorical response variables (see Chapter 13) and the latter on continuous responses. Regression trees are also sometimes called decision trees.

9.3.1 Standard Regression Trees

The standard approach to building regression trees with a continuous response variable starts by assessing all possible binary splits of the observations for each predictor variable, based on a splitting criterion that is an index of impurity, a measure of the heterogeneity of the groups at a split (De'ath & Fabricius 2000). For continuous response variables, the usual splitting criterion is the unexplained or explained SS for the two groups at the split, analogous to using OLS to fit linear models; this is sometimes called the ANOVA method. The first split is based on the predictor that results in the smallest $SS_{Residual}$ or largest $SS_{Between-groups}$. This process is repeated within each of the two groups for all the predictors, again choosing each split based on the predictor that results in the minimum $SS_{Residual}$ or largest $SS_{Between-groups}$. This continues until the final tree, based on the best balance between fit (or predictive error) and complexity, is reached.

For each split, the predictor that determines the split is sometimes called the primary split. Some CART software will also list the other predictors at each split, ordered by how much they changed the SS. There are also surrogate splits, alternative splits that match the primary split as closely as possible but can be used when a primary splitter has missing values for some observations.

While SS about the mean is the usual splitting criterion for continuous response variables, more robust measures such as sums of deviations about the median can also be used (De'ath & Fabricius 2000). Suppose we put regression trees in the context of GLMs. There, the splitting criterion could be a likelihood-ratio statistic for Poisson responses, and fit is measured via deviance. Classification trees with true categorical predictors often use the Gini coefficient for splitting (Boehmke & Greenwell 2019).

Regression trees are mainly used as predictive models. For any observation, the predicted value is the mean of the observations at a leaf – that is, in a terminal group. Predicted values for observations in the one group (leaf) will be the same. This contrasts with the usual linear model, which will have different predicted values for all observations unless they have identical values for all predictors.

The splitting process (tree-building) could continue until each leaf contains a single observation – for example, for the Loyn (1987) data we would have 56 terminal nodes. In this situation, the tree would perfectly predict the response variable's observed values and explain all the variance in the response variable, the equivalent of fitting a saturated linear regression model. Usually, we want the best compromise between tree simplicity (fewer nodes) and goodness-of-fit. One way of achieving this is to adjust (penalize) the goodness-of-fit measure by the number of terminal nodes (or the minimum sample size in any terminal mode) and a complexity parameter (cp). An example of a cp might be specifying a minimum change in r^2, representing the improvement in between-groups SS because of the split; if this cp is not met, then the previous split is the final one on that branch of the tree. The complexity parameter is sometimes termed a stopping criterion.

Once we have chosen a value for cp, we can either:
- build a tree based on this *a priori* cp; or
- build a larger tree with few restrictions and then prune the tree based on our chosen cp value.

While we can manually try different cp values to help find the best tree, it is much more useful to use k-fold cross-validation (De'ath & Fabricius 2000; Kuhn & Johnson 2013), as described for linear regression models in Section 9.2.5. The tree with the smallest average prediction error can be selected based on a specified or software default cp. A modification of this criterion is the one SE rule (Breiman et al. 1984; De'ath & Fabricius 2000; Kuhn & Johnson 2013), where the simplest tree within one standard error of the tree with the minimum prediction error is selected. Note that the various regression tree algorithms might present the prediction error differently, such as residual SS, MS, root MS, or one of these scaled somehow.

Regression trees can also provide measures of predictor importance by aggregating or averaging, over all the trees, the amount each predictor improves the $SS_{Between-groups}$ (or reduces the $SS_{Residual}$) at each split. A predictor's importance is not just determined from those splits where it was the primary splitter but also where it came in the ordered list of predictors for each split and on the surrogate list of predictors. Generally, predictors that split early (higher in the tree) or appear multiple times in the tree will rank as more important (Kuhn & Johnson 2013). Regression trees are based on binary splits, so this measure of importance may cause a different order of predictors than the relative importance measures described in Section 9.1.

Because we have observed and predicted values from regression trees, we can also calculate residuals for each observation and use these residuals as a diagnostic check for the appropriateness of the tree and whether assumptions have been met. Normality of predictor variables is not a concern because only the rank order of a variable governs each split, although transformation of the response variable to alleviate variance heterogeneity may be important (De'ath & Fabricius 2000). It is also useful to watch out for unusually influential observations (outliers). Regression trees may be more robust to collinearity among the predictors than traditional linear regression models, especially as they are mainly used for prediction. However, the binary splitting algorithm may end up choosing between highly collinear predictors at random, making the final predictive tree sensitive to small changes in the data. Methods based on multiple trees (Section 9.3.2) will be more reliable under collinearity.

The regression tree analysis using the data from Loyn (1987) is in Box 9.3. We transformed patch area and distance to the nearest patch to be consistent with our multiple regression model analysis, although transforming predictors makes little difference with regression trees. The first split was between patches with grazing indices from one to four and those with a grazing index of five. This former group was further split into two groups with a log area above and below 1.145 (approx. 14 ha). The smaller patches were further split into those isolated before and after 1964. The final tree is presented in Figure 9.1. Using the one SE rule, a simpler tree that doesn't include the split by year isolated would have been chosen. Note that the predictor importance measure has the same three predictors at the top (grazing intensity, year isolated, log patch area) as we found using relative importance measures from linear models based on partitioning r^2 (Box 9.1). However, their rank order is quite different.

The main advantages of regression trees are that they can handle various response variable types (continuous, categorical, even multivariate), they don't assume linear relationships between the response and the predictors, and they deal with missing data more easily than linear regression models. However, the final regression tree might not be a particularly good predictor for new observations, even after pruning based on cross-validation using the original dataset. One way to address this issue is to combine multiple trees from the one dataset to make predictions, analogous to model averaging described for linear regression models in Section 9.2.4.

9.3.2 Bagging and Boosted Regression Trees

Standard regression tree analysis results in a single tree based on a particular stopping criterion and/or minimizing cross-validated prediction error. Even with cross-validation, the final tree may still have high prediction variance – it may not be a particularly good predictor with new data. With multiple regression models, one way of dealing with this problem was to use model averaging to consider model selection uncertainty and produce a more reliable predictive model. There are also ways of dealing with regression tree uncertainty that produce multiple trees, which are combined to provide a more reliable tree with better predictive capacity.

Bagging (bootstrap aggregation) regression trees incorporate bootstrap resampling to produce multiple trees (Breiman 1996; Kuhn & Johnson 2013). An unpruned regression tree is fitted to each bootstrap sample, each tree is then used to predict values for a new test dataset (or the existing training dataset), and the predicted values from each bootstrapped tree are averaged. Bagging should reduce the prediction error compared to the single tree produced using standard regression tree analysis. A modification of bagging is random forests, where a random sample from the full set of predictors is used to determine each split in each bootstrapped tree. If there are a few very strong predictors in the dataset, the multiple bagged trees will be correlated because the early splits in most trees will be based on these strong predictors. Random forests reduce this correlation among the bootstrapped trees by using a random sample of predictors at each split. Bagging and random forests can both provide measures of predictor importance as described for standard regression trees. One downside of bagging trees and random forests is that because we average the predictions from many trees generated using bootstrap resampling, we can't produce a single descriptive tree as in Figure 9.1. This is not really an issue, as these modifications of standard regression trees focus on improving their predictive ability.

We calculated the prediction error (as RMSE) for the existing observations in the Loyn (1987) dataset, comparing a single unconstrained standard regression tree ($cp = 0$, to match the bagging algorithm, which doesn't constrain the bootstrapped trees), a bagged regression tree, and a random forest tree (both based on 500 bootstrapped samples). The RMSE for the bagged and random forest trees were similar (6.751 and 6.734) and considerably smaller than the single unpruned tree (8.633), indicating the advantage of bagging. However, the standard tree would usually be pruned to improve prediction.

Boosted regression trees (De'ath 2002; Elith et al. 2008; Kuhn & Johnson 2013), or BRTs, also generate multiple trees from the one dataset, but the trees are developed in sequence, with each tree "learning" from the earlier trees to improve the trees' predictive capacity. The algorithm is sometimes called gradient boosting because each step in the learning process reduces some measure of error, such as the mean squared prediction error. The method uses the residuals from previous trees as the response variable for later trees in the learning sequence. Boehmke and Greenwell (2019) describe this as each tree learning from the previous tree's mistake. The algorithm is complex, but the references cited above provide readable introductions. Note that with boosting, unlike bagging or random forests, each tree learns from previous trees, the trees are constrained in size, and they contribute unequally to the final predictions.

In contrast to bagging, where the only decision required by the researcher is how many bootstrapped samples to use, and random forests, where the number of random predictors to choose for each split is additional input, the boosting algorithm requires several *a priori* decisions by the researcher:
• Learning rate, sometimes called the shrinking rate, determines each tree's contribution to the final predictive

---

Box 9.3 ® Worked Example of Regression Trees: Bird Abundance in Remnant Forest Patches

A regression tree for Loyn's (1987) data related the abundance of forest birds in 56 forest fragments to log patch area, log distance to nearest patch, grazing intensity (treated as continuous), altitude, and year of isolation. Transformations of predictors to improve linearity are unnecessary for regression trees (and made almost no difference to this specific analysis), but we kept these predictors transformed to match our previous analyses of these data (Boxes 8.2 and 9.1). We used the ANOVA method of maximizing the between-groups SS for each split and used a change in r^2 of 0.01 as a default *cp* value. No other tree-building constraints were imposed. The residual plot from fitting the regression tree did not reveal any strong variance heterogeneity nor outliers.

The summary of our tree-building was as follows.

Split	*cp*	Mean CV prediction error	SE CV prediction error
0	0.467	1.040	0.134
1	0.232	0.720	0.125
2	0.045	0.441	0.082
3	0.010	0.409	0.078

Note that the mean coefficient of variation (CV) prediction error is scaled, so the first node is approximately 1. Our tree had three splits and four nodes (Figure 9.1) based on a minimum change in r^2 (*cp*) of 0.01; the r^2 for the fit of this tree was 0.744. The first node in the tree was between 43 habitat patches with grazing indices from one to four and the 13 patches with a grazing index of five, resulting in an improvement in r^2 of 0.467 (*cp* column). This former group was further split into two groups, 24 patches with log area less than 1.145 (approx. 14 ha) and 19 patches with log area greater than 1.145, with a *cp* of 0.232. The final split separated the smaller patches into those isolated before 1964 and those after, with a *cp* of 0.045. Any further splits failed to improve r^2 by more than 0.01. Interestingly, the same three predictors (log patch area, grazing, and year isolated) in the regression tree were also in the best regression model based on AIC_C (Box 9.1).

The predictor importance measures, scaled to sum to 100, across the five predictors, were as follows.

Predictor	Importance
Log patch area	37
Year isolated	29
Grazing	20
Altitude	8
Log distance to nearest patch	5

The ranking of predictors is similar to that found by partitioning the r^2 from the fit of traditional linear models in Box 9.1, except that year isolated is ranked higher from the regression tree.

Note:

• Using the 1 + SE rule would have resulted in a simpler tree with only two splits as the mean cross-validated prediction error for this tree is within one SE of the tree with three splits (0.441 − 0.082 = 0.359).
• A minimum CV error was not reached. Halving the minimum change in r^2 to 0.005 resulted in the same tree, but the tree with two splits now had the minimum CV error.

Two predictors (altitude and log distance to nearest patch) did not contribute to the final tree.

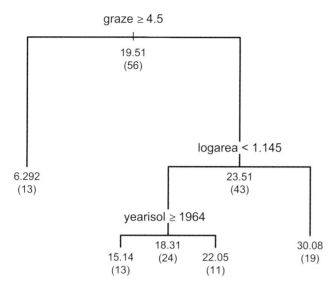

graze ≥ 4.5

19.51
(56)

logarea < 1.145

6.292
(13)

23.51
(43)

yearisol ≥ 1964

18.31
(24)

30.08
(19)

15.14
(13)

22.05
(11)

Figure 9.1 Regression tree modeling bird abundance in forest patches against log patch area, log distance to nearest patch, grazing intensity, altitude, and year isolated for the 56 patches surveyed by Loyn (1987). The criteria for each node are included, with left-hand branches indicating observations with values for that predictor below the cut-off and right-hand branches indicating observations with values for that predictor above the cut-off. The predicted value (mean) and number of observations for each leaf (terminal group) are also provided.

model (Elith et al. 2008). Boosting can cause overfitting and a struggle to find the overall optimal tree (Kuhn & Johnson 2013). The learning rate varies between 0 and 1, and smaller values (e.g. <0.1) tend to be better at avoiding overfitting but require more trees to be developed (Boehmke & Greenwell 2019) and more computation time.

• Tree depth or tree complexity, representing whether the model contains interactions, is commonly set to between one (where each tree is simply a "decision stump" with only two terminal nodes, i.e. a completely additive model) and about eight, depending on the size of the dataset (number of observations and predictors). Elith et al. (2008) suggested that simple trees (two or three nodes) would be most suitable with small samples.

• The number of trees used to develop the final boosted predictions, which will depend on learning rate and tree depth (Elith et al. 2008). Note that BRT software will sometimes have defaults that are probably too small (e.g. 100), and at least 1,000 trees are recommended.

• Bag fraction, the fraction of the dataset – chosen randomly – used at each iteration. This is like the bagging method described above and can be set by the researcher, although it is common to use a fraction of 0.5. Including this random fraction of the data at each iteration is called stochastic gradient boosting.

Ideally, the first three criteria will be optimized simultaneously. Elith et al. (2008), Kuhn and Johnson (2013),

and Boehmke and Greenwell (2019) give examples of this optimization process with suitable code. One approach is to try different combinations of learning rate (e.g. between 0.001 and 0.1) and tree complexity (e.g. between 1 and 5) to see which provides the lowest prediction error and reaches the recommended number of at least 1,000 trees.

As with bagging and random forests, we can't get a single tree diagram with BRTs because we aggregate predictions from many trees. However, we can get partial dependence plots that show each predictor's effect on the predicted response considering the average effects of the other predictors (Elith et al. 2008). We can also get a measure of predictor importance as described for standard and bagged trees and random forests.

We fitted BRTs to the data from Loyn (1987) in Box 9.4. The BRT model that minimized the CV prediction error and reached at least 1,000 trees (with learning rate of 0.005, tree complexity of 2, and bagging fraction of 0.5) had a comparable RMSE to the bagged tree and random forest result.

Finally, bagging, random forests, and boosting will be most beneficial for more complex regression trees with many predictors. The advantages for simpler datasets and simpler trees will be less apparent, as shown with the Loyn (1987) dataset in Box 9.4. Bagging and boosting are general methods that can be applied to a range of statistical models, but biologists most commonly use

Box 9.4 Ⓡ Worked Example of Boosted Regression Trees: Bird Abundance in Remnant Forest Patches

We fitted boosted regression tree models to the Loyn (1987) data relating the abundance of forest birds in 56 forest fragments to log patch area, log distance to nearest patch, grazing intensity (treated as continuous), altitude, and year of isolation. As with the standard regression tree in Box 9.3, we retained the log-transformed predictors of patch area and distance to nearest patch to maintain consistency with previous data analyses. However, predictor transformations are unnecessary for regression trees.

We initially fitted a BRT with a bag fraction of 0.5, a learning rate of 0.01, and a tree complexity of 2 (as this was a relatively small dataset). We also included 10-fold cross-validation. We evaluated the BRT models using the minimum cross-validated prediction error (in this case, $SS_{Residual}$, i.e. deviance). Note that by using a bag fraction <1 and cross-validation, there is a stochastic component to the BRT modeling, so repeated model fits even with the same settings will produce slightly different results. For illustration, we just provide the output from one run for each model below, but in practice multiple runs would be used to get a range of CV prediction errors for each model. The first model had a minimum CV prediction error of 49.28 (±6.88 SE) at 550 trees. Ideally, we would want at least 1,000 trees, so we reduced the learning rate to 0.005 and refitted the BRT. The cross-validated prediction error was now 44.99 (±7.05) at 1,200 trees. We reached our target of using at least 1,000 trees to build the BRT with a lower prediction error, but given the one SE rule, we wouldn't distinguish these two models.

We then tried trees of different complexities using this learning rate of 0.005. All models reached at least 1,000 trees, and again, we wouldn't distinguish between any of these based on the CV prediction error and one SE rule. Given the small sample size and the relatively small number of predictors, more complex trees were difficult to justify, so the tree based on a complexity of 2 and a learning rate of 0.005 seemed reasonable. The RMSE from this model was 6.707, similar to that from the bagged regression tree and random forest models.

The partial dependence plots (Figure 9.2) show that birds are more abundant in larger patches with lower grazing intensity, with some tendency for more birds at higher altitudes. There was little relationship for year isolated and log distance to nearest patch. The variable importance measure (Figure 9.2) showed that log patch area was most important followed by grazing intensity and altitude. Compared to the standard regression tree (Box 9.3) and the partitioning of r^2 from the linear model (Box 9.1), the BRT changes the order so that year isolated is less important than altitude.

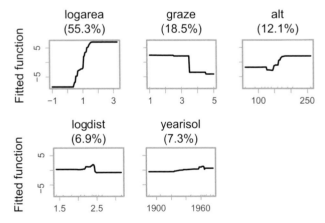

Figure 9.2 Partial dependence plots for log patch area, log distance to nearest patch, grazing intensity, altitude, and year isolated from a boosted regression tree model fitted to the data from Loyn (1987). The BRT had a learning rate of 0.005, tree complexity of 2, and bag fraction of 0.5, and resulted in 1,100 trees. The predicted (fitted) bird abundance is centered, and the relative importance of each predictor (scaled to %) is also provided.

them for improving the predictive capacity of regression and classification trees.

9.4 Key Points

• As linear models, especially multiple regression models, get more complex, biologists often wish to evaluate the relative importance of each predictor.

• Relative importance has been assessed with standardized partial regression slopes, although methods such as LMG and hierarchical partitioning that measure each predictor's contribution to explained variance are preferred.

• Biologists commonly wish to simplify their complex linear models by retaining those predictors that are most important and obtaining more precise predictions from the model.

• Model selection has been based on comparisons of simpler models to a full model (e.g. adjusted r^2), but methods based on information criteria such as the AIC are much preferred.

• Stepwise model selection should be avoided; comparing all possible models (unless the number of parameters is very large) is usually feasible.

• Any model selection process will rank all models from best to worst (e.g. based on AIC), but no single "best"

model should be relied on. Model averaging can produce a model that uses weighted averages of the regression coefficients from all models.

• Whichever model is finally chosen, it should be validated by either testing its predictive capability with new data or splitting the original dataset and developing the model on part of the data and testing it with the other part.

• Regression trees can be a useful and more robust alternative to traditional multiple regression modeling, especially for predictive models, producing a tree-like diagram. Complex regression trees can be improved by bagging, using bootstrap resampling to produce multiple trees that are then averaged, or by boosting, developing multiple trees sequentially so that predictive capacity is improved each time.

Further Reading

Model selection: Burnham and Anderson (2002) is the standard reference for model selection in linear (and generalized linear) models. Bolker (2008) describes model selection, and fit criteria, from an ecological perspective.

Regression trees: Kuhn and Johnson (2013) cover regression trees, including an introduction to boosting and bagging, as well as predictive linear models more generally.

10 Random Factors in Factorial and Nested Designs

10.1 Fixed vs. Random Effects and Mixed Models

In Section 4.2, we introduced two types of predictors in linear models, and their distinction is critical when the predictors are categorical. The most common type is a *fixed predictor*, where all the groups or values of the predictor of interest are included in the analysis. Fixed predictors in biological research are discrete factors such as experimental treatments or groups in sampling programs, or continuous covariates (treated as fixed although often selected randomly along with the response variable). We cannot extrapolate our statistical conclusions beyond these specific groups. If we repeated the study, we would usually use the same groups again. Linear models based on fixed predictors are fixed effects models. The fixed effects are the fixed but unknown model parameters (coefficients) associated with the predictors. Statistical inference focuses on estimating these coefficients, which are group means (or differences between them) for categorical predictors or regression slopes for continuous predictors.

A *random predictor* is nearly always a categorical factor where we are using a random subset of all the possible groups of the factor. We usually wish to make inferences about all possible groups from our sample. If we repeated the study, we would usually take a different sample of groups. Random factors in biology are usually spatial units like sites or blocks or subjects (individual animals or plants) and indicate clustering in the study design where observations are clustered within groups of the random factor. Observations within each random group are not independent. The parameters for random effects are considered random variables rather than fixed constants in the population, and therefore will have a probability distribution. If the random effects are of

biological interest, we estimate their variances (and potentially covariances), termed variance components (VCs). Estimating, and evaluating null hypotheses about, VCs is a key part of fitting models with random effects.

Classifying predictors as fixed or random can be difficult at times (Box 10.1), but it is crucial because it defines the correct statistical model for the data. Decisions should be made before data collection, and the analysis and interpretation should follow the classification.

10.1.1 Designs Applicable to Mixed Models

We generally have two statistical aims for mixed model designs:

- Estimating the effects of the fixed predictors on the response variable, which has two aspects:
  - Accounting for the clustered nature of the data and lack of independence between observations within random groups.
  - Using the random factors to reduce unexplained variation and improve the clarity of our examination of the fixed effects.
- Estimating the variability in the response variable across all the possible groups (clusters) of the random factor(s).

The mixed model literature also distinguishes between random intercept models and random slope models. These models usually apply when we have a multilevel regression design with one or more continuous covariates (Section 8.2), such as measuring body size as a covariate alongside the primary response measure of organism performance or using time as a continuous covariate in repeated measures designs (see Chapter 12). We can now estimate the variance, across all possible groups of the random factor, in the regression coefficients for the response against the covariate. A random intercept

10.1 Fixed vs. Random Effects and Mixed Models

195

Box 10.1 Fixed vs. Random Effects: Sometimes Confusing, But Important

Classifying predictors, particularly categorical ones, as fixed or random is still a subject of some discussion, particularly when we want to draw inferences beyond the groups used in the study. Under these circumstances, our ability to extrapolate from a study sample to a broader population relies on a clear understanding of the relationship between the two. We are most confident when that sample is random. Many authors have proposed that how we select groups of a factor is critical to whether it can be defined as random (e.g. Vittinghoff et al. 2012), and Milliken and Johnson (2009) argued that a factor should be considered fixed unless it was demonstrated that the groups were chosen at random. Bolker (2015) relaxed that requirement slightly, suggesting that the groups used should represent a broader population. Other statisticians have recommended abandoning the distinction between fixed and random effects and just focusing on the relevant coefficients, all treated as random effects, in an appropriate linear mixed model (Gelman & Hill 2006), although this path might create difficulties when trying to predict new situations.

Criteria

A range of criteria is available, and Bolker (2015) provided a good summary table of them. Briefly,

An effect is fixed is where:

- The groups we use are all those of interest.
- We are interested in comparing the effects of some of these specific groups.
- If we did the study again, we would choose the same groups.
- Some authors calculate the ratio of groups used (p) to groups available (P) and use it to determine the form of the statistical model, with $p/P \Rightarrow 1$ for fixed, 0 for random.

A random factor is where:

- The groups have been chosen from some larger population of potential groups.
- The individual groups are not of much interest, but their variance might be.
- We have some interest in generalizing from the groups used to the larger population.
- If we did the study again, it is unlikely that the same groups would be selected.

It Can be Complicated

The same factor might be considered fixed or random, depending on the aims of the study and the hypotheses of interest (e.g. Gelman & Hill 2006; West et al. 2015). For example, consider an ecological experiment repeated at several places. This repetition is desirable either given the "repeatability" crisis (see Chapter 1) or because it allows us to see whether the ecological process of interest is consistent. In this case, we can identify at least three scenarios:

- We can only identify a few locations ("Sites") where it is relevant to repeat the experiment – for example, if the species concerned is rare or a particular habitat is uncommon. In this case, Sites could be a fixed factor because we are studying all sites of interest.
- We randomly pick multiple sites from a population of candidate sites, aiming to estimate the variance in the ecological process. Here, site would be a random effect, as would its interaction with our ecological process.
- A more complicated situation could arise where the sites chosen are not the only ones of interest, but we do not choose them randomly, so they do not allow inferences about a broader population of groups. In a world where resources are unlimited, we'd repeat the experiment enough to treat sites as a random effect (i.e. to estimate the relevant variance sufficiently accurately), but in a more typical, resource-limited situation, we could repeat the experiment under quite different conditions. We might, for example, choose sites very different from each other, though if we did the study again, we may not choose the same extreme situations.

If we pick extreme situations and find little difference in our ecological process, we might be disinclined to investigate the situation further, even though we don't know much about what happens between the extremes. If we find an important difference, our next step might be to repeat the experiment many times to estimate the variance in the particular process or to group our repeats in some explanatory way, such as using several places along a productivity gradient. This two-stage process might save us some effort. Site in this case should be treated initially as a fixed effect, because although these places aren't the only ones of interest, we don't (and can't) use them to make any inferences about other places.

Box 10.1 (cont.)

Time (e.g. months or years) as a factor can be tough to allocate as fixed or random, although it is difficult to envisage a sequence of days, months, or years being a random sample from a population of times to which we could justifiably extrapolate, but it does depend on the time scale of interest. It's hard to see how we'd design a study to estimate inter-annual variation with years as a random effect if, for example, our random numbers suggested collecting data in years 1, 4, 9, 10, and 17. It's probably even harder to envisage a successful grant or thesis proposal using this plan! On the other hand, if we are interested in short-term variation, for example, within a season or a month, it's easier to think how sampling days could be allocated more or less randomly. This time scale has been an issue in environmental monitoring, which often focuses on times just before and after particular human activities, but times are often treated as random (Downes et al. 2002).

The other problem is that even if we estimate temporal variances, we can only make inferences about future events by assuming that nothing is changing, because the "population" of times from which we have sampled is constrained. That is a concern when we know there is ongoing change, such as ecological work in the face of changing climates.

But It Matters

Being clear about which effects are fixed and random is important, even if it is complicated. When an effect is important to us, the fixed–random decision can be viewed as a trade-off. Treating it as fixed means that we might be confident about our assessment of the effect but can't generalize. Alternatively, viewing the effect as random means that we can no longer say much about specific groups and may lose sensitivity for assessing other fixed effects, but we can generalize. Neither option is correct in all situations, and the decision depends on the research aim.

The classification of an effect as fixed or random can alter how we fit and interpret statistical models, and as seen in this chapter and following ones, random effects counterintuitively alter how we assess fixed ones. The changes may not be obvious unless we specify the statistical model and methods for assessing effects in detail. When we are clear how particular effects of interest will be evaluated, we can make sensible decisions about where to allocate resources in our design. If we are confused at the design stage, we may finish up fitting a model for which the sampling design provides reduced sensitivity, and resources have been wasted. It is much better to avoid these situations by thinking carefully before starting.

Our Recommendation

• As with all data collection exercises, think at the planning stage about the statistical models that will be fitted and how to allocate sampling effort best to those models. Doing this should automatically trigger a discussion of whether each predictor is fixed or random.

• Think carefully about whether random effects are present simply to remove "noise" from the data (i.e. taking account of clustering in the data) or if they are of interest.

• Be clear how any random effects might alter your assessment of other effects of interest and design your sampling accordingly.

• If estimating random effects is important, you will need a moderate number of instances of this effect to be confident about it. For example, Bolker (2015) suggested that when a random effect is represented by ≤ 5 instances, that sample's representativeness is so uncertain that the effect may be better treated as fixed.

• It is very difficult to justify a fixed factor, where groups were chosen nonrandomly, being modeled as a random effect simply because researchers want to generalize, because the groups are unlikely to represent any larger population.

• Make your decisions about fixed vs. random prior to collecting data (ideally) and before any analysis and interpretation (definitely!).

• Check your software carefully, as some packages default to fixed effects unless specified otherwise. Expect to declare each effect as fixed or random.

Acting Ethically

There are two temptations to resist during analysis and interpretation.

It is not uncommon to read papers in which models were fitted with all or most factors fixed through the Methods and Results sections, but then the discussion generalizes happily as if effects were random. This is inappropriate because the interpretation is not consistent with the model fitting. It is generally done, we suspect, as a way of

10.1 Fixed vs. Random Effects and Mixed Models

197

Box 10.1 (cont.)

"selling" the results to a broader audience or higher-impact journal. Our sense is that interpreting fixed effects as if they were random is common, but the reverse error of interpreting random effects as fixed is rare.

A second situation is where the change occurs between design and analysis, as a designed random effect is analyzed as fixed. There may be circumstances when this is appropriate, but they are rare, such as:

- The study's aim changed when it was underway, so particular groups of the random effect become of interest.
- Unexpected events limited the number of groups of a random effect. An ecologist might, for example, find that an experiment planned for multiple sites could not be sampled fully because a pandemic limited field work or a bushfire destroyed some sites. In such a case we might decide that there are too few remaining sites for generalization, or the losses were nonrandom. In this case, a particular effect might better be viewed as fixed.

The first temptation should be resisted, but it can be dealt with by referees and readers, who should be vigilant toward mismatches between results and discussion.

The second is more worrying. Our impression is that switches from random to fixed effects do not often occur for the good reasons above, but happen more often when there were design errors, and it is realized that tests for particular effects are weak, or, most worryingly, as a form of *P*-hacking. It is particularly worrying when it happens because a test of an effect is not "statistically significant" with the right mix of fixed and random effects, but becomes so when some effects are reclassified as fixed. Pre-registration of sampling designs and statistical models is the best way to prevent this behavior.

Linear models with only random factors are random effects models. They are most commonly used with nested/hierarchical designs, where the interest is in determining how much variation in the response is attributable to hierarchical spatial or temporal scales (Section 10.5). Other designs that include only random factors are uncommon. Mixed effects models include fixed and random effects and are also common. We will describe some in the next section.

model allows the intercepts to vary across groups of the random factor but uses a common slope. A random slopes (or coefficients) model allows intercepts and slopes to vary across groups. Random slopes are incorporated into the model by including an interaction between the covariate and the random effect, analogous to evaluating homogeneity of slopes in an analysis of covariance, described in Section 8.2.4 (Figure 10.1). The interaction between a fixed categorical factor and a random effect is also assessing whether the fixed effect varies across all possible groups of the random factor, so it can also be termed a random "slope" model.

We will now describe some mixed model designs. We can represent them with factor relationship diagrams, which we use in Figure 10.2 to illustrate the differences between nested, crossed, and partly nested designs (see also Schielzeth & Nakagawa 2013). We will use a study looking at the birthweight of rat pups in litters from different randomly chosen female rats (Pinheiro & Bates 2000; West et al. 2015). Only one litter was recorded for each rat for this design, so rat and litter are interchangeable terms. We will follow the terminology of West et al. (2015) for hierarchical designs. Observation units will be

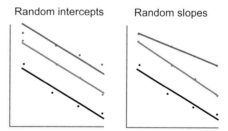

Figure 10.1 Random intercepts and random slopes approaches for a design with a continuous fixed predictor and a random effect with three groups. In the random intercepts model, a single pooled slope is calculated and applied to each group, and then an intercept is calculated. In the random slopes model, a separate slope (and usually intercept) is calculated for each random group.

level 1 and the clusters of those units (random factor groups) will be level 2. Designs with three or more levels are also possible.

10.1.1.1 *Single Random Factor Designs*

The simplest design has a single random factor, litters from female rats representing level 2 groups. Litters

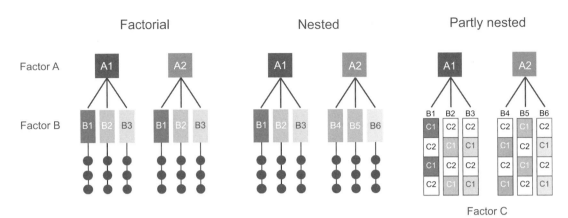

Figure 10.2 Layout of three common multifactor designs, showing differences between factorial, nested, and split-plot (partly nested) arrangements of factors.

(and therefore female rats) are considered a random sample of all possible litters, so they will be treated as a random effect. Our response variable is the weight of each pup in a litter, so individual pups within each litter are level 1 units. Our biological interest is in the variability in mean weight of pups. The analysis of these single random factor designs will be described in Section 10.3.

10.1.1.2 *Nested or Hierarchical Designs*

The previous design was nested because we had pups nested within litters (random groups). Nested designs can easily be extended to multiple levels (Figure 10.3). The levels of nesting nearly always represent random factors, except for the highest level, which can be fixed or random. For example, the rat pup data includes a fixed factor, three doses of a specific compound, to which the rats were allocated in a randomized design. We now have litters (female rats) nested within treatments and pups within litters. These nested designs could also just include random factors, such as randomly chosen genetic lines, randomly chosen rats/litters within each line, and pups within each litter. Our interest now might be in partitioning the variation in the response among these scales. The relative magnitudes of the variances associated with each level (genetic lines, litters, pups) would be of specific interest. Models for nested designs will be described in Section 10.5.

10.1.1.3 *Crossed and Block Designs*

Random factors can also be included in crossed or factorial designs. Our simple single random factor design could be extended by including a categorical factor at level 1, whereby pups from each litter are further classified by groups of a fixed factor such as gender (male vs. female;

see Figure 10.3). Like the crossed designs described in Chapter 7, we can now assess the fixed effect of gender on pup weight and whether this effect is consistent across litters (i.e. the gender by litter interaction, itself also a random effect). These crossed mixed model designs will be covered in Section 10.6. A random factor in crossed designs is often a blocking factor where the groups of the fixed factor are grouped into (usually) spatial blocks, with one or more units of each groups in each block. Blocking can be a useful design strategy to reduce noise.

10.1.1.4 *Split-Plot Designs*

The split-plot or split-unit is a more complex design with at least two fixed factors and a random factor. Here, one factor is applied to the level 2 groups (sometimes termed "plots"), and a second factor is applied to level 1 sub-units ("sub-plots") within each group. Split-plot designs incorporate aspects of crossed and nested designs. In the rat pup dataset, we have the fixed treatment (three groups of a compound) applied to the level 2 groups (rats and their litters, representing plots) and the second fixed factor, gender, applied to the level 1 units, pups within litters (representing sub-plots). These split-plot designs are sometimes termed partly nested as they include both crossed and nested components. They are common in biological research and will be described in Chapter 11.

10.1.1.5 *Repeated Measures and Longitudinal Designs*

A major application of mixed models is analyzing designs where units, often termed "subjects," are recorded multiple times. These designs usually also include fixed factors applied to different groups of subjects. The rat

Figure 10.3 Illustration of the rat example showing more complex nesting. The top left panel shows the basic design with three rats, each with three pups. The panel to the right adds an experimental treatment with three groups (shown by different shading), and rats nested within each treatment. The lower panels add to the complexity; in the lower left panel, male and female pups (shown with different shading) are measured within each litter. The lower right panel has rats with variable litter sizes, allowing litter size to be added as a predictor.

pup dataset didn't have a temporal component, but imagine the weight of each pup was measured at four times to examine growth patterns. The repeated observations within subjects represent level 1 data, and the subjects themselves represent level 2 random groups. These designs can be analyzed using a similar linear model to that for split-plot designs. Repeated measures (RM) designs will be covered in Chapter 12.

10.2 Fitting Linear Models with Fixed and Random Factors

We illustrate the approaches for fitting models with random effects with the stream diatom dataset introduced in Box 6.5, with diatom diversity (response variable) measured from stations (sampling units) along streams in Colorado. The original analysis modeled diatom diversity against the fixed effect of zinc level (four groups). In this analysis, we will make use of the clustered nature of the dataset with the stations coming from six streams that we will treat as a possible random sample from all metal-polluted streams. There were 4–7 stations as level 1 sampling units on each stream (level 2 groups). Stream is a random factor, and our interest is in the variation in mean diatom species diversity across all possible streams in this region. A linear model would include the fixed overall mean diatom diversity, the random effect of any stream, and an error term (Section 10.3.1). Given we have also recorded zinc concentration for each station, we could include the fixed zinc concentration covariate plus the random interaction between zinc concentration and stream, allowing random regression slopes (Section 10.4). Our interest is estimating the fixed effects and also the VCs associated with the random effects. We will emphasize two methods for fitting these models, using OLS ANOVA models and using ML/REML mixed effects models.

10.2.1 Traditional OLS "ANOVA" Models Approach

The traditional OLS ANOVA approach essentially describes and fits mixed models as if all effects were fixed. In our examples, we would fit a single-factor linear model as described in Chapter 6 or an ANCOVA-type model including zinc concentration as a covariate from Chapter 8, except that stream is now interpreted as a random effect.

10.2.1.1 *Estimation and Tests*

The VCs associated with random effects are estimated by equating the mean squares from the ANOVA to their expected values (expected mean squares or EMS). Confidence intervals are based on the χ^2 distribution. Confidence intervals are only approximate for VCs above the residual, and Satterthwaite's adjusted df for linear combinations of mean squares are often recommended as an approximation (Kutner et al. 2005; Milliken & Johnson 2009). Confidence intervals in designs with unequal sample sizes are problematic, although more complex approximations are available for some situations (Hocking 2013; Searle et al. 1992). Tests on VCs are based on constructing Fs derived from the relevant EMS.

Any fixed effects, such as the overall mean in this example, are estimated using OLS as described in previous chapters, with SEs and CIs based on t statistics. Note that the EMS for fixed effects in mixed models will often include terms associated with the random effects, so F

tests may not use the $MS_{Residual}$ as the denominator, in contrast to models where all effects are fixed.

10.2.1.2 *Assumptions and Diagnostics*

The OLS models described in earlier chapters had only one random component, the model error terms. The models in this chapter and the next two have at least one extra random term. The normality and homogeneity of variance assumptions applicable to the error terms also apply to the other random terms in the model. These models are essentially fitted as though all the effects are fixed, so the normality assumption for the random effects is rarely scrutinized. The usual methods for assessing assumptions (e.g. residual plots) are applicable.

In mixed model designs, the complete independence assumed for fixed effects models does not apply. Observations within random groups will be correlated, but we assume that the covariances between pairs of observations are equal within each random group and consistent across any groups of the fixed factor. Combined with equal variances, this pattern of covariances is termed compound symmetry, or a slightly less strict pattern, sphericity. This assumption is likely to be met where there is randomization in the allocation of fixed factor groups to units at the different levels in a hierarchy – for example, different treatment groups of a fixed experimental factor are randomly allocated to units within a block (Section 10.6.1). Where this is impossible, such as in a design where random "subjects" are recorded through time (and time cannot be randomized), compound symmetry is unrealistic. This issue will be considered further in Chapter 12.

10.2.1.3 *Overview*

This traditional OLS approach for fitting mixed models to fully balanced designs with fixed and random factors is straightforward and has served us well for many years. If appropriate sampling and/or experimental design principles have been followed, we have a consistent approach to linear modeling that, for better or worse, requires few decisions by the researcher except to check underlying distributional and the more restrictive assumptions about constant variances and covariances. There are, however, some important limitations on using OLS-based linear mixed models.

• The model structure does not explicitly distinguish fixed and random effects. This distinction only arises in deriving the EMS and the calculation of F statistics. There is still some statistical controversy about the correct derivation of EMS in OLS mixed models with interaction terms (Section 10.6.2).

• This approach is less suited to designs where continuous covariates are recorded along with the response variable at the different hierarchical levels of the designs outlined in Section 10.1. Our interest is whether intercepts and/or slopes vary between the random groups. We could fit a random effects ANCOVA-type model as described above (see also Section 8.2) and determine the EMS based on relevant effects being random instead of fixed. However, this approach is not commonly applied, as the EMS can be tricky to derive (Hector 2015), especially if we have multiple random effects in a nested or crossed design. Treating the intercepts and/or slopes as responses and using a single-factor random effects model might be an option. However, the mixed effects models described in Section 10.2.2 are far more helpful under these circumstances.

• The OLS approach to mixed models isn't very adaptable to unbalanced designs. While point and interval estimates of fixed effects are reasonably robust to unequal sample sizes, this is not the case for random effects, and calculating CIs for VCs is very difficult.

• Observations within the same group of a random factor are likely to be correlated, and the OLS model assumes these correlations (covariances) are constant within each random group. We can use generalized least squares (GLS) to fit some fixed effect models with different variance–covariance structures, and there are methods based on adjusting degrees of freedom to deal with complex correlation structures (Section 12.1.2), but they are somewhat inelegant and inflexible.

• The MS for the random effect will sometimes be less than the residual MS (and their F will be less than 1), resulting in a negative estimated VC. Variances cannot be negative. The usual recommendation is to convert a negative VC estimate to zero (Kutner et al. 2005; Milliken & Johnson 2009). Hocking (2013) and Searle et al. (1992) argued that negative VCs suggest an inappropriate model has been applied or there may be serious outliers in the data, so negative components might be a useful diagnostic tool. Negative values can also occur by chance, particularly with small sample sizes. Using an estimation method like REML that precludes negative VCs is a better approach.

10.2.2 Linear Mixed Effect (or Multilevel) Models Approach

The second approach is variously called linear mixed effects modeling (West et al. 2015), hierarchical linear modeling (Raudenbush & Bryk 2002), multilevel modeling (Snijders & Bosker 2012), or multilevel regression modeling (Gelman & Hill 2006; Gelman et al. 2020). In

biology, linear mixed (effects) modeling is the more common term (e.g. Bolker et al. 2009; Zuur et al. 2009). Mixed effects models are fundamentally different from OLS models because the random effects are identified explicitly in the model, which is fitted with ML or REML.

Mixed effects models are often thought of as multilevel regression analyses (Fitzmaurice et al. 2011; Zuur et al. 2009). Multilevel regression involves two (or more) stages. The first fits a linear model, relating the response variable to one or more fixed covariates or factors, to the observations within each random group of units. With our diatom example, we could estimate the intercept and/or the slope of the relationship between diatom diversity and zinc concentration for each stream. If there are also categorical factors recorded at level 1 (e.g. sex of rat pups), they are included in the model fitted at this stage. For RM designs where time is the covariate (Chapter 12), separate regression models relating the response variable to time would be fitted for each subject.

The second stage treats each random group's regression coefficients (intercepts and/or slopes) as random variables. We model the variance (and covariance) of these regression coefficients (or means) against any associated fixed covariates and factors at level 2 (e.g. litter size and treatment in the rat pup example). In the stream dataset, this second stage could be a random intercept model that allows the intercepts to vary across the streams but assumes that the slope of the regression relating diatom diversity to zinc concentration is the same for all streams. A more realistic scenario is a random intercept and random slope model that allows for varying regression models across groups. If there are higher-level random groups, such as streams being nested within randomly chosen catchments (watersheds), we would have a third stage where the relevant means or regression coefficients from the second stage are modeled across the level 3 groups and any associated fixed covariates or factors. The parameters from modeling the level 1 regression coefficients (intercepts and/or slopes) at level 2 or higher are sometimes called hyperparameters (Gelman & Hill 2006), to distinguish them from the parameters of the level 1 regression models. They are model parameters that are not group- or cluster-specific.

Some points to note:

• Even without a continuous covariate for each observation at level 1, the descriptor "multilevel regression" is relevant as we still fit a linear model. However, most disciplines reserve the term for designs with a continuous covariate within each random group.

• The two steps in the multilevel regression approach are usually incorporated in a single model (Gelman & Hill

2006) that includes fixed and random effects, including their hierarchical structure (level 1, 2, etc.), plus any relevant interactions.

• The model terms in mixed effects models are essentially the same as for the OLS models, except that fixed and random effects are distinguished when coding in statistical software and model fitting is based on ML/REML.

10.2.2.1 Estimation and Tests

In general terms, mixed effects models are fitted using ML or REML, often incorporating GLS (West et al. 2015) or penalized least squares (Bates et al. 2015). Estimation of parameters of mixed effects models, particularly the variances and covariances of the random effects, can be challenging, especially if there are unequal sample sizes or missing cells. Maximum likelihood estimates require "integration over all possible values of the random effects" (Bolker et al. 2009), weighted by their probability of occurring; this is termed the marginal likelihood (Bolker 2015), a modification of the usual likelihood that considers the random effects as random variables. Algorithms such as Fisher scoring and Newton–Raphson (NR), introduced in Box 4.3 and commonly used for fitting GLMs, can also be used for models with random effects. The expectation-maximization (EM) algorithm can also be used for complicated likelihood functions by itself or to provide starting values for the other methods (West et al. 2015). Other, more advanced, algorithms can be used, especially for generalized linear mixed models (GLMMs) (Chapter 13), including penalized quasi-likelihood and the Laplace approximation (Bolker 2015; Bolker et al. 2009). No single algorithm works best in all situations. Different software packages might use different algorithms for estimating mixed effects models and, for more complex models, they may produce different results.

While the fixed and random effects can be estimated simultaneously, it is usually simpler to "profile out" the fixed effects – that is, construct a profile likelihood function for the random effects maximized over the fixed effects (see Box 4.3) – to estimate VCs and their CIs (West et al. 2015). Bootstrapping procedures can also be used to construct CIs on VCs, as can Markov chain Monte Carlo (MCMC) methods based on the posterior distributions of the random effects. Markov chain Monte Carlo estimation is usually part of a Bayesian framework, but can sometimes be applied in a frequentist setting. Profile likelihood, bootstrap, and MCMC CIs are feasible with moderate-sized datasets. Tests of individual VCs (i.e. the H_0 that a VC equals zero) can be based on likelihood-ratio (LR) tests comparing REML models with and without the

random effect of interest. However, the resulting P-values underestimate the evidence against the H_0 as a VC can only be different from zero in one direction so the H_0 is on the boundary of possible parameter values; Bolker (2015) recommends dividing the resulting P-value by 2 if testing the only random effect in a model. Inference from CIs is probably more reliable than that from LR tests for VCs.

Confidence intervals for fixed effects are calculated, as usual, based on the SEs and the t distribution, and tests of H_0s can be based on t statistics or Fs from an ANOVA on fixed effects. Standard errors for fixed effects are biased downwards with ML and REML but can be adjusted for CIs and testing H_0s (West et al. 2015). The Kenward–Roger approximation, based partly on the Satterthwaite method (Section 2.5.6.1), is usually recommended. Models with and without specific fixed effects can be compared using LR statistics or AIC, but the models must be fitted with ML, not REML, for this comparison.

The focus for biologists is usually on getting the random effects structure correct, so we have the most reliable estimates and tests of fixed effects. We are often also interested in the random effects, as they provide insight into the patterns of variability across our hierarchical design. Occasionally, patterns might also suggest other groupings hidden among the random clusters (Winter 2019). We can explore the random effects formally, but there are a couple of points to note:

• The random effects are random variables representing deviations from an overall mean (or intercept; West et al. 2015), rather than true fixed population parameters, so statisticians often refer to predicting rather than estimating effects (Fitzmaurice et al. 2011). For random effects, we are after the best linear unbiased predictor (BLUP) rather than the best linear unbiased estimator (BLUE), and the random effects are sometimes termed conditional modes (Bolker 2015).

• The estimate (or prediction) of each effect considers the data from the other groups of the random factor. Bolker (2015) described each effect as a weighted average of the effect for each group and the overall mean across all the random groups. The random effects are also sometimes termed shrinkage estimators (or predictors) because the effects are shrunk toward the population mean effect ("complete pooling") compared with estimating each effect based only on that group's data ("no pooling"), especially for groups with small sample sizes. Gelman and Hill (2006) provide an excellent overview of shrinkage and pooling in mixed effects models.

One of the strengths of mixed effects models is that we can incorporate different patterns of variances and covariances of the random terms in our model. The D matrix has the VC associated with each random effect in its diagonal and the covariances between random effects in the off-diagonal (West et al. 2015). There are two common structures for this matrix that we can use in our model fitting. One, sometimes termed the variance component structure, sets all covariances to zero, so only VCs of the random effects are estimated. This structure can reduce the complexity of the model and allow better convergence of the ML algorithms, especially with random slope models (Bates et al. 2015). The alternative is an unstructured matrix where variances and covariances are unconstrained. This requires many variance and covariance parameters to be estimated with complex models, which may be impossible except with large datasets.

The residual or R matrix has the variance of the error terms for each observation in its diagonal and their covariances on the off-diagonal. With standard linear models, we assume that the variances of these error terms are constant and the covariances are zero – that is, each observation is independent of the others; this is termed the diagonal structure. However, as we expect observations within the same group of the random factor to be correlated, we can use other R matrix structures, termed R-side effects by Bolker (2015). Different structures for the R matrix are most relevant for analyzing RM designs, and we will describe suitable options in Section 12.1.2. Biologists are often (too often!) dealing with relatively small sample sizes, which don't allow the estimation of too many additional covariance matrix parameters. Other more flexible variance–covariance structures (e.g. unstructured and Toeplitz; Fitzmaurice et al. 2011) are possible. However, they require the estimation of many covariance parameters, which is only possible with medium to large datasets. Not all mixed effects modeling software allows flexible specification of D and R matrices.

Maximum likelihood variance and covariance estimates are biased downwards, and REML estimation is preferred. Restricted ML estimators of variance integrate over the fixed effects (Bolker 2015) – that is, consider the loss of df from the estimation of the fixed effects parameters (Snijders & Bosker 2012). However, ML and REML both have a role in mixed effects models. The former is required for comparing models with different fixed effects using LR tests or AIC values, whereas the latter provides the most reliable variance and covariance estimates.

Some descriptions of mixed effect or multilevel models distinguish between the general mixed effects model we have been describing that includes the fixed and random effects and the implied marginal model (e.g.

West et al. 2015). The marginal model only includes the fixed effects and is sometimes called a population-averaged model – that is, averaged over all the possible groups of each random factor in the design. The error terms of the implied marginal model have a variance–covariance matrix that combines the R and D matrices. The implied marginal model is sometimes used when the estimation algorithm for the original mixed effects model doesn't converge reliably. While omitting all random effects is rarely necessary, the marginal model generates marginal residuals that may be useful for checking assumptions.

10.2.2.2 Assumptions and Diagnostics

Checks for assumptions use the usual techniques such as boxplots and residual plots. Residuals can be conditional (based on the model with fixed and random effects) or marginal (based on fixed effects from the marginal model). Conditional residuals are the usual software default, although West et al. (2015) suggested that the marginal residuals might be more suitable. Variance heterogeneity can be dealt with by the specification of the R matrix and transforming the response may also be useful. The assumption of independence of errors will not be met for most datasets requiring mixed effects models, and some mixed model algorithms default to assuming compound symmetry (Section 10.2.1.2). When this may not be realistic (e.g. RM designs), different covariance structures can be applied by specifying the structure of the R matrix. This will be examined further in Chapter 12.

We also assume that the random effects higher than the residual in more complex designs are normally distributed. This assumption is more important than for OLS models where the models are fitted as though all effects are fixed. Checking this assumption is difficult and usually ignored. Fortunately, Schielzeth et al. (2020) demonstrated via simulations that mixed effects models are very robust to nonnormal distributions of the random effects (and the error terms).

Many mixed effects models include continuous covariates (true multilevel regression models). In these cases, we also assume a linear relationship between the response and the covariate for each random group, and diagnostic tools we have illustrated for regression analyses are particularly useful. We also must be careful about collinearity, just as we cautioned for multiple regression models (Section 8.1.9). The precision of our estimates of fixed effects will be reduced if there are strong correlations among the fixed covariates in our mixed effects models, although again, mixed effects models are robust

(Schielzeth et al. 2020). Centering the covariates will usually reduce collinearity and may help interpretation.

10.2.2.3 Overview

Key attributes of these linear mixed effects (multilevel) models include:

- Fixed and random effects are explicitly distinguished in the model and fitted using ML/REML. Maximum likelihood is better for comparing models with different fixed effects using LR tests or information criteria, but REML is preferred for estimating/predicting variances and covariances of the random effects.
- Clustering in the design is incorporated into the model explicitly, so the analysis considers observations within random groups more correlated than those from different groups. Depending on the software, different variance–covariance (R) structures can be incorporated for the within-group observations.
- It is important to correctly specify the hierarchical design structure of the random effects in the model, so estimation and tests of fixed effects are based on df that represent the correct experimental or sampling units. As always, be suspicious of results from mixed models analyses where the df seem much higher than expected from the design and numbers of true experimental units.
- Restricted maximum likelihood mixed effects models are much more reliable than OLS models for estimating VCs, especially with unbalanced designs.
- Collinearity can be an issue with multiple covariates. Intercepts and slopes can be highly correlated across random groups, and this is exacerbated if there are covariates at two or more levels in the hierarchical design. Centering of covariates can help reduce collinearity.
- Maximum likelihood convergence to a reliable solution is not always possible, especially for complex models or messy datasets with missing values and/or small sample sizes (e.g. Boxes 10.2 and 10.3). Options for improving convergence include simplifying the model by omitting the more complex random effects, setting the correlation between random effects (e.g. the intercept and slope) to be zero, increasing the iteration limit of the ML algorithm, or trying a different algorithm.
- Unfortunately, the appropriate df for the fixed effects can be difficult to calculate, especially for complex models (e.g. random slope models) and unbalanced designs. Therefore, some software won't provide P-values for tests of the fixed effects. This is still an active area of debate and research in linear models.
- For relatively simple, balanced datasets, OLS and ML/REML mixed effect models produce comparable results for the fixed effects. However, the iterative nature of ML/REML estimation and the preclusion of negative VCs with REML can cause differences. The real advantage of mixed effects modeling is when we have covariates (multilevel regression) and unbalanced datasets.
- Terminology for these models can be confusing, including subscripting in formal model equations. The available software can vary in algorithms and estimation methods, provide different options for modeling covariances, and produce warning messages that can be difficult to interpret.

We will explain the structure of linear mixed effects models, including the multilevel regression approach, in Section 10.3 for single random factor designs. For subsequent mixed models in this chapter and Chapters 11 and 12, we won't distinguish between the OLS and mixed effects model statements as both essentially include the same terms, even if estimation and fitting the models is very different.

10.2.3 Modeling Strategies

We emphasized model selection methods for OLS multiple regression models with fixed predictors in Chapter 9. We try to find a model that maximizes the precision of our estimates and predictions while using the fewest predictors. Selection methods included assessing the relative importance of the predictors in a model and directly comparing the fit of models with and without the relevant predictors using statistical tests or preferably information criteria.

For designs with fixed factors or covariates and random factors, researchers might simply fit a single mixed model that includes all fixed effects and relevant interactions, and all random effects and their hierarchical (nested) structure (and possible interactions with some fixed effects for more complex models described in Section 10.6 and Chapters 11 and 12). Conclusions are drawn from the estimates of, and tests on, this model's fixed and random effects.

Model selection or simplification, however, can be important for mixed models for two reasons. First, we might be interested in predicting from our model, especially if the design can be viewed as a multilevel regression. As described in Chapter 9, simpler models might provide more precise and interpretable predictions. For example, in a multilevel regression, conclusions may be more straightforward if there is little evidence for varying slopes, allowing random slopes terms to be omitted. Second, the ML/REML algorithms do not always converge with numerous, possibly correlated, random effects. Overfitting is a risk, so a model with fewer parameters that still adequately explains and predicts the response variable might be desirable.

Model selection is more common with ML/REML-based mixed effects models than with OLS models, and West et al. (2015) describe two alternative strategies. A top-down strategy starts with all possible fixed effects and interactions, termed a loaded mean structure, and a chosen structure for random effects. It then simplifies the model by removing terms that don't contribute to the model's explanatory or predictive capacity. Unimportant random effects are removed first, and then fixed effects are considered. A bottom-up (step-up) strategy starts with a random effects structure and no fixed effects except an intercept (overall mean). The model is built by adding fixed effects at each level starting at level 1 – that is, build the model in a true multilevel fashion. Models with different structures for the R matrix can also be considered. Our preference is to fit the full model and examine all the fixed and random effects or use a top-down strategy.

Removing effects is commonly test-qualified (Colegrave & Ruxton 2017), comparing models with and without the effect of interest, based on F or likelihoods. Information criteria such as AIC_C can also be used to select between models. In both cases, model selection should be based on ML, even though the chosen model is best refitted with REML. We make two additional points. First, our biological knowledge should influence our decision, such as whether a particular random effect is of biological interest or importance. Second, we need to consider the appropriate sampling or experimental units for assessing fixed effects. For example, combining a nested effect with the residual in some designs can mean that one or more fixed effects are no longer tested against the appropriate units of replication (see Section 10.5.1).

10.3 Simple Random Factor Designs

We will illustrate fitting mixed models using a straightforward design with a single categorical random factor, then look at multilevel regression designs including a covariate. Designs with only random factors are not common in biology, but they are helpful for introducing the model and analysis options.

We have outlined the alternative analysis of the Medley and Clements (1998) study to use their six streams as a possible random sample from all metal-polluted streams in the southern Rocky Mountain ecoregion of Colorado. The analyses are given in Box 10.2.

This design has only a single factor, but it can be viewed as hierarchical or nested with random stations (level 1 units) within each randomly chosen stream (level 2 groups), although the term *nested* is usually reserved for designs with more complex hierarchies (Section 10.5).

Consider a design with a random factor A with p randomly chosen groups and n units at level 1 within each group. While it is common practice in the mixed models (especially multilevel regression) literature for model subscripts to represent level 1, 2 etc. from left to right – for example, y_{ij} would be observation i from the jth random group – we will follow our previous chapters where y_{ij} is the jth observation from group i.

For our modified example from Medley and Clements:

$$(\text{diatom species diversity})_{ij} = \mu + (\text{stream})_i + \varepsilon_{ij}.$$

In this model:

μ is the fixed population mean, such as mean diatom species diversity across all possible stations on all possible streams, estimated by the unweighted mean of the sample group means in the design.

ε_{ij} is random or unexplained error associated with replicate j from the ith group of factor A (e.g. station j in stream i). These random error terms are assumed to be normal with a mean of zero and a constant variance within each random group – that is, $N(0, \sigma_\varepsilon^2)$ with σ_ε^2 being the residual or error VC.

The other term is the random effect of any group of factor A (e.g. stream) on the response variable (e.g. diatom diversity). While the individual random effects are usually not of interest, their variance component is. The random effects are assumed to be $N(0, \sigma_\alpha^2)$, with the VC σ_α^2 measuring the variance in mean values of the response variable across all the possible groups of the factor that could have been used. We assume that these random effects and the error terms are independent of each other.

The models used for these designs are sometimes termed intercept-only, "empty" (Snijders & Bosker 2012), "null" (Finch et al. 2014; Raudenbush & Bryk 2002) or "random effects" (Zuur et al. 2009), as we haven't included any fixed factors or covariates recorded for each sampling unit within each random factor group. The only fixed component is the overall mean (or model intercept).

Because of the correlated nature of observations within each random group, we can calculate the intraclass correlation, calculated as the proportion of total variance due to the random effect of factor A:

$$\rho = \frac{\sigma_\alpha^2}{\sigma_\alpha^2 + \sigma_\varepsilon}.$$

The estimate of ρ is obtained by substituting the estimates of the relevant VCs into the equation. The intraclass correlation is not widely reported in biological research, with estimates of the actual VCs preferred.

Box 10.2 Ⓡ Ⓔ Worked Example of Single Random Factor Design: Diatom Communities in Metal-Affected Streams

Here we use the Medley and Clements (1998) data differently from the original analysis in Chapter 6. Now we are comparing diatom species diversity across streams. Streams are treated as a random factor, assuming they represent a random sample of all possible streams in this part of the Rocky Mountains. The design is unbalanced, with 4–7 stations (level 1 units) on each stream (level 2 clusters). We will fit a model that focuses on estimating the diatom diversity VCs for streams and stations within streams (i.e. a null or random effects model).

We used OLS to fit an effects model including the overall mean and the random effect of stream to the diatom diversity data. The residual plot indicated no variance heterogeneity or unusual values. The resulting ANOVA is given here.

Source	df	MS	F	P
Stream	5	0.366	1.41	0.251
Residual	28	0.259		
Total	33			

There is little evidence that the added variance in diatom diversity among streams is greater than among stations within streams.

We estimated the VCs using the ANOVA method. The estimate for streams and the CIs for both components are approximate due to unequal sample sizes, the latter based on the Satterthwaite-corrected df.

Variance component	Estimate	95% CI
Stream	0.019	−0.065 to 0.103
Residual, i.e. between stations	0.259	0.163 to 0.474

Most of the variation in diatom diversity is at the smaller scale of stations within streams compared to among randomly chosen streams. The CI for the streams includes zero.

We also fitted the equivalent model using the mixed effects (multilevel) approach, which should provide better VC estimates. The ML and REML estimates, with profile CIs, were as follows.

Variance component	ML	REML	95% profile CI for REML
Stream	0.009	0.021	0.000–0.174
Residual, i.e. between stations	0.257	0.258	0.159–0.444

The variance pattern was similar to the ANOVA method, although the ML estimate of the streams variance is biased downwards compared to REML and OLS. The profile CIs are more reliable than those based on the ANOVA method.

10.3.1 Traditional OLS Approach

The linear effects model for a design with a single random categorical predictor looks the same as for a single fixed factor:

$$y_{ij} = \mu + \alpha_i + \varepsilon_{ij}.$$

Now α_i represents a random effect instead of a fixed effect. There is no way to designate the effect as random

in the model structure. This is a characteristic of OLS linear models. Our primary interest in fitting this model is estimating σ_α^2, and comparing the relative magnitude of σ_α^2 and σ_ε^2 – for example, is there more variation in diatom diversity at the scale of streams or the smaller scale of stations?

The OLS model fit proceeds identically to the fixed effect case (Section 6.2.2). The ANOVA and EMS are in

Table 10.1 Expected values of ANOVA mean squares and F statistic for a single-factor random effects model

Mean square	EMS	Variance component	F
Between groups			
Unequal n	$\sigma_\varepsilon^2 + \dfrac{\sum_{i=1}^{p} n_i - \left(\sum_{i=1}^{p} n_i^2 / \sum_{i=1}^{p} n_i\right)}{p-1}\sigma_\alpha^2$	$\dfrac{MS_A - MS_{Residual}}{\left(\sum_{i=1}^{p} n_i - \sum_{i=1}^{p} n_i^2 / \sum_{i=1}^{p} n_i\right)(p-1)}$	$\dfrac{MS_{Groups}}{MS_{Residual}}$
Equal n	$\sigma_\varepsilon^2 + n\sigma_\alpha^2$	$\dfrac{MS_A - MS_{Residual}}{n}$	
Residual	σ_ε^2	$MS_{Residual}$	

p is the number of groups and n_i is sample size of group i and n is sample size when equal.

Table 10.1. Just like a fixed effects model, we assume homogeneity of error variances across all possible groups and the expected value of the $MS_{Residual}$ is σ_ε^2. We can use the EMS to provide estimates for the VCs from the model using what is termed the ANOVA, EMS, or method of moments approach (Table 10.1).

Confidence intervals for VCs are based on the χ^2 distribution, and this is straightforward for σ_ε^2 with equal or unequal sample sizes. In contrast, the estimate of σ_α^2 is a linear combination of mean squares and its distribution is usually unknown (Hocking 2013), so we can only calculate an approximate CI based on Satterthwaite's adjusted df (Searle et al. 1992). With unequal sample sizes, the reliability of the Satterthwaite approximation is debatable.

We may also wish to test whether $\sigma_\alpha^2 = 0$ – for example, there is no variance in diatom diversity between all possible streams. The appropriate F statistic is $MS_{Groups}/MS_{Residual}$, just like a fixed effects model (Section 6.2.5).

10.3.2 Linear Mixed Effects (Multilevel) Models

The mixed effects model has essentially the same terms as the OLS model, but the terminology is different and varies among reference sources:

$$y_{ij} = \mu(\text{or } \beta_0) + u_i + \varepsilon_{ij}.$$

Because these models are often used in multilevel regression scenarios, the model is commonly described using regression terminology, with the intercept replacing the mean; the random effect is often represented by the symbol u_i. We still assume the random effects and the error terms are normally distributed with zero means and variances of σ_α^2 and σ_ε^2, and independent of each other.

While these analyses don't include a fixed covariate (see Section 10.4), we can still consider them a multilevel regression model. The first step is to fit a no-predictor, intercept-only, regression model through the level 1 observations for each group – for example, for each stream, we would model diatom diversity for each station against the stream-specific mean diatom diversity and an error term associated with each observation.

The second step is to treat the estimated regression coefficients (group means) from step 1 as random variables and model them at level 2 – that is, fit a model with the random group intercepts (means) as the response variable modeled against the overall fixed intercept or mean and a random effect measuring deviations of each random group mean from the overall mean. These two steps are combined into the single model.

As with the OLS model, our main interest for the random effects and the error term is estimating their VCs using REML, with CIs derived from profile likelihoods. The null hypothesis that σ_α^2 is zero can be assessed by comparing the fit of REML models with and without the random effect of interest using LR tests (but see the caveat about P-values for testing VCs in Section 10.2.2.1).

10.4 Multilevel Regressions

These simple random effects models can be made more complex by recording a fixed covariate for each sampling unit. For example, Bleeker et al. (2017) examined variability in reproductive characteristics of round gobies along the Rhine River in Europe. They had five sampling sites along the Rhine. At each site, they collected multiple fish and measured gonad mass and the covariate of body length. Our focus is on variation in the slopes and intercepts of the gonad mass–body length regressions for each site. The data analyses are given in Box 10.3.

We can fit an extension of the previous model by adding the covariate and associated regression slope, which only allows the intercept to vary among the random groups (random intercepts models) or by also including the interaction between the covariate and the random factor, which allows both intercepts and slopes (random

Box 10.3 Ⓡ Ⓔ Worked Example of Single Random Factor Design with Continuous Covariate: Gobies along the Rhine River

Bleeker et al. (2017) examined variability in reproductive characteristics of round gobies (*Neogonius melanostomus*) along the Rhine River in Europe, where they are a nonnative species. They had five sampling sites along the Rhine correctly treated as a fixed factor in their analyses. However, for our purposes, we will imagine that the aim was to comment on gobies across their whole distribution in this river, and the sites are a random sample from a population of possible sites. At each site, they had 7–22 fish. The response variable was gonad mass (g), and the covariate was body length (mm).

A plot of gonad mass against body length for each site showed consistent positive relationships (Figure 10.4), suggesting a random intercept model might fit similarly to a random slopes model. We initially fitted a full random slopes model. The residual plot showed no unusual values or variance heterogeneity, but we did get a warning about singularity, presumably because some of the VCs were essentially zero and the random intercept effect was perfectly correlated with the random slope effect (see discussion in Bates et al. 2015; Bolker 2015). We centered the covariate to reduce the correlation between random effects and then compared the fit of a REML mixed model that allowed only the intercepts to vary across sites with one that allowed both slopes and intercepts to vary.

Model	AIC$_C$	LL	LR test
Random intercepts	17.194	−4.285	
Random slopes and intercepts	18.732	−3.890	$\chi^2 = 0.790$, df $= 1$, $P = 0.374$

The model fits were not strongly different, with the LR test indicating little improvement by including a random slopes term. We then fitted a REML mixed model with the same slope for each population, only allowing intercepts to vary, and determined the VCs for the random effects, with profile CIs. The overall fixed effect of body length on gonad mass was 0.043 ± 0.012 ($t_{1,36.25} = 3.65, P < 0.001$; Kenward–Roger df), as expected given Figure 10.4. The REML estimates were as follows.

Variance component	Estimate	95% CI
Sites	<0.001	<0.001–0.009
Residual	0.057	0.040–0.079

There was essentially no variation in intercepts among sites, with all the random variance being residual (among fish).

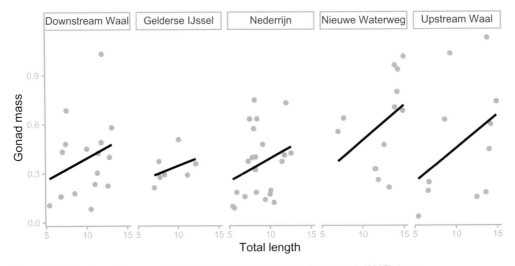

Figure 10.4 Plots of gonad mass vs. fish length for the five sites from the Bleeker et al. (2017) dataset.

coefficients or random slopes model) to vary. Using our worked example from Bleeker et al. (2017), the random slopes model (omitting the subscripts for clarity) would be:

gonad mass = overall mean or intercept + site
+ body length + site × body length + ε.

The random intercept model would be:

gonad mass = overall mean or intercept + site
+ body length + ε.

These models are like ANCOVA models (Box 8.14), except that instead of a fixed effect and covariate, we now have a random effect and a fixed covariate. The first model is equivalent to the heterogeneous slopes model for fixed effects ANCOVA and the second is equivalent to the ANCOVA model assuming equal slopes.

As with ANCOVA models, the covariate can be centered by subtracting the grand mean. Centering results in the random effect measuring variance in the response for the mean value of the covariate rather than when the covariate equals zero (see Box 8.14). Centering might be appropriate if the intercept is outside the data range, as in the fish example.

Now we have a covariate, the multilevel regression context becomes clearer. The random slopes model involves fitting a linear regression model to the level 1 observations within each level 2 group (e.g. site), each with its own intercept and slope. The estimated parameters from these regressions are then treated as random variables modeled across the random level 2 groups. If there is little evidence for the slopes varying from site to site, we can omit the random slopes (interaction) term and refit the random intercepts model.

Note that the random slopes model has two random effects besides the error term:
• The site × body length interaction term measures the deviation between the slope for random group i and the overall slope, and is $N\left(0, \sigma^2_{\text{site} \times \text{body length}}\right)$, measuring the variation in slopes of the regressions of gonad mass against body length across all possible sites.
• The random site term measures the deviation between the mean or intercept for random group i and the overall mean or intercept, and is $N\left(0, \sigma^2_{\text{site}}\right)$, measuring the variation in gonad mass across all possible sites.

We could fit these models using OLS, resulting in an ANOVA that combines aspects of ANCOVA and a two-factor crossed mixed model (Section 10.6 and Table 10.3) with the VCs estimated from the EMS. Including

continuous covariates in OLS random effect models is not straightforward when the designs become complex (Hector 2015), and mixed effects (multilevel regression) modeling is preferred.

The mixed effect models are fitted and the effects estimated as previously described. The first step is usually the decision between the random slopes and the simpler random intercepts models, based on three considerations:
• Plots of the regressions within each group indicate whether a heterogenous or homogeneous slopes model should be used (e.g. Figure 10.4).
• A formal comparison of the two REML models or whether the CI for the random slopes VC includes zero.
• With small datasets, especially with lots of variability in group-specific slopes, the modeling algorithms might not converge to a reliable solution when random slopes are included, necessitating a simpler random intercepts model.

We may also be interested in the overall fixed effect (i.e. slope) of the covariate if one is included, and its ML estimate is standard output from mixed models software, with the t distribution used for CIs. The formal hypothesis that the overall fixed effect of the covariate equals zero is assessed by comparing the ML fit of random intercept models with and without the fixed effect using LR tests or with t or F statistics from ANOVA (both with Kenward–Roger adjustment).

10.5 Nested (Hierarchical) Designs

Nested or hierarchical designs are common in biology (Aarts et al. 2014; Pearl 2014; Schielzeth & Nakagawa 2013). They are usually based around a categorical main factor with sampling or experimental units within each factor group. These units are then subsampled with smaller units, often because the scale of the primary sampling or experimental units is not appropriate for making the observations on the response variable. Consider a design with two levels of nesting under a fixed factor of interest. In the hierarchical design framework, the bottom-level observation units are the level 1 units (termed micro-level), and the sampling or experimental units for the main categorical factor are the level 2 groups (macro-level) (Snijders & Bosker 2012; West et al. 2015). The main factor can be viewed as level 3 in the hierarchy. It is commonly fixed but can be random – for example, when the design is evaluating variation at multiple scales, such as measuring ecological variation at several scales (e.g. Downes et al. 1993) or in some breeding programs (see Schielzeth & Nakagawa 2013). The nested factors are nearly always random.

Continuous covariates can also be included at different levels in the hierarchy. For example, Leonard et al. (1999) measured the attachment strength (response) and size (covariate) of intertidal mussels (level 1 units) from randomly chosen sites (level 2 groups) nested within a fixed factor representing the importance of crab predation (high versus low). The emphasis in these analyses might be on how the intercepts and/or slopes of the regressions of the response against the covariate vary across the level 2 groups or interact with the groups of the main fixed factor. While nested designs with continuous covariates are often the focus of mixed effect (multilevel) models in the social/educational sciences and some biomedical settings, our experience is that they are not that common in biology.

Schielzeth and Nakagawa (2013) distinguished two types of nested designs. The first is a naturally nested design, where the nesting is due to subsampling from the original experimental or sampling units with smaller-scale observation units. The second nesting is a design choice, reflecting the distribution of experimental or sampling units. They viewed this second category as representing situations that could also be approached with a crossed design.

Our experience is that the first type is common, and our view, expressed in Chapter 3, is that it is preferable to summarize or aggregate subsamples into sampling or experimental units before modeling. This step simplifies the modeling and emphasizes the true replication. The second category is more often associated with more complex designs such as those described in Chapters 11 and 12.

We illustrate the first type with two examples. Lozada-Misa et al. (2015) compared two morphologies of the coral genus *Porites* for disease-derived lesions. They had colonies of the branching *P. cylindrica* and of the massive *Porites* spp. and measured multiple lesions per colony. Colony morphology was an observational fixed factor, with colony a random factor nested within morphology, and size of individual lesions was the response variable. This design is fully balanced, and the complete data analysis is shown in Box 10.4.

Newman et al. (2015) compared different phenotypes of a geophytic plant in South Africa. They had multiple populations for each phenotype, and from each population they measured tepal length from multiple individual plants. Phenotype was an observational fixed factor, with population a random factor nested within phenotype. This design was very unbalanced, with unequal numbers of populations per phenotype and flowers per population. The complete analysis is presented in Box 10.5.

10.5.1 Two-Level Nested Designs

Consider a nested design with a main fixed factor (A) with p groups and two levels of nesting, a nested factor (B) at level 2 with q random groups within each group of A, and n units at level 1 within each group of B within each group of A (Figure 10.2). Note that the B groups within each A group are different, as required for a nested design – randomly chosen groups from a population of possible groups within each group of factor A. For example, Lozada-Misa et al. (2015) had $p = 2$ fixed morphology types, with $q = 10$ random colonies within each morphology and $n = 5$ random lesions from each colony within each morphology type. The number of level 2 groups within each factor A group can be different, as can the number of level 1 units within each level 2 group, as with the Newman et al. (2015) example.

Each observation has the subscript ijk indicating observation k from B subgroup j within group i of factor A; the subscript $j(i)$ indicates the jth group of factor B within the ith group of factor A. The population mean for each group of A is μ_i (the average of the means for all possible B groups within each A group), and the mean for each random group of B within each A group is $\mu_{j(i)}$.

The models used to analyze these two-level nested designs are:

Lozada-Misa et al. (2015):

$$(\text{lesion size})_{ijk} = \mu + (\text{morphology})_i + (\text{colony within morphology})_{j(i)} + \varepsilon_{ijk},$$

and

Newman et al. (2015):

$$(\text{tepal length})_{ijk} = \mu + (\text{phenotype})_i + (\text{population within phenotype})_{j(i)} + \varepsilon_{ijk}$$

In these models:
- The response is the size of lesion k from colony j within morphology i or the plant k tepal length in population j within phenotype i.
- μ, as usual, is the overall population mean of the response variable across all combinations of A and B. It is estimated from the unweighted mean of the factor A group means.
- Factor A is commonly fixed (e.g. coral morphology or plant phenotype), so we have a fixed effect for each A group – for example, the effect of morphology on mean lesion size averaging across all lesions and colonies.

Box 10.4 Ⓡ Ⓔ Worked Example of Two-Level Nested Design: Lesions in Corals

Lozada-Misa et al. (2015) compared lesions caused by white syndrome disease in two colony morphologies of the coral genus *Porites*. They collected 10 colonies each of the branching *P. cylindrica* and the massive *Porites* spp. and measured five lesions per colony. Morphology was a fixed factor, with colony a random factor nested within morphology and size of individual randomly chosen lesions as the response variable. Note that in this design, it may be hard to separate effects of morphology from interspecific differences.

Boxplots of lesion size indicated strongly skewed distributions with very different spreads for the two morphologies, and residual plots from the fit of a nested linear model (colony nested within morphology) also showed a strong relationship between residuals and predicted values (colony means). Log transformation improved the distribution and reduced the difference in variances so our analysis, like that of Lozada-Misa et al., will be on log-transformed lesion sizes, although we have used $\log_{10}$ in contrast to their natural logs.

The OLS linear model analysis yielded the following.

Source of variation	df	MS	F	P
Morphology	1	9.820	32.29	<0.001
Colony(morphology)	18	0.304	1.42	0.145
Residual	80	0.214		

The mean log size of lesions was greater in *Porites* colonies with massive morphologies (1.23) than those with branching morphologies (0.61); back-transformed, these values correspond to roughly 17 and 4 cm^2, or, alternatively, the difference between the two log-transformed means (0.6) indicates a four-fold difference. Note that we could achieve a similar result for morphology by modeling mean lesion size for each colony with morphology as the single predictor.

Source of variation	df	MS	F
Morphology	1	1.964	32.29
Residual	18	0.061	

This design is balanced, so we can estimate VCs using OLS and EMS calculations. The CI for colonies within morphology type is based on the Satterthwaite approximation.

Variance component	Estimate	95% CI
Colony(morphology)	0.018	0.005–0.958
Residual, i.e. between lesions	0.214	0.160–0.299

Most of the variation in lesion size was between individual lesions within a colony rather than colonies within morphology type. Note the very wide asymmetrical CI for the colony VC, indicating the conservative nature of the Satterthwaite method and our uncertainty in this variance estimate.

With a balanced design, a linear mixed effects model produced an identical result for the fixed effect of morphology as the ANOVA. We can also evaluate the morphology effect by comparing the ML fit of the full model with a model lacking it.

Model	df	AIC$_C$	LL	Deviance
Reduced	3	161.21	−77.480	154.96
Full	4	142.83	−67.206	134.41

Box 10.4 (cont.)

The log-likelihood χ^2 was 20.549 with df $= 1$ ($P < 0.001$), indicating a strong effect of morphology on log lesion size, the same conclusion as the linear model ANOVA.

The VCs were estimated using ML and REML, the latter with profile likelihood CIs.

Variance component	ML	REML	95% CI
Colony(morphology)	0.012	0.018	0.000–0.067
Residual	0.214	0.215	0.159–0.296

The residual VC matched that for the ANOVA method, and the CI was similar. For the colony VC, the ML estimate is biased downwards (as usual), whereas the ANOVA and REML estimates match. The CI is much narrower than the conservative Satterthwaite approximation from the ANOVA method.

- Factor B is nearly always random, so we have a random effect for each possible B group – that is, the effect of any colony that could have been used for each morphology type or the effect of any population that could have been selected for each phenotype. The random effects are assumed to be $N\left(0, \sigma_\beta^2\right)$.
- The error terms (ε_{ijk}) represent random or unexplained error associated with each observation – for example, the random error in lesion size associated with any lesion within any colony within one of the morphologies. These error terms are assumed to be $N(0, \sigma_\varepsilon^2)$.
- There are two VCs from this model, the variance among all possible groups of factor B within each A group and the error variance.

While factor A is commonly fixed, it may occasionally be random. In this case, the model would include the random effects of A, assumed $N(0, \sigma_\alpha^2)$.

10.5.1.1 OLS Analysis

We can fit this model with OLS, and the ANOVA and EMS are presented in Table 10.2.

Note:
- The ANOVA partitions the total variation into three components: variation associated with the fixed effects of factor A (MS$_A$), the random effects of factor B $\left(\text{MS}_{B(A)}\right)$, and the residual (MS$_{Residual}$).
- The EMS for A includes the VCs for the residual and the random effects of B nested within A. If A is random, then the EMS for A does not include the VC for B.
- Guided by the EMS, the SEs for the fixed effects of A are calculated using MS$_{B(A)}$, not the residual. This makes sense, as the B groups are the true experimental or sampling units for evaluating factor A, with the level 1 units (residual) representing variation within B groups. This has

implications for pooling (model selection) in these analyses (see below).

The standard OLS assumptions apply and are checked in the usual way, especially with plots of residuals.

The VC estimates for the random effects are derived by equating their mean squares to their expected values in the ANOVA (Table 10.2). These VCs are sometimes presented as percentages of the total random variation in our data, and sometimes as percentages of the total variation. Our comments about the challenges for OLS estimation of VCs with unbalanced data from Section 10.3.1 apply here. Only an approximate CI for σ_β^2 based on Satterthwaite's adjusted df can be calculated, and with unbalanced data its reliability is questionable.

The F statistics for testing hypotheses about the fixed factor A (all fixed effects equal zero; all population group means equal) and the random factor $B(A)$ (variance component for B equals zero) are presented in Table 10.2. Again, note that the test for A uses MS$_{B(A)}$ as the denominator. The logic and mechanics of planned and unplanned comparisons for the fixed factor are the same as for simple and crossed categorical predictor models (Sections 6.2.7 and 7.1.10), but ensure your software uses the correct SE based on MS$_{B(A)}$. If A is random, it is tested against the residual.

10.5.1.2 Mixed Effects Model Analysis

With fully balanced designs, the ML/REML estimates and tests of the fixed effects will be similar to the OLS model. For unbalanced datasets, or when a nondefault R matrix is used (e.g. with heterogeneous variances), the degrees of freedom for fixed effects can be difficult to calculate (Bolker 2015), and the Kenward–Roger approximations are recommended for t and F statistics. The null hypothesis for the fixed factor can also be

Box 10.5 Ⓡ Ⓔ Worked Example of Two-Level Unbalanced Nested Design: Flower Characteristics in Different Phenotypes of a Geophytic Plant

Newman et al. (2015) compared flower characteristics of two different phenotypes (short- and long-styled flowers; see their figure 1) of the geophytic plant *Nerine humilis* in South Africa. They had multiple populations for each phenotype (seven or eight short- and three long-style, depending on response variable). From each population, they measured flower characteristics (e.g. tepal length, nectar volume) from 12–37 flowers. Phenotype was fixed, with population random and nested within phenotype; we will analyze tepal length as the response variable.

Boxplots of tepal length based on individual plants (there were too few localities for each phenotype for boxplots on population means) did not indicate skewed distributions, and spreads were similar. Residual plots from a nested linear model (population nested within phenotype) showed no patterns between residuals and predicted values (population means).

The standard OLS linear model analysis resulted in the following.

Source of variation	df	MS	F	P
Phenotype	1	2,363	11.55	0.009
Population(phenotype)	8	205	11.06	<0.001
Residual	236	18.5		

There was strong evidence that mean tepal length was larger for the long-style phenotype, although the test as part of the mixed effects model (below) is more reliable due to the unbalanced design. There was also strong evidence that the added variance in tepal length between populations within each phenotype differed from zero.

We could achieve a similar result (df, F, and P-value) for the effect of phenotype by fitting a model based on mean tepal length per population and phenotype as the only predictor. The result ($F_{1,8} = 10.78$, $P = 0.011$) is not exactly the same because of the unbalanced design.

We can estimate the VCs using the ANOVA method with CIs based on Satterthwaite's method.

Variance component	Estimate	95% CI
Population(phenotype)	7.66	3.75–25.86
Residual, i.e. between plants	18.49	15.99–21.66

Note the very wide (conservative) CI for the VC for population within phenotype. Because the design is unbalanced, VCs are better estimated using ML and REML.

A linear mixed effects model was also fitted. The F test for the fixed effect of treatment, with Kenward–Roger's adjusted df to deal with the unequal sample sizes, resulted in $F_{1,7.97} = 11.027$, $P = 0.011$, a result very similar to the original OLS ANOVA.

We can also use a LR test based on comparing the ML fit of the full model with a model that omitted the treatment effect.

Model	df	AIC	LL	Deviance
Reduced	3	1,453.1	−723.54	1,447.1
Full	4	1,446.4	−719.21	1,438.4

The log-likelihood χ^2 was 8.650 with df = 1 ($P = 0.003$), again indicating an effect of phenotype on tepal length, the same conclusion as from the linear model ANOVA.

The VCs were estimated using ML and REML, the latter with profile likelihood CIs.

Box 10.5 (cont.)

Variance component	ML	REML	95% CI
Population(phenotype)	6.56	8.46	2.58–19.90
Residual, i.e. between plants	18.50	18.50	15.53–22.29

Bootstrap CIs produce similar intervals for both components. The VCs were similar to those from the ANOVA method, with the REML CI for population being more reliable. Again, more of the variation in tepal length is between plants within populations rather than between populations.

Table 10.2 Expected values of ANOVA mean squares and *F* statistics for a three-level nested model where factor *A* is fixed and factor *B* is random and nested within factor *A*

Mean square	EMS	Variance component	*F*
Factor *A*	$\sigma_\varepsilon^2 + n\sigma_\beta^2 + nq\dfrac{\sum_{i=1}^{p} \alpha_i^2}{p-1}$	NA	$MS_A/MS_{B(A)}$
Factor *B(A)*	$\sigma_\varepsilon^2 + n\sigma_\beta^2$	$\dfrac{MS_{B(A)} - MS_{Residual}}{n}$	$MS_{B(A)}/MS_{Residual}$
Residual	σ_ε^2	$MS_{Residual}$	

Factor *A* has *p* groups, there are *q* factor *B* groups within each *A* group, and *n* units within each *B* group within each *A* group. Unbalanced designs (i.e. unequal numbers of units in each *B* group and/or unequal numbers of *B* groups in each *A* group) use similar sample size equations as in Table 10.1, modified for the nested structure.

evaluated using a ML LR test from comparing the full model with the model omitting the fixed effect. The REML estimates of the VCs of the random effects, with profile or bootstrap CIs, are more reliable than OLS estimates, especially with unbalanced datasets.

10.5.1.3 *Pooling and Model Selection in Nested Analyses*

With these two-level nested designs, biologists usually fit a full model and present the estimate and test for the fixed effect plus estimates of VCs. In many situations, the primary focus is the main fixed factor, and there is no interest in the random effects, particularly when the nested structure is simply a result of having the main experimental/sampling units at a different scale to the observation units. Then the fixed factor could be evaluated by averaging the observations within each level 2 unit and fitting a single-factor model using these averages as the response. The estimates and test of the fixed effects will be the same since the fully nested model tests use the means of the level 2 groups to estimate and test *A* anyway. We recommend this option when the nesting reflects subsampling.

Another way to simplify the model is to assess whether to retain the *B(A)* term or pool it with the residual. This

decision is usually test-qualified – that is, the decision is based on a statistically "nonsignificant" result of the test of the null hypothesis (using *F* or log-likelihood statistics) that the VC for *B(A)* equals zero. This approach has sometimes been advocated because it increases the df for the test of the fixed effect. In the OLS model framework, this pooling issue has created some debate in the statistical literature (e.g. Hines 1996; Janky 2000). One real problem with pooling in these hierarchical designs is that it can represent a form of pseudoreplication by creating the illusion that the subsamples are legitimate independent units and artificially inflating the df (Colegrave & Ruxton 2017). Our view is that pooling in these models should usually be avoided and the correct sampling/experimental units for evaluating the highest-level fixed effect should be used. To simplify the models for these nested designs, researchers are much better off averaging the observations at levels below these units, as described above.

10.5.2 More Complex Nested Designs

These designs can be extended to three or more nested factors. For example, Caballes et al. (2016) examined the effects of maternal nutrition treatments (fixed factor) on the larvae of crown-of-thorns seastars. There were female seastars (random) nested within each treatment. Fertilized eggs from each seastar were reared and groups of larvae were placed into glass jars (random); a sample of larvae was removed from each jar after four days and larval morphological characteristics were measured. This design has three random levels of nesting: females within nutrition treatment, jars within females, and individual larvae within jars. The model (omitting the subscripts for clarity) is:

$$\text{(morphology)} = \mu + \text{nutrition} + \text{female(nutrition)} + \text{jars(female(nutrition))} + \varepsilon.$$

The analyses are shown in Box 10.6. As emphasized in the previous section, if the focus is assessing the fixed effect of nutrition, then the data from each female (3 jars and 10 larvae) could simply be averaged and a single fixed factor model fitted using these averages. This would be preferable, as the female seastars are the true experimental units.

10.5.3 Sample Size and Nested Designs

When considering the estimation of the fixed effect of A, and the power to detect any effects, increasing the number of B groups (level 2) within each A group is more important than increasing the number of level 1 units. This has implications for the design of nested experimental and sampling programs. The higher-level "units" in nested designs are often more costly, either because they are more expensive (e.g. whole animals vs. pieces of tissue) or take longer to record (large spatial areas vs. small plots). It is then tempting to take more observations at lower levels in the design hierarchy. It is very important to realize that to increase the precision of the estimate for fixed main effects (and the power of a test), we need to increase the number of groups of the random factor immediately below it. For example, Lozada-Misa et al. (2015) could improve the estimate of the effect of morphology much more by sampling more colonies per morphology rather than by recording more lesions per colony.

Despite this general advice, smaller-scale noise as part of the apparent variation in factor B can still be important. If we examine the EMS from a two-level nested design (Table 10.2), we see that the $\text{MS}_{B(A)}$ includes two components, small-scale (level 1) variation (σ_ε^2) and the true variance between B groups ($n\sigma_\beta^2$). As we increase our subsampling effort (i.e. raise n), we expect the estimate of MS_B to be increasingly dominated by σ_β^2. Therefore, while subsampling at levels below B has no direct effect on the power of the test of A, when there is considerable level 1 variation, taking more subsamples at lower levels will provide some improvement.

Another important aspect of the design of studies using a series of nested random factors is allocating limited resources to each level of sampling. For example, imagine we followed up the study of Lozada-Misa et al. (2015). The number of morphology types is fixed, but how should we allocate our sampling effort to the two nested levels (colonies and lesions) in a new study design?

We use two criteria to decide. First is the precision of the means for each level of the design or, conversely, the variance of these means. Second is the cost of sampling each level in the design. Several texts (Andrew & Mapstone 1987; Snedecor & Cochran 1989; Sokal & Rohlf 2012; Underwood 1997) provide equations for relating costs and variances to determine the optimum sample number at each level. These calculations are generally based on OLS estimates of the VCs at each level and assume balanced designs.

Understanding the averaging that happens as part of testing higher-level fixed effects can also lead to efficiencies in experimental designs. When subsampling is done for cost or logistical reasons, knowing that these subsamples will be averaged at some stage of the process opens up the possibility of doing this averaging at the sample processing stage, rather than during data preparation or model fitting. This process is called compositing and can offer savings (Carey & Keough 2002).

For example, consider an environmental sampling program to measure levels of a contaminant that is a serious concern, even at low levels. We are comparing two land uses, sampled by randomly chosen sites (level 2) within each land use category. Each site is heterogeneous, but soil samples are small, so we need multiple samples (level 1) to characterize each site. The chemical assays are expensive, but visiting sites and collecting them is relatively cheap. We would like more sites, but cost is an issue. With our knowledge of averaging, we could combine the subsamples from each site into a single composite sample, which is then analyzed chemically. The comparison of land use categories is unaffected by this compositing, but we have reduced the overall sampling costs dramatically, potentially allowing more sites with more precise estimates.

10.6 Crossed (Factorial) Mixed Designs

Another important use of mixed models in biological research is analyzing data from factorial designs. Our interest with these designs is mostly in the effects of the fixed factors and whether they are consistent across the

Box 10.6 Ⓡ Ⓔ Worked Example of Three-Level Nested Design: Nutrition of Seastar Larvae

Caballes et al. (2016) examined the effects of maternal nutrition (three treatments: starved or fed one of two coral genera: *Acropora* or *Porites*) on the larval biology of crown-of-thorns seastars. There were 3 female seastars nested within each treatment, 50 larvae reared from each female were placed into each of 3 glass culture jars, and the lengths of 10 larvae from each jar were measured after 4 days. This fully balanced design has maternal nutrition as a fixed factor, with three random levels of nesting: females within nutrition treatment, jars within females, and individual larvae within jars.

We first fitted an OLS model with each term in the hierarchy evaluated against the term immediately below. The residual plot did not reveal any variance heterogeneity nor outliers. The ANOVA was as follows.

Source of variation	df	MS	F	P
Nutrition	2	1.267	20.35	0.002
Female(nutrition)	6	0.062	2.13	0.100
Jar(female(nutrition))	18	0.029	1.67	0.046
Residual	243	0.018		

A mixed effects model showed the same F statistic for the fixed effect. There was strong evidence for an effect of maternal nutrition, with starved larvae shorter than those fed on either coral genus (Figure 10.5).

This is a balanced design, so OLS (95% CIs) and REML (95% profile CIs) VCs are similar.

Variance component	OLS estimate	REML estimate
Female(nutrition)	0.001 (0.000–0.042)	0.001 (0.000–0.004)
Jar(female(nutrition))	0.001 (0.004–0.011)	0.001 (0.000–0.003)
Residual	0.018 (0.015–0.021)	0.018 (0.015–0.021)

The profile CIs, except for the residual, are narrower than those based on the χ^2 distribution. The variance among larvae within each jar is much greater than the variances among jars or females.

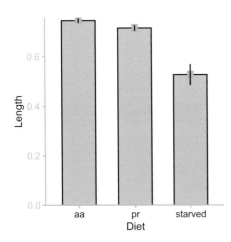

Figure 10.5 Mean size of larval seastars fed in three different ways. The means and the dark error bars were calculated by first averaging larval sizes for each culture jar. The wider gray error bars show those calculated (incorrectly) from individual larval lengths. The smaller size of these error bars, particularly for starved larvae, reflects the misleading larger sample size obtained by using observational (larvae), rather than experimental, units (jars).

random factors – that is, the interactions between fixed and random effects.

10.6.1 Types of Factorial Mixed Designs

We will describe two factorial designs here, although they are not fundamentally different and use pretty much the same models. The next two chapters will describe more complex designs that include both crossed and nested factors or repeated measurements through time.

10.6.1.1 General Factorial Mixed Designs

Random factors can be incorporated into factorial sampling and experimental designs where we have multiple units for each combination of factors. For example, Schlegel et al. (2012) examined the responses of sea urchin sperm to present and future ocean pH. pH group was a fixed effect that was crossed with different individual male sea urchins (random factor; level 2 groups), with multiple recordings (level 1 units) of sperm for each combination of pH and individual urchin. The response variable was sperm swimming speed. While the design focused on the fixed effects of pH, the authors also wanted to assess how much individual urchins varied in their response to changing pH. This design is crossed and has two factors (one fixed and one random), with multiple observations in each combination, and the data analyses are shown in Box 10.7.

10.6.1.2 Randomized Block Designs

Sometimes the random factor is a "blocking" variable and the other factors are the main fixed factors of interest. Consider a design with a single fixed factor. We could set this up as a completely randomized (CR) design with multiple sampling/experimental units for each group. The residual variation represents variation among the units within each group. One possible downside of the CR design is that the background environment might be very patchy in some cases, and we end up with a large residual variance. An alternative is to use a complete blocks (CB) design where we cluster units into blocks (areas that we expect to have more similar background conditions). Then each group of the fixed factor is applied randomly to a unit in each block (Figure 10.6). The analysis now has two factors: the fixed factor crossed with the blocks factor, with a single unit for each factor combination. Blocks are usually treated as a random factor, assuming that they are sampled from a population of possible blocks. If the allocation of units within each block to fixed factor groups is randomized, this design is referred to as a randomized complete blocks (RCB) design.

The total sample size is the same in the CR and CB designs, but in the latter the residual from the CR design is now split into variation due to blocks, representing an attempt to control "nuisance" variables related to the scale of blocking, and the remainder. If blocks explain some variation in the response, the residual will be smaller,

allowing more precise estimates and more powerful tests of fixed effects.

As an example of a single-factor CB design, Wamelink et al. (2014) examined the growth of plants grown in soil collected near the Rhine River with artificial soil representing the composition on the Moon or Mars. Spatial blocks (level 2) were established in a glasshouse, and each block had one small pot (level 1) of each soil type. For each pot, they recorded the total biomass of seedlings. Soil type was the fixed factor and block the random factor. The analyses are given in Box 10.8.

Although "blocks" commonly refers to spatial groupings, they may also be biological groupings, such as organisms of similar size, age, or genetic structure. They can be established in two ways. First, units may be grouped into blocks at a scale chosen by the investigator as part of the study design, as Wamelink et al. (2014) did in their glasshouse experiment. The name of this design reflects its agricultural origin, where the blocks were sections of land under the control of the investigators. Second, blocks may be naturally occurring groups (e.g. pairs of leaves on a shrub in the Walter and O'Dowd (1992) example from the first edition of this book), and their scale is not under the control of the investigator.

The classic CB design described is essentially an unreplicated version of the two-factor crossed design, and the blocks only serve to reduce residual variation. If the interaction between the fixed and the random effects is important, a replicated crossed design must be used instead.

10.6.2 Analysis of Crossed Designs with One Fixed and One Random Factor

Consider a crossed design with a main fixed factor A with p groups crossed with a random B factor with q random level 2 groups and n level 1 units within each A–B combination (Figure 10.2). The groups of B within each group of A are the same, in contrast to the nested design in Section 10.5. For example, Schlegel et al. (2012) had $p = 3$ fixed pHs (A) with $q = 19$ random urchins (B) and $n = 9$–10 random sperm from each pH–urchin combination.

The model fitted to the data from Schlegel et al. (2012) is:

$$\begin{aligned}(\text{swim speed})_{ijk} = {} & \mu + (\text{pH})_i + (\text{urchin})_j \\ & + (\text{interaction between pH and urchin})_{ij} \\ & + \varepsilon_{ijk}.\end{aligned}$$

- The response is observation k from the combination of the ith group of factor A and jth group of factor B (cell ij) – that is, swim speed from sperm k from urchin j at pH i.

Box 10.7 Ⓡ Ⓔ Worked Example of Two-Factor Mixed Crossed Design: Urchin Sperm and Ocean Acidity

We will analyze a modified dataset from Schlegel et al. (2012). They examined swimming speed of sea urchin sperm under three different seawater pH levels (a pH 8.1 control and two more acidic treatments of pH 7.8 and 7.5) and from 19 different randomly chosen individual animals. There were 9 or 10 runs per individual and pH combination, and the response variable was the average sperm swimming speed.

Boxplots and residual plots did not suggest any skewness or markedly unequal variances. We first fitted the traditional OLS factorial linear model (Type III SS).

Source	df	MS	F	P
pH	2	235.2	12.31	<0.001
Individual	18	620.6	42.17	<0.001
pH × individual	36	19.1	1.30	0.119
Residual	509	14.7		

Sample sizes were almost completely balanced, so we estimated the VCs using OLS.

Source	VC	95% CI
Individual	20.18	8.57 to 31.60
pH × individual	0.44	−0.32 to 1.20
Residual	14.71	13.31 to 16.36

There is no evidence for a strong pH × individual interaction – the effect of pH on swimming speed is statistically consistent among the random individuals. There was a strong effect of pH and considerable variation among individuals.

We then fitted linear mixed models to these data. First, we compared the REML fit of mixed models with and without the random pH × individual interaction.

Model	AIC	LL	Deviance	Test
Random intercepts	3,215.6	−1,603.5	3,207.1	
Random "slopes" and intercepts	3,216.4	−1,603.1	3,206.3	$\chi^2 = 0.823$, df = 1, $P = 0.364$

Again, there was no evidence that effect of pH varies across random individuals. This was supported by the profile CI on the interaction VC (estimated with REML), which includes zero.

Source	VC	95% profile CI
Individual	20.554	10.679–40.290
pH × individual	0.445	0–1.495
Residual	14.710	13.042–16.676

The effect of pH is consistent across randomly chosen individuals. Most of the variation in swimming speed is between readings for each individual (residual variance) and between individual urchins (individual variance).

In mixed effects models like this, the interaction between pH and individual is the equivalent of a random slopes term when the fixed predictor is a continuous covariate. Given the very weak interaction, a model selection strategy might drop this effect from the model. The fixed effects would then be evaluated against a residual term that

Box 10.7 (cont.)

combined the interaction and original residual. Our pH conclusions would be the same (full model: $F_{2,35.984} = 12.345, P < 0.001$; reduced model: $F_{2,545} = 15.76, P < 0.001$; Kenward–Roger). The counterargument to pooling, in this case, is that individuals are the experimental unit for assessing pH, and we have "artificially" increased the df by pooling with the residual (see Section 10.5.1).

The reduced model is more straightforward to interpret, so we included contrasts among pH groups as part of this model fit. The control was pH 8.1, so contrasts comparing it to the other two treatments made sense. There is a strong effect of pH on swimming speed, with reductions in speed as pH drops (pH 8.1 vs. 7.8: -1.605 ($\pm$ 0.398), $t_{545} = -4.028, P < 0.001$; pH 8.1 vs. 7.6: -2.154 ($\pm$ 0.399), $t_{545} = -5.400, P < 0.001$). Similar conclusions were obtained for these contrasts if the full model was used, although the SEs and df differed.

The REML VC estimates for this simpler model showed that the variances in swimming speeds are broadly similar between individuals and between readings within each individual (residual). However, we are more confident in our estimate for the residual variance.

Source	VC	95% profile CI
Individual	20.69	10.791–40.392
Residual	15.00	13.307–16.869

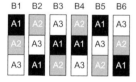

Figure 10.6 Layout of a randomized complete blocks design for an example of a fixed factor with three groups (A1, A2, and A3) and six blocks. The panel at the right shows this design for the example in Box 10.8, where each block is a physical location in the greenhouse, with three pots of seeds, one for each soil type.

- μ is the overall mean as usual, and pH is a fixed factor, as described in earlier chapters.
- Urchins represent a random factor, and this random effect is $N\left(0, \sigma_\beta^2\right)$, measuring the variance in the swim speed across all possible urchins that could have been used, pooling pH groups $\left(\sigma_{urchin}^2\right)$.
- The interaction effect between pH and urchin is $N\left(0, \sigma_{\alpha\beta}^2\right)$, measuring the variance associated with the interaction between pH and individual urchins $\left(\sigma_{pH \times urchin}^2\right)$. Biologically, this interaction term measures whether the pH acts consistently across urchins.
- ε_{ijk} is the usual random or unexplained error, associated here with sperm k from urchin j at pH i.

For CB designs, there is only a single unit for each A and B combination, so Wamelink et al. (2014) had $p = 3$ fixed soil types (A) and $q = 20$ random blocks (B) with $n = 1$ pot for each combination.

The model fitted to their data is:

$$(\text{plant biomass})_{ijk} = \mu + (\text{soil type})_i + (\text{block})_j + \varepsilon_{ijk}.$$

The fixed effect (soil type) and the random effect (block) are defined as for the Schlegel et al. (2012) example. This model is additive, lacking an interaction term. We can never estimate the random interaction effect separately from the error when we have $n = 1$.

10.6.2.1 OLS Analysis

The linear effects model fitted to data from a replicated crossed design with one fixed factor (A) and one random factor (B) is essentially the same as that used when both factors are fixed (Box 7.4). The fixed and random effects are not distinguished in the model nor estimated differently. The ANOVA and EMS are shown in Table 10.3.

Notes:
- The ANOVA partitions the total variation into four components: variation associated with the fixed effects of factor A (MS_A) and the random effects of factor B (MS_B), the AB interaction (MS_{AB}), and the residual ($MS_{Residual}$).
- There are two approaches for determining the EMS. One approach imposes sum-to-zero restrictions on the estimation of the fixed effects and the other doesn't. We present the first approach in Table 10.3, with more details in Box 10.9.

- The EMS for the fixed factor A includes VCs for the residual and the random AB interaction. Consequently, the SEs for the A effects are calculated using MS_{AB}, not the residual, in contrast to where both factors are fixed (Table 7.1).
- Three VCs can be estimated: $\sigma_\varepsilon^2, \sigma_{\alpha\beta}^2, \sigma_\beta^2$. Confidence intervals can be based on the χ^2 distribution for balanced designs only; otherwise, mixed effects models are preferred.

- With unbalanced designs, researchers choose between Type I, II, or III SS (see Section 7.1.7). We have used Type III for our worked examples in this chapter.
- The ANOVA for a CB design is the same, except that the random interaction effect cannot be estimated and is combined with the residual, and only two VCs are relevant. We can only estimate the VC for blocks if we assume there is no blocks × A interaction.

Box 10.8 ® Worked Example of Single-Factor Block Design: Plant Growth on Earth, the Moon, and Mars

Wamelink et al. (2014) compared the growth of 14 plant species grown in soil collected near the Rhine River with simulant regolith representing the soil composition on the Moon and on Mars. We will focus on total biomass of *Sedum reflexum* as the response variable. Twenty spatial blocks were established in a glasshouse, each with three small pots containing five seeds, representing the three soil types (Figure 10.6). Soil type is a fixed factor and block is random. The full layout of the experiment is shown clearly in Wamelink et al.'s (2014) figures 1 and 2.

We fitted a standard OLS model with the fixed effect of soil type and the random effect of blocks. As there is only a single observation in each soil type and block combination, no interaction could be included in this model. Boxplots showed strong skewness and variance heterogeneity, and an "interaction" plot showed an inconsistent size of differences between soil types across blocks, a pattern supported by the residual plot. While a mean–variance relationship could have been modeled using GLS, we transformed the response to logs. The boxplots now showed an even spread across soil types and the interaction plot showed differences between soil types were more consistent across blocks with a much-improved residual plot. The ANOVA (Type III) results from this model were as follows.

Source	df	MS	F	P
Soil type	2	9.695	196.02	<0.001
Blocks	19	0.035	0.71	0.788
Residual	38	0.049		

There was strong evidence for an effect of soil type on log plant growth (Figure 10.7). The OLS estimates for the VCs were as follows.

Source	VC	95% CI
Blocks	0	0–0.001
Residual	0.049	0.035–0.076

Blocks accounted for little variation in log plant growth, suggesting that an RCB design offered no statistical advantages over a CR design.

For comparison, we also fitted a (random intercept) mixed effects model. There was a strong effect of soil type ($F_{2,38} = 217.17, P < 0.001$, Kenward–Roger) with log plant growth lower in Moon soil than Earth soil ($t_{57} = 9.543, P < 0.001$), but greater in Mars soil ($t_{57} = -11.274, P < 0.001$) – see Figure 10.7. The REML VC estimates were as follows.

Source	Var comp	95% profile CI
Blocks	0	0–0.009
Residual	0.045	0.030–0.062

As with the OLS model, the variance due to blocks was essentially zero.

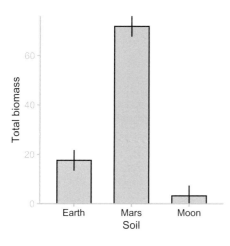

Figure 10.7 Mean plant growth (with SE bars) for simulated soils from three different astronomical bodies, using data from Wamelink et al. (2014).

Table 10.3 Expected values of ANOVA mean squares (Model 1 or Σ-restricted approach) and F statistics for a balanced two-factor crossed mixed model where factor A is fixed and B is random

Mean square	EMS	Variance component	F
Replicated			
Factor A	$\sigma_\varepsilon^2 + n\sigma_{\alpha\beta}^2 + nq\dfrac{\sum_{i=1}^{p}\alpha_i^2}{p-1}$	NA	$\mathrm{MS_A}/\mathrm{MS_{AB}}$
Factor B	$\sigma_\varepsilon^2 + np\sigma_\beta^2$	$\dfrac{\mathrm{MS_B}-\mathrm{MS_{Residual}}}{np}$	$\mathrm{MS_B}/\mathrm{MS_{Residual}}$
$A\times B$	$\sigma_\varepsilon^2 + n\sigma_{\alpha\beta}^2$	$\dfrac{\mathrm{MS_{AB}}-\mathrm{MS_{Residual}}}{n}$	$\mathrm{MS_{AB}}/\mathrm{MS_{Residual}}$
Residual	σ_ε^2	$\mathrm{MS_{Residual}}$	
Unreplicated (e.g. CB)			
Factor A	$\sigma_\varepsilon^2 + \sigma_{\alpha\beta}^2 + q\dfrac{\sum_{i=1}^{p}\alpha_i^2}{p-1}$	NA	$\mathrm{MS_A}/\mathrm{MS_{Residual}}$
Factor B	$\sigma_\varepsilon^2 + p\sigma_\beta^2$	$\dfrac{\mathrm{MS_B}-\mathrm{MS_{Residual}}*}{p}$	$\mathrm{MS_B}/\mathrm{MS_{Residual}}$*
Residual	$\sigma_\varepsilon^2 + \sigma_{\alpha\beta}^2$		

Factor A has p groups, B has q groups, with n units within each combination. For unreplicated designs where B represents blocks (e.g. CB designs) $n=1$ and a nonadditive model is assumed.

* Variance component and F statistic only valid if we assume no interaction $\left(\sigma_{\alpha\beta}^2=0\right)$.

The F statistics for testing hypotheses are presented in Table 10.3. Again, note that we test A using $\mathrm{MS_{AB}}$ as the denominator. The logic and mechanics of planned and unplanned comparisons for the fixed factor are the same as for simple and crossed categorical predictor models (Sections 6.2.7 and 7.1.10), but ensure your software uses the correct SE based on $\mathrm{MS_{AB}}$.

We emphasized in Section 7.1.8 that the main fixed effects from a crossed linear model were difficult to interpret in the presence of a strong interaction. With a two-factor crossed mixed model, the fixed factor includes the VC for the interaction term. Therefore, the F test for the fixed factor actually evaluates the effect of the fixed factor over and above the variation due to the interaction and the

Box 10.9 The OLS Mixed Models Controversy

Return to the model for a factorial design with one fixed and one random factor:

$$y_{ijk} = \mu + \alpha_i + \beta_j + \alpha\beta_{ij} + \varepsilon_{ijk}.$$

To estimate the parameters using OLS, we usually impose two sum-to-zero constraints. The first is that the sum of the effects of the fixed factor A, pooling B, is zero $\left(\sum \alpha_i = 0\right)$. The second is that the sum of the interaction effects across the groups of A is also zero $\left(\sum \alpha\beta_{ij} = 0\right)$. This constraint also defines a covariance between pairs of interaction terms within each group of factor B – that is, any two interaction terms will not be independent within each group of B. Using the Schlegel et al. (2012) example, this model allows for the sperm swim speeds within an individual urchin to be positively or negatively correlated. The version of the mixed model that imposes this constraint originates with Scheffé (1959) and is termed the restricted (or Σ-restricted) model (Kutner et al. 2005; Searle et al. 1992), also Model I in Ayres and Thomas (1990) and the constrained parameters (CP) model (Voss 1999). It is the version most commonly presented in linear models texts and the version we use.

An alternative model (II) does not restrict the interaction terms and is termed, not surprisingly, the unrestricted model or the unconstrained parameters model (Voss 1999). It implies that any two interaction terms are independent, within each group of A and B, and is recommended by several influential authors, including Hocking (2013), Milliken and Johnson (2009), and Searle et al. (1992). Using the Schlegel et al. example, this model assumes that the covariance of sperm swim speeds is the same for each pair of urchins.

The two approaches result in different EMS. Model I results in the values in Table 10.3. Model II has a different expectation for MS_B, which now estimates the residual variance plus the variance due to the interaction and the variance due to the random main effect of B. Note that the difference in the two approaches is only for the random factor, not the fixed factor or the interaction, so only the estimation and test of VC for B is affected.

The most appropriate approach for testing the main effects of factor B has been an issue of considerable debate (Ayres & Thomas 1990; Fry 1992; Hocking 2013; Schwarz 1993; Searle et al. 1992; Voss 1999). Ayres and Thomas (1990) argued that Model II should be applied only after careful assessment of the covariance assumptions – that is, independent interaction effects (but see also Fry 1992). In most biological cases, it is difficult to assess the assumption.

Voss (1999) proposed that the Model I test for factor B is correct no matter which EMS alternative is used. He argued that the H_0 for no main effects of factor B in Model II is actually that $\sigma^2_{\alpha\beta} = \sigma^2_\beta = 0$, which results in the same F test as in Model I. Voss (1999) claimed that this effectively resolved the controversy.

In a more radical approach, Nelder and Lane (1995) proposed that the usual sum-to-zero constraints imposed when using overparameterized effects models are unnecessary and pointed out that if we don't apply them, the EMS for A and B both include the interaction effect. Indeed, the EMS and Fs for each term become basically identical for all combinations of fixed and random factors. Under this model for EMS, which is not conventional, testing fixed main effects is relevant even in the presence of interactions because we are testing for the effect of the fixed factor over and above the interaction.

Hector et al. (2011) argued that one advantage of mixed effects (multilevel) models is that there is better agreement around appropriate tests (see also Galway 2014).

residual. This implies that the fixed main effect can be interpreted even in the presence of an interaction. This also justifies evaluating fixed effects in a CB design even though there is no formal test for the interaction between the fixed effect and blocks. If we do want to evaluate this interaction, there are three approaches:

1. a plot of the response variable values in each cell against factor A groups and blocks (the unreplicated equivalent of the interaction plot described in Section 7.1.11);

2. the usual residual plot might show the residuals changing from positive to negative and back to positive again (Kutner et al. 2005); and

3. Tukey's test for additivity, which more formally tests for this curvilinear relationship in the residuals.

Approaches (2) and (3) only detect simple interactions that involve different magnitudes of fixed effects for each block, but not different directions.

10.6.2.2 *Linear Mixed Effects (Multilevel) Model*

In a multilevel context, the model for the replicated crossed design is a random "slope (coefficient)" model. The interaction measures how the effect of the fixed factor varies across the random factor. The additive model used for a CB design does not include this interaction, so might be described as a random intercept model.

The estimates and tests of the fixed effects will be like the OLS model for fully balanced designs. For unbalanced datasets or nondefault R matrices, the df for fixed effects can be difficult to calculate (Bolker 2015). The Kenward–Roger approximations are recommended for t and ANOVA F statistics. The fixed factor can also be evaluated using a LR test on ML models. The REML estimates of the random effects, with profile or bootstrap CIs, are more reliable than OLS estimates, much more so with unbalanced datasets. For models including the AB interaction, model convergence might be an issue, especially with small sample sizes. Simplifying the model by setting the correlation between the two random effects to zero or omitting the random interaction effect might help convergence. Our comments about pooling random effects with the residual from Section 10.5 apply here as well.

10.6.3 Crossed Designs with Two or More Fixed Factors and One Random Factor

These crossed designs including fixed and random factors can extend to three or more factors. For example, Singh et al. (2016) examined how copulatory traits in male fruit flies are affected by cold stress. Different cold selection treatments (fixed factor) were crossed with multiple ancestral populations (random factor). A second fixed factor was recovery time after being cold-shocked; different flies were used each time. The response variable was mating latency (minutes). The analyses of these data are shown in Box 10.10.

The fully crossed model fitted to these data has two fixed effects (cold selection, recovery time) and one random effect (ancestral population, termed "block" by the authors) plus the three (two random and one fixed) two-way and one random three-way interactions. The VC for the random three-way interaction effect measures how the fixed two-way interaction between selection and recovery varies across populations. The VCs for the random two-way interaction effects are interpreted similarly (e.g. selection × population: how the main effect of cold selection varies across all possible populations, pooling recovery times). The fixed interaction effect measures how the effect of recovery time changes with

selection, pooling across populations. The main effects are interpreted as previously.

A modification of this design is the classical factorial CB design, which would have only a single observation for each combination of the two fixed factors. Van der Geest et al. (2020) studied the mutualistic relationship between the seagrass *Zostera noltei* and lucinid bivalves and how it ameliorated the effects of sulfide stress in a coastal lagoon. They set up a field experiment with two fixed factors, density of bivalves (added vs. control) and pore water sulfide production (organic matter added vs. control), with a single experimental plot for each combination, replicated across randomly chosen locations (blocks). The analyses are shown in Box 10.11.

There are two versions of the model fitted to an unreplicated factorial CB design. Model 1 is the same as for a replicated three-factor crossed design with one factor random, but with only a single observation in each combination, no three-way interaction. Model 2 is a simpler fully additive model. It omits all random interaction effects, which are combined into the residual term, and just includes the fixed main effects and interaction and the random main effect of blocks. Model 2 increases the residual df compared to Model 1, potentially improving the estimate and tests of the fixed effects, but it assumes implicitly that all random interaction effects are unimportant. We prefer starting with the nonadditive Model 1 and possibly moving toward Model 2 by evaluating random effects and omitting those that explain little variation.

The key issue with these OLS models is the derivation of the EMS and how that affects estimation of VCs and tests of the fixed effects (Table 10.4). The EMS for each fixed effect incorporates the random interaction between it and the random factor, so the F for each fixed effect uses the interaction between that effect and the random factor as its denominator. With fully balanced designs, VCs for the random effects are estimated in the usual way, with CIs based on the χ^2 distribution. Variance component estimation using OLS for complex unbalanced designs is very challenging, if not impossible.

The mixed effects models are straightforward extensions of those for two-factor mixed designs. Their main advantage is dealing with replicated but unbalanced designs. REML estimates VCs of random effects and CIs using the profile or bootstrap methods.

10.6.4 Design and Analysis Issues with Crossed Mixed Designs and Their Models

Several issues need consideration when fitting models to mixed factorial designs.

pe.

Random Factors in Factorial and Nested Designs

224

Box 10.10 Worked Example of Three-Factor Mixed Crossed Design: Cold Stress and Fruit Fly Mating

Singh et al. (2016) examined the response of copulatory traits in male fruit flies (*Drosophila melanogaster*) to cold stress. They had two selection treatments (populations selected for cold shock versus control populations, a fixed effect) crossed with five ancestral populations from which experimental flies were derived (random effect, termed "block" by Singh et al.). The data we will analyze are from an experiment that compared files from the two selection treatments that were cold-shocked, allowed to recover, and copulatory traits were measured after 4, 12, and 30 hours. Different flies were used for each period (a fixed effect). This design is fully crossed and the sample size was 35–62 flies in each treatment combination. The response variable was mating latency (minutes).

Boxplots indicated strong skewness and uneven spread of residuals. There were some zeroes, so a log $(y+1)$ transformation was applied before analysis.

The traditional OLS factorial linear model was fitted first, using Type III SS.

Source	df	Denominator	MS	F	P
Selection	1	Selection × population	4.72	36.31	0.004
Period	2	Period × population	11.09	33.61	<0.001
Population	4	Residual	3.21	27.66	<0.001
Selection × Period	2	Selection × period × population	0.03	0.21	0.812
Selection × population	4	Residual	0.13	1.15	0.334
Period × population	8	Residual	0.32	2.74	0.054
Selection × period × population	8	Residual	0.14	1.23	0.276
Residual	1,449		0.12		

Some software will adjust the df (usually Satterthwaite's) for these effects. The selection effect was consistent for the different periods, but mating started earlier for cold shock selected flies (7.2 vs. 9.4 min), and latency declined with increasing period (12.0–7.9–5.9). Means were back-transformed after analysis.

Because of the unequal sample sizes, we will not estimate VCs for the random effects with OLS, but instead fit a mixed effects model using REML.

Random effect	Variance component	95% profile CI
Population	0.009	0.002–0.039
Selection × population	0.000	0.000–0.003
Period × population	0.002	0.000–0.006
Selection × period × population	0.000	0.000–0.003
Residual	0.116	0.108–0.125

The largest VC was the residual (variation between flies in each factor combination), followed by the population variance. All of the interactions involving population had 95% CIs including zero (to three decimal places). This contrasts with the OLS test for period by population, which showed evidence that this differed from zero. The REML estimates are more reliable. The fixed effects tests (with Kenward–Roger adjustment) revealed no interaction between selection and period ($F_{2,7.9} = 0.31$, $P = 0.740$) but strong main effects of period ($F_{2,8} = 35.96, P < 0.001$) and selection ($F_{1,4} = 36.72, P = 0.004$), the same result as for OLS.

Given how small the VCs for the random interaction effects were, an argument could be made to simplify this model by omitting those and re-examining the fixed effects, although our conclusions would not change.

Box 10.11 Ⓡ Ⓔ Worked Example of Two-Factor Crossed Block Design: Seagrasses, Mutualistic Bivalves, and Organic Matter

Van der Geest et al. (2020) set up a field experiment to examine the mutualistic relationship between seagrasses (*Zostera noltei*) and lucinid bivalves (*Loripes orbiculatus*), in particular the extent to which the bivalves and their endosymbiotic bacteria could oxidize harmful sulfides and prevent their accumulation by seagrasses. In a coastal lagoon in the western Mediterranean, they established two fixed factors: abundance of bivalves (added vs. control/background levels) and organic matter (added to increase pore water sulfide production vs. control), with a single experimental plot for each combination replicated across six randomly chosen locations (blocks). The response variable we will analyze was the percentage of sulfur in plant tissue originating from the sediment ($F_{Sulfide}$).

Residuals from the full OLS model looked fine, with even spread and no unusual values. The ANOVA based on Type III SS showed the following.

Source	df	Denominator	MS	F	P
Organic	1	Organic × block	100.711	21.047	0.006
Bivalve	1	Bivalve × block	25.122	5.367	0.068
Block	5	Residual	3.716	0.394	0.835
Organic × bivalve	1	Residual	5.072	0.537	0.497
Organic × block	5	Residual	4.785	0.507	0.763
Bivalve × block	5	Residual	4.681	0.496	0.770
Residual	5		9.442		

Note that the residual is an estimate of the three-way interaction and the true residual, which can't be separated with only a single plot at each combination of fixed factors and block, and the denominator df are only five for the fixed effects tests.

There was no interaction between bivalve addition and organic matter addition, but evidence for a strong main effect of organic matter ($F_{Sulfide}$ means: control = 19.7, added = 23.8, OLS SE = 0.887) and a possible effect of bivalve addition ($F_{Sulfide}$ means: control = 22.8, added = 20.7, OLS SE = 0.887).

We didn't use OLS to get VCs as the Fs for block and its interactions were <1, suggesting negative VCs. Our mixed effects model resulted in a singular fit; essentially, the random effects of block and its interactions with the other factors were trivial, with only the residual having a VC > 0. Singularity was not solved by removing any of the random effects. The fixed effects (Kenward–Roger) tests from this overfitted model produced similar results as the OLS model (Organic × bivalve: $F_{1,5} = 0.897$, $P = 0.387$; Bivalve: $F_{1,5} = 4.442$, $P = 0.089$; Organic: $F_{1,5} = 17.806$, $P = 0.008$).

10.6.4.1 *Number of Random Factor Groups*

The random factor usually represents a spatial (or temporal) scale of replication that affects how we estimate and test the fixed effects. Designs should include a reasonable number (e.g. ≥5; Bolker 2015) of random factor groups. Sometimes we might design an experiment with fixed factors and think it would be useful to know how the fixed effects generalize across space, so we include a random factor representing these spatial units. If we cannot include enough groups of the random factor, we might be better restricting our study to a single group of the random factor (e.g. a single location) and being more confident in our conclusions about the fixed effects.

Simply doing an experiment at two random locations (or with two populations, genotypes, etc.) instead of one will not be very useful. An alternative might be to do the experiment under several conditions chosen to span the range of conditions and treat these different conditions as a fixed effect (see more details in Box 10.1).

10.6.4.2 *Issues with Multiple Random Factors*

Our designs sometimes include multiple random factors. Then, there are problems deriving appropriate F statistics for some terms if we fit an OLS model (see, e.g., Underwood 1997; Winer et al. 1991). One option in this situation is to construct quasi-F statistics by combining mean squares until

Table 10.4 Expected values of ANOVA mean squares (Model 1 or Σ-restricted approach) and F statistics for a balanced three-factor crossed mixed model where factors A and B are fixed and C is random

Mean square	EMS	Variance component	F
Factor A	$\sigma_\varepsilon^2 + nq\sigma_{\alpha\gamma}^2 + nqr\dfrac{\sum_{i=1}^{p}\alpha_i^2}{(p-1)}$	NA	MS_A/MS_{AC}
Factor B	$\sigma_\varepsilon^2 + np\sigma_{\beta\gamma}^2 + npr\dfrac{\sum_{j=1}^{q}\beta_j^2}{(q-1)}$	NA	MS_B/MS_{BC}
Factor C	$\sigma_\varepsilon^2 + npq\sigma_\gamma^2$	$\dfrac{MS_C - MS_{Residual}}{npq}$	$MS_C/MS_{Residual}$
$A \times B$	$\sigma_\varepsilon^2 + n\sigma_{\alpha\beta\gamma}^2 + nr\dfrac{\sum_{i=1}^{p}\sum_{j=1}^{q}(\alpha\beta)_{ij}^2}{(p-1)(q-1)}$	NA	MS_{AB}/MS_{ABC}
$A \times C$	$\sigma_\varepsilon^2 + nq\sigma_{\alpha\gamma}^2$	$\dfrac{MS_{AC} - MS_{Residual}}{nq}$	$MS_{AC}/MS_{Residual}$
$B \times C$	$\sigma_\varepsilon^2 + np\sigma_{\beta\gamma}^2$	$\dfrac{MS_{BC} - MS_{Residual}}{np}$	$MS_{BC}/MS_{Residual}$
$A \times B \times C$	$\sigma_\varepsilon^2 + n\sigma_{\alpha\beta\gamma}^2$	$\dfrac{MS_{ABC} - MS_{Residual}}{n}$	$MS_{ABC}/MS_{Residual}$
Residual	σ_ε^2	$MS_{Residual}$	

Factor A has p groups, B has q groups, C has r, with n units within each combination. For unreplicated designs where C represents blocks (e.g. CB designs) $n = 1$, and the EMS are the same except that the three-way interaction is incorporated with the residual, and VCs for the random effects assume this interaction to be zero.

a suitable numerator and denominator combination is found that tests the hypothesis of interest (Kutner et al. 2005; Winer et al. 1991). Quasi-F values do not follow an F distribution precisely under the H_0, so they are approximate at best. The problem becomes almost intractable if the design is also unbalanced. Mixed effects (multilevel) models might be more useful in this situation, although some software packages limit the number of crossed random effects. Our experience is that multifactor crossed designs with more than one random factor are not common.

10.6.4.3 *Efficiency of Blocking*

It is only worth including a blocking factor in a design if blocks account for a nontrivial amount of variation. Otherwise, the cost from reduced df will outweigh any benefit from the block design. Unless good pilot data are available, we must use our judgment. After using a CB design, we can retrospectively assess whether using blocks was better than a CR design. Lentner et al. (1989) introduced a measure of efficiency of an CB design relative to a CR design. Their measure was monotonically related to the ratio of $MS_{Blocks}/MS_{Residual}$ from the OLS model. If this ratio is greater than 1, a CB design is more efficient than a comparable CR design where the number of observations per fixed factor group equals the

number of blocks. In the Earth–Mars soil example, the blocks F is <1 and VC close to 0, so the CB design probably was not more efficient than a CR design.

10.6.4.4 *Missing Values in CB Designs*

Missing observations are a problem for unreplicated CB designs because a missing observation is a missing cell. The simplest approach to missing observations in CB designs is to omit any block with missing values. Snedecor and Cochran (1989) and Sokal and Rohlf (2012) proposed methods for estimating a missing value based on treatment and block totals. Both methods use the available information from the same treatment and block in estimating the missing value and produce very similar estimated values. The imputation techniques outlined in Section 5.6 are also applicable here.

10.6.4.5 *Incomplete Block and Latin Square Designs*

Very occasionally, we may have an experimental design where blocking would be helpful, but we have fewer units in each block than treatments, so we cannot have every treatment represented in each block. Under these circumstances, the trick is to allocate treatments to blocks so that relevant effects can be estimated and tested, although some interactions must be assumed to be zero. The simplest arrangement is a balanced design where every pair of treatments occurs once (and only once) in one of the blocks. These designs can also be unbalanced so that not every pair of treatments occurs in any block. The definitive reference is Cochran and Cox (1957), but Kirk (2012) and Mead et al. (2012) also describe these designs.

Of course, if there is such a mismatch between the available experimental units and number of treatment combinations, reducing the number of treatments might be a better solution, especially if treatment by block interactions is likely.

Sometimes we want to include two blocking factors in our design to further reduce the unexplained variation in our response variable, for example in an ecological experiment with two different environmental gradients. In this case, we may use a special design called a Latin square (Cochran & Cox 1957; Kutner et al. 2005; Mead et al. 2012). As the name suggests, Latin squares consider the experimental design a square with equal rows and columns. The number of rows/columns is equal to the number of treatments, and treatments are allocated randomly to cells, with the restriction that each treatment is represented once in each row and each column. Row and column are random effects.

10.7 Key Points

• Random predictors, nearly always categorical factors, are important components of many linear models, usually representing random, often hierarchical, spatial replication.
• Random factors are usually included in designs that also include fixed predictors and the resulting models are termed mixed models. The fixed predictors can be categorical factors and/or continuous covariates.
• Research questions with mixed models are about estimating and testing fixed effects and their interactions and estimating the variance associated with the random effects, termed variance components. If continuous covariates are included, models can include interactions between random effects and the fixed covariates (random slopes) or just random main effects (random intercepts).
• Mixed models can be fitted with OLS, with VC estimated by equating the ANOVA mean squares to their expected values; this approach is only reliable with relatively simple, fully balanced, designs and can result in negative estimates of VCs.
• A more general approach to fitting mixed models is using ML methods, sometimes termed linear mixed effects models or multilevel models. This approach can handle a range of designs, including unequal sample sizes, and provides more reliable estimates of VCs using REML.
• The OLS and mixed effects models approaches will produce similar estimates for fixed effects with balanced designs. The usual assumptions (normality, variance homogeneity, etc.) apply to the random terms in these models, although the mixed effects approach can be generalized to other distributions because of the flexibility of ML and can more easily allow nonindependence of errors. For any models including covariates, collinearity should also be checked.
• The common mixed model designs in biology are:
  – Nested or hierarchical designs where we have a series of nested random factors representing a hierarchical design structure, often due to subsampling the true experimental units with smaller observation units; the top-level factor can be fixed or random.
  – Factorial designs where at least one of the factors is random, usually representing spatial variability, and the others fixed; these models include the fixed main effects and interactions and random main effects and random interactions that involve a random factor.
  – A simplified factorial design is one where the fixed factor groups are arranged in, usually spatial, random blocks with one unit for each group in each block, termed a complete blocks design; the blocks effect is considered random and the interaction between the fixed effect and blocks cannot be separately estimated.
  – Split-plot and repeated measures designs are covered in Chapters 11 and 12.

Further Reading

Mixed models and random effects: Most linear models textbooks will introduce random factors and mixed models, although more commonly from an OLS perspective (e.g. see Kutner et al. 2005). There has been a recent proliferation of books on mixed effects models, often emphasizing multilevel regression; West et al. (2015), Zuur et al. (2009), and Faraway (2014) provide approachable introductions from a biological perspective.

Nested designs and models: Underwood (1997), Sokal and Rohlf (2012), and Milliken and Johnson (2009) provide good introductions to analysis of nested designs with a focus on OLS models.

Mixed factorial designs and blocking: Mead et al. (2012) and Milliken and Johnson (2009) describe the principle of blocking in experimental design, with the former describing appropriate OLS models; both cover more complex designs including blocks.

There are fewer books that describe a mixed effects (ML) modeling approach to traditional nested and block/factorial designs with random effects, a gap we hope this chapter has partly addressed.

11 Split-Plot (Split-Unit) Designs
Partly Nested Models

One particular class of sampling or experimental design with both crossed and nested factors is the split-plot (or split-unit) design, initially developed for agricultural settings. The rationale behind these designs is that different fixed factors are applied to sampling or experimental units at different spatial scales (see Figure 10.2). One factor is applied to larger units termed "plots," with two or more plots for each category of this factor; this is sometimes called the between-plots component. Without additional factors, this is simply a single-factor design with the plots as experimental or sampling units. A split-plot design applies a second factor to smaller units (termed "sub-plots") within each plot; this is often called the within-plots component. Using the hierarchical design terminology from Chapter 10, the sub-plots are level 1 units and the plots are level 2 units.

The classic split-plot design has a single sub-plot within each plot for each category of this second factor, but replicate sub-plots for each category can also be used. Note that if the design is unreplicated at the sub-plot level, then within each category of the first factor we have a blocks design, with the second factor crossed with plots (representing the blocks). While representing different designs developed originally for agricultural experiments, the terms plots and blocks essentially have the same meaning in split-plot designs. This terminology becomes particularly confusing when the between-plots component is arranged as a blocks design, with blocks (plots) crossed with, rather than nested within, the between-plots factor (e.g. see the example from Morrissette-Boileau et al. 2018).

These designs can be extended in several ways, some of which will be described later in this chapter:
- The between-plots component can include two or more factors applied to plots arranged in a crossed design.

- The within-plots component can also include two or more factors applied to sub-plots within each plot arranged in a crossed design.
- More complex hierarchies can be included where one factor is applied to plots, a second factor to sub-plots within each plot, and a third factor applied to sub-sub-plots within each sub-plot.
- Continuous covariates can be included at any level, focusing on comparing slopes and intercepts across the plots or sub-plots.

These split-plot designs (and their associated models) are sometimes called partly nested or partly hierarchical designs (and models) because they include both nested and crossed design aspects (e.g., Kutner et al. 2005).

11.1 Simple Split-Plot Designs

The classic split-plot design has one between-plots factor applied to whole plots (level 2 units), and one within-plots factor applied to sub-plots (level 1 units) within each plot. These factors are commonly considered fixed, with the plots and sub-plots deemed random. These designs are quite common in biological research.

Stokes et al. (2014) studied the neurotoxin tetrodotoxin (TTX) in different species of flatworms and its possible roles in predation and defense. One factor (between plots) was flatworm species with individual flatworms as plots nested within species. The second factor (within-plots) was body segment. Each segment represented a "sub-plot" in split-plot terminology, so this design was unreplicated for the segment factor with only a single segment or sub-plot for each body region. The response variable was the tissue TTX concentration. The experimental layout is shown in Figure 11.1 and the data analysis in Box 11.1.

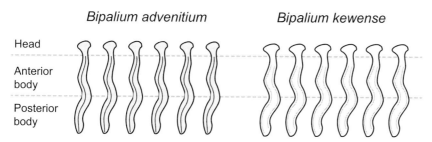

Figure 11.1 Experimental design for the study by Stokes et al. (2014) comparing tetrodotoxin concentrations in different body regions of two species of flatworms. For each worm, the head, anterior body, and posterior body regions are distinguished.

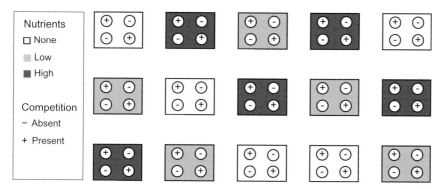

Figure 11.2 Experimental design for the study by Legault et al. (2018) examining effects of nutrients and competition on reed growth. The diagram shows bins with different nutrient levels, in which were placed pots of *Phragmites* plants with and without competitors.

With two factors, a completely randomized (CR) crossed design might have been considered for this study. However, the scales at which the two factors operate are quite different. Comparing species obviously requires individual worms as the sampling units, whereas the comparison of different body regions needs to be done for segments within individual worms. A CR design would require three times as many worms, taking measurements from only a single body region of each flatworm. The CR design may be less effective if there is extensive variation between individual flatworms. If there are animal ethics issues for a particular research question, split-plot and repeated measures (Chapter 12) designs can provide ways of reducing animal use (McDonald et al. 2021).

A second experimental example is from Legault et al. (2018), who examined the effects of nutrient loading on competitive interactions between two species of saltmarsh plants. The between-plots factor was nutrient addition, with bins containing half-strength seawater as the plots. The within-plots factor was competition, applied to sub-plots (small pots) within each bin. The design was replicated for the competition factor at the sub-plot (pot) level. The response variable was the biomass of *Phragmites*. Figure 11.2 shows the layout of the experiment and Box 11.2 has the analysis.

Again, a CR factorial design could have been used for this study, but presumably the researchers decided that the scale of experimental units for manipulating competition needed to be smaller than the scale for manipulating nutrient levels, or they needed to reduce the costs or spatial footprint of the experiment by using fewer bins.

Split-plot designs can also be used when the plots or blocks do not represent spatial units of replication, hence why the term "split-unit" is sometimes used. For example, Westley (1993) used a split-plot design to examine the effects of inflorescence bud removal on asexual investment in the Jerusalem artichoke (*Helianthus tuberosus*). There were four populations of *H. tuberosus*, five individual genotypes nested within each population, and two treatments (normal flowering and inflorescence removed) applied to different tubers from each genotype. Genotypes were plots, population was the between-plots factor, and treatment was the within-plots factor.

11.1.1 Analysis for Simple Split-Plot Designs

We will first describe the linear models for a standard split-plot design with two main factors, A and C, that are commonly fixed, and a random term, B (plots), nested within A but crossed with C. For the flatworm example, A is species, $B(A)$ individual flatworms (plots) within species, and C body segment (C). For saltmarsh plants, A is nutrient addition (A), $B(A)$ is bins (plots) within nutrient addition ($B(A)$), and C is competition. In both examples, A and C form a factorial design. $B(A)$ and C are also crossed because every flatworm has three body segments, and every bin gets both competition treatments. While we will focus on the common design, where factors A and C are fixed, either of these can be random. For example, Keough and Quinn (1998) examined the effects of pedestrian trampling on algal and invertebrate assemblages on rocky intertidal shores. Plots were randomly chosen areas of rocky coast and the within-plots factor was experimental trampling intensity applied to strips of shore (sub-plots) within plots. The between-plots factor

---

Box 11.1 Ⓡ Ⓔ Worked Example of Split-Plot Unreplicated Design: Neurotoxins in Flatworms

Stokes et al. (2014) compared tetrodotoxin concentrations in different body regions of two flatworm species. The between-plots factor was flatworm species (fixed with two groups: *Bipalium adventitium* and *Bipalium kewense*) with individual flatworms (plots) nested within species. The within-plots factor was body segment (fixed with three groups: head, anterior body, posterior body), and each segment represented a "sub-plot." The response variable was the TTX concentration of tissue adjusted for weight (ng/mg). The main research questions were about the fixed effects of species, body segment, and their interaction on TTX concentration, but the analyses also provide information about the variances associated with the random effects of individuals within species and the term that combines the random interaction between individuals within species and body segment and the true residual.

We first fitted the OLS linear model that included fixed main effects of species, body segment, and their interaction, the random effects of individuals within species, and the interaction between individuals within species and body segment. Because this study is unreplicated at the body segment level within each individual, this model will fit the data perfectly, and there will be no residual term. To check assumptions by examining the pattern in the residuals, we first fitted a two-factor crossed model that ignored the hierarchical structure. The residual plot indicated a strong mean–variance relationship, so, like Stokes et al., we log transformed TTX concentration for analysis.

The resulting ANOVA was as follows.

Source	df	MS	F	P
Between-plots				
Species	1	0.809	6.17	0.032
Individual(species)	10	0.131		
Within-plots				
Segment	2	8.575	46.965	<0.001
Species × segment	2	0.037	0.202	0.819
Individual(species) × segment	20	0.183		
Total	33			

There was no evidence of an interaction between species and segment (the pattern of log TTX concentration between segments was consistent across the two species), but strong main effects. There was a much higher TTX concentration in the head than in the two more posterior body segments (Figure 11.3), and *B. adventitium* consistently had more TTX than *B. kewense*.

We also fitted a mixed effects model using REML to these data. Because the design is balanced (same number of individuals for each species), the fixed effects results were similar to those from OLS: segment ($F = 51.840$, $P < 0.001$); species × segment ($F = 0.223$, $P = 0.802$), with only the species result causing some thought for those deciding around 0.05 ($F = 4.889$, $P = 0.051$). Contrasting the segments showed a strong difference between head

Box 11.1 (cont.)

and anterior body ($t_{20} = 7.747$, $P < 0.001$), but a much smaller difference between anterior and posterior body segments ($t_{20} = 1.850$, $P = 0.079$). The mixed model estimates of the variance components with 95% profile CIs were: individual within species (0.000; 0.000–0.050) and individual within species × segment (0.165; 0.090–0.227); remember, this latter variance component (VC) includes the variance due to this interaction plus the true residual variance. The REML estimate of the VC for individual within species was close to zero.

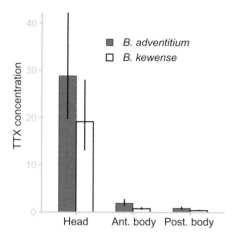

Figure 11.3 Mean tetrodotoxin concentrations (standardized for mass) for different body segments of two flatworm species. Note that the data were log transformed for analysis, but the figure shows data back-transformed to the original scale, accounting for the asymmetrical error bars (standard errors).

was shore, representing three randomly chosen rocky shores along this stretch of coastline.

Consider a design with p groups of A ($i = 1$ to p), q groups of B (plots) nested within each A group ($j = 1$ to q), r groups of C ($k = 1$ to r), and n sub-plots for each C group within each plot ($l = 1$ to n). For flatworms, $p = 2$ species, $q = 6$ individuals within each species, and $r = 3$ body segments. For this unreplicated split-plot design, $n = 1$ – that is, one segment for each region of the flatworm body. For saltmarsh plants, $p = 3$ nutrient treatments, $q = 5$ bins within each nutrient treatment, $r = 2$ competition treatments, and $n = 2$ pots for each competition treatment within each bin. We will talk more about options for models for replicated split-plot designs below.

For most biologists, the focus of these designs is the fixed effects of factors A, C, and their interaction, which we interpret in the same way as for a two-factor crossed ANOVA (Chapter 7). Specific contrasts between groups within these fixed effects will usually be relevant. Estimating the VCs associated with the random effects may also be important, especially if researchers are interested in partitioning variation among different spatial scales – that is, between and within plots.

11.1.1.1 Linear Models for Split-Plot Designs

The classic split-plot design has only a single sub-plot for each factor C group, so $n = 1$ and l can be omitted from the model subscripts for simplicity:

$$y_{ijk} = \mu + \alpha_i + \beta_{j(i)} + \gamma_k + \alpha\gamma_{ik} + \varepsilon_{ijk}.$$

For flatworms:

$$(TTX)_{ijk} = \mu + (\text{species})_i + (\text{individual within species})_{j(i)}$$
$$+ (\text{body segment})_k + (\text{species} \times \text{body segment})_{ik} + \varepsilon_{ijk}.$$

In these models, we have:

y_{ijk} is the TTX concentration in segment k for individual j in species i.

μ is the overall population mean TTX concentration, as usual.

Factors A and C are fixed, so α_i, γ_j, and $\alpha\gamma_{ij}$ are the effects of species i and body segment k on the TTX concentration, as for earlier chapters.

Individuals within species represent random plots, so the random effects $\beta_{j(i)}$ are assumed to be N(0, σ_β^2), with σ_β^2 measuring the variance in mean TTX concentration across all possible individuals that could have been used for each species, pooling segments.

Box 11.2 ℝ 𝔼 Worked Example of Split-Plot Replicated Design: Effects of Nutrient Loading on Competition Between Saltmarsh Plants

Legault et al. (2018) experimentally examined the effects of nutrient loading on the outcome of competitive interactions between two species of saltmarsh plants, the native *Spartina alterniflora* and the exotic *Phragmites australis*, in the northeastern United States. The between-plots factor was nutrient addition (fixed, with three groups: none, low, and high) with five 50-gallon bins containing half-strength seawater as the plots within each nutrient group. The within-plots factor was competition (fixed with two groups: no competition, with competition), with sub-plots being small pots containing one (no competition) or two (with competition) species within each bin. This design was replicated at the sub-plot level, with two pots for each competition group within each bin. We will analyze the biomass (g) of *Phragmites* as the response variable.

We first checked the pattern in the residuals by fitting a simple two-factor crossed model ignoring the hierarchical structure. While the residual plot suggested some mean–variance relationship, we analyzed the untransformed biomass to be consistent with the original paper. The ANOVA from the fit of the full OLS model was as follows.

Source	df	MS	F	P
Between-plots				
Nutrients	2	7,535.4	46.45	<0.001
Bins(nutrients)	12	162.2		
Within-plots				
Competition	1	1,400.4	9.73	0.009
Nutrient × competition	2	142.7	0.99	0.400
Bins(nutrients) × competition	12	144.0		
Residual	30			
Total	33			

We could also have tested null hypotheses about the VCs associated with the random effects of bins within nutrient group and bins within nutrients × competition by using the residual term.

There was no evidence for an interaction between competition and nutrient level, but there were main effects of nutrients and competition, with biomass of *Phragmites* greater without competition and increasing with increasing nutrients (Figure 11.4).

Fitting the mixed effects model resulted in virtually identical Fs and P-values for the fixed effects as the OLS model, as expected, given this design is fully balanced.

There is no evidence for an interaction between nutrients and competition, so we could also contrast each added nutrient group to the no-nutrient group. Both contrasts indicate a strong effect of adding nutrients on biomass (none vs. low: $t_{12} = 6.565$, $P = 0.001$; none vs. high: $t_{12} = 9.395$, $P < 0.001$).

The VCs, estimated using REML with 95% profile CIs were as follows.

Variance component	Estimate	95% CI
Bins(nutrients)	4. 6	0–46.8
Bins(nutrients) × competition	33.4	0–69.9
Residual	77.1	48.4–132.5

Most of the variation was among pots within each bin, and the CIs for both bins within nutrients and bins within nutrients × competition included zero.

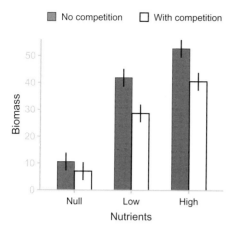

Figure 11.4 Mean biomass of *Phragmites* at three different nutrient levels, and with and without competition from *Spartina*. Error bars are standard errors.

ε_{ijk} is random error, incorporating two components: the random interaction between body segment and individuals $\left(\beta\gamma_{j(i)k}\right)$, measuring the random variation in effects of body segment across the random individuals within each species, and the sub-plot error – that is, random variation between segments for each body region. These two terms cannot be separately estimated when there is only a single sub-plot for each treatment combination (only one segment for each body region in each flatworm), just like in the models for an unreplicated blocks design (Section 10.6). These random error terms are assumed to be $N\left(0, \sigma_\varepsilon^2\right)$, with σ_ε^2 being the residual or error VC.

We can fit these models using OLS, distinguishing the fixed and random effects only through the expected values of the mean squares from the ANOVA.

These split-plot designs can also be analyzed with mixed effect (multilevel) models. The complete model includes essentially the same terms as described for the OLS models above, but is fitted using REML or ML as described in Section 10.2. The model for the flatworm example is:

$$y_{ijk} = \mu + \alpha_i + \gamma_k + \alpha\gamma_{ik} + u_{j(i)} + \varepsilon_{ijk}.$$

The terms are the same as described for the OLS model except that $u_{j(i)}$ is the random effect of individuals within species; this random effect has an associated VC interpreted the same way as the OLS model but estimated differently (and more reliably) with REML. In a multilevel context, the individual-specific mean TTX concentration and individual-specific differences in TTX between body segments (level 1) are modeled against

species for the between-plots (level 2) component. These individual-specific patterns can vary across species (and hence the species by body segment interaction), but we cannot separately identify any individual within species by segment interaction.

11.1.1.2 *ANOVA and Estimates of Effects*

For the OLS model, the fixed effects or contrasts and SEs are estimated as usual, and CIs are based on the t distribution. With these models, make sure your statistical software calculates the SEs and t or F statistics appropriately – for example, the between-plots fixed effects are based on the between-plots error term. Most modern linear models software packages will get them right, but a quick check of the df for the t and F statistics will show if the wrong error term is being used.

We introduced simple main effects to explore interaction terms in factorial models with a single error term (Section 7.1.11). Simple main effects can also be used for the *AC* interaction in these partly nested models. However, deriving an error term for examining *A*'s effects at each group of *C* is not straightforward when there are multiple error terms available in the model. For OLS models, Kirk (2012) and Maxwell et al. (2018) describe the use of pooled error terms for simple main effects.

The OLS partitioning of the SS_{Total} for Y from the fit of the model in Section 11.1.1.1, where *A* and *C* are fixed and plots are random, is shown in Table 11.1.

With mixed effect models, the fixed effects and their SEs are derived using ML as described in Section 10.2. Confidence intervals are based on t statistics with df adjusted using the Kenward–Roger adjustments. Most

Table 11.1 Expected values of ANOVA mean squares (Model 11.1 or Σ-restricted approach) and F statistics for the case of A (p groups) and C (r groups) fixed, but B (q plots) random

For designs that are unreplicated at the sub-plot level, such as the example from Stokes et al. (2014), there is no straightforward F statistic for $B(A)$ or $B(A)C$ in these circumstances. Separate terms for $B(A)C$ and residual are only possible when there is more than one sub-plot for each C group within each plot.

Mean square	EMS	Stokes et al.	F	Legault et al.	F
Between-plots					
Factor A	$\sigma_\varepsilon^2 + rn\sigma_\beta^2 + rqn\dfrac{\sum \alpha_i^2}{p-1}$	Species	$\dfrac{\mathrm{MS_A}}{\mathrm{MS_{B(A)}}}$	Nutrients	$\dfrac{\mathrm{MS_A}}{\mathrm{MS_{B(A)}}}$
Factor $B(A)$	$\sigma_\varepsilon^2 + rn\sigma_\beta^2$	Individual (spec.)	NA	Bins (nutr.)	$\dfrac{\mathrm{MS_{B(A)}}}{\mathrm{MS_{Residual}}}$
Within-plots					
Factor C	$\sigma_\varepsilon^2 + n\sigma_{\beta\gamma}^2 + pqn\dfrac{\sum \gamma_k^2}{r-1}$	Segment	$\dfrac{\mathrm{MS_C}}{\mathrm{MS_{B(A)C}}}$	Competition	$\dfrac{\mathrm{MS_C}}{\mathrm{MS_{B(A)C}}}$
AC	$\sigma_\varepsilon^2 + n\sigma_{\beta\gamma}^2 + qn\dfrac{\sum\sum (\alpha\gamma)_{ik}^2}{(p-1)(r-1)}$	Spec. × seg.	$\dfrac{\mathrm{MS_{AC}}}{\mathrm{MS_{B(A)C}}}$	Nutr. × comp.	$\dfrac{\mathrm{MS_{AC}}}{\mathrm{MS_{B(A)C}}}$
$B(A)C$	$\sigma_\varepsilon^2 + n\sigma_{\beta\gamma}^2$	Residual: combined $B(A)C$ & residual	NA	Bins (nutr.) × competition	$\dfrac{\mathrm{MS_{B(A)C}}}{\mathrm{MS_{Residual}}}$
Residual	σ_ε^2			Residual: pots in each bin–competition group	

Note that for balanced designs (i.e. equal number of plots within each A group), the between-plots component is the same as the ANOVA from a single-factor model fitted to the mean values for each plot (i.e. averaging over the groups of factor C). Variance component estimates are derived by equating mean squares to their expected values, but this becomes more problematic for these more complex hierarchical designs with fixed and random effects – for example, negative VCs are more likely.

mixed effects models software will output an ANOVA based just on the fixed effects. Variance component estimation is based on REML with CIs determined using the profile or bootstrap methods.

11.1.1.3 *Null Hypotheses*

The usual null hypotheses of interest are that each fixed effect (an overall factor effect or effects of individual contrasts within each factor) or random VC equals zero. Using the OLS ANOVA, we can evaluate the overall factor effects: The F statistic for the test that the between-plots fixed effect (A) equals zero uses the plots within $A\left(\mathrm{MS_{B(A)}}\right)$ in its denominator, whereas those for the within-plots fixed effects (C and AC) use the residual. We can also derive t statistics for individual contrasts.

Note that if factor A is random and C is fixed (e.g. see Keough & Quinn 1998), the test for C changes to use the

AC interaction as the denominator of the F statistic. If C is random, there is no F statistic for evaluating the fixed effect of A, and the methods described in Section 10.6.4 should be considered. Testing hypotheses about the VCs using Fs derived from an OLS ANOVA is only possible when we have separate estimates of the B(A) × C and residual terms –that is, only with design replicated at the sub-plot level. Variance components are better handled with REML using mixed effects modeling.

With the mixed effects model, the fixed effects can be tested using t or F statistics from the ANOVA incorporating just the fixed effects, based on Kenward–Roger adjusted df. Alternatively, likelihood-ratio (LR) tests can be used based on comparing ML models with and without the fixed effect. Likelihood-ratio tests based on REML can also be used for testing hypotheses about VCs, although there are issues with the resulting P-values (Section 10.2.2.1).

11.1.1.4 Split-Plots with Sub-plot Replication

For a split-plot design with replication at the sub-plot level, we can estimate an additional random interaction term, the interaction between plots and C, measuring whether the effect of factor C varies among random plots within each group of factor A. There are different options for analysis in this situation, depending on how we view this interaction term.

First, we could simply average the data for the two pots within each competition group and fit the previous models to these means. For the Legault et al. (2018) data, the model would be:

$$(\text{biomass})_{ijk} = \mu + (\text{nutrients})_i + (\text{bin within nutrients})_{j(i)}$$
$$+ (\text{competition})_k + (\text{nutrients} \times \text{competition})_{ik} + \varepsilon_{ijk}.$$

This approach is the same as discussed for two-level nested designs. The effect of the main fixed factor could be evaluated by averaging the observations within each level 2 sampling or experimental unit and fitting a single-factor model using these averages as the response variable (Section 10.4).

Second, we could use the full dataset but still fit the previous models – that is, we exclude the $B(A)$ by C interaction, so we don't allow the effect of C to vary among random plots within each A group. This model combines the plots(A) by C and residual terms into a single error term.

Third, we could fit a full model that separately estimates the random $B(A)$ by C interaction and the residual (among replicate sub-plots within each C group) – that is, we include a term in the model allowing the effect of C to vary across the randomly chosen plots in each A group. This model would be (note the l subscript allowing for $l > 1$ sub-plots):

$$(\text{biomass})_{ijkl} = \mu + (\text{nutrients})_i + (\text{bin within nutrients})_{j(i)} +$$
$$(\text{competition})_k + (\text{nutrients} \times \text{competition})_{ik} +$$
$$(\text{bin within nutrients} \times \text{competition})_{j(i)k} + \varepsilon_{ijkl}$$

There are now three random effects: bin within nutrients, bin within nutrients × competition interaction, and the error terms. In a multilevel context, we are modeling the overall mean biomass and the effect of competition on biomass within each bin (level 1) against nutrient group for the between-plots (level 2) component.

The decision to omit the random plots by C interaction effect when replicate sub-plots are included in the design might be based on biological grounds (this random interaction is not of interest or considered unlikely to be important) or the mixed modeling algorithm might not

converge unless this term is excluded. In either case, it would be sensible to evaluate it before omitting the term, by seeing if the CI includes zero or by comparing fits of models with and without this random term (see Box 11.2). This model comparison might be one step in a model-building process to produce the most parsimonious model, retaining only important terms (see Section 11.1.4). Two further points to note:

• It is also important to structure these models correctly in statistical software. If the plots are specified as the only random effect, most software will combine the $B(A) \times C$ interaction and the residual into a single residual term, with increased degrees of freedom.

• It is important to distinguish a design with separate individual sub-plots for each C group within each plot from one where we subsample individual sub-plots. For example, we should average data from multiple plants from an individual pot (as Legault et al. 2018 did), because they do not represent independent applications of factor C. However, data from separate pots within each bin can be used because each one represents a separate competition "arena." The sub-plots (pots) are not independent from each other, being clustered in bins, but the model structure accounts for this.

11.1.2 Assumptions

Statistical inference for the OLS and mixed effects models assumes a normal distribution of random effects, and emphasis is usually on the error terms (and usually the response variable). Still, as we emphasized in several earlier chapters, the normality assumption is not critical. The other important assumptions are related to homogeneity of variances and independence of the residuals, and we will consider the OLS and mixed effects models separately.

The OLS model strictly assumes equal variances and zero covariances (i.e. independence) of the error terms for between- and within-plots components. This is potentially an issue for the latter as we know that observations within the same group of a random factor (i.e. plots) are not independent. However, the F statistics and P-values for the within-plots components of an OLS model are reliable if the pattern of variances and covariances meets a pattern called compound symmetry (or a related pattern termed sphericity); this means that the variances among the C groups are equal and the covariances between the C groups are equal, but no longer need to be zero. This condition is likely to be met if the allocation of C groups to sub-plots within each plot is randomized. We will discuss this issue in more detail in Chapter 12 in the context of repeated measures designs, where covariances are unlikely to be equal because observations made on times closer together will be more correlated than those from times further apart.

Mixed effects models also assume homogeneity of variances, but take into account the nonindependence of the within-plots observations by separately specifying and estimating the random and fixed effects in the model. Additionally, many mixed model algorithms allow the researcher to specify a structure for the residual (within-plots) variance–covariance (R) matrices (see Section 10.2.2) and compare the fit of models with different structures.

Residuals are, as always for linear models, important for checking assumptions and identifying unusual/influential observations; a plot of residuals against predicted values is the most useful approach. We could plot the level 1 residuals for assessing the reliability of within-plots inference and then separately plot the level 2 residuals for the between-plots component. We prefer to keep things simple and just plot the level 1 residuals from the final fitted model. If we only have one sub-plot for each C group in each plot, the OLS model will be saturated (a perfect fit) and all the residuals will be zero; looking at residuals from fitting a model ignoring plots is probably the best approach in this situation (see Box 11.1).

11.1.3 Unbalanced Split-Plot Designs

Unbalanced split-plot designs can arise because of different numbers of plots in each group of the between-plots factor and because of different numbers of sub-plots in each C group within each plot (assuming the number of sub-plots is greater than one). We first introduced the rat pup dataset from Pinheiro and Bates (2000) in Section 10.5, and it is a good example of an unbalanced split-plot design, although it is not usually described in split-plot terms (e.g. see West et al. 2015). There were three drug treatments as the between-plots factor and the plots were female rats and their subsequent litters (level 2) assigned to each treatment. The within-plots factor was the sex of each pup, and pups were the sub-plots (level 1). The response variable was pup weight, and there was also a level 2 covariate (litter size). The design was unbalanced for both plots (some rats didn't survive) and sub-plots (litter size varied, as did the relative numbers of males and females in each litter). The analysis of these data is given in Box 11.3.

We can still fit the OLS model with unbalanced designs, but as we have highlighted previously, the F statistics and P-values are more sensitive to the homogeneity of variances assumption, and we have the usual issue of choosing between Type I, II, or III SS. Additionally,

the estimation of VCs is much more difficult, and reliable CIs are impossible. The mixed effects approach is more appropriate with unbalanced designs, especially if VCs are a priority. However, a decision about SS type still needs to be made if we test fixed effects based on an ANOVA; we use Type III in Box 11.3.

11.1.4 Model Building

The usual practice for biologists is to fit the full OLS model and interpret the terms from the ANOVA table. This full model could be simplified by comparing it to appropriate reduced models omitting specific terms using F statistics. For example, suppose the full model from a design with multiple sub-plots for each C group within each plot does not fit the data markedly better than one that omits the random $B(A) \times C$ interaction effect. In that case, we might be comfortable using the simpler model with a residual term with more df.

Model building and simplification tend to be more common with mixed effects models (Section 10.2.3). For example, West et al. (2015) simplified the model they used for the rat pup data analysis by first testing whether the random litter within treatment effect could be removed (they didn't include a litter within treatment by sex interaction in their full model) and then considered omitting the fixed treatment by sex interaction. Models with different structures for the residual variance–covariance matrix can also be compared – for example, allowing within-group variances to differ.

11.2 More Complex Designs

Split-plot designs can be made more complex in many ways. Some balanced repeated measures designs described in the next chapter can also be analyzed by the same partly nested OLS models used for split-plot designs. The repeated measures factors (usually representing some set of temporal measurements) are analogous to the within-plots factors. We will illustrate these more complex designs here and in the next chapter.

11.2.1 Additional Between-Plots Factors

Additional factors (usually crossed) can be included in the between-plots component of these designs, so we have between-plots factors A and B crossed, plots (C) nested within $A \times B$, and within-plots factor D applied to sub-plots within each plot. For example, Al-Janabi et al. (2016) examined the effects of ocean acidification and warming (OAW) and nutrient levels on the growth of different genotypes of the early life stage of a seaweed. The plots were seawater tanks nested within each

Box 11.3 ℞ Worked Example of Split-Plot Unbalanced Design: Growth of Rat Pups

The rat pup dataset from Pinheiro and Bates (2000) was introduced in Chapter 10. The between-plots factor was drug treatment (fixed with three doses: control, low dose, high dose) and the plots were 10 female rats and their subsequent litters (level 2) assigned to each treatment. The pups from a litter produced by each female were sexed and weighed, so pups were the sub-plots (level 1) with sex (fixed: male vs. females) as the within-plots factor and pup weight as the response variable. Litter size (number of pups) was recorded as a continuous covariate for each litter (level 2), as we might reasonably expect larger litters to have smaller pups on average. The design was unbalanced for both plots (only 7 out of 10 rats survived in the high-dose group) and sub-plots (litter size ranged between 2 and 18 pups). Additionally, one litter (number 12) had only female pups, so it was omitted from the analysis to avoid a missing cell.

Given the imbalance in the design, we focused the analysis on mixed effects models rather than a traditional OLS model. Initial boxplots of pup weight by treatment group and sex revealed no marked nonnormality or differences in spread. First, we will ignore the covariate of litter size. Our initial model included the fixed effects of treatment and sex and their interaction, and the random effects of litter(treatment), the interaction between litter(treatment) and sex, and the residual. The residual plot from this model suggested some unequal variance, and there was one unusual observation. Nonetheless, we will analyze untransformed pup weights.

To simplify the model and match the analysis of West et al. (2015), we first evaluated the random effect interaction between litter(treatment) and sex. The VC was small (0.016 and CI included zero), and a REML comparison of this model with one that omitted this interaction resulted in a likelihood χ^2 statistic of 0.628 and $P = 0.428$. There was no evidence that the effect of sex varied among litters within each treatment, so we omitted this term and refitted the model. For the fixed effects, using Kenward–Roger Type III adjustments, we see the following.

	MS	df_{num}	df_{den}	F	P
Treatment	0.369	2	23.1	2.26	0.127
Sex	7.817	1	295.6	47.84	<0.001
Treatment × sex	0.072	2	295.0	0.44	0.645

There was no treatment by sex interaction nor a strong main effect of treatment, but male pups were markedly heavier than females irrespective of treatment. Using Type II SS did not change our conclusions. The VC for litter within treatment was 0.253 (95% profile CI: 0.131–0.437); a LR comparison of the current model with one that omitted the litter within treatment indicated we should retain this term.

Now we will re-analyze these data, including the level 2 covariate, litter size. The model is the same as above, except that litter size is included. Note that we didn't include an interaction between litter size and treatment, so we are assuming that the slopes of the relationships between mean pup weight and litter size are consistent across treatment groups.

	MS	df_{num}	df_{den}	F	P
Treatment	1.760	2	22.2	10.80	<0.001
Sex	7.673	1	300.9	47.07	<0.001
Litter size	5.419	1	25.9	33.25	<0.001
Treatment × sex	0.075	2	299. 8	0.46	0.632

There was a strong relationship between pup weight and litter size, and there is now evidence for a main effect of treatment (pooling sex) on mean pup weight once we account for litter size.

There were differences between our analysis and that of West et al. (2015), who took a model-building approach. First, they did not omit litter number 12. Second, they further refined their model by omitting fixed effects that were not statistically significant – in this case, the treatment by sex interaction. Third, where the software allowed it, they fitted a model with a heterogeneous variance structure between the treatment groups. This model was a better fit than one that assumed homogeneous variances.

combination of OAW and nutrients. The sub-plots were sandstone cubes containing germlings from one of many sibling groups (populations) of the seaweed. One cube for each sibling group was placed within each tank. Al-Janabi et al. (2016) quite reasonably treated sibling group, the within-plots factor, as a random effect, representing all the possible sibling groups that could have been generated. More commonly, the within-plots factor is fixed in these designs, and we will illustrate the analysis as if sibling group was fixed. Tanks (plots) were also random as usual. The response variable was relative growth rate and the analysis of these data is shown in Box 11.4.

Including additional between-plots factors is a relatively simple extension of the standard split-plot models described earlier. When all factors except plots are fixed, the model should include the fixed main effects of A, B, and D and their interactions plus the random effect of plots within AB against which A, B, and AB are assessed, and an error term against which D and its interactions with A and B are assessed. If there are multiple sub-plots for each factor D group, the model can also include the random interaction between plots within AB and D estimated separately from the error term; this random interaction is a measure of how the effect of factor D varies across the random plots within each AB combination.

Box 11.4 ⓡ ⓔ Worked Example of Split-Plot Design with Two Crossed Between-Plots Factors: Nutrients and Future Climate Effects on Seaweeds

Al-Janabi et al. (2016) examined the effects of OAW and nutrient levels on the growth of different sibling groups of the early life stage of a seaweed, *Fucus vesiculosus*. There were two fixed and crossed between-plots factors: OAW (ambient pH and temperature vs. future pH and temperature) and nutrient levels (ambient nutrients vs. doubled nutrients). The plots were "benthocosms," large tanks in an outdoor facility in Germany; three tanks were nested within each combination of OAW and nutrients. The sub-plots were small sandstone cubes containing germlings from one of 16 sibling groups (populations) of *F. vesiculosus* produced by mating male and female plants collected from the wild. One cube for each sibling group was placed within each mesocosm. As discussed in the text, we will treat sibling groups as a fixed effect. The response variable was the mean relative growth rate (percentage change in area) of a subset of individual germlings on each cube after eight weeks.

We assume the one missing observation was missing at random, and we used mixed effects modeling rather than OLS for the analysis. We fitted a model that included the three fixed effects and their interactions plus the random effect of mesocosms nested within OAW and nutrients. The residual plot did not indicate any variance heterogeneity or unusual observations. The ANOVA for the fixed effects based on Type III SS and the Kenward–Roger method (note the adjusted df for the within-plots effects) was as follows.

Effect	MS	df	F	P
OAW	11.36	1, 8	32.10	<0.001
Nutrients	4.76	1, 8	13.47	0.006
OAW × nutrients	3.08	1, 8	8.72	0.018
Sibling group	1.88	15, 119.03	5.30	<0.001
OAW × sibling	0.27	15, 119.03	0.76	0.717
Nutrients × sibling	0.41	15, 119.03	1.16	0.314
OAW × nutrients × sibling	0.51	15, 119.03	1.45	0.134

There was no evidence that the effect of sibling group interacted with OAW or nutrient level. There was a strong effect of sibling group, and evidence that the effect of nutrients depended on the group of OAW – nutrients had little effect on growth at ambient OAW conditions, whereas for warmer and more acidic conditions not adding nutrients reduced growth rates compared to adding nutrients (Figure 11.5). Note these ANOVA results differ somewhat from those reported by Al-Janabi et al. (2016), as they used different software, adjusted the df with Satterthwaite's method, did not include the three-way interaction in their analysis, and treated sibling group as a random effect; nonetheless, the broad conclusions were similar, focusing on the interaction between nutrients and OAW.

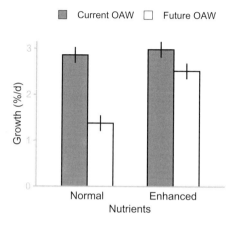

Figure 11.5 Mean daily growth of *Fucus* juveniles at two different nutrient levels and under temperature and pH levels associated with current vs. future seawater conditions. Error bars are SEs.

11.2.2 Additional Within-Plots Factors

Extra within-plots/subjects factors can also be included in these designs in two ways. First, we could have a split-split-plot design where a third factor is applied to smaller units within each sub-plot. For example, Ferrarezi et al. (2019) examined the effects of three fixed factors on grapefruit production. The first factor was the production system (screenhouses vs. open-air plots) and was applied to plots. The second factor was planting system (in-ground vs. potted) applied to sub-plots within each production system plot. The final factor was rootstock (Sour orange vs. Cleopatra crossed with Flying dragon) compared within each sub-plot within each production system plot. We don't cover split-split-plot designs here, but more detail can be found in Mead et al. (2012) and Kirk (2012).

Second, two or more factors could be applied to the sub-plots within each plot in a crossed design. For example, Hey et al. (2020) did a field experiment to examine the effect of artificial light at night (ALAN) on the growth of a wildflower. The between-plots factor was ALAN, with independent plots of ground for each ALAN treatment. The sub-plots were small pots, and there were two within-plot factors arranged in a crossed

design: planting density and soil moisture. Each combination of density and moisture had a single pot so the design is unreplicated at the sub-plot level. The response variable was total biomass (root and shoot) per plant in each pot, averaged across the plants within the higher-density pots. The analysis of these data is shown in Box 11.5

The model fitted in this example would include the fixed effects of ALAN, density, moisture and their interactions, and the random effects of plots within ALAN, plots within ALAN by density, and plots within ALAN by moisture. The error term is a composite random term combining the plots within ALAN by density by moisture and the true sub-plot error; these cannot be estimated separately with only a single sub-plot for each density and moisture combination in each plot.

11.2.3 Including Continuous Covariates

These split-plot designs can also include continuous covariates recorded at the plot (level 2) or sub-plot (level 1) levels, or other levels in more complex hierarchies. The rat pup data described in Section 11.1.3 included a level 2 covariate, litter size (i.e. number of pups) recorded at the plot (litter) level as we might expect pups from bigger

Box 11.5 Ⓡ Ⓔ Worked Example of Split-Plot Design with Two Crossed Within-Plots Factors: ALAN Affects Wildflowers

Hey et al. (2020) studied the effect of ALAN on the growth of a wildflower species (*Asclepias syriaca*) in an outdoor field experiment. They used a split-plot design for their experiment. The between-plots factor was ALAN, with five plots receiving artificial light and five control plots with the same setup but only receiving ambient light. The sub-plots were 11-L small pots, and there were two within-plot factors arranged in a crossed design: planting density (one or three plants per pot) and soil moisture (weekly addition of water vs. no watering). There were four pots in each plot, so each combination of density and moisture had a single pot; the design is unreplicated at the sub-plot level. The response variable was total biomass (root and shoot) per plant in each pot, averaged across the plants within the higher-density pots.

We first checked the pattern in the residuals by fitting a simple three-factor crossed model, ignoring the hierarchical structure. There was evidence for a mean–variance relationship, with one outlier (an unwatered pot with a single plant in an artificially lit plot; standardized residual close to 4). Like the original paper, we transformed the response to log(biomass + 1), which improved the residual plot and reduced the outlier's influence.

We initially fitted a model that included the interactions between the within-plots fixed effects and random plots, resulting in separate error terms for *F*s associated with the within-plot fixed effects.

The fit of the OLS partly nested model resulted in the following ANOVA table.

Source	df	MS	F	P
Between-plots				
ALAN	1	0.012	0.33	0.582
Plots (ALAN)	8	0.037		
Within-plots				
Density	1	0.002	0.21	0.661
ALAN × density	1	0.002	0.17	0.693
Plots (ALAN) × density	8	0.010		
Moisture	1	0.039	4.38	0.070
ALAN × moisture	1	0.026	2.93	0.125
Plots (ALAN) × moisture	8	0.009		
Density × moisture	1	0.003	0.29	0.605
ALAN × density × moisture	1	0.021	2.12	0.183
Plots (ALAN) × density × moisture	8	0.010		

There was no evidence for an interaction between ALAN and density or moisture and no main effect of ALAN. There was some evidence plant biomass was greater for plants that received supplemental water. The *F*s and *P*-values for the fixed effects (based on the Kenward–Roger adjustment) from the fit of the mixed effects model were only slightly different from above for some within-plots effects, but our conclusions were not changed. While the focus of these designs is mainly on the fixed effects, the REML estimates of the VCs for the random effects are provided for completeness.

Component	Variance	95% Profile CI
Plots (ALAN)	0.0066	0.0000–0.0184
Plots (ALAN) × density	0.0004	0.0000–0.0079
Plots (ALAN) × moisture	0.0000	0.0000–0.0063
Plots (ALAN) × density × moisture	0.0094	0.0037–0.0134

Box 11.5 (cont.)

A simpler model combines the interactions between the within-plots fixed effects and random plots into a single random effect (residual). There was little difference in the fit of this model compared to the full mixed effects model above (REML: LR $\chi^2 = 0.021$, df = 2, $P = 0.941$), so the simpler model is probably fine. The within-plots fixed effects are now tested based on 24 df (rather than 8), and while the P-values change slightly, our conclusions would be the same. We could also fit the simpler model using OLS by omitting the interaction between plots (ALAN) and the within-plots fixed effects.

In their analysis, Hey and colleagues included a covariate (herbivore damage), only included two-way interactions between ALAN and moisture and ALAN and density, and used Satterthwaite's adjustment or the tests of fixed effects. They found a "statistically significant" effect of moisture ($P = 0.032$), but otherwise their broad conclusions were the same.

litters to have smaller weights. The analysis of these data, including the covariate, is also in Box 11.3.

11.3 Key Points

• A common design in biology is the split-plot or split-unit design, characterized by having different fixed factors applied to units at different scales.

• The typical split-plot has one between-plots fixed factor applied to units termed plots and a second within-plots fixed factor applied to smaller-scale units within each plot, termed sub-plots.

• These designs can be replicated at the sub-plot level, with two or more sub-plots for each group of the within-plots factor, or unreplicated, with a single sub-plot for each within-plots factor group.

• The mixed models fitted to these designs include the fixed effects of the two factors and their interaction, plus the random effects of plot (nested within the between-plots factor) and sub-plots and any interactions between the fixed and random effects.

• These models can be fitted with OLS for fully balanced designs or with mixed effects models based on ML/REML.

• In addition to the usual assumptions, these models assume that the pattern of variances and covariances (correlations) for the within-plots components meet a condition known as compound symmetry so the correlations among error terms are equal. This condition is likely to be met if the within-plots factor groups are allocated randomly to sub-plots.

• Split-plot designs can be extended to include multiple between-plots factors and/or multiple within-plots factors, and/or continuous covariates at either the between-plots or within-plots levels.

Further Reading

Split-plot (unit) designs and their linear models are well described by Kutner et al. (2005; termed partly nested), Mead et al. (2012), and Milliken and Johnson (2009), although all with an OLS focus.

12   Repeated Measures Designs

The previous chapter introduced the split-plot (split-unit) design, where factors are applied at two or more different scales (e.g. plots and sub-plots). Another common situation in biological research is where the same experimental or sampling units are recorded repeatedly through time or exposed to different experimental treatments through time. These designs are often termed "repeated measures" (RM) because the response variable is measured from the same units at different times or under different conditions. They have similarities in structure to a split-plot design, and the same linear model can sometimes be fitted to both. Repeated measures designs, and their analysis with OLS-based linear models, originated in psychological research, where the units are often called "subjects." The experimental treatments applied to each subject or times at which subjects were recorded represent within-subjects factors. The subjects or units are commonly also allocated to different groups of one or more between-subjects factors.

Additional terminology is sometimes used in psychological research for these designs:

- Subjects × treatments designs have the order of within-subjects factor groups randomized for each subject. The within-subjects (RM) factor is a set of treatments that can be ordered independently of time – for example, a set of drugs applied to experimental animals.
- Subjects × trials designs do not allow the order of within-subjects factor groups to be randomized. The RM factor is usually time.
- Groups × treatments or groups × trials designs extend the first two designs by including one or more between-subjects factors. These designs are the commonest in biology.

The main reason for using RM designs is that we are interested in the response of units/subjects through time or across the within-subjects treatments and whether the effect of between-subjects factors is consistent or not through time. RM designs can also be used to manage limited resources. For example, Allen and Marshall (2014) studied the effects of egg size on larval characteristics of a marine tubeworm. They allocated eggs from a group of large or small female tubeworms to 10 jars for each egg size. These eggs were fertilized, and after the embryos hatched, the proportion of larvae that had settled and metamorphosed was recorded for each jar every 3 days for 15 days. A completely randomized (CR) design would require separate, independent jars for each tubeworm and time combination, and 100 total jars, so using RM is more resource-efficient. Reducing the total number of animals might also be an ethical consideration; McDonald et al. (2021) discuss these issues for studies involving uptake of radioisotopes.

In the RM context, jars are subjects, egg size is the between-subjects factor, and day is the within-subjects factor (which we could also treat as a continuous covariate). This design has similarities to a split-plot design: subjects replace the term plots. Instead of a within-plots factor applied to sub-plots within each plot, we now have a within-subjects factor applied or recorded from each subject through time. Using our multilevel terminology, the level 1 data are the repeated measurements through time, with subjects being the level 2 units, usually with their own level 2 factor(s) or covariate(s).

There are specific challenges with analyzing and interpreting RM designs. Observations taken from the same units through time will not be independent. We have already pointed out that observations from the same groups of a random factor are not independent, and subjects in RM designs, just like plots in split-plot designs, are nearly always random. It is also likely that observations closer together in time will be more correlated than those further apart, so the covariance structure can be complex. We will address this issue in Section 12.1.2.

Two other difficulties apply when the within-subjects factor represents experimental treatments applied by the investigator (e.g. drugs given to experimental animals). The first is the problem of carryover effects, where the response to one treatment may be affected by the preceding treatment in the sequence. This difficulty can be solved only by ensuring that the time interval between treatments is long enough for the "subjects" to recover. The second problem is the order or sequence effect, where measurements early in a sequence may differ from those later, irrespective of treatment. We can alleviate this problem by randomizing the order in which a subject receives each treatment. Where the within-subjects factor is not a set of treatments but simply represents time itself, carryover, and order or sequence effects are usually implicit in the hypotheses being tested.

Even though the RM (within-subjects) factor can be experimental treatments, we will refer to this factor as "time" for the rest of this chapter for simplicity. There are three common statistical approaches for analyzing these RM designs:

• Fitting a linear model using OLS with an emphasis on the resulting ANOVA. The models we fit are the same as those described in Chapter 10 for block designs, where subjects are equivalent to blocks, and Chapter 11 for split-plot designs, where subjects are equivalent to plots. Here, we have between- and within-subjects factors, just like we had between- and within-plots factors.
• Producing summary statistics for each subject for the within-subjects factor (e.g. means, or intercepts/slopes if the within-subjects factor can be treated as a continuous covariate) and then modeling these summary statistics against any between-subjects factor(s) with the usual OLS approach.
• Using mixed effects (multilevel regression) modeling based on ML and REML. We specify the fixed and random effects (subjects and any interactions with subjects). This approach may allow us to model different correlation structures through time.

When we have a factor (e.g. time) measured on a quantitative scale, we could include it in our linear model as a categorical or a continuous predictor. If the latter, the emphasis is then on the linear regressions between the response variable and time for each subject. We can investigate how the slopes and intercepts of those regressions vary among the fixed factors and across random subjects. This approach also helps when we have missing data (e.g. some subjects are missing observations for one or more times, or subjects are recorded at different times). We can fit a linear regression for each subject irrespective of having the same number of observations or the same

times. If the relationships are nonlinear, we can also model polynomial regressions (e.g. quadratic). Another advantage of treating time as continuous is that we can fit random slopes mixed effects models by including a random time by subjects interaction effect. However, if the number of times is relatively small, say <5, or the relationships between the response and time for the different subjects can't be modeled with a polynomial or by transformation, then fitting time as a categorical factor is more appropriate. When time is categorical, even if the design allows a time by subjects interaction, the number of random effects created will often be too many for the model to be fit unless we have a large sample size.

12.1 Simple Repeated Measures Designs

A simple RM design has no between-subjects factors. For example, Steiger et al. (2008) used the burying beetle *Nicrophorus vespilloides* to study the Coolidge effect, the decline in a male's interest in mating with the same female compared to novel females. Individual male beetles were presented with the same female beetle four times, and then a novel female on the fifth occasion, making this a RM design. The within-subjects factor was the order of presented females (fixed with five groups). Order of mating was not treated as a continuous covariate as the focus was on differences between groups, especially the final group (novel female) versus the others. In multilevel terminology, the RM are the level 1 observations, and the beetles are the level 2 units. However, these units are not grouped or clustered (there is no between-subjects factor), so the data are not truly multilevel. Box 12.1 shows the analysis of these data.

We will use a second example, from Chadha et al. (2019), to illustrate a quantitative RM factor treated as a continuous covariate. These authors established a CR experiment studying the effects of different soil water-holding capacities (WHCs) on an agricultural weed. Plants were grown individually in pots allocated to different WHCs, a between-subjects factor. We will use a subset of the data from only one treatment group (so no between-subjects factor) in which the leaves on each plant were counted weekly for nine weeks. Time was the within-subjects (RM) fixed factor, and individual plants were the random subjects. These data are analyzed in Box 12.2.

12.1.1 Analysis of Simple Repeated Measures Designs

The simple RM design is similar to an unreplicated complete blocks (CB) design (Section 10.6.1). There are q

Box 12.1 Ⓡ Ⓔ Worked Example of Repeated Measures Design with Single Within-Subjects Factor: Testing the Coolidge Effect in Beetles

Steiger et al. (2008) studied the Coolidge effect, the decline in males' interest in mating with the same female compared to novel females, using the burying beetle *Nicrophorus vespilloides*. Eighteen male beetles were presented with the same female beetle four times, and then a novel female on the fifth occasion. There was no evidence that physical exhaustion affected time to mating as a separate control group of males were presented with novel, unmated females five times in succession and there was no change in time to mating. The within-subjects factor was the order of presented females, treated as a fixed factor with five groups. The response variable recorded was time to mating.

The OLS model for this design includes the fixed mating order and random beetle effects, with the error being a combination of true error within each mating order and individual combination and any interaction between mating order and beetle. The residuals from this model and boxplots of mating time against mating order showed uneven spread with skewed distributions for the first and fifth matings, and an outlier in the direction of skewness in the latter group. Like Steiger et al. (2008), we will analyze log-transformed mating time.

The ANOVA (Type III SS) from the fit of this model was as follows.

Source	df	MS	F	P	GG-P
Mating order	4	2.091	6.318	<0.001	0.002
Beetles	17	0.395	1.195	0.293	
Residual	68	0.331			

The Greenhouse–Geiser ε was 0.678, so the adjusted P-value for the mating order effect is slightly higher. There was evidence against the null hypothesis of no effect of mating type, with Figure 12.1 indicating an effect of mating order on log mating time, with mating time increasing steadily from the first to the fourth mating, and then dropping sharply for the fifth mating with the novel female. Note that the F and P-value for the random effect of beetles are only valid if we assume no mating order by beetle interaction –that is, the effects of mating order are the same for each beetle.

The OLS estimate (with 95% CIs) for the variance components for random beetles was 0.013 (-0.036 to 0.061), and the residual was 0.331 (0.255 to 0.450). There was little variation between individual beetles, but again, interpretation of this variance component relies on the absence of a mating order by beetle interaction.

We can also analyze this design by fitting a mixed effects model that includes the fixed effect of mating order and the random effect of individual beetles. The fixed effects ANOVA F statistic and P-value (based on Kenward–Roger adjustment to df) were similar to the OLS model. The REML estimates for the variance components were the same as the OLS estimates, although the 95% profile CIs were slightly different with the lower limit for the beetles constrained to zero (beetles: 0–0.094; residual: 0.229–0.437).

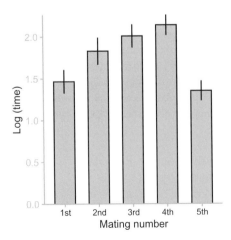

Figure 12.1 Log time (mins) to mating for burying beetles against mating order. The bars show means for each mating position with standard error bars. Note that the error bars do not incorporate the correlated data structure.

Box 12.2 ℝ 🄴 Worked Example of Repeated Measures Design with Single Quantitative Within-Subjects Factor: Plant Growth Through Time

Chadha et al. (2019) set up a CR experiment studying the effects of different soil WHCs on various characteristics of the weed *Lactuca serriola*. Plants were grown individually in pots allocated to one of four WHCs (100%, 75%, 50%, and 25%). There were seven plants for each WHC, although we only used data from the 100% treatment group. The leaves on each plant were counted weekly for nine weeks. Time was the within-subjects (RM) fixed effect, and individual plants were the random subjects. With nine weeks and reasonably linear trends through time for each plant (Figure 12.2), it made sense to treat time as a continuous covariate. We did not center time, so intercepts represent the number of leaves for week 0.

The OLS linear (ANCOVA) model was fitted, including the interaction term allowing different linear slopes for each plant. Examination of the residuals did not show strong patterns in the variances related to means. The resulting ANOVA (Type III SS) from the model fit was as follows.

Source	df	MS	F	P
Time	1	4,999.1	271.69	<0.001
Plant	6	21.8	1.35	0.253
Time × plant	6	18.48	1.14	0.352
Residual	49	16.1		

Since plant was a random effect, the F for time uses the time by plant interaction in the denominator. There was little evidence against the null hypothesis of no time by plant interaction, and Figure 12.2 shows that the slopes of number of leaves against time were very consistent across plants. The OLS estimate of the interaction variance component (VC) was 0.038 (95% CI: −0.326 to 0.401; includes zero), representing <1% of the total random variation. As with fixed effects ANCOVA models, it is common to drop the unimportant heterogeneous slopes term – that is, the interaction was pooled with the residual, and the model refitted; the F for time is now 305.78 (df = 1, 55, $P < 0.001$; note the increase in denominator df from 6 to 55) and remaining VCs were: plant = 11.33 (95% CI: −3.56 to 26.30; includes zero) and residual = 16.10 (95% CI: 11.23 to 25.00).

We also estimated a mixed effects model, including time as a fixed covariate, plant as a random effect, and random slopes by incorporating the time by plant interaction effect. The fit of this model was singular, so we set the correlation between the two random effects (intercepts and slopes) to be zero. Again, we compared this model's fit to the random intercept model (by omitting the random slopes term). The AIC_Cs of the two models were similar (random slopes: 376.81; random intercepts: 375.28), and the log-likelihood ratio χ^2 was 0.828 (df = 1) with a P-value of 0.363, indicating the random intercepts model fits almost as well as the random slopes model (supporting the results from the OLS model and Figure 12.2). The random intercept model showed a strong effect of time (Kenward–Roger $F_{1,55} = 305.78$, $P < 0.001$; overall slope = 3.45 with SE = 0.197) and the estimates of the VCs (with 95% profile CIs) were: plants (11.33; 2.77–38.24) and residual (16.35; 11.33–23.83).

Had we kept the random slopes model, the P-value for time is still very small, although the F is now based on 1 and 10.28 df (instead of 1 and 55), reflecting the "cost" of including the random time by plant interaction effect in the model.

Finally, we examined whether an AR(1) correlation structure might improve our random intercept mixed model's fit. The number of leaves in any week is cumulative, so we might expect that weeks closer together would be more correlated than those further apart. Including an AR(1) correlation structure through time within each plant did not change the model ML fit much (AIC_Cs were almost identical with a log-likelihood ratio χ^2 of 2.265 (df = 1) with a P-value of 0.132).

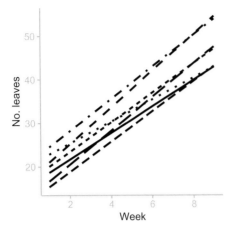

Figure 12.2 Changes in the number of leaves through time for the weed *Lactuca serriola*. Each line is an OLS linear regression representing one plant. The data are taken from the 100% WHC soil treatment of Chadha et al. (2019).

randomly chosen sampling or experimental units termed subjects (equivalent to random blocks, factor *B*). Each subject is recorded *p* times or subjected to a set of *p* treatments (factor *A*), the within-subjects fixed factor. The main difference between this design and a CB design is that the fixed factor is "applied" to whole subjects rather than smaller units within each block. In the Steiger et al. (2008) example, the fixed factor is mating order (*p* = 5), and the subjects are individual male beetles (*q* = 18). In the Chadha et al. (2019) example, time (weeks, *p* = 9) could be treated as a factor or a quantitative covariate, and subjects are individual plants (*q* = 7). When we treat time as a continuous covariate, we have a value for the response variable and the covariate for each of our subjects at each time.

12.1.1.1 *Linear Models for Simple Repeated Measures Designs*

The linear models for a simple RM design are the same as for an unreplicated single-factor blocks design. Using the Steiger et al. (2008) example:

$$(\text{time to mating})_{ij} = \mu + (\text{mating order})_i + (\text{beetle})_j + \varepsilon_{ij}.$$

In this model, the response variable (the mating time from mating *i* of beetle *j*) is modeled against the overall population mean (or intercept) mating time, the fixed effect of the *i*th mating order, pooling all possible individual beetles, and the random effect of beetle *j* assumed to be $N(0, \sigma^2_{\text{beetle}})$ where the VC σ^2_{beetle} measures the variance in the time to mating across all possible beetles, pooling mating orders. The error term, ε_{ij}, represents the random interaction between mating order and beetle and the error due to differences among random observations for each

mating order and beetle combination. As with the model fitted to an unreplicated block design, we cannot estimate these two sources of variation separately. These error terms are assumed to be $N(0, \sigma^2_\varepsilon)$.

When we treat the within-subjects factor (e.g. time) as a continuous covariate, we fit a model similar to the ANCOVA model in Section 8.2, except that instead of a fixed factor and a fixed covariate, we now have a random factor (subjects) and a fixed covariate. The overall mean is replaced by the intercept of the regression of the response against time over all subjects. This model allows the intercepts and slopes of the subject-specific regression lines to vary by including the interaction between the covariate and the random subjects (equivalent to the heterogeneity of slopes term from an ANCOVA model). From the Chadha et al. (2019) example, we model the number of leaves from any plant against:

- the intercept of the regression of the number of leaves against time;
- the fixed effect (slope) of the regression of the number of leaves against time, pooling all possible individual plants;
- the random effect of any plant, pooling times;
- the random effect of the interaction between plant and time, measuring how much the slopes of the regressions through time for each plant vary from each other.

Both random effects are assumed to be normal with a mean of zero, and they have VCs of σ^2_{plant}, measuring the variance in the mean number of leaves across all possible plants, and $\sigma^2_{\text{plant} \times \text{time}}$, measuring the variance in the slopes of the relationship between the number of leaves and time across all possible plants, respectively. By treating time as a continuous covariate, we can separately estimate the random plant × time interaction effect. The error term represents the error associated with each observation – that is, the component of each observation not due to the random plant or the specific time with which it is associated. These error terms are, as usual, assumed to be $N(0, \sigma_\varepsilon^2)$.

We can think of this model in a multilevel regression context. We are fitting a linear regression model for each level 2 subject (plant), each with its own intercept and slope. The estimated parameters from these regressions are then modeled across the level 2 subjects (plants). If there is little evidence for the slopes varying from plant to plant, this model can be simplified by omitting the plant × time interaction term (random slopes term) and refitting the model with just the random effect of plant and the fixed effect of time. Now, the regression model within each subject (plant) has a common slope, and only the intercept varies from plant to plant.

With the RM factor treated as a continuous covariate, it may be appropriate to center the covariate. Centering results in the plant term measuring variance in the number of leaves for the mean value of time rather than the number of leaves at time zero. If the intercept (i.e. the number of leaves when time is zero) is outside the data range, this may make more biological sense. Centering can also alleviate issues with collinearity in more complex models with multiple interaction terms.

We fit the models and estimate the fixed effect parameters (with SEs and CIs) with OLS, where fixed and random effects are not distinguished in the model fitting. The variances associated with the random effects are derived from the expected values of the mean squares from an ANOVA. The ANOVA and F statistics from the OLS fit for the Steiger et al. (2008) example, where mating order is categorical and beetles are random, are the same as for the unreplicated block design in Table 10.3. For the Chadha et al. (2019) example with time as continuous, the ANOVA is analogous to that from a replicated two-factor mixed model (Table 10.3), except now, as a continuous covariate, week has df = 1 and the fixed effect of time is tested against the time × plants interaction.

Alternatively, we can analyze these models with a mixed effects model. For the fixed effects, df-based Kenward–Roger adjustments are recommended, as are profile (or bootstrap) CIs for REML-estimated VCs.

When the RM factor is treated as a continuous covariate, the decision to fit the simpler random intercept-only model instead of the random slopes model is usually based on:
• an examination of the individual regression lines for each subject (e.g. Figure 12.2);

• an assessment of whether the interaction between time (e.g. week) and random subjects (e.g. plants) is considered important, based on the CI for its VC or a comparison of the two models.

There are two other important issues when including random slope terms in mixed effect models:
• The two random effects (slopes and intercepts) may be correlated. That correlation may be useful biological information, such as plants that produce leaves faster through time may also have more leaves to start with. We may wish to set the correlation to be zero in our modeling (i.e. a VC D matrix – see Section 10.2.2) if our model fitting doesn't converge, as this reduces the number of random effects to be estimated.
• Similarly, we may have to omit the random slopes term from our model to help achieve convergence, especially with small sample sizes.

12.1.2 Assumptions for Simple Repeated Measures Models

The assumptions for OLS and ML/REML mixed effects linear models described in Section 10.2, including normality and homogeneity of model error variances, apply to these RM analyses. When the within-subjects factor is included as a continuous covariate, we also assume the relationship between the response and the covariate is linear for each subject. Some nonlinearities can be managed by including quadratic polynomial terms in the model or transforming the response (e.g. to logs). Plots of residuals against predicted values are, as always, important diagnostic tools to check for homogeneity of variances and outliers. For mixed effects models, conditional residuals based on the model including the fixed and random effects are most commonly used. When the within-subjects factor is a continuous covariate, some software will also provide influence diagnostics (e.g. Cook's D) for the within-subjects regressions.

12.1.2.1 *Independence and Covariance Structures*
The other key issue with models for RM designs is the lack of independence of model error terms. We have pointed out that observations (and their error terms) within groups of a random factor (blocks, plots, clusters, etc.) will be correlated. Additionally, when we have repeated measurements through time, observations closer together are likely to be more correlated than those further apart. This contrasts to a block or split-plot design, where we can usually randomize the allocation of groups of the fixed factor to sub-units within each block or plot, so we wouldn't expect a particular pattern of correlations.

$$\begin{array}{ccc}\text{Generalized} & \text{Diagonal} & \text{Compound symm.}\end{array}$$

$$\begin{bmatrix} \sigma_1^2 & \sigma_{12}^2 & \sigma_{13}^2 \\ \sigma_{21}^2 & \sigma_2^2 & \sigma_{23}^2 \\ \sigma_{31}^2 & \sigma_{32}^2 & \sigma_3^2 \end{bmatrix} \begin{bmatrix} \sigma^2 & 0 & 0 \\ 0 & \sigma^2 & 0 \\ 0 & 0 & \sigma^2 \end{bmatrix} \begin{bmatrix} \sigma^2 & \sigma_c^2 & \sigma_c^2 \\ \sigma_c^2 & \sigma^2 & \sigma_c^2 \\ \sigma_c^2 & \sigma_c^2 & \sigma^2 \end{bmatrix}$$

Figure 12.3 Variance–covariance matrix showing a generalized version and two options that are used commonly, assuming complete independence (diagonal) and for compound symmetry. σ_{12}^2 represents a covariance between times 1 and 2, and for the compound symmetry case, σ_c^2 represents a constant covariance between all pairs of times.

The most convenient description of the pattern of correlations is via the variance–covariance matrix. In the simple design with a categorical within-subjects factor, the rows and columns refer to the groups of the within-subjects factor (e.g. mating order), the diagonal represents the constant variance of the response for each mating order, and off-diagonals represent the pairwise covariances between the observations for different mating orders (Figure 12.3).

OLS Models

While inference in OLS models assumes equal variances and zero correlations for the errors within groups, the P-values associated with the t or F statistics for within-subjects fixed effects (e.g. mating order) are still reliable if the correlations meet a pattern known as compound symmetry. This means the variances are the same, and the pairwise covariances are the same (Figure 12.3). With RM designs involving time, this scenario is unlikely. However, the compound symmetry condition is too restrictive – the F statistics and P-values are still reliable if a slightly more lenient condition known as sphericity is met, where the variances of the differences between within-subjects groups are equal. If these assumptions are violated, the P-values associated with the within-subjects effects are less reliable and overestimate the evidence against the null hypothesis. Mauchly's test is a formal, but oversensitive, test for sphericity.

Unfortunately, different variance–covariance structures cannot be incorporated into OLS models, so we have limited options for dealing with violations of compound symmetry or sphericity. We could fit the model with generalized least squares (GLS), allowing a range of variance–covariance structures. However, GLS models don't allow random effects, so they are difficult to use for more complex hierarchical designs with random factors, like those we consider in the rest of this chapter.

If the focus is on interpreting the P-values associated with the within-subjects effects, we can adjust the df for the relevant F by how much the variance–covariance matrix departs from compound symmetry and sphericity. This departure is measured by the epsilon (ε) parameter. When sphericity is met, $\varepsilon = 1$; the closer ε is to zero, the more the sphericity assumption is violated. We can estimate ε from the sample variance–covariance matrix and the most common and recommended variant is the Greenhouse–Geisser ε. An alternative is the Huynh–Feldt ε. The df are adjusted downwards based on the estimate of ε, making the tests more conservative.

Another approach that does not assume sphericity is to use the differences between pairs of within-subjects groups (e.g. between pairs of mating orders) as multiple response variables in a multivariate analysis of variance (MANOVA; see Chapter 15). Statistical software often provides this analysis when running a classic RM ANOVA and evaluates whether the differences have a population mean vector equal to zero.

Mixed Effects Models

Mixed effects models provide more flexibility for dealing with the lack of independence between repeated observations. The REML-estimated variances and covariances (D matrix) of the random effects and the model error terms (residuals; R matrix) can be used to weight the estimation of fixed effects (Section 10.2). The better our estimate of the variances and covariances associated with the random effects, the better our estimates of the fixed effects and their standard errors (West et al. 2015). We introduced different structures for the D matrix in Section 10.2 (see also Section 12.1.1), so we will focus on four commonly used structures for the R matrix here (Fitzmaurice et al. 2011; Gueorguieva 2017; West et al. 2015):

• The simplest structure specifies independence of observations from the same subject – that is, that the covariances equal zero, and the only parameter to be estimated is the constant variance $\left(\sigma_\varepsilon^2\right)$. This is sometimes termed the diagonal structure (Figure 12.3).

• As described above for OLS models, the compound symmetry structure has equal variances and equal covariances. Only two parameters need to be estimated, the constant variance and the constant covariance (Figure 12.3). Constant covariances are unlikely with repeated measurements on the same subjects.

• The first-order autoregressive structure, AR(1), is commonly used for RM and longitudinal datasets. The variances are assumed to be constant, but the covariances depend on how far apart the observations are, with the correlations between observations decreasing exponentially as observations become further apart. There are still only two parameters to be estimated: the constant variance and the correlation parameter, which changes by a power term measuring how many repeated measurements separate the observations. The AR(1) structure is particularly useful for RM designs with equally spaced time intervals.

- The Toeplitz structure again assumes equal variances but specifies different correlations depending on how far apart the observations are, rather than a simple exponentially decreasing correlation as in AR(1). The number of parameters to be estimated depends on the number of times (t): the constant variance plus $t - 1$ covariances. Like the AR(1) structure, a special case of the Toeplitz, it is best with equally spaced time intervals.

Three other structures are important to mention. First, an unstructured variance–covariance matrix allows all the variances and covariances to be estimated. While offering the greatest flexibility, the number of parameters to be estimated quickly gets very large as the number of times increases, and the parameters may not be estimable for the sample sizes commonly found in biological research. Second, to overcome the requirement of equally spaced time intervals for the AR(1) and Toeplitz structures, a modification of the standard AR(1) structure is the exponential where the correlation changes based on the actual difference in times rather than just the separation in the number of repeated measurements. Third, these variance–covariance matrices can also be structured in most software to allow different variances through time (variances often increase if the trend through time is also positive).

The choice between the different R matrix structures should be based on biological knowledge and the comparison of models' fit with different structures. However, note that not all software algorithms allow different R matrix structures to be applied.

12.1.3 Missing Observations

With simple RM designs, missing observations for a subject can occur in two circumstances:
- when one or more recordings are missed, but subsequent recordings are made;
- when subjects drop out or die during the study, the measurements' sequence is shorter for some subjects than for others. Drop out is common in health research but can occur in other biological research.

When we have missing observations in RM and longitudinal datasets, the data are often called incomplete. How we deal with missing observations depends on whether they are missing completely at random (MCAR) or just missing at random (MAR) – see Section 5.6 and Fitzmaurice et al. (2011). Missing at random observations, where the missing values depend on the other observed data but not on the specific data that could have occurred, is the more realistic scenario. Under these circumstances, ML estimation is reliable (West et al. 2015).

Where the within-subjects factor is categorical, there are only two options for dealing with missing observations: The subject with the missing observation(s) is removed from the analysis or imputation is used to replace the missing observation(s).

When the within-subjects factor is continuous, we essentially fit linear regressions for each subject. For observations MAR it doesn't matter if the regressions for different subjects are based on different numbers of time points. The subjects don't even have to be recorded at the same times.

There are caveats to the flexibility with missing observations treating time as continuous. First, if many subjects have very few time points or some subjects have only a single time point recorded, reliable estimation of slopes can be difficult. Second, if there is also a between-subjects factor, we need to be careful that the between-subjects treatments do not influence the pattern of missing observations. For example, if the between-subjects factor affects survivorship, the treatment effects will be confounded with differences in observation numbers. It would be important to analyze survivorship in these circumstances and be cautious with interpreting the analysis on the original response.

12.2 More Complex Repeated Measures Designs

The simple designs in the previous section are not that common in biological research. Biological designs usually include one or more between-subjects predictors. Often, these are single-factor or multifactor crossed designs, but where the sampling or experimental units are recorded through time. We will describe these designs now.

12.2.1 One Between-Subjects Factor

In their simplest form, we have one between-subjects factor with random subjects nested within each group, and these subjects are recorded over multiple times or treatments. For example, Garcia et al. (2015) studied the effect of the appetite-regulating hormone leptin on appetite and mating preferences in toads. The basic design was a simple two-group single-factor experiment where female toads were allocated to a treatment group receiving leptin or a control group. After the treatment period, each toad was presented with crickets, and the response variable was the cumulative number of attacks by each toad over five three-minute intervals. Treatment (leptin versus control) was the fixed between-subject factor, and toads were the subjects. The within-subjects fixed factor was time. These data are analyzed in Box 12.3.

The within-subjects factor can also represent different treatments, rather than just a time sequence.

The linear model for a RM design with a fixed between-subjects factor and a fixed within-subjects factor, both categorical, is the same as for an unreplicated split-plot design. Using the Garcia et al. (2015) example, omitting subscripts for convenience:

$$\text{number of attacks} = \mu + \text{treatment} + \text{toad (treatment)} \\ + \text{time} + \text{treatment} \times \text{time} + \varepsilon.$$

In this model, the response variable is modeled against:
- the overall population mean number of attacks;
- the fixed effects of the treatment (pooling over toads and times), time (pooling over toads and the two treatments), and the treatment × time interaction (how much the difference between the treatments changes across time);
- the random effect of toads within treatments assumed to be N(0, σ_{toad}^2) with the VC σ_{toad}^2 measuring the variance in the cumulative mean number of attacks across all possible toads that could have been used within each treatment;
- the random error term, ε, incorporating two components: the random interaction between time and toads, measuring the random variation in effects of time across the random toads within each treatment, and the random variation between observations for individual toads at each time. These two terms cannot be separately estimated, as there is only a single observation for each time–toad combination (only one observation for each time for each individual toad), just like in the models for an unreplicated split-plot design. These error terms are assumed to be N$(0, \sigma_{\varepsilon}^2)$.

This model can be fitted with OLS or as a mixed effects (multilevel) model, with checks of assumptions based on residual plots as described in the previous section. The OLS fit assumes compound symmetry for the covariance structure within subjects and we use Greenhouse–Geisser adjusted df for the F statistics for these within-subject terms. The ANOVA from the OLS fit is essentially the same as for the split-plot design, with the F for treatments using toads with treatments for the denominator and the Fs for time and treatment × time using the residual.

When the within-subjects factor is treated as a continuous covariate (e.g. time), the OLS model becomes tricky because of the hierarchical structure. It becomes a nested ANCOVA-type model, and it is much easier to use the mixed effects (multilevel regression) approach. We can allow slopes of the relationship between the number of attacks and time to vary across toads within each treatment by including a toad within treatment × time interaction term separate from the residual. The decision to use a random slopes model (that includes this interaction) instead of the simpler random intercepts model (that doesn't include the interaction) is based on the same criteria outlined in Section 12.1.1.1 for the model with no between-subjects effects. Including a random slopes term may not be possible with small sample sizes because the ML/REML algorithms won't converge.

12.2.2 Two or More Between-Subjects Factors

The between-subjects component of these designs can be more complex than a single factor. The multifactor designs discussed in previous chapters can be extended to include recordings on sampling or experimental units over time. For example, Skrip et al. (2016) studied how egg-laying order in zebra finches affected the eggs' mass. Mating pairs were allocated randomly to two diet groups (supplemented with high antioxidant food, no supplement) and two exercise groups (additional flight exercise, no additional exercise). Egg mass was the response variable from pairs that laid between two and five eggs. The between-subjects component was a two-factor (diet and exercise) crossed design with pairs as subjects and egg order (within each pair) as the within-subjects factor, treated as a continuous covariate. These data are analyzed in Box 12.4.

The fitted model includes the fixed effects of diet and exercise and their interaction, the fixed effects of the within-subjects factor egg order, plus its interactions with diet and exercise, and the random effects of pair within each diet and exercise combination. If we treated egg order as a categorical factor with four groups, the model would have a residual term representing the combination of true residual plus the random effect of the interaction between pair and egg order. In our analysis (Box 12.4), we treated egg order as a continuous covariate to partly deal with the incomplete dataset (not all pairs laid five eggs, and some observations were missing) and used the mixed effects (multilevel regression) approach. The model included the same terms as above and the option of including a random slopes term representing the interaction between egg order and pair within treatments.

Box 12.3 ℝ 🄴 Worked Example of Repeated Measures Design with Single Between-Subjects Factor and Single Within-Subjects Factor: Effect of an Appetite-Regulating Hormone on Toad Hunger

Garcia et al. (2015) studied the effect of the appetite-regulating hormone leptin on appetite and mating preferences in the spadefoot toad *Spea bombifrons*. Eighteen female toads collected from the wild were allocated randomly to a treatment group, which received a daily subcutaneous injection of leptin for six days, or a control group, which received saline injections with the same frequency. One hour after the day 6 injections, each toad was presented with approximately 50 crickets. The response variable was the cumulative number of attacks by each toad over three-minute intervals for 15 minutes. Treatment (leptin versus control) was the fixed between-subject factor, and toads were the subjects. The within-subjects fixed factor was time with five groups (3, 6, 9, 12, and 15 minutes after introducing crickets).

Our first analyses will treat time as a categorical factor with the five groups; we treat time as a continuous covariate in a second analysis below. We fitted the "split-plot" model using OLS. Our exploratory analyses using boxplots and a residual plot from fitting a simple factorial treatment by time model showed some mean–variance relationship but not marked, so we, like Garcia et al. (2015), analyzed the original cumulative number of attacks. We also included linear and quadratic contrasts for time and its interaction with treatment, as the times were equally spaced.

The resulting OLS "repeated measures" ANOVA was as follows.

Source	df	MS	F	P	GG-P
Between-subjects					
Treatment	1	1,646.9	8.59	0.010	
Toad(treatment)	16	191.7			
Within-subjects					
Time	4	881.2	91.81	<0.001	<0.001
Linear	1	3,493.6	364.02	<0.001	
Quadratic	1	27.3	2.85	0.096	
Treatment × time	4	144.7	15.08	<0.001	<0.001
Linear	1	572.5	59.65	<0.001	
Quadratic	1	0.7	0.07	0.792	
Residual	64	9.6			
Total	99				

The Greenhouse–Geisser ε (0.371) indicated that compound symmetry was unlikely to be met, although using the adjusted P-values for the within-subjects effects makes no difference to our conclusions. There was strong evidence against the null hypothesis of an interaction between treatment and time (including the linear trends). Figure 12.4 shows that cumulative attacks increased through time for both groups, but at a much lower rate in toads that received leptin, consistent with suppressed appetite.

We also fitted a mixed effects model using REML, including toads as a random effect. The Type III ANOVA for the fixed effects (Kenward–Roger adjustment did not change df as the design was balanced) was the same as for the OLS fit, and the linear trend was different for the two treatment groups ($t_{64} = 7.723$, $P < 0.001$; note $t^2 = F$ from OLS ANOVA above). The REML estimates of the VCs (with 95% profile CIs) were: residual = 9.60 (6.26–12.05); toad within treatment = 36.41 (17.18–69.20). There was considerable variation in cumulative attacks between toads within each treatment.

Our initial plots indicated differences in the variances between treatment and time combinations. We fitted a mixed effects model that allowed the variances to differ between these groups (we could also allow variances just to differ between treatments or between times if our exploratory analyses suggested the variance heterogeneity was mostly due to differences in main effect groups). This second model (AIC$_C$: 518.67) was a better fit (based on ML) than the original model assuming variance homogeneity (AIC$_C$: 530.29; model comparison LR = 39.16 with df = 9, $P < 0.001$). Refitting this second model with REML changed the Fs and P-values for the fixed effects (treatment:

Box 12.3 (cont.)

$F_{1,16} = 7.04$, $P = 0.017$; time: $F_{4,64} = 77.06$, $P < 0.001$; treatment $\times$ time: $F_{4,64} = 13.325$, $P < 0.001$), but not our main conclusion of a strong interaction between treatment and time. This model is probably more appropriate if your software allows different variance structures to be specified with mixed models.

We also analyzed these data treating day as a continuous covariate, although with only five time measurements for each toad it is marginal whether time is best treated as a categorical or continuous predictor. Ordinary least squares ANCOVA-type mixed models that include complex hierarchical structures are tricky to fit and interpret, so we focused on mixed effects models. We fitted a model that allowed for random intercepts and random slopes of the cumulative attacks against time relationships for individual toads and compared this to a model that only had random intercepts. A plot of the linear relationships for each toad suggested some variation in slopes across toads, and estimates of slopes ranged from 0.367 to 3.300. The ML comparison of these two models produced a LR statistic based on df = 2 of 57.48 ($P < 0.001$), indicating the random slopes model is the better fit (random slopes AIC_C: 466.78; random intercepts AIC_C: 519.50).

Examining the fixed effects from a random slopes model is tricky because of the difficulty of calculating the df, and different software and algorithms can produce different results. In this case, the model showed evidence for a treatment $\times$ time interaction (ANOVA based on Type III SS: $F_{1,16} = 18.35$, $P < 0.001$). Note that fitting a random intercept model would not have changed our conclusions about the fixed effects, with the F for the treatment $\times$ time interaction increasing to 61.502, with 1 and 70 df. The main difference between these two models is the REML estimation of VCs (the correlation between the two random effects in the random slope model was very low at -0.044, indicating that toads with steeper slopes did not have consistently higher or lower starting number of attacks).

Component	Random slope (95% CI)	Random intercept (95% CI)
Toad within treatment	13.94 (5.12–28.46)	36.47 (17.07–69.10)
Toad within treatment by time	0.31 (0.14–0.61)	NA
Residual	2.84 (1.99–4.25)	9.31 (6.64–12.78)

The original OLS analysis indicated that the assumption of compound symmetry of the within-subjects variance–covariance matrix was not met. We tried to fit a random slopes model with an AR(1) covariance structure but could not get convergence (other software/algorithms might have more success). Out of interest, we compared the fit of a random intercepts model with and without an AR(1) structure; this indicated that including an AR(1) structure did improve the fit of the model (AIC_C: AR(1) = 476.03, default = 519.50; LR = 45.82 with 1 df, $P < 0.001$), although the previous random slopes model was probably a better option.

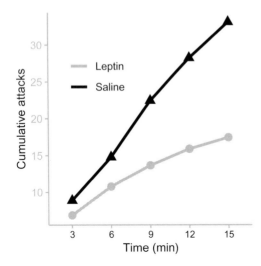

Figure 12.4 The mean cumulative number of attacks on crickets by toads over 3-minute intervals for 15 minutes, for toads injected with leptin or a saline control. Data are from Garcia et al. 2015. We omitted error bars because they don't reflect the relevant variances.

Box 12.4 ℝ 🅴 Worked Example of Repeated Measures Design with Two Crossed Between-Subjects Factors and a Single Within-Subjects Factor: Effect of Laying Order on Egg Mass in Birds

Skrip et al. (2016) studied the effects of dietary antioxidants and exercise on the allocation of nutrients to eggs in zebra finches. We will focus on one aspect of their study: how the order of laying affected the mass of the eggs. Mating pairs of finches were allocated randomly to two diet groups and two exercise groups. The finches laid between one and nine eggs, but few laid more than five and pairs that laid only one egg don't provide enough information to fit a regression between order and egg mass. We analyzed egg mass as the response variable from pairs that laid between two and five eggs. The between-subjects component was a two-factor (diet and exercise) crossed design with pairs as subjects, and egg order (within each pair) was the within-subjects factor.

We could treat egg order as a categorical factor, but the dataset was incomplete as some pairs laid fewer than five eggs, and there were also some missing recordings in the group with the control diet and no exercise, so either the pairs or egg orders with missing values would be omitted. The sample size was already small, with three groups with $n = 5$ and one group with $n = 3$, so omitting pairs would exacerbate this problem. Therefore, we used a mixed effects model with egg order as a continuous covariate. These kinds of decisions are not uncommon for biological experiments with small sample sizes.

We fitted a mixed effects model with random intercepts and slopes for the finch pairs within each diet and exercise combination. We compared the fit of this model to one with just random intercepts (and constant slopes) using ML. A plot of the individual regressions of egg mass against order suggested some variability in slopes, with mostly positive (increasing egg mass for later eggs) or flat slopes. Comparing the two models produced a LR statistic based on 2 df of 2.72 ($P = 0.257$), indicating the random intercepts model is comparable to the more complex random slopes model (random slopes AIC_C: -136.48; random intercepts AIC_C: -133.73).

The analysis of fixed effects (with Kenward–Roger adjustment) from the random intercept model showed the effect of egg-laying order was not consistent among the diet and exercise groups, particularly varying with diet (diet × exercise × order interaction: $F_{1,58.80} = 3.66, P = 0.061$; exercise × order: $F_{1,58.80} = 5.47, P = 0.023$; diet × order: $F_{1,58.80} = 21.06, P < 0.001$). Egg mass increased more rapidly with order for birds that had supplemented diets, especially those that were also exercised (Figure 12.5). If interest was also in the random effects, VCs (with 95% CIs) for random pairs (0.009; 0.003–0.015) and residual (0.006; 0.004–0.008) were similar.

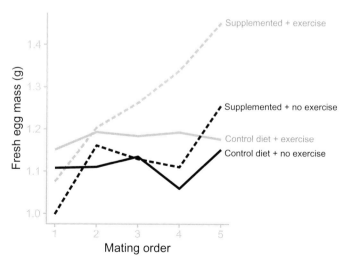

Figure 12.5 Egg masses of zebra finches, showing changes in mean egg mass with mating order for four treatments (with or without dietary supplement and with or without additional exercise). Data from Skrip et al. (2016); note that this figure differs from that shown in their paper, as we used a subset of their data. We omitted error bars because they don't reflect the relevant variances.

Box 12.5 ® Ⓔ Worked Example of Repeated Measures Design with Single Between-Subjects Factor and Two Crossed Within-Subjects Factors: Birds on Billabongs

Parkinson et al. (2002) examined the differences in the abundance of birds between different habitats on a floodplain. Habitat was a between-subjects factor, and there were three groups: permanent billabongs, temporary billabongs, and nonwetland habitat. Four sites (subjects) were chosen within each habitat group. The authors were also interested in whether the pattern between habitats was consistent over months (austral summer: November, December, January, February) and time of day (morning and evening chorus), so each site was sampled morning and evening on one day each month; month and time of day were the crossed within-subjects factors. The relative abundance of birds was the response variable.

A simple crossed linear model, ignoring the hierarchical nature of sites nested within habitats, was fitted to check the pattern of the residuals. There was a strong relationship between the mean and the variance for bird abundance. While Parkinson et al. used a fourth root transformation, there were no zeroes in the data, so we applied a log transformation, as we have consistently for strongly skewed data in this book. Note that instead of a log transformation, we could fit a mixed model that allowed the group variances to differ (e.g. Box 12.3).

We first fitted the standard "split-plot" model using OLS. This model included the crossed fixed effects of habitat, month, and time and their interactions and the random effect of site within habitat and its interactions with month, time, and month × time. The expected mean squares meant that habitat was tested against site within habitat, month and habitat by month was tested against site within habitat by month, time and habitat by time was tested against site within habitat by time, and the three-way interaction was tested against site within habitat by month by time. The resulting "repeated measures" ANOVA was as follows.

Source	df	MS	F	P	GG-P
Between-subjects					
Habitat	2	1.971	36.597	<0.001	
Site(habitat)	9	0.054			
Within-subjects					
Month	3	0.020	0.717	0.550	0.531
Habitat × month	6	0.120	4.326	0.003	0.006
Site(habitat) × month	27	0.028			
Time	1	0.244	9.273	0.014[†]	
Habitat × time	2	0.009	0.358	0.708[†]	
Site(habitat) × time	9	0.026			
Month × time	3	0.059	2.534	0.078	0.121
Habitat × month × time	6	0.024	1.034	0.425	0.411
Site(habitat) × month × time	27	0.023			
Total	99				

[†] Adjusted P-values not produced as time only had two groups.

Greenhouse–Geisser ε values of 0.853 for month and habitat × month and 0.538 for month × time and habitat × month × time were used to adjust P-values for within-subjects effects. There was evidence against the null hypothesis of no effect of time, which was consistent across habitats and months. There was also evidence against the null of no effect of habitat, with some evidence that this effect differed between months (habitat by month interaction), mostly driven by more birds in permanent than temporary wetlands in February, in contrast to the earlier months (Figure 12.6).

Box 12.5 (cont.)

With only four months and two times of day, neither within-subject factor could be treated as a continuous covariate. The dataset was complete, so the advantages of fitting a mixed effect model are limited to more reliable estimates of the VCs for the random effects. The df, Fs, and P-values from the ANOVA for the fixed effects, even with Kenward–Roger adjustment, were the same as for the OLS fit. The REML estimates for the VCs with profile 95% CIs were as follows.

Component	REML estimate	95% Profile CI
Site within habitat	0.003	0–0.010
Site within habitat by month	0.002	0–0.009
Site within habitat by time	0.001	0–0.007
Site within habitat by month by time (residual)	0.023	0.011–0.026

If we choose to simplify the model by omitting unimportant random effects, we could compare models with and without the site(habitat) by time and site(habitat) by month random effects.

Term(s) removed	AIC_C	LR χ^2 compared to full
None – full model	−18.37	
Site within habitat by time	−22.29	0.081, $P = 0.776$
Site within habitat by month	−22.07	0.304, $P = 0.581$
Both	−25.95	0.311, $P = 0.856$

The simpler model without either random effect fitted almost as well as the original full model. There were small changes to the estimates of the within-subjects fixed effects and their SEs; the df (63) were now based on the combination of the random effects, and the Fs and P-values therefore changed. Our broad conclusions remained the same.

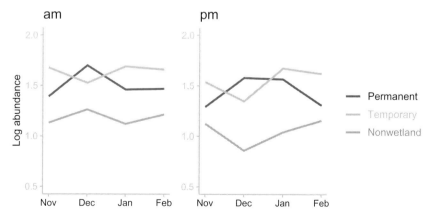

Figure 12.6 Mean log abundance of birds during visual censuses of billabongs of three types (data from Parkinson et al. 2002). Each billabong was surveyed at four different times over spring–summer and in the morning and afternoon. We omitted error bars for clarity and because they don't reflect the relevant variances.

12.2.3 Two or More Within-Subjects Factors

While less common in biological research, RM designs can include multiple within-subjects factors, usually arranged in a crossed design. For example, Parkinson et al. (2002) set up a sampling design to investigate the use of different floodplain habitats by birds. They had three habitat types (between-subjects factor: permanent wetland, temporary wetland, forest) with four sites (subjects) within each category. They recorded the number of birds at each site at two times (morning and evening) in each of four months (November, December, January, February). Month and time of day were the two crossed within-subjects factors. These data are analyzed in Box 12.5.

For the Parkinson et al. (2002) example, we would fit a model that includes the fixed effects of habitat, month, and time of day, plus their interactions, and the random effects of site within habitat, site by month, site by time of day, and site by month by time of day. Depending on the software used, one or a combination of the random interactions of site within habitat by month and/or time might be combined into the residual term.

12.2.4 Model Selection in Complex Repeated Measures Designs

Our preference with these models is to retain the random effects that represent the design structure, such as effects associated with nested random factors, to ensure the variation among the appropriate units is used for evaluating the fixed effects and we are not inflating the df. However, as with nested and split-plot designs in the previous chapters, it is common practice with these complex RM designs for models to be simplified by omitting unimportant random effects as determined from a formal model comparison or biological knowledge. With the example from Parkinson et al. (2002), a comparison of models with and without the random effects of site within habitat by month or by time of day showed that neither of these random effects made much difference to the model fit (Box 12.5), so many researchers would simplify the model by leaving these terms out (pooling them with the residual).

12.3 Key Points

- Repeated measures designs, where the same units are recorded through time, are particularly common in biology.

- The commonest version of this design has similarities to a split-plot, where we have one between-subjects fixed factor applied to units termed subjects and a second within-subjects fixed factor being recordings through time or another set of treatments applied through time to the subjects. The fitted model has similar terms to the split-plot model, with fixed main effects of the between- and within-subjects factors and their interaction, and the random effects of subjects and the interaction between subjects and the within-plots factor. Either OLS or mixed effects models can be used, the latter being better for unbalanced designs (see next point) and for VCs.

- Time (or quantitative treatments through time) can also be treated as a continuous covariate; now we have a multilevel regression model with the slopes and/or intercepts of the linear relationship between the response and time being modeled across the random subjects and across the groups of the between-subject fixed factor. This approach handles missing observations through time as slopes and intercepts can be derived even if the number of times and their intervals vary between subjects.

- Repeated measures designs are unlikely to satisfy the assumption of compound symmetry because observations and error terms from times closer together are likely to be strongly correlated than those further apart. Ordinary least squares models rely on more conservative df for within-subjects terms to deal with this. Some mixed effects model software allows the correlation structure to be included in the model fitting.

- As with split-plot designs, these RM designs can be extended to include multiple between-subjects factors and/or multiple within-subjects factors, with the latter often better treated as continuous covariates.

Further Reading

Repeated measures analyses, especially from an OLS perspective, are covered in detail in many experimental design texts in psychology, including Kirk (2012) and Tabachnick and Fidell (2019). Milliken and Johnson (2009) include two chapters on these designs and their analyses with OLS models, and West et al. (2015) focus on mixed effects (ML) models for these designs with biological examples.

13 Generalized Linear Models for Categorical Responses

In Chapter 4 we introduced GLMs, which were characterized by:

• a random component representing the response variable and its probability distribution from the exponential family (Box 4.2);

• a systematic component representing the linear combination of predictors;

• a link function $[g(\mu_i)]$ describing how the predictors relate to the mean response; the link function can be viewed as a transformation of the response to make the relationship with the predictors linear. The three commonest link functions are identity (models the mean response), log (models the log of the mean response), and logit (models the log of the odds of one outcome in a binary response).

Estimation and model fitting are based on ML. Inference (CIs and tests) about individual model parameters is based on the Wald statistic or a likelihood-ratio (LR) statistic comparing models with and without the parameter of interest. While it is difficult to generally specify a threshold, the LR statistic is more reliable for smaller sample sizes. The linear models we have described so far are special cases of GLMs where the model error terms (and commonly the response variable) are assumed normal and the identity link function relates the predictors to the mean response.

We start this chapter by describing regression-type models where our response variable is binary or a proportion (binary observations that are grouped or result from an underlying binary process), or a count that has no upper limit but can't be negative. The former models are based on the binomial distribution for the response and termed logistic regression models. The latter are based mostly on the Poisson distribution and termed Poisson regression models, although other distributions such as

the negative binomial might sometimes be needed. The link function is the logit or the log-link, respectively. We'll start with at least one of the predictor variables being continuous, so we have regression-type models. We then consider methods for goodness-of-fit and statistical inference (CIs and hypothesis tests) for these logistic and Poisson regression models. The key assumptions and diagnostic checks are then covered, with a focus on overdispersion and zero-inflation.

While we will mostly use models based on binomial and Poisson distributions, plus beta and negative binomials for overdispersed data, GLMs can use several other distributions in the exponential family. The gamma and inverse Gaussian distributions can be especially useful for modeling positively skewed continuous variables and can be used with a variety of link functions.

When all predictors are categorical, we have a situation familiar to many biologists where counts are classified into a two (or more)-dimensional ("contingency") table, and the focus is on assessing independence of the rows and columns. We describe traditional analyses of these tables and show how these analyses do not work for more complex datasets. The best approach is a type of GLM called a loglinear model where the modeled response variable is actually the counts in each cell of the table.

Chapters 10–12 described linear mixed models, an extension of linear models to include fixed and random predictors. We can similarly extend linear mixed models to discrete responses, termed generalized linear mixed models (GLMMs). Understanding GLMMs provides entry into a wealth of advanced linear models, capable of dealing with a wide range of unusual biological circumstances. Finally, we introduce generalized additive models (GAMs), regression-type models based on smoothing functions rather than linear models.

13.1 Logistic Regression

One of the most important GLMs used in biological research is based on the binomial distribution and the logit link function, modeling the log of the odds of one of the two outcomes. These models are used in two common situations:

- The response is binary, and each observation consists of one of the two outcomes (e.g. success or failure, present or absent, survive or not survive). These data occur commonly when the predictor is continuous and are termed ungrouped (Hosmer et al. 2013). Dunn and Smyth (2018) describe this analysis as a Bernoulli GLM. Interpreting goodness-of-fit and model diagnostics (e.g. residuals) is more difficult with binary data (Section 13.5).

- Each data point is the proportion of one of two possible outcomes; the sample size on which each proportion is based is also part of each observation. This is called grouped data and is common with a categorical predictor, so multiple observations have the same predictor value. Dose–response studies that we describe in Box 13.3, where we record the response of multiple organisms (singly or in groups) at each dose group, are an example of grouped data

When we have at least one continuous predictor, we have the classic logistic regression analysis. Categorical predictors can be included in these models (as dummy variables), just as we described in Chapters 6 and 8 for linear regression.

With a binary response and one or more categorical predictors, the dataset essentially becomes a contingency table with counts in each combination of the response and predictor categories; loglinear models that focus on modeling the counts in each category combination can be used (Section 13.7). However, loglinear modeling does not automatically distinguish one variable as a response variable, so sometimes logistic models with a binary response and dummy variables for the factors are more appropriate. This distinction between models with continuous vs. categorical predictors mirrors the historical approach of treating ANOVA and linear regression as distinct, but it is more sensible to link them under a GLM framework.

13.1.1 Binary Response with a Single Continuous Predictor: Simple Logistic Regression

Consider a situation where we have a single binary response and a single continuous predictor. For example, in Chapter 4 we introduced the work by Polis et al. (1998) studying the presence or absence of lizards (response) on small islands of a range of shapes in the Gulf of California. Recall we constructed a linear model relating the probability of "success" (π), in this case that an island has lizards, to the continuous predictor, the perimeter:area ratio (PA) of an island. Rather than model this probability directly, which would require constructing a nonlinear model, we build a GLM using the logit link function and binomial distribution, so we model the log of the odds of success. For this example, this model takes the form:

$$\text{logit}(\pi_i) = \ln\left(\frac{\pi_i}{1 - \pi_i}\right) = \beta_0 + \beta_1 x_i$$

$$\text{Log(odds of island having lizards)} = \beta_0 + \beta_1(\text{PA}).$$

This is commonly called a logistic regression model (or sometimes a logit model). Like a simple linear regression model, we have two parameters – the slope (β_1) and the intercept (β_0) – and use ML to estimate them (see Section 13.4). The intercept, β_0, is the log of the odds of lizards being present when $\text{PA} = 0$. Its interpretation is similar to the intercept of the linear regression model (Chapter 6), and it is rarely of biological interest unless the predictors are centered. The slope parameter, β_1, indicates the change in the log odds for a one-unit increase in the predictor.

We can convert β_1 to an odds ratio (OR), which measures how much the odds of success change with a single unit change in the predictor variable:

$$\text{Odds ratio} = \frac{\text{Odds}_{x+1}}{\text{Odds}_x} = e^{\beta_1}.$$

For the Polis et al. (1998) data, the estimated logistic regression coefficient (b_1) is an estimate of how much the log odds of *Uta* occurring on an island (compared to not occurring) would change for an increase in the PA ratio of one unit. When the predictor is categorical, or we are dealing with contingency tables without specifying a response variable, ORs are more useful (Section 13.7).

13.1.2 Categorical Predictors in GLMs

Where categorical predictors are included, you should know how your statistical software codes dummy variables, as interpretation of the model coefficients and ORs changes for different methods. With reference coding, one group of the categorical predictor is a reference, and the effects of the other groups are relative to it. Alternatively, effects coding compares each group logit to the overall logit (Hosmer et al. 2013).

Box 13.1 Ⓡ Ⓔ Worked Example of Single-Predictor Logistic Regression: Presence/Absence of Lizards on Islands

Polis et al. (1998) identified lizards as potential influences on spider populations on islands in the Gulf of California and collected data to model the presence/absence of *Uta* lizards against the ratio of perimeter to area (PA), as a measure of input of marine detritus, for 19 islands in the Gulf of California. We fitted a GLM of the presence of *Uta* (binary) against PA ratio. *Uta* occurred on 10 of the 19 islands and the data are plotted in Figures 4.4 and 13.1. One island (Cerraja) with and two (Bahia Animas Norte and Sur) without *Uta* were separated from the others in their category and were potentially influential given the small sample size. The question of main interest was the nature of any relationship between the presence of *Uta* (i.e. the odds that *Uta* occurred relative to not occurred) and the PA ratio of an island. Inferentially, this is the H_0 that the slope of the logistic regression model, β_1, equals zero.

The ML parameter estimates were as follows.

Parameter	Estimate	ASE	Wald statistic	P
β_0	3.606	1.695	2.127	0.033
β_1	−0.220	0.101	−2.184	0.029

There was evidence from the Wald test that $\beta_1 \neq 0$, supported by the LR $\chi^2 (P < 0.001)$. The OR for PA was estimated as 0.803 with 95% CI 0.659–0.978. For a one-unit increase in PA, an island has a 0.803 chance of having *Uta* compared to lacking *Uta*, a decrease in the odds of having *Uta* of approximately 20%. The plot of predicted probabilities from this model is shown in Figure 13.1, clearly showing the logistic relationship.

The plot of deviance residuals against predicted values is difficult to interpret with binary observations; the more useful plot of quantile residuals suggested no particular issues. Influence diagnostics showed that Cerraja (*Uta* present and PA of 25.92) was more influential than the rest on the model fitting. It had the largest quantile residual (1.924) and Cook's D near 1 (0.843). Omitting this observation changes the P-value for the Wald slope test to 0.098, but the P-value from the more reliable LR χ^2 is still very small ($P < 0.001$).

Goodness-of-fit statistics were calculated to assess the model. The Hosmer–Lemeshow statistic indicated that the model was a reasonable fit and Tjur's (2009) coefficient of discrimination indicated that about 52% of the uncertainty in the log of the odds of *Uta* on islands could be explained by PA ratio.

Statistic	Value	df	P
Hosmer–Lemeshow ($\hat{C}$)	7.898	8	0.443
Tjur's D	0.521		

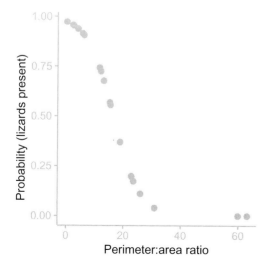

Figure 13.1 Scatterplot of the predicted probabilities from logistic regression model of the presence of *Uta* in relation to the perimeter:area ratio on 19 islands in the Gulf of California (Polis et al. 1998).

13.1.3 Binary Response with Multiple Predictors: Multiple Logistic Regression

Logistic regression can be extended easily to situations with multiple predictors. For example, Mayekar and Kodandaramaiah (2017) studied influences on a plastic trait, pupal color, of a tropical butterfly species. We use their data to model pupal color (green vs. brown) against two continuous and one categorical predictor. The analysis of these data is shown in Box 13.2.

13.1.3.1 *Logistic Model and Parameters*

The full multiple logistic regression model for the butterfly example (with sex requiring a single dummy variable) is:

$$\log(\text{odds of pupa being green vs. brown}) = \beta_0 + \beta_1(\text{pupation time})$$
$$+ \beta_2(\text{weight}) + \beta_3(\text{sex}) + \beta_4(\text{pupation} \times \text{weight}) + \beta_5$$
$$(\text{pupation} \times \text{sex}) + \beta_6(\text{weight} \times \text{sex}) + \beta_7(\text{pupation} \times$$
$$\text{weight} \times \text{sex}).$$

This model is fitted using ML. Interpreting the model terms is like that for multiple linear regression models, including both continuous and categorical predictors. This example includes a categorical predictor, so interpreting the interaction terms including sex is like an ANCOVA model, described in Section 8.2 – the weight by sex term is assessing how much the slope of the relationship between log odds of a pupa being green and weight changes between male and female pupae. Interactions between continuous predictors indicate that the slope of log odds against one predictor changes for different values of a second predictor. Simple slopes that examine how the relationship between the response and one predictor changes at specified values of a second predictor can be adapted for GLMs. Collinearity is also likely between main effects and their interactions, and centering continuous predictors often helps.

If interactions are not considered important or of interest, we can fit a main effects logistic regression model:

$$\log(\text{odds of pupa being green vs. brown}) =$$
$$\beta_0 + \beta_1(\text{pupation time}) + \beta_2(\text{weight}) + \beta_3(\text{sex}).$$

These effects are interpreted similarly to a multiple linear regression without interactions – for example, the effect of pupation time on the log odds of a pupa being green, independent of weight and sex.

13.1.4 Nominal and Ordinal Multinomial Response Variables

The approaches described above are not restricted to binary responses, and they can readily be expanded to response variables with several discrete categories. Biologists might see these variables as part of behavioral studies where multiple choices are presented or in ecological studies where, for example, selection of prey by predators is measured. These response variables are very common in biomedical areas and are a major part of marketing surveys, so an extensive literature covers their analysis. Agresti (2019) provides a clear outline of the most common ways of fitting models to these data. The analysis depends on whether the variable is nominal or ordinal.

13.1.5 Proportion Response

Generalized linear models based on the binomial distribution can also be used to fit models to observations that represent the proportion of one outcome versus its alternative, so a binomial distribution is appropriate. These proportions are constrained to between 0 and 1. For example, as part of a study on behavioral thermoregulation in birds, Ryeland et al. (2017) recorded the proportion of time individuals of different bird species spent resting with their heads back and modeled this against several predictors. This design also included a random effect, so we will present the analysis in Section 13.8. Another common use of logistic models with proportion data is analyzing dose–response curves, where the response can be the proportion of organisms responding/ surviving to different doses of some substance (Box 13.3).

13.2 Count Responses: Poisson Regression

Biologists often wish to model counts (e.g. number of organisms in a sampling unit, numbers of cells in a tissue section). Counts often have a Poisson distribution, where the mean equals the variance, so we use the log-link function. The broader term for these GLMs is loglinear models.

To illustrate a Poisson GLM, recall the research by Fill et al. (2021), who studied the effect of duff (leaf litter) on the post-fire ecology of wiregrass in a section of pine savanna (Section 4.7.2.2). We will model numbers of culms against basal area and duff treatment using each plant as the unit of analysis. The analyses of these data are presented in Box 13.4.

The full Poisson model was:

$$\log(\text{mean culms}) = \beta_0 + \beta_1(\text{treatment}) + \beta_2(\text{basal area}) +$$
$$\beta_3(\text{treatment} \times \text{basal area}).$$

Box 13.2 ℝ 🅔 Worked Example of Multiple-Predictor Logistic Regression: Determinants of Butterfly Color

Mayekar and Kodandaramaiah (2017) studied potential determinants of pupal color (green vs. brown; see their figure 1) of a tropical butterfly. A laboratory population was established, and newly hatched larvae were placed in a growth chamber set at either 60% or 85% relative humidity. Resulting pupae were recorded for color, along with time to pupation, pupal weight, and sex. We will use their data to model green vs. brown pupae (binary response) against the two continuous (time to pupation, pupal weight) and one categorical (sex) predictor. We only used data from the low-humidity treatment as brown pupae were rare at high humidity.

Like Mayekar and Kodandaramaiah (2017), we initially modeled the log of the odds of green vs. brown pupae against all three predictors and their two- and three-way interactions. There was evidence of strong collinearity as the main effects were highly correlated with their interaction terms (very large variance inflation factors [VIFs]). Centering the continuous predictors alleviated this and the model based on centered predictors indicated no strong interactions (Wald test P-values between 0.332 and 0.886). The LR test comparing the full model with the no-interaction model also indicated little difference in fit ($P = 0.556$ with 4 df).

We then fitted the simpler no-interaction model using uncentered predictors, as collinearity was no longer an issue without interactions (all three VIFs <2). Weight did vary somewhat between the sexes and this may explain the large SE for weight. There were no unusually influential observations (no Cook's D values close to 1). The model results with Wald statistics (LR χ^2 P-values were similar) were as follows.

Parameter	Estimate	ASE	Wald statistic	P
Intercept	−3.223	2.332	−1.382	0.168
Time to pupation	0.221	0.063	3.504	<0.001
Weight	0.440	6.352	0.069	0.945
Sex	0.257	0.435	0.590	0.555

Only time to pupation, adjusting for weight and sex, affected color, with green pupae more likely with longer pupation times (OR = 1.247 with 95% CI: 1.102–1.411). Not surprisingly, the CIs for the other ORs included 1. Note that we could have further simplified our model by omitting the effects of sex and weight, as Mayekar and Kodandaramaiah (2017) did through a full model selection process based on AIC$_C$ values, but we have kept all main effects to illustrate the classic multiple logistic regression model.

While the Hosmer–Lemeshow statistic (6.877 with 8 df and $P = 0.550$) did not indicate any lack of fit of the model, Tjur's coefficient of discrimination was only 0.043, suggesting the model did not have strong explanatory power.

The model parameters are estimated using ML and it is actually a GLM equivalent of the ANCOVA model (Section 8.2). The interaction term assesses how much the slope of log mean number of culms vs. plant basal area is affected by the treatment. If this interaction explains little, it could be dropped, and an additive model with main effects of treatment and basal area would be used. The parameter for the main effect of the continuous predictor represents a multiplicative effect on the response for a one-unit change in the predictor (holding the other predictors constant). For example, a one-unit change in basal area results in an e^{β_2} change in mean number of culms.

The biggest issue in biological research is that counts (especially abundances of animals and plants in nature) often have variance greater than the mean, termed over-dispersion. A Poisson distribution is no longer appropriate, and we need to consider alternatives like the negative binomial or quasi-Poisson (Section 13.6.2). In contrast to binomial GLMs, an alternative to fitting a Poisson GLM is to transform the response variable and fit a standard linear model. A square root will often normalize the distribution for a true Poisson where the variance equals the mean (see Section 5.4). The square root transformation is less effective with overdispersion, and there are two other options:

• A log transformation, usually adding a small constant, like 1, to deal with zero values (see discussions and different opinions in Ives 2015; Morrissey et al. 2020; O'Hara & Kotze 2010; Warton et al. 2016). We now

Box 13.3 ® ⓔ Dose–Response Curves: You Might Already Use GLMs or GLMMs without Knowing It

One task familiar to many biologists is the construction and interpretation of dose–response curves, in which we model the probability of a response, such as an organism dying or a gene being upregulated, as a function of one or more predictor variables (e.g. a toxic substance). These analyses are particularly common in environmental science, to predict chemical risks, and environmental regulators use them widely. Dose–response relationships can be modeled as a GLM.

The feature of many dose–response relationships is that they are not linear; most commonly, they are sigmoidal. For a hypothetical pollutant, we may see no detectable effect at very low concentrations (and some may even be beneficial) then a range of concentrations over which increasing numbers of organisms respond, then the response saturates, possibly because all organisms have responded (or, in much toxicity testing, died!).

As with many "standard" approaches, the initial steps were based around limitations to computational power and a desire to make analysis possible for a range of scientists. In the case of binary dose–response relationships, the first approach was probit analysis (Bliss 1934). It involved a simple transformation of the response variable in which the proportion responding (*p*) was transformed to its probit, using an inverse normal distribution. The probit is essentially the *z*-score – the value of a standard normal distribution for which the cumulative probability is *p*. It then had a linear relationship with the dose, and the slope and intercept could be calculated with OLS (Figure 13.2). In practice, 5 was often added to the probit to keep numbers positive (Finney 1971).

A more recent approach has been to use logistic regression, using the logit transformation to linearize the relationship. In the most common use for these methods, estimating chemical toxicity, doses are often applied as a log-series to ensure that the full range of toxicity is seen. The dose is then log transformed, and this step can be done *a priori*, followed by back-transforming any critical dose, or it is done internally in many standalone ecotoxicology packages, where it is labeled log-logistic regression.

There have been several extensions of these basic approaches, including dose–response models where there is no threshold response – that is, the response increased steadily from the smallest dose – but probit and logit regressions are used widely.

The most common use of these regressions is to calculate critical doses. In environmental regulation, the most common is the concentration at which there is a 50% response – the LC50 and EC50 (lethal and effective concentrations, respectively).

What's the Difference Between Probit and Logistic Regression?

Not much! Both approaches are GLMs. We have covered the logistic regression approach, with its logit link function. A probit regression, while often listed as a separate statistical tool, is also a GLM, based on the inverse normal distribution – that is, we model the expected value of $Y = \Phi^{-1}(p)$. In both cases we can then construct a linear model linking the response to one or several predictors. The two approaches are closely related (Agresti 2019), and produce very similar results.

We illustrate the results using the original dataset used by Finney, examining the toxicity of rotenone to a species of aphid (Martin 1942). Figure 13.2 shows the dose–response relationship, and the probit and logit transformed data.

Ordinary least squares estimates for probit and logit data gave similar r^2 values (0.98) and LC50 values. Historically, OLS was used, but current approaches use ML. We show those results online:

- For probit data, the regression equation is probit = 2.839 + 4.136(logDose). For Finney's probit values (with +5 added), $p = 0.50$ has a probit of 5, so we can solve the regression equation for $y = 5$ to get 0.686, which we could back-transform to 4.86 mg/L.
- For logit data, the equation is −2.087 + 3.042(logDose). $p = 0.5$ corresponds to a logit of 0, so we solve the equation for $y = 0$ to get an identical 0.686.

Which One to Use?

Probit analysis predates logistic regression and has supported simple toxicity testing. This was in part because of its computational simplicity. There is a range of other, more sophisticated approaches to modeling dose–response relationships, but for "basic" threshold responses, logistic regression is a better option. Adopting this model provides a link to the increasing literature on logistic regression.

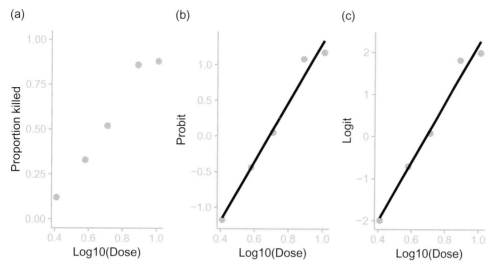

Figure 13.2 Analysis of toxicity data using logistic and probit regression. The panels show the original dataset used by Finney (1971) developing probit regression, and show survival of aphids with increasing concentrations of rotenone. Panel (a) shows the original raw data, showing a sigmoid relationship (with an additional data point of zero mortality at a dose of 0). Panels (b) and (c) show probit and logit transformed data, both of which make the relationship linear. Linearity is helped in this case by log transforming the rotenone concentration, even though in this example the range of concentrations is not as broad as is commonly used in toxicity studies.

model the mean of the log-transformed response, rather than the log of the mean response.

• A fourth root (double square root) transformation is sometimes recommended for multivariate analyses of species abundance data (see Section 14.4).

13.3 Goodness-of-Fit for GLMs

Generalized linear models are fitted using an iterative reweighted least squares algorithm (e.g. Newton–Raphson; see Box 4.3) that finds parameter estimates that maximize the log-likelihood function. Two common measures of goodness-of-fit are used with GLMs. The first is the (residual) deviance, based on the log of the ratio of the maximized likelihood of the model of interest (L_1) and the maximized likelihood of the saturated model (L_S), one with as many parameters as data points:

$$-2\log(L_1/L_S) = -2(L_1 - L_S),$$

where L_1 is the maximized log-likelihood of the model of interest and L_S is the same for the saturated model (see Section 4.5.2). This is the ML/GLM equivalent of

$SS_{Residual}$. It is used more broadly for GLM model comparison and testing hypotheses about parameters (see below). When analyzing contingency tables, the deviance is often called the G^2 statistic.

The second measure of goodness-of-fit is the Pearson χ^2 statistic based on comparing observed and fitted values. The statistic is determined by calculating for each observation the difference between the observed and fitted values divided by the square root of the fitted value. These are then summed over all observations. This is the classic Pearson χ^2 statistic (Section 13.7).

In both cases, lower values indicate that the model fits the data better. The Pearson χ^2 and the deviance statistics approximately follow a χ^2 distribution if the fitted (expected) values are not too small. When some fitted values are <5, the χ^2 statistic is more reliable than the deviance (Agresti 2015), but Dunn and Smyth (2018) suggest slightly more lenient conditions. For binomial models with ungrouped (binary) data, neither statistic will have approximate χ^2 distributions and neither statistic is particularly useful as a goodness-of-fit measure (Agresti 2015; Dunn & Smyth 2018), although the deviance can still be used for comparing models. Hosmer et al. (2013) describe a solution to the

Box 13.4 Ⓡ Ⓔ Worked Example of Multiple-Predictor Poisson Regression (Overdispersion and Zero-Inflation): Post-Fire Ecology of Wiregrass

Fill et al. (2021) studied the effect of duff (leaf litter) on the post-fire ecology of wiregrass (*Aristida beyrichiana*) in a section of pine savanna. They sampled 99 plants in an area of 0.1 km², recorded plant basal area, and allocated each plant to one of three treatments: high duff, low duff, low duff with added pinecones. They then burnt the area and five months later counted the number of aerial stems, called culms, on each plant. We will model counts of culms per plant against basal area (cm²) and duff treatment using each plant as the unit of analysis.

We initially modeled log mean numbers of culms against basal area and treatment using a GLM with a Poisson distribution. There was a strong interaction between treatment and basal area (LR $\chi^2 = 89.733$, df = 2, $P < 0.001$) and the AIC was 1,891.3 (AIC$_C$ was similar). The data were very overdispersed, with the residual deviance much greater than the residual df and the dispersion statistic was 21.121.

To manage the overdispersion, we fitted a negative binomial model, as did Fill et al. The dispersion statistic was now around 1. The residual plots looked okay, and there were no strongly influential observations. The AIC for the negative binomial model (627.85) indicated a much better fit than the Poisson. However, the interaction effect was different in the negative binomial (LR $\chi^2 = 3.238$, df = 2, $P = 0.198$), suggesting the interaction in the Poisson model may have been driven primarily by the overdispersion. After removing the interaction term from the negative binomial model, there was no effect of treatment (LR $\chi^2 = 0.456$, df = 2, $P = 0.796$), but culm number increased with basal area (LR $\chi^2 = 14.296$, df = 1, $P < 0.001$) and log mean number of culms increased by 0.003 (SE: 0.0007) for an increase in basal area of 1 cm². Conclusions were the same even if the interaction term was retained in the model. The fit of the two models is shown in Figure 13.3.

While there was no indication of zero-inflation compared to a negative binomial distribution (ignoring treatments, the proportion of observed zeroes was 0.22 and the proportion predicted from a negative binomial was 0.26), we will use this example to illustrate ZIP (zero-inflated Poisson) models.

We used a negative binomial additive model because convergence was difficult with the ZIP model that included the interaction in the binomial component. The first model had treatment and basal area as predictors for the negative binomial component, but just included an intercept for the binomial part, assuming the predictors do not influence the initial split of zeroes. The second model included both predictors in both components of the model. The two models had very similar AICs, close to the original additive negative binomial GLM (ZIP1: 629.089; ZIP2: 628.231; additive NB: 627.084), suggesting zero-inflated models were not better fits than a negative binomial. Both models were consistent in their conclusions for the Poisson component (no evidence of treatment effect but strong effect of basal area), and neither predictor affected the binomial component in the second model.

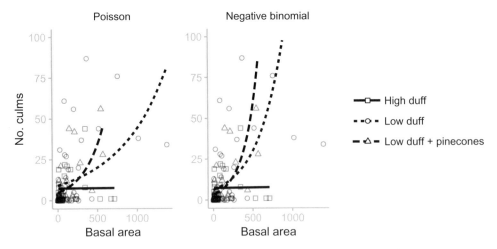

Figure 13.3 Poisson and negative binomial models fitted to the relationship between number of culms and basal area of wiregrass subject to three treatments (data from Fill et al. 2021).

problem of testing goodness-of-fit for continuous predictors in logistic regression by grouping observations so the minimum expected frequency of either of the binary outcomes is not too small. The Hosmer–Lemeshow statistic, also termed the deciles of risk (DC) statistic, is derived from aggregating the data into 10 groups. The grouping is based on either each group having one-tenth of the ordered predicted probabilities, so the groups have equal numbers of observations, or the groups being separated by fixed cut points (e.g. first group having all probabilities ≤ 0.10, etc.). Both grouping methods produce a statistic ($\hat{C}$) which approximately follows a χ^2 distribution with df as the number of groups minus two.

There are also GLM analogs of the r^2 used as a measure of explained variance in OLS regression. Menard (2000) discussed a range of measures for logistic regression and Tjur (2009) recommended a measure termed the coefficient of discrimination (D). It is the difference between the average predicted (fitted) values for success and failure and is equivalent to the r^2 described for linear models:

$$D = \bar{\bar{\pi}}_1 - \bar{\bar{\pi}}_0.$$

Since then, several other options have been proposed and evaluated, not just for logistic regression, but for a range of GLMMs (Ives 2019; Nakagawa & Schielzeth 2013; Nakagawa et al. 2017; Tjur 2009).

Finally, information criteria such as AIC and AIC_C, which we introduced in Sections 4.5.3 and 9.2.1, can also be used to assess model fit. The AIC for a model is:

$$-2(L_1 - p)$$

where L_1 is the maximized log-likelihood of the model of interest and p is the number of parameters in the model. The adjustment of AIC for small samples (AIC_C) is described in Table 9.1. An advantage of AIC for comparing models is that the models don't have to be nested.

13.4 Inference for Parameters in GLMs

Standard errors of the parameter estimates in binomial and Poisson GLMs are derived as for linear model parameters based on the sampling distributions of each parameter being asymptotically normal. Confidence intervals are also calculated as for linear model parameters except that the standard normal distribution is often used instead of the t distribution. Alternatively, profile CIs as described for mixed effects models in Section 10.2.2.1 might be

more appropriate for complex models with numerous parameters. For logistic models, a CI for the OR can also be derived by exponentiating the upper and lower CIs for β_1 (Kleinbaum et al. 2013).

The null hypotheses of interest in binomial and Poisson GLMs are that the parameter associated with each term in the model equals zero – for example there is no relationship between the presence/absence of *Uta* and the PA ratio of an island or log mean parasitoid abundance is independent of moth abundance for a specific treatment. These H_0s can be evaluated in one of two ways:

- The Wald test, which is the parameter estimate divided by its SE and compared to the standard normal (z) or t distribution, considering sample size. This is a ML/GLM equivalent of a t test for testing hypotheses about a single linear model parameter.
- A test based on the log of the ratio of the likelihood of the full model (including the parameter of interest) to that of the reduced model (omitting the parameter of interest). This log-likelihood-ratio statistic $[-2(L_F - L_R)]$ is essentially the difference between the deviances of the two models. It is compared to a χ^2 distribution, assuming the expected values are not too small. This is a ML/GLM equivalent of an F for testing hypotheses about linear model parameters by comparing full and reduced models. In the same way that we could produce an ANOVA table for comparing OLS models, we can also produce an analysis of deviance table for comparing GLMs.

In contrast to some normal-based linear models where $t^2 = F$, the Wald and LR tests are different for GLMs. The Wald test tends to be less reliable, lacks power for smaller sample sizes, and is difficult to interpret with categorical predictors that require two or more dummy variables, so the LR statistic is preferred.

13.5 Assumptions and Diagnostics for Binomial and Poisson GLMs

Inference for GLMs assumes that the response variable's probability distribution is adequately described by the random component and probability distribution chosen. The key is that the relationship between the mean and the variance in our data matches that specified by the chosen probability distribution. We also assume that the observations are independent. Nonindependent observations are often the result of random factors in our design and can be handled by including one or more random effects in the model (GLMMs; Section 13.8), just like we did for standard linear models.

As well as assessing the model's overall fit, it is also important to evaluate the contribution of each observation, or group of observations, to the fit. In OLS linear models, we have emphasized the importance of residuals, the difference between each observed and fitted or predicted value. There are two types of residuals for observations in GLMs, which derive from the measures of goodness-of-fit described in Section 13.3.

The first is the deviance residual, the square root of an observation's contribution to the model's residual deviance, termed the unit deviance:

$$\text{sign}(y_i - \hat{\mu}_i) \sqrt{d_i},$$

where for observation i, $\hat{\mu}_i$ is the predicted mean value, and d_i is the unit deviance.

The second is the Pearson residual, an observation's contribution to the model's Pearson χ^2 statistic. This is the raw residual (the observed minus the predicted value) standardized by dividing by the square root of the variance of the predicted value. For a logistic model based on grouped data:

$$\frac{y_i - n_i \hat{\pi}_i}{\sqrt{[n_i \hat{\pi}_i (1 - \hat{\pi}_i)]}},$$

where for observation i, n_i is the sample size for that proportion and $\hat{\pi}_i$ is the predicted odds of one outcome (e.g. success). Note that because the Pearson χ^2 goodness-of-fit statistic is difficult to interpret for ungrouped binary data with continuous predictors (Section 13.3), Pearson residuals are also of little value. For a Poisson model:

$$(y_i - \hat{\mu}_i) / \sqrt{\hat{\mu}_i},$$

where for observation i, $\hat{\mu}_i$ is again the predicted mean value. The denominator here is the SD of the count, which for a Poisson is the SD of the mean, hence $\hat{\mu}_i$ appearing in numerator and denominator.

Dunn and Smyth (2018) also described the quantile residual with the same normal cumulative probability distribution as the response, and recommended it for use with truly discrete probability distributions such as binomial, because plots of the residuals are easier to interpret (see Box 13.1 for an illustration).

If the conditions for the deviance and χ^2 statistics to follow a χ^2 distribution hold, both residuals will follow a normal distribution. These residuals are commonly standardized by dividing by their SE, so they have N(0, 1) (see Table 6.4 for OLS residuals). The GLM residual plots are like those from OLS models, residuals against predicted values and/or against each predictor, and again, we are looking for an even spread with no unusually large residuals.

Diagnostics for the influence of an observation are like those for OLS models (Section 6.5). They include (1) leverage, measured in the same way as for OLS regression and (2) an analog of Cook's statistic, Dfbeta, based on either the deviance, Pearson or quantile residual, and leverage ([Agresti 2019], also termed $\Delta\beta$ [Hosmer & Lemeshow 1989]). This statistic measures the standardized change in the estimated GLM model coefficients when an observation is deleted. We can also calculate the change in χ^2 or deviance goodness-of-fit with deletion. Influential observations should always be checked, and our recommendations from earlier chapters apply here.

When there are multiple predictors in the model, strong collinearity can affect statistical inference for model parameters just as it did for OLS regression models. While not necessarily reducing the model's predictive value, collinearity will inflate the SEs of the estimates of the model coefficients and can produce unreliable results (Hosmer et al. 2013). The methods we described in Section 8.1.9 are applicable here, especially examining a correlation matrix between the continuous predictors. Variance inflation factors (or tolerances) can also be calculated for each predictor by simply fitting the model as an OLS linear regression.

Other aspects of OLS multiple linear regression also apply to GLMs with multiple predictors. For example, we can produce partial regression plots to visualize the relationship between each predictor and the response, independent of the other predictors. Note that because these plots are based on residuals, they can be difficult to interpret for binary responses.

13.6 Overdispersed Data

It is common with biological count data for the variance to exceed the mean. This is termed overdispersion and makes Poisson distributions inapplicable. This is often manifest as a distribution where many observations are zero, more than predicted from a Poisson distribution with that mean. Such a distribution is also called zero-inflated, and we look at these data in Section 13.6.3. Overdispersed distributions often occur as an effect of latent variables – variables not measured but important, or measured but not included in the model structure. As a simple example, consider ecological surveys that count the numbers of a burrowing shrimp in patches of seagrass, with the seagrass hypothesized to protect the shrimp from mobile predators. Our data have a range of patches with variable numbers of shrimp and many patches in which no shrimp were found. One plausible biological model for these data might be that we have measured seagrass density, which does indeed predict the number of shrimp, but only when the sediment is suitable for shrimp to burrow. The zeroes in the data come from two populations – suitable habitat where, by chance, no shrimp occur, and seagrass patches where sediment is unsuitable. Our model of shrimp numbers vs. seagrass density captures only one of the biological processes influencing shrimp.

Using a Poisson distribution for overdispersed data will result in SEs of estimated parameters smaller than they should be, and statistical tests of hypotheses will have inflated Type I error rates. We will also be fitting a statistical model that does not reflect the underlying biology.

13.6.1 Identifying Overdispersion
We can identify overdispersion in two main ways:
• Compare the observed distribution to that predicted by the relevant distribution (in this case a Poisson, but overdispersion can also occur for other distributions) using a quantile–quantile plot (Section 5.1.1.4).
• Calculate the dispersion statistic as the Pearson χ^2 goodness-of-fit statistic divided by the residual df (number of observations − number of model parameters). Values >1 indicate overdispersion, whereas for Poisson data it should be 1. This statistic is also an estimate of the dispersion parameter of our response variable, termed the Pearson estimator (Dunn & Smyth 2018). The dispersion parameter can also often be estimated by fitting a quasi-Poisson model (see below).

13.6.2 Correcting for Overdispersion
We can adjust our model in a few ways to account for the overdispersion.

13.6.2.1 *Quasi-likelihood (Quasi-Poisson) Models*
Because overdispersion reduces the SEs of our parameter estimates from a Poisson GLM, a simple solution would be to correct those SEs by multiplying by some measure of overdispersion. This approach is termed quasi-likelihood modeling, and quasi-Poisson when we use it with a Poisson GLM. Essentially, we multiply the SEs by the estimate of the dispersion parameter, usually $\sqrt{(\chi^2/df)}$ (Agresti 2015). Some software allows GLMs to be specified with quasi-Poisson (or quasi-binomial) and will estimate the dispersion parameter as part of the model fitting. Note that the quasi-Poisson is not a distribution; we simply fit a Poisson (using a log-link) with modified SEs. The parameter estimates will be the same as a Poisson GLM; only the SEs are changed.

There are some drawbacks with the quasi-likelihood approach. Zuur et al. (2013) argued that overdispersion is likely to affect the parameter estimates and their SEs, so

only correcting the latter might not deal with overdispersion properly. Dunn and Smyth (2018) also pointed out that because quasi-likelihood models are not true probability models, we cannot determine an AIC value (quasi-AICs can be determined; Bolker et al. 2009), and quantile residuals are no longer applicable (although deviance residuals are still okay). Also, LR statistics to compare different nested quasi-Poisson models are questionable, and F statistics based on differences in deviances should be used (Zuur et al. 2009). Whether a quasi-Poisson model should be chosen instead of, say, a negative binomial when overdispersion is present is often based on more subjective assessments of residual plots.

13.6.2.2 *Negative Binomial Models*
We could use a more appropriate probability distribution, such as the negative binomial (Table 4.2). The negative binomial is a mixture distribution, a modification of the Poisson that allows the mean to vary randomly following a gamma distribution (Pekár & Brabec 2016). The relationship between the mean and the variance in a negative binomial can be presented in several ways. Zuur et al. (2013) define the variance as $\mu + \mu^2/k$, where smaller k indicates greater overdispersion. In contrast, Hilbe (2007) and Agresti (2019) define the variance as $\mu + D\mu^2$, where larger D values indicate greater overdispersion. Software for fitting negative binomial models often produces an estimate of k or D.

Negative binomial GLMs are usually based on the log-link, although the identity link can be used. The model is fitted using ML to estimate the model parameters and k or D. The fit can be compared to a Poisson model using either AIC (or AIC_C) values or a LR statistic; for the latter, the P-value should be halved from what we would get if we assumed the LR statistic followed a χ^2 distribution with df = 1 (Agresti 2015). Residual plots are also important for checking the fit, and Dunn and Smyth (2018) recommend using quantile residuals.

Hilbe (2007) provides a comprehensive account of regression using negative binomial distributions, and Rhodes (2015) describes them in biology (though largely ecology). We illustrate the analysis of overdispersed data with a negative binomial in Box 13.4 using the wiregrass example (Fill et al. 2021).

13.6.2.3 *Including Observation-Level Random Effects*
When fitting GLMMs (Section 13.8), one option to deal with overdispersion besides using a negative binomial distribution is to include a random effect representing individual observations, termed an observation-level random

effect (OLRE). Harrison (2014) showed via simulations that including an OLRE worked well unless the overdispersion resulted from zero-inflation (Section 13.6.3). Zuur et al. (2012, 2013) describe OLRE with worked examples. We illustrate the use of OLRE in Box 13.9.

13.6.3 Too Many Zeroes (Zero-Inflated Data)

Biological count variables often include zero values. Zuur et al. (2012), Zuur and Ieno (2016), and Blasco-Moreno et al. (2019) described two main types of zeroes in ecological data. False zeroes result from observer or study design errors and could have been nonzero observations – for example, an individual was misidentified or the sampling duration was too short. True zeroes are where there haven't been any errors made, such as there being zero individuals in a sampling unit. Zuur et al. (2012) and Zuur and Ieno (2016) describe true zeroes as linked to the predictors included in the statistical model, while false zeroes won't be. They are also sometimes termed structural zeroes.

Count variables with more zeroes than expected from a Poisson distribution are zero-inflated. Zeroes can sometimes be the most common value, such as ecological sampling of rare taxa. This can cause overdispersion, but the usual solution of fitting a mixture model based on a negative binomial distribution may not be appropriate if there are many zeroes and the distribution is bimodal with one mode at zero and a second mode at some nonzero value (Agresti 2015).

Zero-inflated data often arise when multiple processes are involved, one determining whether a value is zero and a second determining the actual nonzero number. Mixture models explicitly incorporate these two processes, and Rhodes (2015) emphasizes that they fit the data better and have the added benefit of stimulating us to think about the biological processes responsible. In our shrimp example above, we should think about what makes sediment suitable for shrimp, and, in suitable sediments, how seagrass density influences shrimp numbers.

There are two common approaches for handling zero-inflated count data. The first is termed a hurdle or two-part model (Agresti 2015; Zuur & Ieno 2022) or zero-altered (ZA) model (Zuur et al. 2012). The first part is to fit a binomial model to the probability of an observation being a zero or not, and the second part is to model the nonzero counts with a truncated (to exclude zeroes; see below) Poisson (ZAP) or negative binomial (ZANB) model.

The second approach is to use ZI models. These models combine a distribution that has all its mass at zero (essentially a binomial model for the probability an observation is zero) and a Poisson (ZIP) or negative binomial

(ZINB) component to model the counts including zeroes. The main difference is that ZI models allow zero counts in the Poisson or NB component, but ZA models don't. The ZI approach conceptually distinguishes false zeroes (the binomial component) from true zeroes in the Poisson or NB component. However, in practice we don't know whether zeroes are true or false. We assume that zeroes that fit the Poisson or NB component model are more likely to be true zeroes, whereas those that don't and are identified with the first part of the analysis are more likely to be false zeroes. As Blasco-Moreno et al. (2019) emphasize, we should minimize false zeroes with robust sampling designs. The choice between a Poisson and NB for the count part of a ZI model is based on whether there is overdispersion beyond that caused by excess zeroes.

For both ZA and ZI models, the two parts may or may not use the same predictors. We would usually include the full set of predictors for the Poisson or negative binomial part. For the binomial part, the two most common options are to assume that the predictors don't affect the initial split of zeroes (so include only an intercept) or use the full set of predictors.

Zero-inflated data are widespread in ecological research, and Zuur et al. (2012) and Zuur and Ieno (2016, 2022) provide a thorough coverage of them. We illustrate ZI models in Box 13.4 with the wiregrass data.

13.6.4 Zero-Truncated Data

Some biological count data cannot follow a Poisson distribution because zeroes are impossible. For example, counts of paternal genotypes in a brood of offspring must be a positive integer, as must counts of group size. The problem with this distribution and the recommended solution depends on the response variable values. If the mean is large, the probability of Y being zero may be small, so the Poisson distribution may still be a good fit. As the mean gets smaller, zero values become much more likely using a Poisson distribution, and the mismatch between the data and the Poisson distribution makes parameter estimates unreliable. Here, we can fit a truncated Poisson distribution in which probabilities are adjusted to account for the absence of zeroes, as in the ZAP approach above. We can then proceed with the GLM.

Zuur et al. (2012) describe the analysis of zero-truncated data in considerable detail. They also describe the equivalent truncated negative binomial distribution.

13.6.5 Binomial Overdispersion

Overdispersion can also occur when modeling grouped data with binomial distributions – that is, variance $> \mu(1-\mu)$. Commonly recommended solutions are to use (1) quasi-binomial models, (2) OLREs in a GLMM, or (3) beta-binomial mixture models. In the beta-binomial distribution, we consider the proportion of one outcome (e.g. success) of a binomial to follow a beta distribution. The beta distribution applies to data between 0 and 1 and can represent a variety of distribution shapes depending on the values of its two parameters (α and β). It allows greater variance than a regular binomial, so it is suited for overdispersed binomial data (Zuur et al. 2013). It is not a true exponential distribution, so quasi-likelihood estimation may sometimes be more reliable than standard ML (Agresti 2015). Harrison (2015) compared beta-binomial models to OLREs for overdispersed binomial GLMMs (see Section 13.8) and found they were each suited to different sources of overdispersion. He generally recommended beta-binomial models and suggested that if using OLREs, parameter estimates should be compared to the beta-binomial model.

13.7 Contingency Tables

A common form of categorical data analysis in the biological sciences analyzes contingency tables. These tables involve the cross-classification of sampling or experimental units by two or more categorical variables, with counts or frequencies of units in each combination of the variables, termed a cell, analogous to the factorial designs described in Chapter 7.

Generally, contingency tables are analyzed so that no variable is considered a definite response variable. For example, in the illustration introduced in Chapter 4, Morehouse et al. (2016) cross-classified bear cubs according to whether they were "problem" bears or well-behaved ones and whether their mothers were problem or good bears (Table 4.2). These two variables could not be distinguished as response or predictor because a research question could be around predicting future risks (by knowing parents) or identifying causes of problems in cubs. In other situations, one variable in our table might be envisaged as a response to the other variables – for example, numbers of animals responding or not responding (one variable) to experimental treatments (second variable). This becomes particularly important when interpreting the multiple types of independence in more complex tables (Section 13.7.2).

The key question of interest with contingency tables is usually about independence of the effects of the variables on the cell frequencies, such as whether bear cubs are good or bad is independent of parental behavior. The traditional analysis compares the observed and expected (assuming independence of the variables) frequencies in

each cell and uses a χ^2 (or similar) statistic to evaluate independence. We can also fit a GLM where we model the cell frequencies or counts as the "response" against the classification variables as "predictors" (using a log-link and Poisson distribution) and evaluate the interaction term(s) by comparing the fit of full and reduced models. These models are usually called loglinear models. While the two approaches will result in similar conclusions for simple (e.g. two-variable, termed two-way) tables with reasonable sample sizes, the loglinear GLM approach is preferred for more complex tables and also allows random effects to be included to deal with hierarchical designs and/or correlated observations as GLMMs (Section 13.8). Standard loglinear models are based on Poisson distributions, but, as we construct GLMs, we can use other distributions (and link functions) to deal with issues like overdispersion. Agresti (2019), for example, describes the fitting of negative binomial distributions for contingency tables.

13.7.1 Two-Way Tables

Contingency tables where sampling or experimental units are cross-classified by two variables (rows and columns) are termed two-way tables. They are often described as $R \times C$, with R rows (categories of one variable) and C columns (categories of the second variable). The data from Morehouse et al. (2016) are a 2×2 table (good and bad mothers crossed with good and bad cubs) and are analyzed in Box 13.5. A second example is Teng et al. (2020), who used a survey of domestic cat owners to examine whether reported cat body condition scores (three categories) were related to cat begging behavior (four categories), with the number of survey responses in each cell of the 3×4 table. These data are analyzed in Box 13.6. Analysis of $R \times C$ contingency tables, especially 2×2 tables with small sample sizes, has a long and somewhat controversial history in statistics (see Agresti 2019; Choi et al. 2015).

The general layout for a two-way table (cross-classification of two variables) is illustrated with a 2×2 table in Table 13.1.

We will use A ($i = 1$ to R categories) and B ($j = 1$ to C categories) as labels for the two variables (to follow the factorial designs from Chapter 7); no particular significance should be ascribed to which variable is A and which is B. For a 2×2 table, $R = C = 2$. The observed frequency in each cell is n_{ij}, and the joint probability that an observation occurs in any cell is π_{ij}. We also have marginal totals (e.g. the total in row 1 is n_{1+}) and *marginal probabilities* (e.g. the probability that an observation occurs in row 1 is $\pi_{1+} = \pi_{11} + \pi_{12}$); these marginal

Box 13.5 Ⓡ Worked Example of 2 × 2 Two-Way Contingency Table: Problem Bears, Cubs

Morehouse et al. (2016) used genetic analysis to determine the parentage of bear cubs, and cross-classified cubs and their parents as causing problems around humans (see Table 4.2). The χ^2 statistic suggested that cub behavior was not independent of the mother's behavior ($\chi^2 = 4.403, \mathrm{df} = 1, P = 0.036$). Standardized residuals are tabulated below.

	Problem mother	Good mother
Problem cub	1.657	−0.569
Good cub	−1.092	0.376

There was an excess of problem cubs whose mother was also problematic and fewer good cubs when the mother was problematic than expected if the mother and cub behavior was independent. The odds of a problem mother having a problematic cub are 1.67, compared to the odds of a well-behaved mother having a problem cub, which are 0.36.

The estimated OR is 4.63 (95% CI: 1.003–21.367) – that is, the odds of being a problem cub are 4.63 times greater if the mother showed problem behavior than if she didn't. Even though the CI was wide, it (just) does not include 1, indicating that problem cubs are associated with mothers showing similar problematic behavior.

One out of four cells had an expected value <5, so we might be concerned about the reliability of the χ^2 test for independence. The Fisher exact test resulted in a P-value of 0.050, although we note that neither row nor column totals were fixed (see Section 13.7.1), and the P-value from the mid-P exact test was 0.058. If we were relying on the traditional 0.05 significance level for drawing conclusions about the null hypothesis of independence, these results are equivocal.

We can also assess independence of mother and cub behavior using loglinear models. The saturated model includes the mother behavior × cub behavior interaction and both main effects, and the reduced model omits the interaction. The comparison of the fit of these two models indicated some evidence for lack of independence – that is, an interaction (LR $\chi^2 = 4.005, \mathrm{df} = 1, P = 0.045$). Note the test statistic and P-value are slightly different from the original χ^2 statistic based on observed and expected frequencies.

probabilities are the probabilities that an observation occurs in a particular row or column.

We can distinguish between three types of contingency tables (Agresti 2013):
- No marginal totals are fixed, and the table represents the cross-classification of a random sample of units (e.g. the bear cub example).
- Either the row or column totals are fixed. For example, one variable might represent experimental treatments that have specified (and equal) sample sizes.
- The grand total is fixed, an uncommon situation in biological research.

While a loglinear model (Poisson GLM) can be used in all three situations (Dunn & Smyth 2018), there is some debate about the appropriateness of other tests of independence, especially the Fisher exact test (see below), to each type of table.

The question of independence – that is, the observations come from a population of units in which the two variables (rows and columns) are independent of each other in terms of the cell frequencies – is often expressed

as no association, or interaction, between the two variables. For example, is the nuisance behavior in grizzly bear cubs associated with similar behavior in their parents (Box 13.5) and is being overweight in domestic cats associated with their begging for food (Box 13.6)?

13.7.1.1 Test for Independence

The most common approach to assessing independence is using a χ^2 test by calculating the expected frequencies in each cell based on independence of the two variables. An expected cell frequency is simply the product of the probability of an observation occurring in that cell and the total sample size (n_{++}):

$$f_{ij} = n_{++} \, \pi_{ij}.$$

If rows and columns are independent, the joint probability for a specific cell (π_{ij}) is simply the product of the corresponding marginal probabilities: $\pi_{i+}\pi_{+j}$.

We can then assess whether the values predicted using the assumption of independence match well with the observed ones:

Box 13.6 Ⓡ Worked Example of R × C Two-Way Contingency Table: Cat Owners' Attitudes

Teng et al. (2020) analyzed the results of a survey of domestic cat owners in Australia. The survey focused on factors (e.g. cat demographics, owner attitudes and demographics, etc.) that might affect the prevalence of overweight and obese cats. They related nearly 1,400 survey responses of owner-assessed body condition score (BCS with five categories: very underweight [1], somewhat underweight [2], ideal [3], chubby/overweight [4], and fat/obese [5]) to a range of categorical predictors with a multivariate multinomial GLM. We will use one aspect of their data to construct a contingency table relating the BCS, reduced to three categories (1 and 2, 3, 4 and 5) to cats' begging behavior (four categories: never, sometimes, often, always). The final table is given here.

	BCS 1 and 2	BCS 3	BCS 4 and 5
Never	21	266	56
Sometimes	55	527	162
Often	14	106	69
Always	10	47	44

The BCS were not independent of begging behavior ($\chi^2 = 55.928, \mathrm{df} = 6, P < 0.001$). The standardized residuals were as follows.

	BCS 1 and 2	BCS 3	BCS 4 and 5
Never	−0.783	1.978	−2.913
Sometimes	0.131	0.702	−1.259
Often	0.074	−2.092	3.497
Always	0.984	−2.688	4.003

More owners scored their cats in the overweight and obese category when the cats begged for food often or always and fewer when the cats rarely begged than would be expected if begging and BCS score were independent. Odds ratios can only be calculated for 2 × 2 tables, so we focused on the ORs of being overweight versus ideal for pairwise comparisons of each begging category against never begging. The ORs with 95% CIs (estimated using ML) were as follows.

	OR	95% CI
Sometimes vs. never	1.457	1.045–2.057
Often vs. never	3.082	2.031–4.702
Always vs. never	4.424	2.677–7.344

The odds of being overweight compared to ideal increase as the frequency of begging behavior increases. These ORs are lower than those presented by Teng et al. from their multinomial model for these same comparisons, but the trend was the same.

We can also assess independence of condition score and begging using loglinear models. The saturated model includes the two-way interaction and both main effects and the reduced model omits the interaction. The comparison of the fit of these two models indicated strong evidence for lack of independence – that is, an interaction (LR $\chi^2 = 53.259, \mathrm{df} = 6, P < 0.001$) – a similar result to our original test for independence using the standard χ^2 statistic based on observed and expected frequencies.

Table 13.1 General data layout for a 2×2 contingency table

		Variable B		Marginal totals and probabilities
		B1	B2	
Variable A	A1	n_{11}	n_{12}	n_{1+}
		π_{11}	π_{12}	π_{1+}
	A2	n_{21}	n_{22}	n_{2+}
		π_{21}	π_{22}	π_{2+}
Marginal totals and probabilities		n_{+1}	n_{+2}	n_{++}
		π_{+1}	π_{+2}	Grand total

$$\chi^2 = \sum_{i=1}^{R} \sum_{j=1}^{C} \frac{\left(n_{ij} - f_{ij}\right)^2}{f_{ij}}.$$

We compare this test statistic to a χ^2 distribution with $(R-1)(C-1)$ df to evaluate the null hypothesis of independence.

13.7.1.2 Odds and Odds Ratios

Odds and odds ratios are important summary measures of association or lack of independence in contingency tables, more than for logistic regression models. They can be calculated simply and unequivocally for 2×2 tables and used in larger tables by subdividing them into relevant sets of 2×2 tables (see Box 13.6). We calculate the odds of one of the two possible categories (outcomes) of one variable for each category (j) of the other variable:

$$\frac{\pi_j}{1 - \pi_j},$$

where π_j is the probability of one of the two outcomes and one minus π_j is the probability of the other outcome.

In the grizzly example, the odds of a cub being problematic are 1.37 when the mother was a problem and 0.36 when the mother had no issues.

The OR (θ) is simply the ratio of the odds of one outcome of A for one category of B to the odds of that outcome for another category B. The OR is a population parameter:

$$\theta = \frac{\pi_1/(1 - \pi_1)}{\pi_2/(1 - \pi_2)}.$$

For simple tables like this, an unconditional ML estimate of the OR can be calculated easily:

$$\hat{\theta} = \frac{n_{11} n_{22}}{n_{12} n_{21}}.$$

If the two variables are independent, the OR will not change across rows or across columns, and it will deviate from 1 as associations become stronger.

The sampling distribution of ORs is usually very skewed, especially for small sample sizes (Agresti 2019). To calculate the SE and CI we need to log transform the OR, which results in its sampling distribution being approximately normal:

$$\text{ASE}\left(\log \hat{\theta}\right) = \sqrt{\frac{1}{n_{11}} + \frac{1}{n_{12}} + \frac{1}{n_{21}} + \frac{1}{n_{22}}}.$$

Confidence intervals for the OR are best calculated on the log OR and back-transformed. The 95% CI is $\pm z_{0.95} \text{ASE}\left(\log \hat{\theta}\right)$, where z is the critical value from a standard normal distribution. Ruxton and Neuhäuser (2012) called the method above the Woolf method. They reviewed approaches for determining CIs, including generalizations based on the Fisher exact test (see below). They recommended a method based on the score test as the least conservative.

Note that the OR is zero or undefined if any observed counts in the 2×2 subset table are zero. One approach to this problem is to add a small amount, most commonly 0.5, to each cell. Agresti (2013) describes some alternatives.

13.7.1.3 Residuals

An important step in interpreting lack of independence in contingency tables is examining the pattern of the residuals, defined as the difference between the observed and expected (predicted) values $\left(n_{ij} - f_{ij}\right)$. Each cell will have a residual. Absolute residuals are difficult to compare when the frequencies vary – a residual of five is more substantial as a deviation from an expected value around 10 than when the expected frequencies are around 100. We can standardize each residual by dividing by the square root of the expected value:

$$\frac{n_{ij} - f_{ij}}{\sqrt{f_{ij}}},$$

and the χ^2 statistic is simply the sum of these squared residuals.

Another way of standardizing a residual is to divide it by its estimated SE:

$$\frac{n_{ij} - f_{ij}}{\sqrt{f_{ij}\left(1 - p_{i+}\right)\left(1 - p_{+j}\right)}},$$

where p_{i+} is the proportion of the total row observations in that row and p_{+j} is the proportion of the total column observations in that column. These are also standardized Pearson residuals from the fit of a loglinear model (see below). Large residuals indicate large deviations from independence, and the sign indicates more or less observed than expected under the assumption of independence of A and B. For the grizzly data, the standardized residuals showed more problem cubs with problem parents and fewer cases where the behavior of parents and offspring was mismatched (Box 13.5).

13.7.1.4 Small Sample Sizes

The χ^2 statistic is based on categories' frequencies and can only be compared to the continuous χ^2 distribution if the sample size is big enough. The notion of big enough is somewhat arbitrary, with one common recommendation being that no more than 20% of the cells have expected frequencies less than five.

Yates correction for continuity was developed to improve the accuracy of the χ^2 test for 2×2 tables with small frequencies, but is now regarded as unnecessary (Agresti 2013) because of the availability of "exact" tests.

Fisher's exact test was designed for 2×2 tables with fixed marginal totals. It does not use the χ^2 distribution to assess the H_0 of independence but instead answers the question, "Given our fixed marginal totals, what is the probability of obtaining the observed cell frequencies and all cell frequencies further away from the expected?" Fisher's exact test is probably conservative when the marginal totals are not fixed, especially for small tables. Other exact tests are available, and Ruxton and Neuhäuser (2010) recommended the Fisher–Freeman–Halton test, especially the mid-P version, for $R \times C$ tables and the Fisher–Boschloo test for 2×2 tables. Permutation methods based on the standard χ^2 test are also available.

Another solution to small observed frequencies for tables with more than two rows or columns is to collapse or combine categories. For example, there may be come categories with small frequencies that could meaningfully be combined into a single category with adequate frequencies.

13.7.1.5 Loglinear Models

The best method for analyzing contingency tables, especially more complex tables, is with loglinear models. They treat the cell frequencies as counts distributed as a Poisson random variable, and as with other GLMs, we estimate their parameters using ML. We introduce these models with the two-way tables and expand to more complex cases below.

The basic approach is straightforward, and we will illustrate it with our two worked examples. Under the assumption of independence of A and B, recall that the expected frequency in any cell is the product of the total sample size and the probabilities of being in that cell's row and that cell's column – that is, $f_{ij} = n_{++}\pi_{i+}\pi_{+j}$. We can convert this to a familiar linear model by logging of both sides, to give $\log f_{ij} = \log(n_{++}) + \log(\pi_{i+}) + \log(\pi_{+j})$. We can express this as a loglinear model to model the expected frequencies in each cell, assuming independence of A and B:

$$\log f_{ij} = \lambda + \lambda_i^A + \lambda_j^B,$$
$$\log f_{ij} = \lambda + \lambda_i^{\text{mother}} + \lambda_j^{\text{cub}},$$
$$\log f_{ij} = \lambda + \lambda_i^{\text{begging}} + \lambda_j^{\text{BCS}},$$

where:

λ is the mean of the logs of all the expected frequencies (the equivalent of an intercept);

λ_i^A is the effect of category i of variable A – that is, the effect of mother bears being problematic or not on the log expected frequency of cubs in each cell, or the effect of begging frequency on the log expected frequency of owner responses rating their cat in each cell; and

λ_j^B is the effect of category j of variable B – that is, the effect of cubs being problematic or not on the log expected frequency of cubs in each cell, or the effect of BCS category on the log expected frequency of owner responses in each cell.

These are essentially main effects models, and the main effect terms model the row and column totals; a larger λ means that the expected frequencies will be larger for that variable – that is, that row or that column (Agresti 2019). These parameters are also deviations from the mean of all the log expected frequencies, just like linear effects models described in Chapters 6 and 7.

Table 13.2 Observed frequencies for three-way contingency tables from the McGowen et al. (2020) study of vision genes in cetaceans

PSG	Lineage	Gene type Vision	Nonsensory	Totals
Yes	Ingroup	66	146	212
Yes	Outgroup	77	144	221
No	Ingroup	30	21	51
No	Outgroup	19	23	42
Totals		192	334	526

Main effects are not usually of biological interest, as the main question with contingency tables is about the independence of A and B – that is, the interaction between them. We can fit a more complex model that includes an interaction between A and B:

$$\log f_{ij} = \lambda + \lambda_i^A + \lambda_j^B + \lambda_{ij}^{AB}$$

This saturated model fits the observed frequencies perfectly.

Like with any GLM, we assess the independence of A and B by comparing the ML fit of the models with (full model) and without (reduced model) the interaction term using the log-likelihood-ratio statistic. Because the full model is also the saturated model for a two-way table, the LR statistic is also the deviance.

The lack of independence in loglinear models can be interpreted using ORs and residuals, just as described above. Pearson and deviance residuals are standard output from loglinear modeling routines in most statistical software.

13.7.2 Three-Way and Higher Tables

An obvious extension of two-way contingency tables is the addition of a third variable in the cross-classification. The three variables are labeled A (I categories), B (J categories), and C (K categories). We've switched from R(ows) × C(olumns) to I × J × K here because there's no obvious name for table dimensions beyond the first two. For example, McGowen et al. (2020) used a phylogenetic dataset on cetaceans to count the genes falling into each cell of three cross-classified variables: whether the genes were under positive selection (PSG) or not, sensory or nonsensory, and from ingroup or outgroup lineages from a phylogenetic tree. Their 2 × 2 × 2 table is shown in Table 13.2, and the analysis is in Box 13.7.

As a second example, Sinclair and Arcese (1995) cross-classified wildebeest carcasses from the Serengeti by three variables, to determine whether cause of death (predation, nonpredation) was associated with bone

marrow type (indicating the health status of the animal) and sex. This example is available online.

Three-way contingency tables are often analyzed by breaking them into smaller two-way subset tables:
- Partial or conditional tables represent the association between pairs of variables (one of which might be the "response" variable if one exists) at each category of the third variable. With the cetacean gene example, we can consider separate tables of PSG cross-classified with gene type for each lineage separately or PSG cross-classified with lineage for each gene type separately.
- Marginal tables represent the association between two variables, pooled over the categories of the third variable – for example, PSG cross-classified with gene type across both lineages.

These smaller two-way tables reflect different types of dependence between variables in three-way tables. Let's use the cetacean example, where A is positive selection, B is gene type, and C is lineage:
- Complete dependence is where all the variables are associated with each other in some way.
- Conditional (partial) independence is where there is no association between two of the variables (A and B) conditional on the category of the third variable (C). Conditional dependence means A and B are associated at each category of C.
- Joint independence is where two variables (A and B) are associated, but both are independent of the third variable (C).
- Marginal independence means no association between two of the variables (A and B) pooling across the categories of the third variable (C). Marginal dependence means A and B are associated pooling C.
- Complete independence means there are no associations (no dependence) between any of the variables.

Our preferred approach for analyzing three-way contingency tables is based around loglinear models. They are commonly fitted in a hierarchical fashion, so the inclusion

Box 13.7 ℝ Worked Example of Three-Way Contingency Table: Cetacean Sensory Genes

As part of a phylogenetic study on the adaptation of cetaceans to an aquatic lifestyle, McGowen et al. (2020) counted the number of genes in each cell of three cross-classified variables: sensory (hearing, vision) vs. nonsensory genes, ingroup (whales, dolphins, etc.) vs. outgroup (hippopotamus) lineages from their phylogenetic tree, and whether the genes were under positive selection (PSG) or not. The research question was primarily whether sensory genes were more under positive selection than nonsensory ones, and whether this association differed between lineages in a way linked to the divergence of cetaceans and hippopotamuses. We will focus just on the analysis of vision genes; the $2 \times 2 \times 2$ table is presented in Table 13.2.

Comparing the fit of the saturated loglinear model with the model that omits the three-way interaction showed there was no strong evidence for the interaction (LR $\chi^2 = 2.357$, df $= 1$, $P = 0.123$; Figure 13.4). This means the association between PSG and sensory/nonsensory genes is consistent between phylogenetic groups. The ratio of the odds of being positively selected for the vision versus nonsensory genes is the same for both lineages. This was confirmed by the Breslow–Day test ($\chi^2 = 2.353$, df $= 1$, $P = 0.125$).

McGowen and colleagues focused their conclusions on the presence or absence of the three-factor interaction using the Breslow–Day test. We undertook further analysis of conditional independence for each of the two-way interactions by comparing the model with all three interactions to ones without each interaction.

Model[*]	Interaction tested	LR χ^2	df	P
PSG × gene type + lineage × gene type	PSG × lineage	1.109	1	0.297
PSG × lineage + lineage × gene type	PSG × gene type	12.392	1	<0.001
PSG × lineage + PSG × gene type	Lineage × gene type	0.026	1	0.871

[*] Main effects included.

The only two-way association showing nonindependence is PSG × gene type – that is, the odds of being positively selected actually decline between nonsensory and vision categories (OR = 0.444; 95% CI: 0.281–0.699), but this lack of independence is consistent across lineages (no three-way interaction). The odds of being positively selected don't change between lineages (confirmed by the Cochran–Mantel–Haenszel [C-M-H] test: $\chi^2 = 1.080$, $P = 0.299$) and the association between lineage and gene type is probably not of much biological interest.

The standardized Pearson residuals from the fit of the model that omits the PSG × gene type interaction were as follows.

PSG	Gene type	Lineage	
		Ingroup	Outgroup
Yes	Vision	−2.000	−0.715
Yes	Nonsensory	2.638	0.937
No	Vision	0.981	0.310
No	Nonsensory	−1.294	−0.409

There were more positively selected nonsensory genes and fewer positively selected vision genes in the ingroup than expected if PSG and gene type were independent.

The assessment of model fits based on AIC were as follows.

Model[*]	Deviance (G^2)	df	AIC
Saturated model	0		61.75
PSG × lineage + PSG × gene type + lineage × gene type	2.36	1	62.11
PSG × gene type + lineage × gene type	3.44	2	61.19

Box 13.7 (cont.)

(cont.)

Model[*]	Deviance (G^2)	df	AIC
PSG × lineage + lineage × gene type	14.75	2	72.50
PSG × lineage + PSG × gene type	2.38	2	60.14
Gene type + PSG × lineage	14.75	3	70.50
Lineage + PSG × gene type	3.44	3	59.19
PSG + lineage × gene type	15.81	3	71.56
PSG + lineage + gene type	15.81	4	69.56

[*] All lower-order terms included, for example, PSG × lineage includes main effects of PSG and lineage.

All models that omit PSG × gene type (implying independence between being positively selected and gene type) are worse fits than models that include this term, highlighting the importance of this association (lack of independence) on model fit.

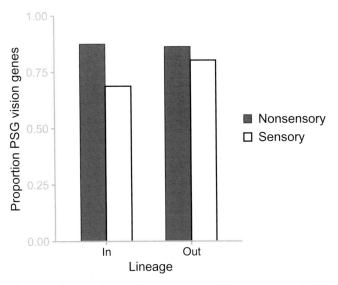

Figure 13.4 Representation of three-way interaction in the McGowen et al. (2020) study of vision genes in cetaceans. The figure shows how the proportion of vision genes (factor *A*) varies with lineage and the type of gene. The similarity between the two pairs of bars reflects the lack of a complex interaction.

of a higher-order term automatically requires the presence of all lower-order terms with those variables – for example, the model with the three-variable interaction automatically includes all two-way interactions and main effects. Similarly, a model which omits one or more two-way interactions also must omit the three-way interaction. The different types of independence described above are evaluated by comparing the fit of models with and without specific terms, mostly interaction terms, just as we could assess independence in a two-way table by comparing models with and without the two-way interaction.

We can also use ORs (based on 2 × 2 subset tables), standardized residuals, and other descriptive methods to examine any lack of independence identified with our modeling.

Loglinear modeling of three-way tables usually starts with the most complex model and proceeds by removing interaction terms that don't contribute to the model's fit. Each of these interaction terms represents one form of association or dependence among the variables.

The most complex model we can fit for a three-way table is a saturated model:

$$\log f_{ijk} = \lambda + \lambda_i^A + \lambda_j^B + \lambda_j^C + \lambda_{ij}^{AB} + \lambda_{ik}^{AC} + \lambda_{jk}^{BC} + \lambda_{ijk}^{ABC}.$$

For example, this saturated model for the cetacean gene example (McGowen et al. 2020) is (subscripts omitted for convenience):

$$\log f_{ijk} = \lambda + \lambda^{PSG} + \lambda^{lineage} + \lambda^{gene\ type} + \lambda^{PSG \times lineage}$$
$$+ \lambda^{PSG \times gene\ type} + \lambda^{lineage \times gene\ type} + \lambda^{PSG \times lineage \times gene\ type}.$$

As with two-way models, λ is the mean of the logs of all the expected frequencies (the equivalent of an intercept). The other terms represent the three-way interaction, three two-way interactions, and three main effects, analogous to the three-factor linear model we described in Section 7.2.

13.7.2.1 Three-Way Interaction

The first step in our analysis is to examine the three-way interaction. A three-way interaction implies complete dependence where the two-way interaction effects are different for each category of the third variable. We can think of this in terms of ORs for two-way subset tables, where the OR for each $A \times B$ table is different for each category of C. For example, the ratio of the odds of a vision gene being positively selected compared to a non-sensory gene changes between lineages. In the wildebeest example, the change in odds of dying due to predation for each pairwise comparison of marrow types is different for males versus females (or for each sex is different for the different marrow types). We assess this interaction by fitting a model that omits this term:

$$\log f_{ijk} = \lambda + \lambda_i^A + \lambda_j^B + \lambda_j^C + \lambda_{ij}^{AB} + \lambda_{ik}^{AC} + \lambda_{jk}^{BC}.$$

This model, sometimes termed the uniform or homogeneous association model, assumes there is no three-way interaction – that is, the association between A and B is independent of C, the association of A and C is independent of B, and the association of B and C is independent of A. This model is compared to the saturated model to evaluate the interaction term.

When the presence of a three-way interaction indicates complete dependence, such as in the wildebeest example (online), there are a few ways we can proceed:
• Examine the standardized residuals from the model without the three-way interaction term to see which cells were causing the lack of independence. Unfortunately, patterns in standardized residuals can be difficult to interpret for complex tables.
• Examine the association of pairs of variables for each category of the third, analogous to simple interaction tests in three-factor linear models. This can include calculating ORs by breaking each table into a series of 2×2 tables.

• If our "response" variable has only two categories, we can use an interaction plot with the proportion of counts in one category of the "response" variable on the y-axis, just like interaction plots for factorial linear models (Section 7.1.8). This is essentially rewriting the model as a logistic regression (Section 13.7.4).

The Breslow–Day test is another test for whether the associations (or ORs) for any 2×2 subset table are the same for each category of the third variable. It is essentially a χ^2 goodness-of-fit statistic calculated assuming there is no three-way interaction.

If there is little evidence for a three-way interaction, as in the cetacean genes example, we can look at the two-way interactions.

13.7.2.2 Conditional (In)dependence

An absence of a three-way interaction implies that the associations (or ORs) between two variables are the same for each category of the third variable. For example, the ratio of the odds of a sensory gene being positively selected compared to a nonsensory gene is the same for both lineages. There can still be an association between the PSG and gene type, but that association is consistent across lineages.

We can now evaluate each two-way interaction term by comparing the fit of the model, including the three two-way interactions to a model which omits a specific interaction. To evaluate the PSG × gene type interaction, we compare these two models:

$$\log f_{ijk} = \lambda + \lambda^{PSG} + \lambda^{lineage} + \lambda^{gene\ type} + \lambda^{PSG \times lineage}$$
$$+ \lambda^{PSG \times gene\ type} + \lambda^{lineage \times gene\ type},$$

$$\log f_{ijk} = \lambda + \lambda^{PSG} + \lambda^{lineage} + \lambda^{gene\ type} + \lambda^{PSG \times lineage}$$
$$+ \lambda^{lineage \times gene\ type}.$$

If these two models fit similarly, PSG and gene type are conditionally independent for the two lineages. A difference in fit of these two models indicates conditional (partial) dependence where there is an association between PSG and gene type that is consistent between lineages.

We can make the equivalent model comparisons for the other two-way interaction terms. The analysis for the cetacean gene dataset (Box 13.7) showed that PSG and lineage were conditionally independent, but there was an association between PSG and gene type for each lineage.

Interpreting these two-way interaction terms doesn't make much sense if there is a three-way interaction because the two alternatives for each two-way term are conditional independence (all ORs equal 1) or conditional dependence consistent across the categories of the third variable. This is analogous to trying to interpret two-way interactions in a factorial linear model in the presence of a strong three-way interaction. The only exception might be if we have such a large sample size that biologically trivial interactions might be statistically "significant"; under these circumstances, we might still examine two-way interactions.

If our evaluations of one or more of the two-way interactions indicate conditional dependence, our approach is similar to interpreting a three-way interaction. We examine standardized residuals and ORs by breaking the full table into 2×2 subsets because when the table dimensions are $2 \times 2 \times K$, we can calculate conditional (partial) ORs for each set of partial tables. In the cetacean gene example, there was conditional dependence between PSG and gene type, and because the full table was $2 \times 2 \times 2$, ORs were straightforward to calculate: the odds of being positively selected for vision genes versus the odds of being positively selected by nonsensory genes for each lineage separately. With conditional independence between PSG and lineage, the ORs for each gene type would not differ from 1.

Another way of testing for conditional independence in $2 \times 2 \times K$ tables is the C-M-H test, which basically tests the null hypothesis that the conditional ORs between A and B equal 1 for all categories of C. It is really only appropriate when there is no three-variable ($A \times B \times C$) interaction (Agresti 2013). The C-M-H statistic is converted to a χ^2 and compared to a χ^2 distribution. A generalization for $I \times J \times K$ tables where I and J are >2 is available in some software. For the cetacean gene example, the PSG $\times$ lineage association C-M-H statistic $= 1.080$ with $P = 0.299$, agreeing with the loglinear model comparison. The C-M-H test also allows a form of meta-analysis to combine the results from several independent 2×2 tables.

13.7.2.3 Joint Independence

If A and B are associated with each other but both are independent of C, it means we have a model that includes only one two-way interaction term ($A \times B$) plus the three main effects. For example, to assess whether being positively selected is associated with lineage, but both PSG and lineage are independent of gene type, we would fit the following model:

$$\log f_{ijk} = \lambda + \lambda^{PSG} + \lambda^{lineage} + \lambda^{gene\,type} + \lambda^{PSG \times lineage},$$

and compare it to the homogeneous association model that includes all two-way interactions (assuming no

three-way interaction). We can make the same comparison to assess joint independence for the other two-way interactions.

13.7.2.4 Marginal Independence

Marginal tables are two-way tables completely ignoring the third variable – for example, the frequencies for A by B pooling categories of C. Marginal independence is independence between the two variables in the marginal table. For the cetacean gene example, one marginal table would be PSG crossed with lineage, pooling gene type. We can also calculate marginal ORs from the marginal table.

A test for marginal independence will not necessarily result in the same conclusions as a test for conditional (or partial) independence; this is known as Simpson's paradox. In biological research, the focus is more often on conditional rather than marginal tests.

13.7.2.5 Complete Independence

The effects of the individual variables represent complete independence and no two- or three-way associations. For example, proportions of positively selected genes are independent of lineage and gene type. Complete independence would be assessed by comparing the homogeneous association model to one that only includes main effects. We wouldn't do this in our worked example because of the two-way interaction in the cetacean gene study.

13.7.2.6 Hierarchical Loglinear Modeling

We pointed out that one measure of fit for a GLM is the deviance. For contingency table modeling, the deviance is often called G^2. We can create a modified analysis of deviance table (like an ANOVA table), which gives the difference in G^2 between hierarchical models and P-values for tests that specific terms equal zero. Many software packages provide an automated approach to hierarchical loglinear modeling. Comparing the fit of models back to the saturated model will not necessarily be appropriate for assessing some forms of independence, so we recommend that you think carefully about the choice of full and reduced models for testing specific model terms.

We can also use information criteria for comparing the fit of different loglinear models so that unnecessarily complex models are penalized. The AIC is commonly used, with smaller values indicating better fit.

Our experience is that most biologists analyzing three-way contingency tables with loglinear models focus on the three-way interaction and/or the individual two-way interactions – that is, specific hypotheses about complete dependence and conditional (in)dependence, rather than joint or mutual independence.

13.7.3 More Complex Tables

We can extend the loglinear model to include four or more variables. The approach does not change, but the complexity increases. The complexity is reduced somewhat when we fit models hierarchically. Still, for four predictors, we have one model with the four-way interaction present, one with all (four) possible three-way interactions, four models with three three-way interactions, six with two of the three-way interactions, and four with three of these interactions. Then we consider models involving the six pairwise interactions…!

Even with hierarchical modeling, we fit many models and try to identify the best-fitting or test for specific forms of independence. The latter is particularly challenging because there may be many pairs of models we could use for some effects. For example, if we label our factors A, B, C, D, and we want to assess the interaction between A, B, and C, we could compare (among others):

- $\ldots ABC + ABD + ACD + BCD$ vs. $\ldots + ABD + ACD + BCD$;
- $\ldots ABC + ABD$ vs. $\ldots ABD$;
- $\ldots ABC + ACD + BCD$ vs. $\ldots ACD + BCD$.

Each pair of models differs only in the presence of ABC, so each comparison is appropriate for assessing this effect, but they will not produce identical answers because they each involve separate parameter estimation. There is no reason to choose one over another, so we may need to develop a model-fitting strategy that specifies the models to be a starting point (e.g. the hierarchical set), identifying all possible pairs of models suitable for assessing a particular effect, and a decision about how we will use these pairs (e.g. by model averaging, Section 9.2.4).

Agresti (2019) suggests an exploratory phase of model fitting at which only three models would be fitted: the model with the four-way interaction omitted, a model with all three factors omitted, and one with all two factors omitted. Each model allows an assessment of whether all interactions of a particular order can be ignored. The fit of each model would be used as a guide to more detailed model fitting. For example, if the model with three-way interactions fitted adequately, but the model with only two-way terms did not, then a more detailed investigation of the different three-way effects would be the next step.

13.7.4 Loglinear vs. Logistic Models for Tables

Suppose we are dealing with contingency tables where one variable can be considered a response, and this variable has only two categories. There, we could fit a logistic regression model where the response is the proportion of observations in one category with the other variables as predictors. We could model the proportion of genes under positive selection and lineage and gene type as two categorical predictors. Here, the two-way interaction between lineage and gene type is the equivalent of the three-way interaction in our original loglinear model.

It is worth noting that the simple calculations for a two-way contingency table described in Section 13.7.1 can also be viewed as fitting a linear model. We constructed a model with the two independent variables and calculated predicted values under this model using marginal probabilities. In the models described earlier in this chapter and others, we have used ML to generate the model with parameter values that provide the best fit. For two-way tables, the ML estimates correspond to the set of row and column totals in the table, allowing us, in effect, to estimate our model using simple calculations and without sophisticated software. For higher-dimensional tables, we must return to ML estimation of joint probabilities.

13.8 Generalized Linear Mixed Models

We saw in Chapters 10–12 that linear mixed effect models (LMEs or LMMs) represent an extension of linear models to include random effects. Similarly, GLMs can also include random effects to become GLMMs. For example, Ryeland et al. (2017) used video to record the proportion of time individuals of various species of shorebirds spent in the backrest position while roosting. They used a binomial model with four fixed continuous predictors describing environmental conditions, group size, and distance to the observer. This would be a standard binomial GLM except that more than one bird was sometimes monitored in each video recording (bout), so video bout was included as a random effect. The resulting model is now a binomial GLMM with a logit link. The analyses of these data are given in Box 13.8.

The full model for the Ryeland data is:

$$\log \text{(odds of bird in backrest vs.not)}$$
$$= \beta_0 + \beta_1(\text{temperature}) + \beta_2(\text{group size})$$
$$+ \beta_3(\text{wind speed}) + u(\text{video bout}).$$

This model is an additive random intercept model. We assume that relationships between odds of backrest and each of the three fixed predictors are also consistent across random video bouts – that is, there were no random slope terms in the model.

Box 13.8 Ⓡ Ⓔ Worked Example of a Generalized Linear Mixed Model: Birds "Backrest Roosting"

Ryeland et al. (2017) studied the roosting behavior of several species of shorebirds. They recorded the proportion of time (number of minutes as a proportion of total minutes in a video bout) individuals of various species spent in the backrest position while roosting. They used a binomial model with a logit link for proportions with four fixed predictors recorded for each video bout: ambient temperature, wind speed, size of group focal bird was in, and distance focal bird was from the observer. We will analyze the data for a single species, the sharp-tailed sandpiper (*Calidris acuminata*). This would be a standard binomial GLM except that more than one bird was sometimes recorded in each bout, so bout was included as a random effect since birds closer together may be correlated in their behavior. The resulting model is a binomial GLMM.

The fit of the GLMM with the four fixed predictors and video bout as a random effect resulted in an AIC value of 224.002 (AIC_C was similar: 226.402). The residual plot looked reasonable, the one video bout (#19) with an unusually large wind speed (17.5) was not influential and there was no evidence for strong collinearity among the predictors. Only the effect of temperature was important (LR $\chi^2 = 17.43$, df $= 1$, $P < 0.001$) with the log of the odds of a bird being in the backrest position declining by -0.249 (SE: 0.055) with each 1 °C rise in temperature.

However, there was evidence of some overdispersion; dispersion statistics for GLMMs are tricky because of the challenge of deciding on the number of parameters (all effects or just fixed effects), but following Zuur et al. (2013), we calculated a dispersion statistic of 2.234 indicating mild overdispersion. As Ryeland et al. (2017) did, we then added an OLRE to handle the overdispersion and refitted the model. The OLRE model was a better fit (AIC: 175.97; AIC_C: 179.27) than the original GLMM. The residual plot was OK and, again, the only strong effect was temperature (LR $\chi^2 = 18.20$, df $= 1$, $P < 0.001$), with the log of the odds of a bird being in the backrest position declining by -0.289 (SE: 0.064) with each 1 °C rise in temperature.

The alternative to using OLRE for binomial overdispersion is to fit a beta-binomial model. This model had an AIC of 178.7 (AIC_C: 181.946), so slightly worse than the OLRE model. Again, the only strong effect was that of temperature (LR $\chi^2 = 19.09$, df $= 1$, $P < 0.001$), with the log of the odds of a bird being in the backrest position declining by -0.193 (SE: 0.048) with each 1 °C rise in temperature.

Note that Ryeland et al. (2017) quite reasonably used a model selection approach to simplify their models. For the sharp-tailed sandpiper, their final model only included temperature and had lower AIC values than the full model with all predictors.

While the logic of adding random effects to GLMs is essentially the same as for normal-based linear models, GLMMs are not straightforward to fit. Bolker et al. (2009) and Bolker (2015) described various methods/algorithms for estimating parameters of GLMMs with ecological examples. Penalized quasi-likelihood estimation is commonly used but has the disadvantage of not providing true likelihoods, so AIC-type measures are not applicable; the Laplace approximation is an alternative that provides true likelihoods. MCMC methods, usually in a Bayesian framework, are increasingly used for complex GLMMs. Bolker et al. (2009) and Bolker (2015) also discuss some difficulties with statistical inference, especially deriving the df associated with fixed effects and the reliability of resulting *P*-values. The other issues we have discussed for GLMs, such as overdispersion, zero-inflation, etc., are just as important for GLMMs, and are often handled the same way.

13.9 Generalized Additive Models

Generalized additive models are nonparametric modifications of GLMs where some predictors are included in the model with nonparametric smoothing functions instead of assuming a linear relationship (Hastie & Tibshirani 1990; Wood 2017). For example, Cabanellas-Reboredo et al. (2019) studied the mass mortality of a Mediterranean bivalve (*Pinna nobilis*) caused by a protozoan endoparasite, using information from scientific surveys and citizen science observations. They fitted a model relating the presence/absence of the disease at a site to salinity and temperature. These data are analyzed in Box 13.9.

A standard binomial GLM using the logit link could have been fitted:

$$\log (\text{odds of disease being present})$$
$$= \beta_0 + \beta_1 (salinity) + \beta_2 (temperature).$$

Cabanellas-Reboredo et al. (2019) wanted more flexibility in the relationship between the response and the predictor variables, so they fitted a binomial GAM:

log (odds of disease being present)
$= \beta_0 + f_1(salinity) + f_2(temperature),$

where f_1 and f_2 are nonparametric functions estimated using a smoothing technique.

With a response variable and p predictors, a GAM can be written as:

$$g(\mu) = \beta_0 + \sum_{j=1}^{p} f_j X_j.$$

In this equation, we have summarized the systematic component as a sum of products between smoothing parameters and predictors. Not all the predictors need to be linked to the response via a smoothing function, as a combination of linear and smoothing functions can be used. Different smoothing functions can also be used for different predictors, although usually the same smoother is used.

While the logic of replacing linear functions in a GLM by smoothing functions in a GAM is conceptually straightforward, two key challenges arise (Wood 2017): (1) what smoothing functions to use; and (2) deciding how smooth they should be.

Box 13.9 ⓡ ⓔ Worked Example of Generalized Additive Model: Disease and an Endangered Bivalve

Cabanellas-Reboredo et al. (2019) studied the spread of a disease in a large bivalve (*Pinna nobilis*) in the Mediterranean caused by a protozoan endoparasite. They collated observations of dead or unwell bivalves from many sites using information from scientific surveys and citizen science contributions. They only used observations from sites that their dispersal models indicated the disease could have spread to. They focused on relating the presence of the disease at a site to salinity and temperature.

We could model these data with a logistic regression model with temperature and salinity as predictors; there was no indication of collinearity between these predictors. There were effects of temperature (LR $\chi^2 = 38.316$, 1 df, $P < 0.001$) and salinity (LR $\chi^2 = 8.597$, 1 df, $P = 0.003$), with disease more likely to be present at higher temperatures and salinities. The quantile residuals showed strong heterogeneity and the model was not a great fit (Tjur's D was only 0.230 and the H-L goodness-of-fit statistic had a P-value < 0.001).

A more flexible approach used by Cabanellas-Reboredo et al. was to fit a GAM. Using thin-plate regression splines for both smoothers with default basis numbers (10) and a logit link, the results were as follows.

Term	Estimated model df	Residual df	χ^2	P
Temperature	4.90	5.88	18.13	0.005
Salinity	2.91	3.59	22.57	<0.001

The intercept was 2.67 (SE: 0.50). The explained deviance was 71.1% and the AIC was 82.05, indicating this model was a better fit than the original GLM (AIC = 188.28). The smoothing parameter was estimated using UBRE instead of cross-validation as the variance of a binomial distribution was specified as 1. Residual plots with binary responses are difficult to interpret, but there were no obvious outliers and a q–q plot of deviance residuals looked okay.

There was evidence that the log of the odds of disease being present was related to temperature; the probability of disease increased with temperature up to about 17 °C and then remained reasonably constant (Figure 13.5). There was also an effect of salinity, with the probability of disease peaking at about 38 psu. Note that the confidence bands on the smooths get wide at the extremes, especially for salinity, making predictions less certain.

For comparison, we also fitted a second model with cubic regression splines for both predictors.

Term	Estimated model df	Residual df	χ^2	P
Temperature	5.12	6.16	17.16	0.010
Salinity	3.06	3.75	17.24	<0.001

The conclusions were similar; the explained deviance (71.4%) and the AIC (82.22) were almost identical to the thin-plate model. The smooth plots for each predictor were difficult to distinguish from the thin-plate model.

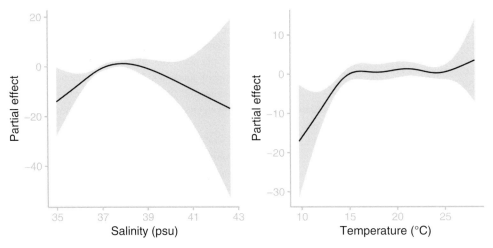

Figure 13.5 GAMs (thin-plate splines) with 95% CIs (shaded areas) fitted to data of Cabanellas-Reboredo et al. (2019), showing partial effects of salinity and temperature on the log of the odds of a bivalve being diseased.

The local (weighted) regression smoothing described in Section 5.1.2 can be used in GAMs. Essentially, locally weighted sums of squares (LOESS) breaks the predictor range into segments (sometimes termed spans or windows) and separate models such as linear or polynomial regressions are fitted to the observations within each segment. Because the segments overlap, the predicted values form a smooth curve. With LOESS smoothing, the proportion of observations within each overlapping segment needs to be specified, and most GAM software that offers LOESS will have a default (e.g. 0.5). LOESS-based GAMs have become less popular in recent years, with splines now being the preferred approach.

Splines are basically piece-wise regressions fitted to adjacent subsets of the data, defined along a predictor variable. Splines comprise basis functions, which allow the smoother to be represented as a linear model or combination of separate linear models. One way to conceptualize basis functions is to think of a single cubic polynomial regression that includes an intercept and three predictors (linear, quadratic, and cubic: x, x^2, and x^3) so there would be four basis functions. The final smoother represents the sum of the basis functions. There are many different splines available, but we will briefly describe three often used by biologists when fitting GAMs: regression splines, penalized regression splines, and thin-plate regression splines.

Regression splines are the simplest. The predictor is partitioned into sections, with the boundary between two sections termed a knot. A linear, quadratic or most commonly cubic polynomial is fitted within each section. The successive polynomials are connected at the knots by imposing conditions on each side of the knot using first-and second-order derivatives (Zuur 2012). The number of basis functions (sometimes termed k) depends primarily on the number of knots (but also on the polynomial chosen), and the smoothness/wiggliness of our smoother depends on the number of knots we use: the more knots, the more wiggly our smoother, although often the differences are quite small (Zuur 2012). Some spline-fitting software is more efficient when the number of basis functions and/or the number of knots is specified. It is, however, tedious to find the optimal number of knots (e.g. comparing models with different knot numbers with AIC), so penalized regression splines are used and are the default regression spline in some GAM software.

Penalized regression splines take a different approach to managing the wiggliness of our smoother. Rather than changing the number of knots, we include a penalty function in our model fitting that penalizes excessive wiggliness with a smoothing parameter (sometimes termed λ), a similar principle to the way we penalized overly complex models in model selection in Chapter 9. We can still specify the number of knots (or basis functions) although the number doesn't matter much if it is not too small; some GAM software provides routines for checking whether the number of basis functions might be inadequate. We also need to set the smoothing parameter, and there are a few different ways for estimating it. If the dispersion of the response distribution is not fixed (e.g. a normal distribution), cross-validation can be used, whereby the smoother is estimated by omitting one observation at a time and seeing how well each model predicts the deleted observation – that is, calculating a residual. The optimum smoothing parameter minimizes the sum of these squared "deleted" residuals; the generalized cross-validation (GCV) score estimates the prediction error where the lower the GCV the better. If the dispersion is specified (e.g. binomial or Poisson), then an unbiased risk estimator (UBRE) can be used. GAM software will usually select the most appropriate estimation method.

Thin-plate splines are particularly useful for models with multiple predictors. They have the advantage of not requiring the location or number of knots to be specified and also penalize excessive "wiggliness" in part based on the smoothing parameter. Wood (2017) described them as almost an ideal smoother. Their downside is their computational cost, as there are as many parameters as there are observations. The computational cost can be partly addressed by using thin-plate regression splines that change the penalizing function, simplifying computation. The underlying mathematics is not for the faint-hearted, and details can be found in Wood (2017). Thin-plate regression splines are the default in some more advanced GAM software.

Originally GAMs were fitted using local scoring based on a backfitting algorithm that iteratively fits a smoothing function, determines the partial residuals, and smooths these residuals (Hastie & Tibshirani 1990). More recent algorithms for fitting GAMs based on splines use (penalized) ML. However, we could apply the penalty function to OLS if we are dealing with normally distributed errors. Like GLMs, GAMs need a link function defined and a probability distribution for the response variable that implies a probability distribution for the model error terms. Smoothing functions don't produce parameter estimates in the way we are used to with GLMs. Still, the smoothing function can be plotted, and predicted values and residuals (partial for multiple covariates) determined and also plotted.

Inference for GAMs is not as straightforward as for GLMs. Degrees of freedom for smooth functions are tricky; software will often provide estimated (effective) and residual (reference) df for each smoother. Goodness-of-fit statistics such as r^2 and percentage explained deviance can be derived. Model comparisons can use AIC or, for two nested models (one with and one without the parameter of interest), LR tests just like with GLMs. P-values for smoothing function evaluating the null hypothesis that the smoothing function is zero (i.e. there is no relationship, based on the smoother type chosen) can be derived from approximate F or χ^2 statistics and are based on Bayesian CIs – see Wood (2017) for details.

There are two components to checking assumptions and model suitability with GAMs. First, we need to do the usual checks of residuals, looking out for unusual patterns

and outliers. Modern GAM software will provide options for generating relevant plots, but remember the challenges of interpreting residuals and goodness-of-fit with binary responses (Box 13.9). Second, we should check that the number of basis dimensions for each smoother is adequate to model the complexity in the data. Many biologists just use the default value in their GAM software (e.g. $k = 10$); this usually just sets the upper limit to the basis dimensions. Wood (2017) recommends checking whether the estimated df are close to the basis dimension used and, if they are, increasing the basis dimension.

This brief introduction doesn't do justice to the flexibility and usefulness of GAMs for modeling nonlinear relationships in biology. There are also extensions to mixed models with random effects (GAMMs).

13.10 Key Points

• Generalized linear models allow fitting of linear models for biological response variables that follow any of a wide range of exponential distributions.
• When the response is normal or is transformed to be normal, GLMs, and often OLS, can be used.
• Biologists mostly use models based on binomial and Poisson distributions, plus beta and negative binomials for overdispersed data.
• Generalized linear models are fitted using ML, and their fit assessed using changes in log-likelihood (deviance), χ^2 statistics, and information criteria, but there are several other measures.
• Residuals remain important for checking assumptions, but they are harder to interpret for binary response variables.
• Binary response variables are analyzed using logistic regression, a GLM that uses a logit link function.
• Response variables that are counts generally follow a Poisson distribution, and the GLM uses a log-link function.
• Overdispersion is common in biological data, where the variance exceeds the mean. Using a Poisson distribution for overdispersed data affects parameter estimates and hypothesis tests.
• Overdispersion requires careful thought about the nature and causes, and there are several approaches that may be relevant.
• A special case of GLMs is the situation where data are cross-classified, with no clear response variable. These contingency tables are analyzed by modeling the counts in each cell, using a log-link function – loglinear models. Interactions between the classifying factors are usually of most interest.

- Generalized linear models can include fixed and random predictors as outlined in Chapters 10–12, and when both types are present we have GLMMs. Understanding GLMMs provides entry into many advanced linear models.
- Finally, we introduce GAMs, regression-type models based on smoothing functions rather than linear models.
- Approaches to model fitting and estimation are more diverse for GLMs, GLMMs, GAMs, and contingency tables than for linear models fitted by OLS, so analysis requires careful thought.

Further Reading

GLMs: Dunn and Smyth (2018) is recommended for a wide range of GLMs, and is a good source of information on less commonly used distributions.

Logistic regression: Hosmer et al. (2013) and Agresti (2019) give thorough coverage, including response variables with several, rather than only two, categories.

Overdispersed data: Zuur et al. (2012) and Zuur and Ieno (2016, 2021) describe analyses of overdispersed, zero-inflated, and zero-truncated data, as do Dunn and Smyth (2018) and Bolker (2015).

Contingency tables: Agresti (2019) offers detailed coverage, including a good account of the move to loglinear models.

GLMMs: Agresti (2015) and Gelman and Hill (2006) provide readable introductions to GLMMs, and Bolker et al. (2009), Bolker (2015), and Zuur et al. (2009, 2013) provide detailed coverage with numerous ecological examples and software code.

GAMS: The two key references for GAMs are Hastie and Tibshirani (1990) and Wood (2017), the latter focusing on more recent approaches using penalized splines and thin plates. We also recommend Zuur (2012) for a less technical description with numerous worked examples from ecological research.

14 Introduction to Multivariate Analyses

A multivariate dataset includes more than one variable recorded from multiple sampling or experimental units, sometimes called objects. If these objects are organisms, the variables might be morphological or physiological measurements; if the objects are ecological sampling units, the variables might be physicochemical measurements or species abundances. We have already considered multivariate data by fitting linear models to datasets with multiple predictors. The multivariate analyses we will discuss in the remaining chapters deal with multiple response variables or multiple variables that could be response variables, predictor variables, or a combination of both.

For each dataset, p variables are recorded for each of n objects. We will introduce multivariate analyses in this chapter using a dataset from a study of wildlife use of underpasses. Clevenger and Waltho (2000) surveyed 11 underpasses ($n = 11$ objects) in Banff National Park in Alberta, Canada, and recorded two sets of variables. The first was the counts of seven wildlife species (black bears, cougars, wolves, etc.; $p = 7$) at each underpass over four years. The second set was each underpass' physical, landscape, and human activity characteristics ($p = 14$).

Multivariate analyses can address a range of questions. We may wish to model a suite of response variables against one or more predictors. While we can model each response separately, interpretation can be difficult when the responses are correlated with each other, and our biological question might be more about the combined responses (e.g. as a measure of the structure of a multispecies community in ecology) than the individual responses. For example, Clevenger and Waltho (2000) were interested in how the suite of wildlife using an underpass related to its physical, landscape, and human activity characteristics. We might want to see how the objects cluster together based on the suite of variables (e.g. how do the underpasses group together in terms of wildlife use), or we might use the suite of variables to classify objects into groups (e.g. use morphological and/or genetic variables to classify individuals into species groups). These different questions are usually associated with graphical procedures that illustrate the relationships among objects (two- or three-dimensional plots of objects are sometimes called ordination plots by ecologists) and which variables are mostly driving those relationships.

We will focus on three broad approaches to multivariate analysis, two of which are shown in Figure 14.1:
- We can derive new variables (termed components, factors, or functions, depending on the analysis). Each is a linear combination of the original variables that consolidates the original variance in as few derived variables as possible. These components can be used as response or predictor variables in further modeling and plotted to show relationships among objects. Ecologists sometimes call these R-mode analyses because they are based on associations (covariances or correlations) among the variables.
- An alternative is to calculate a measure of similarity or dissimilarity, sometimes termed a resemblance or distance measure, between pairs of objects based on the values for all the variables. These dissimilarities can be a response variable in further modeling and represented graphically to show the relationships between objects. These are sometimes termed Q-mode analyses.
- Finally, we can analyze each variable separately with univariate linear models and then combine these univariate analyses to draw conclusions about the suite of variables.

This chapter will introduce the basis of the first two approaches to multivariate data analysis, with their application to specific statistical methods described in this and later chapters.

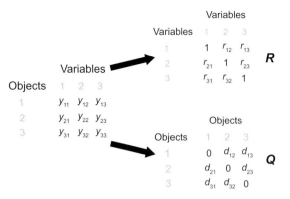

Figure 14.1 Distinction in the initial steps between R- and Q-mode analyses. A data matrix of n rows by p columns is converted to a p by p matrix of associations (e.g. correlations) between variables or an n by n matrix of dissimilarities between objects.

14.1 Distributions and Associations

In a univariate context, we have described the exponential distributions we can use in linear modeling, such as normal, binomial, Poisson, etc. Statistical inference based on these models assumes that our model's response variable and/or error terms follow the chosen distribution, particularly the relationship between the mean and variance. We can also describe distributions for multivariate data. For example, the product of the joint probability densities of normally distributed variables results in the multivariate normal distribution. Some of the inferential methods for multivariate analyses in this and the following two chapters assume multivariate normality, where all variables and linear combinations of variables are normally distributed (Tabachnick & Fidell 2019). As with univariate normality, this assumption is usually not critical. Other multivariate distributions are possible (e.g. multivariate Poisson), although less commonly used in multivariate analyses.

There are also multivariate analogs to univariate distributions' location (mean) and spread (variance). One measure of the center of a multivariate distribution is the centroid. In multivariate space, where each dimension is a variable, the sample centroid is the point represented by the univariate sample means of each variable (Figure 14.2). The population centroid is not usually presented as a single value, but is used to describe the center of a multivariate normal distribution and detect multivariate outliers (Section 14.6).

When we have multiple variables recorded from the same sampling or experimental units, we must consider the variance of each variable and the covariances between variables. We introduced variance–covariance matrices in

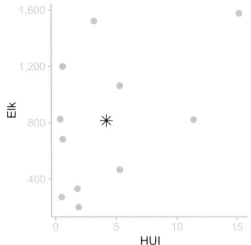

Figure 14.2 Scatterplot of elk numbers against the human activity index (HUI) for the 11 underpasses from Clevenger and Waltho (2000). The centroid, the point represented by the mean of HUI and elk numbers, is shown with a star.

Chapter 10 in the context of mixed effects models and in Chapter 12, where we recorded units repeatedly through time. In multivariate data, we convert the raw data matrix of p variables by n objects to a $p \times p$ variance–covariance matrix ($\mathbf{C}$), where the main diagonal contains the variances for each variable, and the other entries are the covariances between pairs of variables (Box 14.1). We derive $\mathbf{C}$ either via a SS and sums-of-cross-products matrix or directly from the raw data matrix $\mathbf{Y}$ if each variable is centered (to a mean of zero), by $\mathbf{Y}'\mathbf{Y} / (n-1)$, where $\mathbf{Y}'$ is the transpose of the centered raw data matrix.

We can summarize the variability of a multivariate dataset based on the variance–covariance matrix in two ways (Jackson 2003):

- The determinant of a square matrix is a single number summary of the matrix. The determinant of the variance–covariance matrix ($|\mathbf{C}|$) represents the generalized variance of the matrix.

- The trace of the variance–covariance matrix ($\mathrm{Tr}(\mathbf{C})$) is the sum of the diagonal values – that is, the sum of the variances of the centered individual variables.

We can also standardize these covariances by dividing by the SDs of the two variables involved to produce correlations and thus a correlation matrix ($\mathbf{R}$), where r_{12} is the correlation coefficient between variables 1 and 2, etc. (Box 14.1). Note the main diagonal consists of 1s because the correlation between each variable and itself is 1. Covariances and correlations are measures of association between variables. Other measures of association include the χ^2 statistic, discussed in Chapter 13 as a measure of association for contingency tables.

14.2 Linear Combinations, Eigenvectors, and Eigenvalues

14.2.1 Linear Combinations of Variables

One of the fundamental techniques in multivariate analyses is to derive linear combinations of the variables that summarize the variation in the original dataset (Box 14.1). Basically, we are "consolidating" (*sensu* Tabachnick & Fidell 2019) the variance from a data matrix into a new set of derived variables, each of which is a linear combination of the original variables. For n objects and p original variables:

$$z_{ik} = c_1 y_{i1} + c_2 y_{i2} + \ldots + c_j y_{ij} + \ldots + c_p y_{ip}.$$

Here, z_{ik} is the value of the new variable k for object i, y_{i1} to y_{ip} are the values of the original variables for object i, and c_1 to c_p are weights or coefficients that indicate the contribution of each original variable to the linear combination. Depending on the analysis, these new variables are termed discriminant functions, canonical functions or variates, principal components, or factors. This linear combination is analogous to a regression equation. For some analyses, the linear combination may include an intercept:

$$z_{ik} = \text{intercept} + c_1 y_{i1} + c_2 y_{i2} + \ldots + c_j y_{ij} + \ldots + c_p y_{ip}.$$

This form is common when the variables are not standardized to zero mean and unit variance; if they are, the intercept becomes zero and the previous equation is appropriate.

The derived variables are extracted so that the first explains most of the variance in the original variables. The second explains most of the remaining variance after the first has been extracted. The third explains most of the remaining variance after the first and second have been extracted, etc. The new derived variables are independent of (i.e. uncorrelated with) each other. The number of new derived variables is the same as the number of original variables (p), although the variance is usually consolidated in the first few derived variables.

14.2.2 Eigenvalues

Eigenvalues, also termed characteristic or latent roots (λ_1, λ_2, ..., λ_p), represent the amount of the original variance explained by each of the p derived variables. These eigenvalues are population parameters, and we estimate them using ML to produce (l_1, l_2, ..., l_p) and can also

determine their approximate SEs. Note from Box 14.1 that if we use a covariance matrix with centered variables, the sum of the eigenvalues is equal to the trace of the original covariance matrix. If we use a correlation matrix with centered and standardized variables, the sum of the eigenvalues would equal the trace of the correlation matrix – that is, the sum of the variances of the original standardized variables. We have simply rearranged the variance in the association matrix so the first few derived variables explain most of the variation present (between objects) in the original variables. The eigenvalues can also be expressed as proportions or percentages of the original variance explained by each new derived variable (component).

14.2.3 Eigenvectors

Eigenvectors (characteristic vectors) are lists of the coefficients or weights showing how much each original variable contributes to each new derived variable. In general terms, the eigenvectors contain the c_j weights, but these coefficients can be scaled in different ways, so they are often represented as u_j, v_j, or w_j in matrix descriptions of multivariate analyses – see Box 14.1. The eigenvectors are commonly scaled so the sum of squared coefficients equals 1; other forms of scaling are possible. We estimate the coefficients with ML and can determine approximate SEs. These linear combinations can be solved to provide a score (z_{ik}) for each object for each new derived variable. Note there is the same number of derived variables as there are original variables (p). The new derived variables, each with an eigenvector of coefficients and an eigenvalue, are extracted sequentially, so they are uncorrelated.

14.2.4 Derivation of Components

We can derive the new variables (components) with matrix algebra in two ways. We can use an eigenanalysis (spectral decomposition) of a $p \times p$ square matrix of associations among variables (e.g. **C** or **R** matrices) or an SVD of the $n \times p$ original data matrix. The two approaches produce equivalent results if there is a match between the association matrix used and the standardization of variables in the data matrix. One of the biggest problems facing biologists trying to become familiar with multivariate statistical techniques is the bewildering terminology. Different textbooks use different terms for the same property and different labels for the relevant matrices. We have tried to summarize these two approaches for extracting components from a multivariate dataset in Box 14.1, following the terminology of Jackson (2003) where possible.

Box 14.1 ®🄴 Worked Example of Deriving Components (Eigenvectors): Wildlife Use of Underpasses

There are two strategies for extracting eigenvectors (components) and their eigenvalues from a multivariate dataset of n objects by p variables. First, we can use an eigenanalysis (spectral decomposition) of a $p \times p$ association matrix between variables. Second, we can use a singular value decomposition (SVD) of the $n \times p$ data matrix, with variables standardized as necessary. The SVD is more generally applicable, although most biologists are more familiar with obtaining eigenvectors and eigenvalues from a covariance or correlation matrix.

Consider the matrix ($\mathbf{Y}$) of raw data from Clevenger and Waltho (2000), who recorded the numbers of seven taxa of wildlife for 11 underpasses in Alberta, Canada. Normal text is raw counts, italicized is centered data, and bold is standardized data.

Underpass	Black bear	Grizzly bear	Cougar	Wolf	Deer	Elk	Moose
A	10	0	5	1	554	825	1
	−7.55	−0.64	−5.64	−27.27	334.82	10.55	0.82
	−0.53	**−0.41**	**−0.51**	**−0.60**	**2.05**	**0.02**	**2.023**
B	20	0	29	7	42	201	0
	2.46	−0.64	18.36	−21.27	−177.18	−613.46	−0.18
	0.17	**−0.41**	**1.65**	**−0.47**	**−1.09**	**−1.27**	**−0.45**
C	43	0	3	3	294	331	1
	25.46	−0.64	−7.64	−25.27	74.82	−483.46	0.82
	1.78	**−0.41**	**−0.69**	**−0.56**	**0.46**	**−1.00**	**2.02**
D	37	2	30	28	253	1199	0
	19.46	1.36	19.36	−0.27	33.82	384.55	−0.18
	1.36	**0.87**	**1.74**	**−0.01**	**0.21**	**0.80**	**−0.45**
E	13	0	7	3	215	1062	0
	−4.56	−0.64	−3.64	−25.27	−4.18	247.55	−0.18
	−0.32	**−0.41**	**−0.33**	**−0.56**	**−0.03**	**0.51**	**−0.45**
F	8	0	0	5	21	467	0
	−9.55	−0.64	−10.64	−23.27	−198.18	−347.46	−0.18
	−0.67	**−0.41**	**−0.96**	**−0.51**	**−1.22**	**−0.72**	**−0.45**
G	0	0	4	1	61	1576	0
	−17.5	−0.64	−6.64	−27.27	−158.18	761.55	−0.18
	−1.22	**−0.41**	**−0.60**	**−0.60**	**−0.97**	**1.58**	**−0.45**
H	4	0	4	5	338	1522	0
	−13.55	−0.64	−6.64	−23.27	118.82	707.55	−0.18
	−0.95	**−0.41**	**−0.60**	**−0.51**	**0.73**	**1.47**	**−0.45**
I	8	0	20	77	288	821	0
	−9.55	−0.64	9.36	48.73	68.82	6.55	−0.18
	−0.67	**−0.41**	**0.84**	**1.07**	**0.42**	**0.01**	**−0.45**
J	34	5	15	146	291	683	0
	16.55	4.36	4.36	117.73	71.82	−131.46	−0.18
	1.15	**2.79**	**0.39**	**2.60**	**0.44**	**−0.27**	**−0.45**
K	16	0	0	35	54	272	0
	−1.55	−0.64	−10.64	6.73	−165.18	−542.46	−0.18
	−0.11	**−0.41**	**−0.96**	**0.15**	**−1.01**	**−1.12**	**−0.45**

Eigenanalysis

We will illustrate the eigenanalysis of a matrix of associations between variables ($\mathbf{Y'Y}$). This might be a matrix of variances and covariances, $\mathbf{C}$, among p variables based on n objects. Using the example from Clevenger and Waltho (2000), we have the following.

Box 14.1 (cont.)

	Black bear	Grizzly bear	Cougar	Wolf	Deer	Elk	Moose
Black bear	205.67	12.12	59.32	209.74	345.79	−2,855.17	1.79
Grizzly bear	12.12	2.46	6.06	58.81	42.67	11.18	−0.13
Cougar	59.32	6.06	123.66	152.41	4.17	−188.32	−1.33
Wolf	209.74	58.81	152.41	2,058.02	970.15	−2,933.64	−5.25
Deer	345.79	42.67	4.17	970.15	26,584.96	20,110.11	40.96
Elk	−2,855.17	11.18	−188.32	−2,933.64	20,110.11	23,3219.67	−47.29
Moose	1.79	−0.13	−1.33	−5.25	40.96	−47.29	0.16

We could also use a correlation matrix (**R**).

	Black bear	Grizzly bear	Cougar	Wolf	Deer	Elk	Moose
Black bear	1.00	0.54	0.37	0.32	0.15	−0.41	0.31
Grizzly bear	0.54	1.00	0.35	0.83	0.17	0.02	−0.20
Cougar	0.37	0.35	1.00	0.30	0.00	−0.04	−0.30
Wolf	0.32	0.83	0.30	1.00	0.13	−0.13	−0.29
Deer	0.15	0.17	0.00	0.13	1.00	0.26	0.62
Elk	−0.41	0.02	−0.04	−0.13	0.26	1.00	−0.24
Moose	0.31	−0.20	−0.30	−0.29	0.62	−0.24	1.00

Basically, we then derive two matrices, **L** and **U**, so that:

L = **U**′**RU**, although we could just as easily derive **L** and **U** from **C**.

U is an $n \times p$ matrix whose columns contain the eigenvectors (characteristic vectors), the coefficients of the linear combinations of the original variables. The elements of each eigenvector k are u_{jk}, the coefficient for the jth variable in the kth eigenvector. Note that we clearly need to have constraints imposed on the coefficients within each eigenvector, otherwise simply increasing the absolute sizes of the coefficients could increase the variance explained by each new variable. The simplest and most used constraint restricts the sum of squared coefficients to 1. Eigenvectors that are independent and scaled to unity are termed orthonormal. Additional scaling options for the eigenvectors are available to make the variances of the eigenvectors similar (Jackson 2003) – for example, $v_{jk} = \sqrt{l_k}u_{jk}$, so the eigenvectors are in a **V** matrix, and $w_{jk} = u_{jk}/\sqrt{l_k}$, so the eigenvectors are in a **W** matrix.

Our **U** matrix from the wildlife data, based on the correlation matrix, is as follows.

	Eigen1	Eigen2	Eigen3	Eigen4	Eigen5	Eigen6	Eigen7
Black bear	−0.448	−0.329	0.287	−0.195	0.586	−0.476	0.013
Grizzly bear	−0.573	0.043	−0.201	0.256	0.278	0.618	−0.324
Cougar	−0.380	0.167	0.021	−0.817	−0.324	0.194	0.128
Wolf	−0.536	0.114	−0.145	0.430	−0.377	−0.290	0.517
Deer	−0.090	−0.539	−0.537	−0.082	−0.346	−0.273	−0.462
Elk	0.144	0.193	−0.752	−0.190	0.460	−0.068	0.352
Moose	0.102	−0.723	0.039	−0.007	−0.005	0.437	0.524

Each column is an eigenvector (u_k where $k = 1$ to p), the values representing the coefficients or weights for that linear combination of the original variables. For example, the linear combination comprising eigenvector 1 is:

Box 14.1 (cont.)

(-0.448) Black bear $+ (-0.573)$ Grizzly bear $+ (-0.380)$ Cougar+
$\qquad (-0.536)$ Wolf $+ (-0.090)$ Deer $+ (0.144)$ Elk+
$\qquad (0.102)$ Moose,

where the values of each variable are standardized because we used the correlation matrix to extract the eigenvectors; if we had used the covariance matrix, the values would just be centered. These linear equations are often termed components or factors (Chapter 15) and represent new variables derived from the original variables. Note that each variable contributes differently to each component (different coefficients or weights) and that these coefficients will depend on the units of each variable and whether standardizations are used. These linear equations can be solved to produce a component score (z_{ik}) for each object or observation for each component. For example, the score for component 1 for underpass 1 based on standardized variables (remember, we used the correlation matrix in this example) is:

$(-0.448)(-0.526) + (-0.573)(-0.406) + (-0.380)(-0.507)+$
$\qquad (-0.536)(-0.601) + (-0.090)(2.053) + (0.144)(0.022)+$
$\qquad (0.102)(2.023) = 1.006.$

L is a $p \times p$ matrix whose diagonal contains the eigenvalues $l_1, l_2, \ldots, l_p$ (estimates of $\lambda_1, \lambda_2, \ldots, \lambda_p$, the latent or characteristic roots) of **R** (or **C**). The eigenvalues measure the variance explained by each of the eigenvectors. The number of eigenvalues is the same as the number of rows and columns in **C** and **R** and, therefore, the same as the number of original variables (p). The eigenvalues (diagonal of **L**) are as follows.

Eigenvector	Eigenvalue	Percentage variance
1	2.483	0.355
2	1.792	0.256
3	1.331	0.190
4	0.811	0.116
5	0.429	0.061
6	0.088	0.013
7	0.067	0.010

The trace of the **L** matrix, the sum of its diagonal elements (eigenvalues), is the sum of the variances of the original standardized variables. The sum of the eigenvalues from an eigenanalysis of a covariance matrix would equal the sum of the variances of the centered variables. The matrix **L** represents, therefore, a reorganization of the variances of the variables from the original data matrix. Each eigenvalue is associated with each eigenvector, and it is clear that the eigenvectors are extracted in order of decreasing proportions of the total variance.

More formally determination of the eigenvalues involves solving the characteristic equation:

$$|R - l\mathbf{I}| = 0,$$

where **I** is an identity matrix of equivalent dimensions to **R**. The resulting polynomial (pth degree) in l is used to obtain $l_1, l_2, \ldots, l_p$.

Singular Value Decomposition

The SVD of an $n \times p$ data matrix is based on the product of the characteristic vectors of a matrix of associations between variables, the characteristic vectors of a matrix of associations between objects, and their characteristic roots (eigenvalues, which are the same for both association matrices). Suppose **Y** is a matrix of standardized data (as used for the correlation matrix above). In that case, **Y'Y** is the correlation matrix between variables (matrix **R** above), and

Box 14.1 (cont.)

$\mathbf{YY}'$ is the correlation matrix between objects (note these would be covariance matrices for centered data). The characteristic roots (eigenvalues) of these two matrices are the same.

The SVD of $\mathbf{Y}$ is:

$$\mathbf{Y} = \mathbf{ZL}^{1/2}\mathbf{U}',$$

where $\mathbf{L}$ contains the eigenvalues, $\mathbf{U}$ is a $p \times p$ matrix containing the eigenvectors of $\mathbf{Y}'\mathbf{Y}$ as defined above, and $\mathbf{Z}$ is an $n \times p$ matrix of eigenvectors of $\mathbf{YY}'$ and are also the component scores for objects scaled by the square root of the eigenvalues. Note that we now have the square root of the eigenvalues because we are dealing with the original variables rather than covariances or correlations (Jackson 2003). If $\mathbf{Y}$ contains centered data, then $\mathbf{L}$ and $\mathbf{U}$ will be the equivalent to that from the eigenanalysis of the covariance matrix. If $\mathbf{Y}$ contains standardized data, then $\mathbf{L}$ and $\mathbf{U}$ will be the equivalent to that from the eigenanalysis of the correlation matrix. Note that we can determine the original variables (centered and standardized if appropriate) from the matrix of component scores and vice-versa when all components are extracted.

The advantage of using SVD is that extraction of eigenvectors and their eigenvalues is a one-step process and SVD can also be applied to association matrices that are not square, such as chi-square matrices from contingency tables as used in correspondence analysis (Chapter 15). The advantage of eigenanalysis is that the choice of matrix (e.g. covariance vs. correlation) will automatically center or standardize the data.

The usual derivation of components is from an association matrix of covariances or correlations between variables (Box 14.1) – that is, an *R*-mode analysis. We can calculate scores for the derived variables (components) for each. We could also derive components from matrices representing covariances or correlations between objects, and the derived variables (components) are linear combinations of the objects. We can then calculate component scores for each variable, and this is termed a *Q*-mode analysis, although *Q*-mode analyses are more commonly based on dissimilarities between objects. These two sets of component scores are related via matrix algebra, and we can obtain component scores for objects from the eigenvectors of the variables and vice-versa (Jackson 2003).

14.3 Multivariate Distance and Dissimilarity Measures

The methods described in the previous section are *R*-mode analyses. We can also conduct *Q*-mode analyses using a measure of similarity or dissimilarity between objects. These (dis)similarity measures are sometimes termed resemblance measures or (dis)similarity indices or coefficients.

Similarity indices measure how alike objects are, such as how similar sampling or experimental units are in terms of species composition or how alike specimens are in morphology or genetic structure. Dissimilarity indices measure how different objects are and should represent multivariate distance. If each variable is represented by an axis (or dimension), the multivariate distance is how far apart the objects are in multidimensional space. Many indices can be presented as either similarity or dissimilarity (1 − similarity), although we will focus on the dissimilarity versions. Dissimilarity indices are also called distances and are calculated for every possible pair of objects. We usually represent the dissimilarities between objects as a matrix, converting an $n \times p$ data matrix to an $n \times n$ dissimilarity matrix. Like the covariance and correlation matrices described in Section 14.2, dissimilarity matrices are identical above and below the diagonal, which will be zeroes, indicating zero dissimilarity between an object and itself.

Certain properties can categorize different dissimilarity indices, and a common distinction is between metric and semimetric (or nonmetric) versions (Legendre & Legendre 2012). Both versions have three important properties in common: (1) a lower bound of zero when objects have the same variables with the same values; (2) they can only take positive values when objects differ in at least one variable; and (3) symmetry, whereby the dissimilarity between objects 1 and 2 is the same as between 2 and 1. Metric measures also satisfy the triangle inequality: the sum of dissimilarities between objects 1 and 2 and between 2 and 3 cannot be less than between 1 and 3, so they can be directly represented in Euclidean space, unlike semimetric measures. While semimetric measures are commonly used for abundance data, we have to use indirect methods (e.g. multidimensional scaling; Chapter 16) to represent the relationships between objects in multidimensional space.

There is a broad range of indices of dissimilarity between objects, with many used in ecology and environmental science. Dissimilarities can also be calculated for other types of data and data used for the range of "O-mics" – genomics, metabolomics, proteomics, etc.

14.3.1 Dissimilarity Indices for Continuous and Count Variables

We define some of the commonly used indices in Box 14.2 and describe them briefly below, mainly in the context of ecological data; we evaluate them in Section 14.3.4.

14.3.1.1 Metric Measures

Some software will, by default, "normalize" some measures, particularly Euclidean and Manhattan indices, by dividing by the number of variables that contribute to the distance measure. This is only relevant if you wish to compare dissimilarities between datasets with different numbers of variables.

Euclidean

This is based on simple geometry as a measure of the distance between two objects in multidimensional space. It uses the (square root of the) sum of squared differences for each variable. It is bounded by zero for two objects with the same values for all variables and has no upper limit, even when two objects have no variables in common with positive values. A modification of Euclidean is Hellinger's distance, based on Hellinger-transformed values (see Section 14.5).

Manhattan (or City Block)

This is the sum of the absolute differences in the value of each variable between two objects. It has properties similar to Euclidean distance and will be dominated by variables with large values.

Minkowski

Euclidean and city block are both versions of the more general Minkowski index.

Canberra

This is the Manhattan index, except that the difference for each variable is divided by the sum of the variable values before summing across variables. To ensure it has an upper limit of 1, we standardize it by the number of

variables that are greater than 0 in both objects. The Canberra index is less influenced than the Manhattan by variables with very large values.

Chi-square

This dissimilarity index, implicit in correspondence analysis (Greenacre 2010), is applicable when the variables are counts, such as species abundances. It is based on differences between objects in the proportional representation of each species, also adjusted for species totals.

14.3.1.2 Semimetric
Bray–Curtis

This index has a long history and has been "rediscovered" numerous times (Legendre & Legendre 2012). It is a modification of the Manhattan in which the sum of differences between objects across variables is standardized by the sum of the variable values across objects, also summed across variables. Equivalently, it can be calculated as 1 minus twice the sum of the lesser value of each variable when it is greater than 0 in both objects, standardized by the sum of the values of all variables in both objects. It ranges between 0 (same variables and values in both objects – completely similar) and 1 (no variables in common with positive values – completely dissimilar). It is sometimes called percent dissimilarity or percent difference. It is well suited to abundance data because it ignores variables with zeroes for both objects.

It is sometimes mistakenly called the equivalent of a different dissimilarity index, Czekanowski's coefficient. There has been debate as to whether Bray–Curtis implies the use of a double (by objects and by variables) standardization (Somerfield 2008; Yoshioka 2008a, b), although most ecologists separate the use of this measure from any decisions about standardization (see Section 14.5).

Kulczynski

This index, also termed the quantitative symmetric measure, was introduced to biologists by Faith et al. (1987). Like Bray–Curtis, it ranges between 0 and 1 and has similar properties.

14.3.2 Dissimilarity Indices for Dichotomous (Binary) Variables

Another group of dissimilarity indices has been developed for variables measured on a binary scale (e.g. presence and absence). Two simple measures of dissimilarity are

Box 14.2 Worked Example of Dissimilarity Measures for Continuous and Binary Variables: Wildlife Use of Underpasses

Consider two objects (1 and 2), and p variables recorded from each object, such as abundances of p species from each sampling unit. The same variables are recorded from each object (even if some variables have zero values for an object).

For continuous variables, y_{1j} and y_{2j} are the values of variable j in objects 1 and 2, min $\left(y_{1j}, y_{2j}\right)$ is the lesser value of each variable when it is greater than zero in *both* objects, and q is the number of variables that are zero for objects 1 and 2. For example, y_{1j} and y_{2j} might be the abundances of species j in sampling units 1 and 2, Σ min $\left(y_{1j}, y_{2j}\right)$ is the sum of the lesser abundance of species j when it is present in both sampling units, p is the number of species and q is the number of species that are missing (zero values) from both samples.

For binary variables (e.g. presence/absence), let a be the number of variables with nonzero values in both objects, b is the number of variables with nonzero values in object 1 but not 2, and c is the number of variables with nonzero values only in object 2.

The formulae presented below are from Faith et al. (1987) and Legendre and Legendre (2012).

Dissimilarity	Formula
Continuous variables	
Minkowski	$\left(\sum_{j=1}^{p} \left|y_{1j} - y_{2j}\right|^{\lambda}\right)^{1/\lambda}$
Euclidean ($\lambda = 2$)	$\sqrt{\sum_{j=1}^{p} \left(y_{1j} - y_{2j}\right)^{2}}$
Manhattan (city block; $\lambda = 1$)	$\sum_{j=1}^{p}\left|y_{1j} - y_{2j}\right|$
Canberra	$\dfrac{1}{p - q} \sum_{j=1}^{p} \dfrac{\left|y_{1j} - y_{2j}\right|}{y_{1j} + y_{2j}}$
Bray–Curtis	$1 - \dfrac{2\sum_{j=1}^{p} \min\left(y_{1j}, y_{2j}\right)}{\sum_{j=1}^{p}\left(y_{1j} + y_{2j}\right)} = \dfrac{\sum_{j=1}^{p}\left|y_{1j} - y_{2j}\right|}{\sum_{j=1}^{p}\left(y_{1j} + y_{2j}\right)}$
Kulczynski	$1 - \dfrac{\left(\dfrac{\sum_{j=1}^{p} \min\left(y_{1j}, y_{2j}\right)}{\sum_{j=1}^{p} y_{1j}} + \dfrac{\sum_{j=1}^{p} \min\left(y_{1j}, y_{2j}\right)}{\sum_{j=1}^{p} y_{2j}}\right)}{2}$
Chi-square	$\sqrt{\sum_{i=1}^{n} \sum_{j=1}^{p} y_{ij} \sum_{j=1}^{p} \dfrac{\left(\dfrac{y_{1j}}{\sum_{j=1}^{p} y_{1j}} - \dfrac{y_{2j}}{\sum_{j=1}^{p} y_{2j}}\right)^{2}}{\sum_{j=1}^{p} y_{ij}}}$
Binary variables	
Jaccard	$1 - \dfrac{a}{a + b + c}$
Sørensen	$1 - \dfrac{2a}{(2a + b + c)}$

To illustrate some of these dissimilarity measures, we have calculated the dissimilarity between the first three underpasses based on abundances of seven taxa of wildlife from Clevenger and Waltho (2000). We have used the

Box 14.2 (cont.)

original variables and variables standardized to proportion of object totals and standardized to proportion of variable totals (weights taxa equally).

Dissimilarity between underpasses	Euclidean	Chi-square	Bray–Curtis
A vs. B			
Raw data	807.61	1.22	0.69
Standardized by object totals	0.29	1.05	0.26
Standardized by variable totals	0.59	2.19	0.79
A vs. C			
Raw data	559.23	0.47	0.38
Standardized by object totals	0.12	0.40	0.10
Standardized by variable totals	0.21	0.61	0.19
B vs. C			
Raw data	285.70	1.18	0.45
Standardized by object totals	0.36	1.04	0.30
Standardized by variable totals	0.57	2.11	0.73

Euclidean and chi-square do not have upper limits, so the dissimilarity can exceed 1 when applied to unstandardized data. Note that the rank order of the three pairwise dissimilarities is the same under different standardizations for Euclidean and chi-square, but not for Bray–Curtis.

We converted the abundances to presence–absence to illustrate the use of Jaccard and Sørensen indices, the latter equivalent to Bray–Curtis on binary data.

Dissimilarity between underpasses	Jaccard	Sørensen
A vs. B	0.167	0.091
A vs. C	0.000	0.000
B vs. C	0.167	0.091

Note that underpasses A and C are identical in taxon presence/absence and the dissimilarity between underpasses A and B is the same as between B and C.

Jaccard's coefficient and Sørensen's coefficient, the latter weighting double presences more and being identical to the Bray–Curtis measure for dichotomous variables.

14.3.3 General Dissimilarity Indices for Mixed Variables

Gower's coefficient is a measure, usually presented as a similarity, useful for situations that include a mixture of continuous and categorical variables. Gower's coefficient handles a mixture of variable types by calculating similarity for each variable separately (using appropriate coefficients for binary and continuous variables), then averaging those similarities. Legendre and Legendre (2012) provide details.

14.3.4 Choosing Dissimilarity Indices

Some multivariate analyses imply a particular dissimilarity index (e.g. chi-square for correspondence analysis). Other analyses give the researcher a choice of dissimilarity, and that decision is determined by the data and how the dissimilarities are used. When variables are measured on similar scales and have no zero values – for example, environmental or morphological measurements – Euclidean, Manhattan, or Canberra are good measures of dissimilarity between objects. If the scales of measurement vary among different variables, the data should be standardized (to zero mean and unit variance) before calculating these dissimilarities.

Much of the debate about suitable dissimilarity indices has been in ecological settings where the variables are abundances (i.e. counts) of numerous taxa, and the research question is about differences in ecological community composition and structure. Legendre and Legendre (2012) and Clarke et al. (2006) provide detailed coverage of criteria for choosing among the many dissimilarities. We will emphasize a subset of these, focusing on the distinction between metric measures such as Euclidean or chi-square dissimilarity and semimetric measures such as Bray–Curtis or Kulczynski:

• Metric measures have the advantage of being directly represented on plots showing the relationship between objects (termed ordination plots; see Section 14.8), whereas semi- and nonmetric measures require indirect plotting methods.

• Many ecologists consider that two sampling or experimental units with no taxa in common are maximally dissimilar from an ecological perspective, so a suitable dissimilarity index should reach a maximum value in this situation. Bray–Curtis and Kulczynski meet this criterion, but Euclidean and chi-square do not.

• Another criterion related to zero counts is double-zero asymmetry, where the absence of a taxon in both units does not contribute to their (dis)similarity. In contrast, the identical nonzero abundances of a taxon mean that the two units are more similar. Most ecologists would consider that the absence of a taxon in two units tells us little about the ecological difference between those units (Legendre & De Caceres 2013). Chi-square, Bray–Curtis, and Kulczynski are asymmetrical, whereas Euclidean is not.

• Chi-square has also been criticized for giving rare taxa too much weight in the dissimilarity calculation. However, Greenacre (2013b) argued that rare taxa appear as outliers when chi-square is used as part of correspondence analysis, but they won't necessarily be influential. The influence of rarer taxa in multivariate analyses can also be controlled by standardizations and/or transformations of the abundance data (Section 14.6).

• Biologists should consider whether differences in relative abundances (shape) or absolute abundances (size) of taxa represent different aspects of ecological communities. Greenacre (2010, 2017b) pointed out that measures like Bray–Curtis combine both, whereas chi-square, as used in correspondence analysis, only measures differences in shape. For example, Bray–Curtis would describe two units with the same proportions of taxa but different absolute abundances as being different, whereas they would be the same as measured by chi-square as part of a correspondence analysis.

Other assessments of the relative merits of different dissimilarity indices for taxon abundance data test how well the dissimilarities, and their associated analyses, recover simulated ecological or environmental gradients (e.g. Austin 2013; Faith et al. 1987) or capture changes through time or space in β diversity (e.g. Anderson & Walsh 2013; Legendre & De Caceres 2013). The assessments generally recommend Bray–Curtis (or Kulczynski) rather than Euclidean or chi-square for detecting patterns in ecological communities based on taxon abundance data. Bray–Curtis has to some extent become the default for biologists using multivariate analyses that allow a choice of dissimilarity index, with chi-square also commonly used as the basis for correspondence analysis and related methods.

14.4 Data Transformation and Standardization

Transformations can be important for multivariate procedures based on eigenanalysis because covariances and correlations measure linear relationships between variables, so transformation of measurement variables to logs will often improve linearity and increase the efficiency with which the eigenanalysis extracts the eigenvectors.

Count (abundance) variables are nearly always positively skewed, often with lots of zeroes, and transformations based on logs or different powers (e.g. square or fourth root) can normalize these variables and reduce the influence of variables with high values (e.g. very abundant taxa) in multivariate procedures based on dissimilarity indices. The issue with zero values and log transformations has been discussed (Section 5.4), and a constant must be added or square or fourth root transformations used. The square root transformation is effective for Poisson-distributed variables, but count data in ecology are commonly overdispersed, so the fourth root is generally more effective. Transformations that normalize count data may help for linear models on dissimilarities that assume there is no relationship between multivariate means and variances (see Section 16.3.3).

Clarke et al. (2014) recommended square or fourth root transformations as they will reduce the influence of very abundant taxa and give rare taxa more weight (see also standardizations below). One difficulty with this approach is that the effect of the transformation will depend on the underlying distributions of the variables (e.g. taxon abundances). Therefore, reduction of the influence of very abundant taxa will be inconsistent.

Standardizations work differently from transformations by adjusting the data, so means and/or variances, ranges, or totals for each variable are the same. For example:

- Centering the data subtracts the mean from each observation for each variable, resulting in all variables having a zero mean. We introduced centering as a way of dealing with collinearity in linear models with interactions in Chapter 8. Eigenanalysis of a covariance matrix extracts components from centered data.
- Standardizing the data divides the centered observations by the SD for each variable. All variables have a mean of zero and an SD (and variance) of 1. Eigenanalysis of a correlation matrix extracts components from standardized data.
- Variables can also be standardized by dividing each value by the largest value, so each observation is expressed relative to the maximum value of that variable across all objects. This essentially puts the data for each variable in the range 0–1. Another standardization is dividing each value by that variable's total, so each value is expressed as a proportion of the total.

The first two standardizations of variables are important if variables are measured in very different units or scales because otherwise those variables with larger values or larger variances will often be more influential on the results of analysis than variables with smaller values or smaller variances. Standardization of variables is essential if the variables are measured in very different units.

For taxon abundances, standardizing by maximum or total converts the data to relative values for each taxon; this removes differences in absolute abundance (size) and focuses only on relative abundances (shape), like the principle behind the chi-square measure. Without this standardization, rare taxa often contribute little to dissimilarities – of course, this may be the most biologically sensible interpretation. A related transformation is Hellinger's, which converts each taxon abundance to the square root of its proportion of the total for that taxon.

Double standardization standardizes both objects and variables. The double standardization best known to ecologists is the Wisconsin standardization: taxa are standardized by their maximum value and sampling units are standardized by unit totals (Borcard et al. 2018).

Just as variables can be standardized, objects (e.g. sampling or experimental units) can also be standardized, so the value for any variable for each object is expressed relative to the maximum or total value for that object in the whole data matrix. For abundance data, this standardization is very important if the size of the units, and hence the total number of individuals, varies because it removes any effect of different total abundances in different sampling units – that is, all sampling units are considered to have the same total abundance across all taxa.

Finally, converting abundance data to presence and absence might be considered an extreme combination of transformation and standardization. There are specific dissimilarity measures for such binary data (see Section 14.3.2).

It is often useful to analyze the same data with different standardizations, particularly in ecological research. For example, comparing the results of an analysis using raw data with one using sample-standardized data will indicate what influence different total abundances in samples have. Raw data versus taxon-standardized data will illustrate the most abundant species' influence (simply leaving out different combinations of rarer species will provide similar information). Finally, to remove all effects of abundance, we can analyze just presence–absence data.

14.5 Standardization, Association, and Dissimilarity

Measures of association used for eigenanalysis in Section 14.2 have implicit standardizations. Covariances measure the linear relationships between centered variables, whereas correlations measure the linear relationships between standardized (zero mean and unit variance) variables. The choice of association matrix on which to base subsequent multivariate analyses depends on whether differences in variances between variables represent important biological information you don't wish to lose. Standardizations are also important for dissimilarity indices. Some dissimilarity indices are implicitly standardized and are unaffected by data standardizations (Austin 2013; Faith et al. 1987; Legendre & Legendre 2012). Some become identical after data standardization – for example, Bray–Curtis, Kulczynski, and Manhattan are identical for count data if objects are standardized to the same total abundance. Others, such as Bray–Curtis, and Kulczynski, produce nonsensical values when standardization is to zero mean (centering) or zero mean and unit variance (because of negative values). Standardizing by the largest value or total is a better option for these measures if you wish to reduce the influence of very abundant variables (e.g. taxa).

14.6 Screening Multivariate Datasets

Exploratory data screening is just as important for multivariate datasets as for univariate ones because their complexity means that visual inspection of the raw data is likely to miss unusual patterns or observations.

A first step with multivariate datasets is to screen the individual variables, particularly using boxplots, to check distributions and outliers, and scatterplots or scatterplot matrices (SPLOMs) to examine the pairwise relationships between variables. Extracting linear combinations is more efficient with stronger linear relationships between variables, so scatterplots are very helpful for deciding whether transformations might improve linearity and reduce the impact of outliers.

Some of the multivariate analyses comparing groups we describe in Chapters 15 and 16 make assumptions about multivariate normality and/or homogeneous variances and covariances, the latter being more important. There are graphical ways of examining multivariate dispersion, and we will describe them with the relevant analyses in the next two chapters.

Multivariate outliers are more difficult to detect because they may not be univariate outliers for individual variables. A multivariate outlier is an object with an unusual pattern of values for the variables (Tabachnick & Fidell 2019). One quantitative measure of multivariate "outlyingness" is Mahalanobis distance, which is the square of the distance from each object to its centroid (Figure 14.2), adjusted according to the variances and covariances. While P-values can be determined for each distance by assuming they follow a chi-square distribution, we recommend just scanning the distances to see if any are unusually large. We can also identify unusual objects from a two- or three-dimensional ordination plot, if some objects are markedly further away from the rest.

The options for multivariate outliers are similar to those for single variables. Transformations can also reduce the influence of outliers if they are extreme values in a positively skewed distribution. Sometimes we decide that an object is so unusual it is unlikely to be part of the population of objects we wish to describe or make inferences about. Then, we might delete that object from the analysis. Objects that differ greatly from the others may, however, represent important biological information, so we should always think carefully about outliers.

Occasionally, we will have missing observations in our dataset. The approaches for dealing with them depend on why data are missing (Section 5.6). The simplest approach is deleting the entire object (row; casewise deletion) or variable (column; listwise deletion) with the missing value. This is often the software default for multivariate analyses. Pairwise deletion can also be used if the analysis is based on pairwise associations between variables (e.g. correlations) or dissimilarities between objects. Deletion is only relevant when data are missing completely at random (MCAR).

A better approach is to use imputation methods to estimate the missing value. Replacing the missing value by the variable or object mean or median, or the predicted value from a regression model, is out of favor because it

ignores the uncertainty associated with the replacement value. The expectation-maximization (EM) method for single imputation, and multiple imputation (MI), deriving multiple datasets with different substituted values, are now considered best practice. Both methods have been commonly used where the focus is on estimating model parameters. For example, MI averages the parameter estimates from fitting the same model to the multiple imputed datasets. An alternative approach for multivariate analyses like ordinations is to average the imputed datasets and then run the multivariate analysis on this averaged data. Clavel et al. (2014) evaluated different MI methods (including an EM version) by creating different missing data patterns from a full morphological dataset (cranial measurement variables for crocodile specimens) and comparing the ordinations of the full dataset with the missing data and MI method combinations. They recommended FCS (fully conditional specification) and EM algorithms, although these methods are limited to specialist software packages. They also provided a method to graphically compare ordinations arising from MI datasets, estimating MI uncertainty.

14.7 Introduction to Multivariate Analyses

We have described two approaches to the initial analysis of multivariate datasets. First, consolidating the variance among the variables into fewer derived components (linear combinations of the original variables); and second, calculating dissimilarity measures between objects. These two approaches support many of the common multivariate analyses used by biologists. These analyses have common aims.

First, to describe the relationships among objects based on the values of the different variables. Ecologists often use the term ordination for this type of analysis. The pattern among the objects can be represented with a tree-like dendrogram showing clusters or a two- or three-dimensional ordination (sometimes termed scaling) plot (see Figure 14.3) that can be constructed in two ways:

• directly from the metric dissimilarities (e.g. Euclidean or chi-square) implicit in eigenanalysis methods where the axes represent the object scores for the first two or three eigenvectors; or

• indirectly by representing the actual dissimilarities, which can be semi- or nonmetric, between objects on the plot.

Second, to identify the variables contributing most to the pattern among the objects – for example, the taxa contributing most to the similarity between sampling units or the morphological features that best distinguish different species.

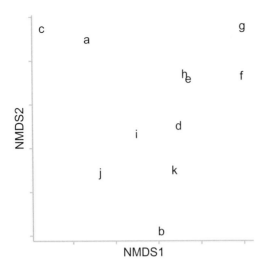

Figure 14.3 Ordination plot of 11 underpasses based on multidimensional scaling (MDS) of a Bray–Curtis dissimilarity matrix derived from proportional abundances of seven wildlife taxa (data from Clevenger & Waltho 2000). Underpasses closer together are more similar in proportional taxon composition. See Chapter 16 for MDS details.

Third, to relate the pattern among objects to predictors, including fitting linear models to the dissimilarities. Constrained ordination uses the predictor variables in the ordination process by constraining the object scores in the ordination plot based on the relationships with the predictors. Unconstrained ordination derives the ordination independently of the predictors and looks at the relationship between the objects and the predictors as a separate part of the analysis.

In Chapter 15, we will describe analyses based on deriving linear combinations of variables (components), including ordination procedures such as principal coordinates analysis and correspondence analysis. We will also look at discriminant function analysis and multivariate analysis of variance (MANOVA), where we have pre-existing groups, and the components are extracted in a way that maximizes the between-group differences.

In Chapter 16, we will focus on analyses based on dissimilarities. Ordination analyses include principal coordinate analysis, distance-based redundancy analysis, and multidimensional scaling. We can also relate one set of dissimilarities to other sets using Mantel tests and

generalized dissimilarity modeling. We can also test hypotheses about group differences using analysis of similarities (ANOSIM) or fit more sophisticated linear models using permutational analysis of variance (PERMANOVA).

14.8 Key Points

- Multivariate datasets in biology are commonly one or more sets of variables, in columns, recorded for units (objects), represented by rows. The variables can be continuous or categorical factors.
- We can extract summary variables, termed components, from a matrix of covariances or correlations among the original variables using either an eigenanalysis or an SVD.
- These components are linear combinations of the original variables and can be used to derive a component score for each object so they can be plotted, as well as measure the contribution or loading of each variable.
- A different approach is to calculate dissimilarities between pairs of objects based on variable values; a wide range of metric and semimetric dissimilarities are available with the Euclidean commonly used for measurement variables and Bray–Curtis or chi-square commonly used for count variables.
- If variables are measured in different units or have very different variances, they may need standardization before analysis. Count variables such as taxon abundances may also be standardized to reduce the influence of taxa with very large numbers or to make the counts proportional relative to row or column totals.

Further Reading

Eigenanalysis: Most multivariate statistics books will describe eigenanalysis, and we have found Jackson (2003), Legendre and Legendre (2012), and Tabachnick and Fidell (2019) provide a good balance between readability and statistical detail.

Distances and dissimilarities: Again, Legendre and Legendre (2012) provide thorough coverage.

Data screening and standardizations: Tabachnick and Fidell (2019) include a chapter on screening multivariate datasets and Clarke et al. (2014) and Legendre and Legendre (2012) cover standardizations and transformations for multivariate analyses from an ecological perspective.

15 Multivariate Analyses Based on Eigenanalyses

In this chapter, we examine multivariate analyses based on linear combinations of variables, termed components (or factors or functions), and whose derivation by eigenanalysis we described in Section 14.2. The main aim of these multivariate analyses is to reduce the original variables to fewer new derived variables that adequately summarize the original variables and can be used for further analysis. A second aim is to reveal patterns in the data, especially among objects, that were not visible when analyzing each variable separately. One way to detect these patterns is to plot the objects in multidimensional space, the dimensions being the new derived variables. This process is sometimes termed scaling, where the distance between objects in multidimensional space represents their biological dissimilarity. Ecologists often use "ordination" instead of scaling, particularly for analyses that arrange sampling or experimental units in terms of species composition or environmental characteristics. Ordination is sometimes considered a subset of gradient analysis. Direct gradient analysis displays sampling units directly in relation to one or more underlying environmental characteristics. Indirect gradient analysis shows sampling units in relation to a reduced set of variables, usually based on species composition. It then relates the pattern in sampling units to the underlying environmental characteristics.

15.1 Principal Components Analysis

Principal components analysis is one of the most common multivariate statistical techniques, and it is also the basis for others. For n objects, PCA transforms p variables into p new uncorrelated variables $(z_1, z_2, z_3, \ldots, z_p)$ called principal components or factors. The components are extracted so each component is uncorrelated with those preceding it, and the variance among the original variables is consolidated as much as possible in the first few

components. For example, Wu et al. (2021) sampled 300 sites representing different land-use types near China's Three Gorges Reservoir. They took a mixed soil sample at each site and determined the concentrations of 10 metals along with five soil characteristics (e.g. pH, soil organic carbon). We can use these data to examine patterns among sites and between land-use types in reduced multivariate space defined by the first few components and determine the variables mostly driving that pattern (Box 15.1).

15.1.1 Deriving Components
15.1.1.1 Axis Rotation

The simplest way to understand PCA is in terms of axis rotation (Legendre & Legendre 2012). Consider the relationship between copper and manganese from the Wu et al. (2021) dataset (Figure 15.1). The distance between each object is their Euclidean distance. Principal components analysis can be viewed as a rotation of the original axes, representing the two variables, after centering to the mean of copper and the mean of manganese, so that the first "new" axis explains most of the variation and the second axis is orthogonal (right angles) to the first. The first new axis is principal component 1, and the second is principal component 2. The first component is actually a "line-of-best-fit" that is halfway between the least squares estimates of the linear regression models of copper on manganese and manganese on copper. It is the estimate of the MA Model II regression (Section 6.1.7) and is the line represented by the correlation between the two variables (either raw or centered). If the variables are standardized (to zero mean and unit standard deviation), the first principal component represents the reduced major axis (RMA) regression. The second component is completely independent of, or uncorrelated with, the first. This rotation preserves the Euclidean distances between objects.

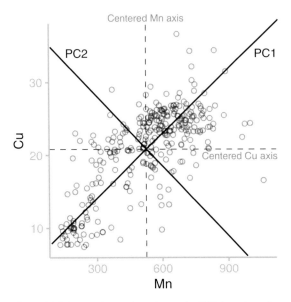

Figure 15.1 Geometric rotation of axes in PCA based on the correlation between soil copper and manganese across 300 sites in China, from Wu et al. (2021).

15.1.1.2 *Decomposing an association matrix*

When there are more than two variables, it is difficult (or impossible) to represent the rotation procedure graphically. In practice, the components are extracted from a sum of squares and cross-products matrix or, more commonly, a covariance (centered variables) or correlation (standardized variables) matrix using the methods described in Section 14.2. Extracting components from a covariance matrix is equivalent to a singular value decomposition (SVD) on centered variables, whereas using a correlation matrix is the equivalent of an SVD on standardized (to zero mean and unit variance) variables. Choice of matrix will be discussed in Section 15.1.4. There will be p principal components, each a linear combination of the original variables:

$$z_{ik} = c_1 y_{i1} + c_2 y_{i2} + \ldots + c_j y_{ij} + \ldots + c_p y_{ip},$$

where z_{ik} is the value or score for component k for object i, y_{i1} to y_{ip} are the values of the original variables for object i, and c_1 to c_p are weights or coefficients, termed loadings, that indicate how much each original variable contributes to the linear combination forming this component. Although p components can be derived, we hope that the first few summarize most of the original variables' variation. The scores for each object on each component are sometimes called z-scores.

The score for object i on PC1 from Wu et al. (2021; see also Box 14.1) is:

$$z_{i1} = c_1(\mathrm{Ca})_i + c_2(\mathrm{Cr})_i + c_3(\mathrm{Cu})_i + c_4(\mathrm{Fe})_i + c_5(\mathrm{Mn})_i \\ + c_6(\mathrm{Pb})_i + c_7(\mathrm{Zn})_i + c_8(\mathrm{Cd})_i + c_9(\mathrm{SOC})_i + c_{10}(\mathrm{pH})_i \\ + c_{11}(\mathrm{N})_i + c_{12}(\mathrm{P})_i + c_{13}(\mathrm{K})_I.$$

15.1.2 Interpreting the Components

The matrix approach to deriving components produces two important pieces of information (see Boxes 14.1 and 15.1). First, the eigenvectors contain the estimates of the coefficients or loadings for each variable (the c_js above). Eigenvector 1 contains the coefficients for each variable for principal component 1, eigenvector 2 for component 2, etc. The eigenvectors are usually scaled so the sum of squared coefficients for each eigenvector equals 1, although other scalings are sometimes used. The further each coefficient is from zero, the greater the contribution that variable makes to that component. It is sometimes more helpful to convert these coefficients or loadings to simple correlations (using r) between the components (i.e. component scores for each object) and the original variables. They are determined by scaling the eigenvector coefficients for each variable by the square root of the eigenvalue for a given component ($\mathbf{V}$ or $\mathbf{W}$ matrices in Box 14.1). High loadings indicate that a variable is strongly correlated with (strongly loads on) a particular component. Note that the signs of coefficients/loadings are arbitrary; a 180° rotation of the objects in multidimensional space would reverse the positive and negative signs.

Second, estimates of the eigenvalues (sometimes called characteristic roots, λ_k) provide relative measures of how much variation between the objects, summed over the variables in the dataset, is explained by each principal component. The components are extracted so that the first explains the maximum variation, the second explains the maximum amount of that unexplained by the first, etc. If there are associations between the variables, the first two or three components will usually explain most of the variation in the original variables, so we can summarize the patterns in the original data based on fewer components (variable reduction). In the analysis of the data from Wu et al. (2021), the first three components comprised >80% of the original variation (Box 15.1). If the original variables are uncorrelated, PCA will not extract components that explain more variation than the same number of original variables. The sum of all the eigenvalues equals the total variation in the original dataset, the sum of the variances of the original variables. Principal components analysis rearranges the variance in the original variables, consolidating it in the first few new components.

It is important to remember that our data usually represent a sample of objects, with observations for each

Box 15.1 Ⓡ Ⓔ Worked Example of PCA: Soil Characteristics and Land Use in China

Wu et al. (2021) took soil samples at 300 sites near the Three Gorges Reservoir in China. Each site was classified into one of three categories of land use: orchard ($n = 75$), dry land ($n = 98$), and paddy field ($n = 127$). They also measured the concentrations of 10 metals (Ca, Cd, Cr, Cu, Fe, Mg, Mn, Ni, Pb, and Zn; mg/kg) and five soil characteristics (pH, concentrations of N and P in mg/kg, and percentage of soil organic carbon [SOC] and K) for each site. Wu et al. (2021) did a PCA on the metals and used the components in further analyses. Using PCA, we will instead examine the pattern among sites and land-use categories based on the metals and the other characteristics (15 variables in total).

Preliminary checks of the variables showed some skewness for Ca and K but no strongly nonlinear relationships, so all variables were left untransformed. Summary statistics for each variable were as follows.

Variable	Mean	SD
Ca	0.747	0.290
Cd	0.221	0.069
Cr	64.388	10.821
Cu	20.824	5.882
Fe	3.232	0.665
Mg	1.124	0.417
Mn	527.030	198.194
Ni	25.793	7.629
Pb	25.525	2.308
Zn	68.223	18.215
pH	6.256	0.827
SOC	0.787	0.209
N	888.335	243.648
P	627.639	253.080
K	2.064	0.419

We used a correlation matrix for the PCA because the variables were measured on different scales and had very different variances, which we did not want to influence the analysis.

	Ca	Cd	Cr	Cu	Fe	Mg	Mn	Ni	Pb	Zn	pH	SOC	N	P	K
Ca	1.00														
Cd	0.41	1.00													
Cr	0.50	0.48	1.00												
Cu	0.67	0.52	0.84	1.00											
Fe	0.69	0.46	0.87	0.90	1.00										
Mg	0.81	0.49	0.79	0.88	0.92	1.00									
Mn	0.72	0.37	0.59	0.72	0.78	0.79	1.00								
Ni	0.68	0.51	0.90	0.89	0.93	0.92	0.72	1.00							
Pb	0.20	0.49	0.65	0.63	0.63	0.47	0.34	0.61	1.0						
Zn	0.74	0.62	0.76	0.89	0.88	0.89	0.73	0.85	0.59	1.00					
pH	0.69	0.32	0.31	0.35	0.35	0.48	0.38	0.43	0.02	0.39	1.00				
SOC	−0.24	0.25	0.02	−0.06	−0.11	−0.17	−0.34	−0.10	0.31	0.01	−0.10	1.00			
N	−0.07	0.43	0.25	0.18	0.15	0.08	−0.18	0.17	0.50	0.24	−0.04	0.72	1.00		
P	0.47	0.30	0.25	0.47	0.36	0.43	0.40	0.32	0.20	0.56	0.15	−0.04	0.04	1.00	
K	0.71	0.48	0.74	0.81	0.84	0.92	0.73	0.87	0.49	0.84	0.45	−0.13	0.09	0.41	1.00

Box 15.1 (cont.)

Bartlett's chi-square statistic was 5,542.9 ($P < 0.001$), indicating that at least the first component differed from the rest. The first three components with eigenvalues >1 were (with percentage variance and nonparametric bootstrap CIs) as follows.

Component	Eigenvalue	Percentage variance (cumulative)	Bootstrap mean percentage	(95% CI)
1	8.57	57.12 (57.12)	57.21	(54.72–59.95)
2	2.37	15.81 (72.94)	15.92	(14.68–17.28)
3	1.09	7.29 (80.23)	7.43	(6.42–8.63)

These three components explained >80% of the total variance, although only the first two components had eigenvalues greater than expected by chance from the broken-stick model (Figure 15.2).

The coefficients for each variable in the first three eigenvectors (i.e. components) are as follows.

Variable	PC1	PC2	PC3
Ca	−0.27	0.23	−0.35
Cd	−0.21	−0.25	−0.33
Cr	−0.29	−0.11	0.25
Cu	−0.32	−0.03	0.13
Fe	−0.32	0.00	0.21
Mg	−0.33	0.09	0.02
Mn	−0.28	0.23	0.08
Ni	−0.32	−0.01	0.15
Pb	−0.21	−0.36	0.29
Zn	−0.32	−0.05	−0.05
pH	−0.17	0.18	−0.62
SOC	0.02	−0.56	−0.25
N	−0.06	−0.58	−0.16
P	−0.17	0.05	−0.25
K	−0.31	0.06	0.03

The loadings of each variable to each component, with bootstrap means and 95% CIs in brackets, are as shown.

Var.	PC1	PC2	PC3
Ca	−0.79 (−0.79: −0.85 to −0.73)	0.35 (0.35: 0.24 to 0.46)	0.37 (0.35: 0.23 – 0.43)
Cd	−0.61 (−0.61: −0.69 to −0.52)	−0.38 (−0.38: −0.49 to −0.26)	0.34 (0.31: 0.13 to 0.47)
Cr	−0.86 (−0.86: −0.88 to −0.82)	−0.17 (−0.17: −0.25 to −0.09)	−0.26 (−0.23: −0.35 to −0.03)
Cu	−0.94 (−0.94: −0.95 to −0.92)	−0.04 (−0.05: −0.10 to 0.01)	−0.14 (−0.13: −0.19 to −0.05)
Fe	−0.95 (−0.95: −0.96 to −0.94)	0.00 (0.00: −0.06 to 0.06)	−0.21 (−0.20: −0.25 to −0.09)
Mg	−0.96 (−0.96: −0.97 to −0.95)	0.137 (0.14: 0.09 to 0.18)	−0.02 (−0.02: −0.07 to 0.04)
Mn	−0.80 (−0.80: −0.85 to −0.75)	0.35 (0.35: 0.28 to 0.42)	−0.09 (−0.08: −0.18 to 0.02)
Ni	−0.94 (−0.95: −0.96 to −0.93)	−0.01 (−0.01: −0.08 to 0.05)	−0.15 (−0.13: −0.24 to 0.04)
Pb	−0.62 (−0.62: −0.70 to −0.53)	−0.56 (−0.56: −0.64 to −0.47)	−0.31 (−0.29: −0.41 to −0.17)
Zn	−0.95 (−0.95: −0.98 to −0.90)	−0.08 (−0.08: −0.15 to −0.02)	0.05 (0.05: −0.08 to 0.20)

Box 15.1 (cont.)

(cont.)

Var.	PC1	PC2	PC3
pH	−0.49 (−0.49: −0.57 to −0.40)	0.27 (0.27: 0.10 to 0.43)	0.64 (0.63: 0.30 to 0.81)
SOC	0.08 (0.08: −0.07 to 0.21)	−0.86 (−0.86: −0.90 to −0.81)	0.26 (0.25: 0.13 to 0.35)
N	−0.18 (−0.18: −0.30 to −0.05)	−0.86 (−0.88: −0.91 to −0.85)	0.17 (0.16: 0.05 to 0.26)
P	0.49 (0.49: −0.58 to −0.40)	0.08 (0.08: −0.09 to 0.25)	0.26 (0.20: −0.35 to 0.74)
K	−0.91 (−0.91: −0.93 to −0.89)	0.09 (0.09: 0.03 to 0.16)	−0.03 (−0.02: −0.08 to 0.05)

The signs of the coefficients and therefore the correlations within each component are arbitrary (Section 15.1.2). The pattern is similar for eigenvector coefficients and loadings.

Generally, each variable tends to correlate highly with only one component, although some have moderate correlations with two (e.g. Pb on 1 and 2, pH on 1 and 3). The 10 target metals and K have the highest correlations with component 1, SOC, and N, followed by Pb for component 2 and pH for component 3; the CIs are relatively narrow for all these except pH on component 3.

All correlation residuals were ≤0.13, except for pH against P (0.28), suggesting the correlations based on the first three components were reasonably close to the original correlations.

A scores plot just for the sites is shown in Figure 15.3 alongside a correlation biplot (scaling = 2) of the first two components. There is no obvious separation of the three land-use categories. It is clear that component 1 represents the metals and component 2 SOC and N.

To improve the interpretability of the loading of the variables on the components, we also tried a varimax rotation. The main change was that pH now only loaded strongly on rotated component 3, and Pb had a higher loading on rotated component 1.

We also extracted the components based on a covariance matrix to illustrate the influence of differences in variances when using a covariance matrix instead of a correlation matrix for a PCA. The first two components now explain a higher proportion (84%) of the total variance. The eigenvalues are considerably larger than for the correlation matrix because the variables are not standardized to unit variance.

Component	Eigenvalue	Percentage variance
1	75,711	46.4
2	61,451	37.6
3	25,896	15.9

The loadings are now covariances rather than correlations, and their pattern among variables is quite different from that based on a correlation matrix. Now component 1 is dominated by Mn and P, followed by N, the variables with the largest variances in that order. Our preference with these data would be to use a correlation matrix because we would not want the large differences in variances (partly related to the measurement scales) to contribute to our interpretation of components.

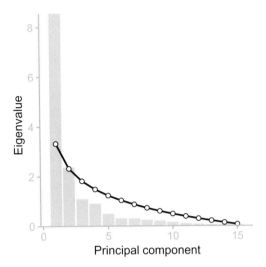

Figure 15.2 Scree (bars) and broken-stick (line) plot from the PCA based on a correlation matrix for the data from Wu et al. (2021).

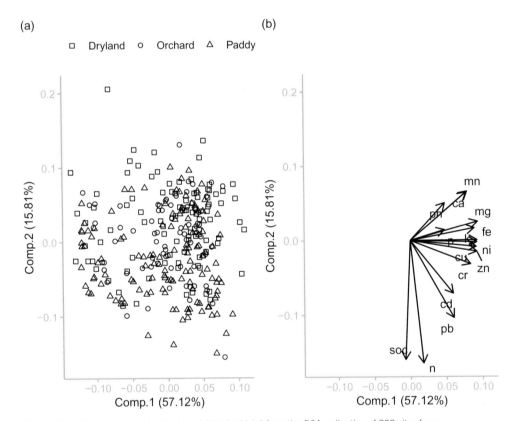

Figure 15.3 Object score plot (left) and variable plot (right) from the PCA ordination of 300 sites from three habitat types in China from Wu et al. (2021). These two panels would usually be combined for a biplot, but are separated to make the figure legible.

variable, from a population of objects; for example, the sites in the Wu et al. (2021) dataset are a sample from all possible sites in this region of China. Therefore, the coefficients and loadings for each variable on each component, and the eigenvalues (explained variance) for each component, are also sample estimates of population parameters. Providing SEs for these estimates or CIs for the parameters is crucial for interpreting the results from a PCA (Bjorklund 2019). Standard ML methods can be used for SEs and CIs (Jackson 2003), but not all PCA software offers them. It is probably best to use bootstrap methods (Jackson 2003; Peres-Neto et al. 2003, 2005) and we provide nonparametric (i.e. not assuming a specific distribution) bootstrap CIs for the Wu et al. (2021) PCA in Box 15.1.

15.1.3 How Many Components to Retain?

A key decision with PCA is deciding how many components to retain and interpret. There are many approaches to these decisions, sometimes called stopping rules (see reviews by Bjorklund 2019; Peres-Neto et al. 2003, 2005). We will introduce some in this section, examine the interpretability of the components, and retain those providing biologically interpretable information.

15.1.3.1 Eigenvalue = 1 Rule

This rule, also called the Kaiser–Guttman criterion, simply says keep any component with an eigenvalue >1 when the PCA is based on a correlation matrix. The logic here is that the total variance to be explained equals the number of variables (because using a correlation matrix standardizes the variables to a mean of 0 and standard deviation of 1), so by chance, each component would have an eigenvalue of 1. In the soils analysis, 3 out of the 15 possible components were kept (Box 15.1). In datasets with many variables, especially those with no strong consolidation of the variance in the first few components, there might be numerous components with eigenvalues greater than 1, and PCA results and biplots (see below) are difficult to interpret.

15.1.3.2 Scree Diagram

We can also examine the scree diagram, which plots the eigenvalues for each component against the component number. We are looking for an obvious break (or elbow) where the first few components explain most of the variation, and the remaining components don't explain much more of the variation (Figure 15.2). A rule of thumb is to keep all components up to and including the first in that remaining group. Our experience is that scree diagrams don't offer more interpretability than simply examining the successive numerical eigenvalues for each component.

15.1.3.3 Broken-Stick Criterion

If we assume the total variance is partitioned randomly among the components, then the pattern in the eigenvalues will follow a broken-stick distribution for which the theoretical equation is known (Borcard et al. 2018; Legendre & Legendre 2012; Peres-Neto et al. 2005). Components are retained if their eigenvalue exceeds that predicted by the broken-stick model (Figure 15.2).

15.1.3.4 Tests of Eigenvalue Equality

There are statistical tests for equality of a set of successive eigenvalues derived from a covariance matrix, such as Bartlett's and Lawley's tests (Jackson 2003; Peres-Neto et al. 2005). We might use one of these to evaluate the null hypothesis that the eigenvalues of the components not retained are equal. Bartlett's test is most common (sometimes called a test of sphericity) and available in most statistical software as part of correlation or PCA routines. The test statistic is compared to a χ^2 distribution. We usually test sequentially, first testing that the eigenvalues of all components are equal (Bartlett's test is then a test of sphericity of a covariance matrix – see Chapter 10). If the evidence does not support this hypothesis, we then test the equality of eigenvalues of all components except the first, and so on, stopping once the evidence indicates the remaining components are equal. Bartlett's and Lawley's tests were initially developed for use with covariance matrices, but can also be used with correlation matrices as a conservative estimate of the number of components to be kept (Jackson 1993).

15.1.3.5 Other Methods

Jackson (1993) and Peres-Neto et al. (2005) used extensive simulations to assess various stopping rules, including those just described and others based on applying bootstrap and randomization methods to eigenvectors and eigenvalues. Peres-Neto et al. (2005) recommended a two-step procedure that first uses Bartlett's test to evaluate whether the first component differs from the rest and followed by one of their other recommended test statistics. However, definitive recommendations were difficult because each method's performance depended on the correlation structure of the variables. Like Bjorklund (2019), Peres-Neto et al. (2005) emphasized the need to interpret component loadings and object scores with appropriate (i.e. bootstrapped) SEs or CIs.

15.1.4 Which Association Matrix to Use?

The choice between the covariance or correlation matrix for a PCA is important. The covariance matrix is based on mean-centered variables. It is appropriate when the variables are measured in comparable units, and differences in variance between variables make an important contribution to biological interpretation. The correlation matrix is based on variables standardized to zero mean and unit variance. It is necessary when the variables are measured in very different units or we wish to ignore differences between variances.

Most PCA software defaults to a correlation matrix, although all offer the covariance matrix. Our experience is that most biologists use the correlation matrix but rarely consider the implications of analyzing variables standardized to zero mean and unit variance. For example, a PCA using the soil data from Wu et al. (2021) is best based on a correlation matrix because of the different units (percentage, concentrations, pH, etc.) and variances of the variables. In contrast, we might compare the results from the two matrices if our variables were species abundances, and we wanted to see if the different patterns of variance in abundance between species across objects were important. We argued in Section 14.5 that analyzing data with different forms of standardization can assist in interpretation (see also Austin 2013). The message when using PCA is that the choice of matrix depends on how much we want different variances among variables to influence our results.

15.1.5 Simplifying Component Structure

Ideally, we would like each variable to load strongly on only one component, and the loadings (correlations) to be close to ± 1 (strong correlation) or 0 (weak correlation). What we sometimes get is much messier, with some variables loading strongly on a couple of components and multiple variables with moderate loadings (0.4–0.6). There are at least two ways to simplify the interpretation of components by having fewer variables load more strongly on each component (Jackson 2003; Jolliffe & Cadima 2016). The first is to apply a second rotation of the components after the initial PCA. Rotation can be of two types. Orthogonal rotation keeps the rotated components uncorrelated with each other and includes methods such as varimax, quartimax, and equimax. As the name suggests, varimax rotation maximizes the variance of component loadings and is the most common orthogonal rotation method used with PCA. Oblique rotation produces new components that are no longer orthogonal. It is more complex and not commonly used in biological research.

The PCA on the soil data from Wu et al. (2021) showed that a secondary varimax rotation did not make much difference, with only a couple of variables changing their loadings. This was probably because most variables already had strong loadings with at least one component. Situations with numerous variables with moderate

loadings on multiple components are more likely to benefit from a secondary rotation.

A second approach to simplifying components is putting some additional constraint on the coefficients and loadings (see Box 14.1). Jolliffe and Cadima (2016) describe a method based on the least absolute shrinkage and selection operator (LASSO) sometimes used in linear regression. In a PCA, the sum of the coefficients of each eigenvector is constrained to be less than a specific tuning parameter; the smaller the parameter, the more loadings closer to zero. Different tuning parameter values can be tried to balance component simplicity and losing explained variance from the simplification.

15.1.6 PCA Assumptions and "Fit"
15.1.6.1 Assumptions
Because it uses covariances or correlations to measure variable association, PCA is more effective when there are linear relationships between variables. Nonlinear relationships are common between biological variables, and under these circumstances PCA will be less efficient at extracting components. Transformations can often improve the linearity of relationships between variables, or robust measures of correlations, such as Spearman's rank correlation, could be used. While PCA is based on linear relationships between variables, we still need to be careful about correlations so high (e.g. >0.95) that we have near-perfect collinearity and redundancy in our variables.

There are no distributional assumptions associated with the ML estimation of eigenvalues and eigenvectors and determining component scores (the descriptive use of PCA). However, calculation of CIs and tests of hypotheses about these parameters do assume multivariate normality. Bootstrap SEs and CIs and randomization tests are preferred. Outliers can also influence the results from a PCA, and we will describe the use of residuals to detect outliers below.

Principal components analysis can run into problems if there are more variables than objects ($n < p$). In these cases, the number of nonzero components needed to explain all the original variance is determined by the number of objects rather than variables (Jolliffe & Cadima 2016). A PCA can still be run, depending on the software. Principal components analysis is an important data analysis tool in disciplines like genomics and proteomics, where the number of variables can be in the thousands, much greater than the number of sampling/experimental units (Abegaz et al. 2019). With enormous datasets, computing speed and memory may become limiting, and faster, more efficient methods have been developed.

Like all multivariate analyses, missing data are a real problem. The default setting for PCA routines in most statistical software is to omit whole objects that contain missing observations. Unless the sample size (number of objects) is large and the objects with missing values are a random sample from the complete dataset, then multiple imputation is more appropriate for dealing with missing observations (see Sections 5.6 and 14.7).

15.1.6.2 *PCA Fit: Residuals*

Residual analysis is useful for PCA, just as for linear models. If we retain fewer than p components, some of the information in the original data will not be explained by the components. This is analogous to the residual in linear models. Two types of residuals are useful when assessing the results from a PCA.

First, we can measure the difference between the observed correlations or covariances between the variables and the predicted (reconstructed) correlations or covariances based on the components we extract – this is termed the residual correlation or covariance matrix (Tabachnick & Fidell 2019). We want these correlation residuals to be as small as possible.

Second, we can focus on residuals associated with the objects. With fewer than p components, the original data (with variables usually standardized to unit variance) can be represented as a multivariate mean (centroid) plus a contribution due to the retained components plus a residual. This residual measures the difference between the observed value of a variable for an object and the value for that object predicted by our PCA with $<p$ components. The sum (across variables) of squares of the residuals, often termed Q (Jackson 2003), can be derived for each object. If the variances differ between the variables and some objects have much larger values for some variables, then the residuals, and Q-values, for those objects will probably be larger for a PCA based on a covariance matrix than one based on a correlation matrix.

Large values of Q for any observation indicate a potential outlier and that the components we have retained do not adequately represent the original dataset for that object. Q-values can be compared to an approximate Q sampling distribution to determine P-values (see Jackson 2003). However, formal statistical testing seems not very useful when exploring a multivariate dataset for unusual values – just check unusual values relative to the rest. These objects can be further examined to see which variable(s) contribute most to the large Q-value – that is, which variables have the large residuals.

We should also mention score outliers, observations further from the origin (the zero–zero point in an ordination plot) than the bulk of the observations. These observations are not necessarily a problem and may represent important biological information as to why that object differs from the rest of the plotted components.

15.1.7 Ordination and Biplots for PCA

The eigenvectors can be used to calculate a new score (z-score) on each component for each object by solving the linear combination (see Box 14.1). For ease of interpretation, these scores are commonly standardized so the variance for each component is 1. The objects can then be positioned on a scatterplot based on their scores with the first two or three principal components as axes (e.g. Figure 15.3); this is a score plot. These plots are centered on the origin point, representing a score of zero on each component because the variables were centered. The position of an object along each component (i.e. on a line drawn at right angles to the component) indicates whether it will have higher or lower values for those variables positively correlated with this component. The distances between the objects are an approximation of their Euclidean distances; objects closer together are more similar in their variable values than objects further apart. If three components are retained, we can either do a 3D plot or three pairwise plots.

We can also plot loadings where each original variable is plotted based on its loading on each eigenvector. Loadings plots by themselves are not commonly used. Instead, biplots, which combine the scores and the loadings, are standard practice. Points represent objects based on their scores, and variables are represented by vectors (arrows) originating at the origin (Figure 15.3). Scaling the objects and variables so they can both be plotted on the same reduced multidimensional space is tricky, and two scaling options are common (Borcard et al. 2018; Legendre & Legendre 2012; Zuur et al. 2007):

• Scaling 1 is where the object scores on each principal component and the eigenvectors scaled to the same unit length are plotted. The distances between objects are approximations of their original Euclidean distances. This makes it easier to interpret the relative positioning of the objects in our plot and is called a distance biplot.

• Scaling 2 is where we plot the object scores scaled to unit variance and the eigenvectors scaled by the square root of their eigenvalues. Now the distances are not approximating Euclidean distances but Mahalanobis distances, which take the covariances among variables into account. This option makes it harder to interpret the pattern among objects but easier to interpret the variable vectors, and is called a correlation biplot.

If the focus of the PCA is on the pattern among the objects, distance biplots are preferred. If the focus is on the variables, correlation biplots are best. Our experience is that the two types of biplots don't differ a lot with many datasets.

Interpretation of biplots can be tricky, especially the variable vectors. In a correlation biplot, the angle between two vectors indicates the correlation between those variables – the smaller the angle, the stronger the positive correlation. Vectors in opposite directions indicate strong negative correlations. The vector lengths are a measure of that variable's contribution to the pattern among objects in that direction. If the PCA was based on a correlation matrix, these lengths are proportional to the variable loadings. It is important not to interpret the positions of the objects in relation to the ends of the variable vectors.

15.1.8 Principal Components Regression

Principal components regression (PCR) is sometimes suggested as a solution to the problem of collinearity in multiple regression. If there are strong correlations among the predictor variables, we can do a PCA on the predictors, at least centered and perhaps standardized, to extract the components. We can then fit a linear regression model that uses all the orthogonal components as the predictors. This is termed principal components regression.

We can also recalculate regression coefficients in terms of the original centered or standardized variables based on the relationship (Jackson 2003):

$$\mathbf{b} = \mathbf{U}\mathbf{b}_z.$$

Here, $\mathbf{b}$ is a matrix of regression coefficients on the original standardized variables, $\mathbf{b}_z$ is a matrix of regression coefficients on the principal components (derived using a correlation matrix), and $\mathbf{U}$ is the matrix of eigenvectors from the PCA on the predictor variables (see Box 14.1). This equation simply states we can obtain regression coefficients in terms of the original variables from the product of the regression coefficients for the principal components and the eigenvectors from the PCA. Using eigenvectors from the $\mathbf{U}$ matrix scales the coefficients so the sum of squared coefficients equals 1 (Box 14.1).

The SE of the regression coefficient for the kth principal component is (Jackson 2003):

$$s_{b_k} = \sqrt{\frac{\mathrm{MS_{Residual}}}{l_k}},$$

where $\mathrm{MS_{Residual}}$ is from the linear regression on the p principal components. The SEs are inversely proportional to the eigenvalues, and the regression coefficients for the early components will have SEs smaller than later components.

If all p components are used, the regression coefficients for the original variables extracted from the PCR will be the same as those from the regression on the original (centered or standardized) variables. The PCR model has the advantage that the predictors (the components) are uncorrelated with each other, so the SEs are not inflated by collinearity, even though predictions will have the same precision as the original regression model. If fewer than p components are used, the extracted regression coefficients will differ from those on the original (centered or standardized) variables. These new coefficients will be biased, the bias increasing as we retain fewer components.

Chatterjee and Hadi (2012) provide a clear example of the calculations involved in PCR. Despite its attractiveness to overcome collinearity in multiple linear regression models, there are limitations to PCR. Hadi and Ling (1998) point out that the components that explain most of the variance in the predictors – that is, the first few derived components – might not be the most important in explaining the variance in the response variable in a multiple regression model. The choice of components to use in PCR should be based on their contributions to the $\mathrm{SS_{Regression}}$, not just their eigenvalues from the original PCA. Principal components regression is also sensitive to outliers, and we recommend the usual diagnostic checks for multiple regression models we described in Section 8.1.6.

15.1.9 Factor Analysis

Factor analysis (FA) is a dimension-reduction method closely related to PCA. Principal components analysis is focused on the variance of the original variables, extracting components that successively explain it. The retained components explain part of the covariance or correlation matrix, and the remainder is residual (zero if we retained all components). In contrast, FA only extracts factors from the shared variance (i.e. covariance) between the variables and tries to explain correlations. Common factors are those that share variability between at least two variables, and the communality is a measure of the variance in each original variable explained by the common factors; communality ranges between 0 (the variable is essentially uncorrelated with the others) and 1 (the variable shares all its variance with others). We also have unique factors that represent the variance of each variable not shared with others.

The mechanics of FA are pretty much the same as for PCA. However, the procedure is more complex because we need to estimate both common factors and the variability associated with the unique factors. Jackson (2003) describes different approaches to estimation of factors. Some (e.g. ML, generalized least squares [GLS]) require the number of common factors to be specified in advance, in contrast to others, such as principal factor analysis, where the matrix of correlations between the variables is modified so that the diagonal contains estimates of the communalities. Spectral decomposition is then applied to this new matrix to extract eigenvectors and eigenvalues.

The common factors can be thought of as estimates of latent variables, the true variables causing the correlation structure in the data. Structural equation modeling (also termed latent variable analysis or causal modeling) is a technique that combines FA with multiple regression so the response and predictor variables may be measured variables or common factors (Tabachnick & Fidell 2019). When only measured variables are used, we have multiple regression modeling, and the possible causal relationships between response and predictor variables can be displayed in various ways (e.g. path diagrams). When we have factors on either side of our regression model, we have structural equation modeling, and the path diagrams are more sophisticated.

We recommend Jackson (2003) and Tabachnick and Fidell (2019) for good overviews of the differences between PCA and FA (and structural equation modeling), although neither has a biological perspective. Factor analysis is not commonly used in biological research, probably because biologists are trying to extract a few new variables that explain most of the variability in the original variables and use these new variables in ordination plots or as response/predictor variables in linear models. Principal components analysis is more appropriate than FA for these purposes.

15.2 Correspondence Analysis

Correspondence analysis (CA) has a long history in data analysis (Legendre & Legendre 2012) and has been "reinvented" a few times. It was partly developed as a method for decomposing contingency tables of counts (see Section 13.7) into a few summary variables and representing ("mapping") the lack of independence between rows and columns of the contingency table as a low-dimensional plot. It is based on a raw data matrix of counts, classified by n rows (objects) and p columns (variables) – for example, n sampling units by p taxa, with the cell entries being the number of individuals of each taxon in each unit. In Section 13.7, we described the χ^2 statistic for independence in a two-way contingency table of counts. Large values of this statistic indicate lack of independence between rows and columns – that is, the proportion of counts in different columns depends on the row and vice-versa. The main purpose of CA is to summarize the lack of independence between rows (objects) and columns (variables) of a contingency table as a few derived variables, sometimes called principal axes. The maximum number of

derived variables is the minimum of $(n-1)$ and $(p-1)$, although only two axes are usually used for plotting.

We will illustrate the use of CA in Box 15.2 with the dataset from Lemmens et al. (2015), who collected biotic data (abundances of different taxa of fish and invertebrates and cover of aquatic plants) from artificial ponds in Belgium. Our analysis will focus on counts of invertebrate families to represent the ponds and the families jointly in low- (preferably two-) dimensional space.

15.2.1 Deriving the Axes

Correspondence analysis proceeds by a transformation of the observed counts into their contributions to the overall χ^2 statistic; these contributions are the standardized residuals for each cell in the table (Section 13.7.1.3). The counts are now relative to their respective row and column totals, termed relativization or profiling, and the data matrix is sometimes termed a **Q** matrix. When the variables are abundances of taxa, differences in overall abundances ("size") will be much less important, and the focus will be on differences in relative proportions ("shape") of different taxa (Greenacre 2010; see also Section 14.3).

This transformed data matrix is then analyzed in one of two ways, just as we did with PCA. First, we can use an SVD to derive eigenvectors and eigenvalues for each axis (Box 14.1). Second, we can convert the matrix of transformed counts into one of two association matrices, between variables or between objects. These matrices are also called sums of squares and cross-products (SSCP), a form of covariance matrix that doesn't divide by $n-1$. We use eigenanalysis of the association matrix to extract the same eigenvectors and eigenvalues as via the SVD approach.

Because the eigenvectors for objects and variables are extracted jointly, after transforming the counts to contributions to the χ^2 statistic, the eigenvalues associated with the principal axes for rows and columns are the same. The sum of these eigenvalues equals the overall χ^2 statistic divided by the total frequency and is called total inertia, a measure of lack of independence. The eigenvalues are interpreted similarly to those from a PCA, with the percentage of the total inertia explained by the successive axes usually presented. The first axis should explain a high proportion of the lack of independence between objects and variables. The axes are extracted in CA so the correlation between variable and object scores is as high as possible. The axes are also orthogonal. We can use methods similar to PCAs to decide how many axes to retain, particularly the Kaiser–Guttman and broken-stick criteria (Section 15.1.3).

The eigenvectors provide loadings for both objects and variables on each axis. In contrast to PCA, these

Box 15.2 ⓡ ⓔ   Worked Example of CA: Invertebrates in Artificial Ponds

Lemmens et al. (2015) did a detailed study of various biotic communities in artificial ponds in Belgium. They sampled 28 ponds that represented distinct types of management, a combination of fish farming strategies (no fish, farming young fish, low-intensity management, no management), and drainage frequencies (>10 years ago, occasional, annual). They also quantified taxon abundances for fish, zooplankton, and macroinvertebrates (different families and species within some groups) and covers of submerged, floating, and emergent vegetation. The macroinvertebrate dataset only included 23 ponds, and we will use these data to illustrate CA by examining the ordination of the macroinvertebrate community (abundances of families).

The χ^2 statistic for the independence of ponds and taxa is 24,485 (616 df, $P < 0.001$), and total inertia is 1.306. The CA extracted 22 eigenvectors (number of ponds minus one), and the first two explained approximately 40% of the total inertia.

Axis (component)	Eigenvalue	Percentage inertia
1	0.365	27.9
2	0.156	12.0
3–22	0.785	60.1
Total inertia	1.306	100.0

While axis 1 dominated the explained inertia, axes 2–7 had higher eigenvalues than the broken-stick model predicted. Nonetheless, we will plot just the first two axes, as is common with CA. The CA correlation (scaling 2) biplot of ponds and families is shown in Figure 15.4. There is not a strong pattern related to management type along principal axis 1, although no-management ponds O1, O5, and O6 separate out together along the second axis. These three ponds were associated with families Nepidae and Lepidoptera, although these taxa were not abundant and the biplot may overestimate their influence. The remaining ponds and families were clustered together on the plot.

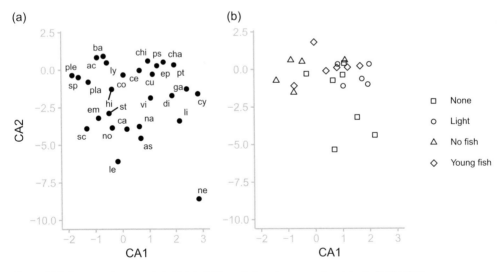

Figure 15.4 Correspondence analysis biplot of 23 ponds based on macroinvertebrate family abundances from Lemmens et al. (2015). Panel (a) shows variable scores and panel (b) shows objects, with different symbols indicating the different management types. The two-letter codes denote macroinvertebrate families, and details are in the original paper.

loadings are not often presented as part of a CA, but are used to derive scores for objects and variables for the ordination plot or biplot, which is usually the focus of CA in biology, especially when the data are taxa abundances across sampling units.

15.2.2 Ordination and Biplots for CA

The eigenvectors are used to determine a score for each principal axis for each object and variable, which are used for the ordination plot or biplot (e.g. Figure 15.3). Commonly, objects (e.g. sampling units) and variables (e.g. taxa) are plotted together as a joint plot, both being represented by points (a "point–point" plot). As with biplots in PCA, there are numerous options for scaling (or standardizing) the object and variable eigenvectors and subsequent scores. The distances between objects or between variables in the CA biplot represent χ^2 distances, depending on which scaling method is used. The three most common methods are (Legendre & Legendre 2012):

- Scaling 1, where the objects are at the centroid of the variables and the focus is on the arrangement of objects. The distances between objects represent their χ^2 distances.
- Scaling 2, where the variables are at the centroid of the objects and the focus is on the arrangement of variables. The distances between variables represent their χ^2 distances.
- Scaling 3 tries to scale object and variable scores comparatively with a method "halfway" between the first two. The downside is that neither distances between objects nor variables are proportional to their χ^2 distances.

In the context of data being taxon abundances in sampling units, the CA relativizes the abundances by totals for the particular taxon and sampling unit. The χ^2 distances between objects in the plot represent differences in relative taxon abundances (i.e. composition or "shape"), and the χ^2 distances between taxa in the plot represent differences in their relative frequencies across sampling units (Borcard et al. 2018). Greenacre (2010) described a modification of CA based on unrelativized counts (i.e. raw data), so distances between taxa are Euclidean and between sampling units are similar to Bray–Curtis dissimilarities.

Interpreting the CA biplot of object and variable scores differs from a PCA biplot. For scaling method 1 or 2, objects and variables that occur together on the plot indicate that the object has a large proportion of that variable or that the variable is more likely to occur in that object than other objects (Borcard et al. 2018). In the ecological setting, proportional representation of taxa (the relativization) and χ^2 distance has been criticized for giving too much weight to rarer taxa in the ordination plot, which may not be ecologically meaningful.

Greenacre (2013b) suggested this is due to rarer taxa being outliers in the plot but not necessarily that influential on the results of the CA. He proposed an alternative scaling method called the contribution biplot (see also Greenacre 2013a) that better reflects the relative contributions of different taxa.

Examining the CA plot with a matrix of residuals from the independence model for the contingency table will be helpful since we can see which cells have large deviations from expected values. We would expect combinations of objects and variables with large positive deviations to be near each other on the plot, whereas combinations with large negative deviations should be in opposite quadrants. This can be difficult to spot on very large tables.

15.2.3 Reciprocal Averaging

An alternative method for CA is termed reciprocal averaging (RA). This iterative procedure calculates object scores for the first axis as a weighted average of variable scores and vice-versa. At each step, the object and variable scores are rescaled so they are comparable. Final scores are obtained when the scores change little between iterations, and convergence is usually quick. The process is then repeated for the second axis. The RA procedure is tedious and produces similar scores (given rounding error) to the much more efficient matrix approach to CA when the two methods are used with the equivalent scaling. Most software for CA uses an eigenanalysis approach rather than RA.

15.3 Use of PCA and CA with Ecological Abundance (Count) Data

There are particular issues when using PCA and CA to produce an ordination of sampling units (e.g. sites) based on counts of multiple taxa (Austin 2013; Legendre & Legendre 2012; ter Braak & Smilauer 2015). While PCA is a valuable method when dealing with measurement variables, especially when there are linear relationships among the variables, this is not the case when using Euclidean distance (the implicit distance measure in PCA) between sites based on taxon abundances. When the underlying environmental gradient is long, Euclidean distance can represent two sites as being similar because they are both missing the same taxa (the "double-zero" problem; see Section 14.3.4). This results in a horseshoe pattern in the ordination plot where sites at the end of a gradient should be furthest apart on the plot, but because of numerous double zeroes, these sites are folded in to appear closer together. While this makes PCA a questionable method for ordinations on abundance data, a double standardization (each abundance divided by the maximum for that taxon and then divided by the total for that site) could remove the horseshoe distortion (Austin 2013).

Ordination plots with CA represent χ^2 distances and so don't have the double-zero problem. Nonetheless, what should theoretically be represented as a linear arrangement of sites on the ordination plot sometimes actually looks like an arch. This arch pattern may reflect the true ecological dissimilarities between sites at the ends of an environmental gradient (Wartenberg et al. 1987), but some ecologists have argued this pattern is a distortion. The causes behind this arch distortion have been a matter of considerable debate in the ecological literature (e.g. Jongman et al. 1995; Legendre & Legendre 2012; ter Braak & Smilauer 2015). Possible explanations are that the underlying χ^2 distance measure in CA (1) doesn't reach a maximum value when two sites have no species in common and (2) is measuring differences in proportional representation of species between sampling units, so it tends to weight rarer species higher in the calculation of dissimilarity than their overall abundance warrants (Legendre & Legendre 2012, but see Greenacre 2013b; McCune & Grace 2002). Another possible reason is that the second CA axis may be due to spurious polynomial axes that contribute to the χ^2 distance (ter Braak & Smilauer 2015) when there is only a single environmental gradient in reality.

One solution to the arch effect proposed by Hill and Gauch (1980) is detrended correspondence analysis (DCA). Detrending breaks the first axis up into several segments, the number determined by the user, and rescales the second axis, so its average is the same for all segments. Detrending is applied to the RA algorithm, with rescaling at each iteration. While this method effectively removes the arch effect, different numbers of segments used in the detrending process can affect the results (Jackson & Somers 1991). Following Legendre and Legendre (2012), we do not recommend DCA in most circumstances.

15.4 Constrained (Canonical) Multivariate Analysis

Principal components analysis and CA are sometimes described by ecologists as unconstrained ordination methods. The final ordination plot of objects (sampling units, sites, etc.) is determined based solely on the abundances (or presence–absence) of the taxa in the analysis. Relationships with any additional explanatory variables, such as the units' physical, chemical, or environmental characteristics, are analyzed separately after the original ordination. These methods can be modified to constrain the original ordination of objects based on taxa by the

(usually linear) relationships with a set of additional variables. These analyses are called constrained or canonical ordinations.

15.4.1 Redundancy Analysis
With PCA, we were interested in reducing dimensions of a single objects × variables dataset. Sometimes, we may have two such datasets and are interested in the relationship between the two sets; this extends a simple bivariate correlation analysis to a multivariate setting. One approach to this analysis is canonical correlation analysis, where we extract linear combinations of variables (components) from the two sets of variables, so the first component for one set has the maximum correlation with the first component from the second set. The components are termed canonical variates, and the first component from each set forms one pair of canonical variates, the second component from each set forms a second pair, etc. The number of canonical variates, and therefore pairs, is the number of variables in the smallest set. Tabachnick and Fidell (2019) provide a detailed introduction to canonical correlation analysis, and Legendre and Legendre (2012) put the method in an ecological context.

Canonical correlation analysis is not commonly used in biological research because we usually distinguish one of the datasets as response variables and the other as predictors. Therefore, an obvious extension of canonical correlation analysis would be to distinguish response and predictor variables and develop a multivariate model whereby we predict a linear combination of response variables from a linear combination of predictor variables – that is, a multivariate version of multiple regression analysis. The predictor variables need not be continuous. An important application of redundancy analysis (RDA) is when the predictors are dummy variables representing categories of categorical factors and their interactions (Legendre & Anderson 1999a). The proportion of the total variance in the response variables that can be explained by (predicted from or extracted by) a linear combination of the predictor variables is termed redundancy (Tabachnick & Fidell 2019). Besides a multivariate version of multiple regression, RDA can be viewed as a constrained PCA, where the extraction of eigenvectors from the response variable matrix is constrained to be a linear combination of the set of predictors.

Lemmens et al. (2015) related biological data (counts of invertebrate families) from their sample of artificial ponds to multivariate sets of management and environmental variables using RDA to constrain the ordination of the biotic data by linear combinations of different sets of predictors. The RDA for these data is in Box 15.3.

Box 15.3 ® ® Worked Example of Constrained Ordination (RDA and CCA): Invertebrates in Artificial Ponds

We described in Box 15.2 the study by Lemmens et al. (2015), who sampled various biotic communities in artificial ponds in Belgium that were classified into one of four management types. In addition to the biotic sampling, they recorded 15 environmental variables (e.g. depth, salinity, pH, etc.) for each pond and covers of submerged, floating, and emergent vegetation. They used RDA to comprehensively examine the patterns in the various biotic communities (ordination) constrained by either management variables, drainage strategy, environmental variables, or the fish community. We will focus on the macroinvertebrate community (abundances of families) and do two RDAs, one constrained by the four management categories and a second constrained by the continuous environmental variables; we did not include management type with the environmental variables in one analysis as some of the predictors varied with management type (i.e. collinearity). We will use a covariance matrix for the invertebrate data and a Hellinger transformation that converts raw abundances to proportional abundances. This provides a comparison to the canonical correspondence analysis (CCA) we will do as a second analysis below. We treated the environmental variables somewhat differently to the more comprehensive analysis in Lemmens et al. (2015) to simplify the number of predictors in our analysis, so our results will differ from theirs. We removed water transparency, oxygen concentration, percentage cover of submerged vegetation, and suspended solids due to correlations (>0.6) with other variables, and log-transformed chlorophyll a (the researchers log-transformed all variables except pH). The remaining variables were standardized to zero mean and unit variance. We used Type III permutation F tests for null hypotheses and present correlation (scaling $= 2$) triplots.

RDA

The first analysis only included the four management categories as predictors. The constrained proportion of the variance explained by management is about 26%, with the remaining 74% unconstrained. There were three constrained components (representing the three dummy variables for management, with no management as reference), with the first explaining 62% of the constrained variance and the second another 31%. There were 19 unconstrained components. The adjusted r^2 was only 0.143, so management category did not explain much of the taxon composition pattern among ponds. The F tests showed evidence against the null hypothesis of no effect of management type ($P = 0.003$; 3,19 df), and the largest effect was between no-management and no-fish ponds. The triplot showed that most of the no-management ponds separated from the no-fish ponds, with the remaining management types intermingled (Figure 15.5). The taxa were harder to interpret, although the Planorbidae and Baetidae were more associated with the no-fish ponds.

The second analysis included 11 of the environmental variables. The constrained versus unconstrained proportions of the variance were 58% and 42%, respectively. Of the 11 constrained components, the first two explained 57% of the constrained variance. The adjusted r^2 was again low (0.168). The F tests on individual predictors indicated log chlorophyll a ($P = 0.047$), surface area ($P = 0.057$), and percentage cover of reeds ($P = 0.076$) are related most strongly to the patterns in macroinvertebrate families. The triplot showed the three predictors along the first axis, with percentage reed cover associated with no-fish ponds (Figure 15.5).

CCA

The CCA using management category as the predictor showed about 25% of the total inertia was constrained, similar to the RDA. The first constrained axis explained about 66% of the constrained inertia, with another 20% attributable to the second axis. As with the RDA, there was an effect of management type ($P = 0.004$; 3,19 df), with no-fish management most different from no-management. The triplot did not show marked separation of ponds along the first axis (Figure 15.5).

The second CCA on the environmental variables showed about 65% of total inertia was attributable to the constrained axes, with about 51% of the constrained inertia on the first two axes. Surface area ($P = 0.010$), percentage emergent vegetation ($P = 0.013$), log chlorophyll a ($P = 0.021$), depth ($P = 0.036$), and percentage reeds ($P = 0.040$) showed relationships with the macroinvertebrate ordination along the first axis, with reeds associated with no fish and the other variables with the remaining management categories (Figure 15.5).

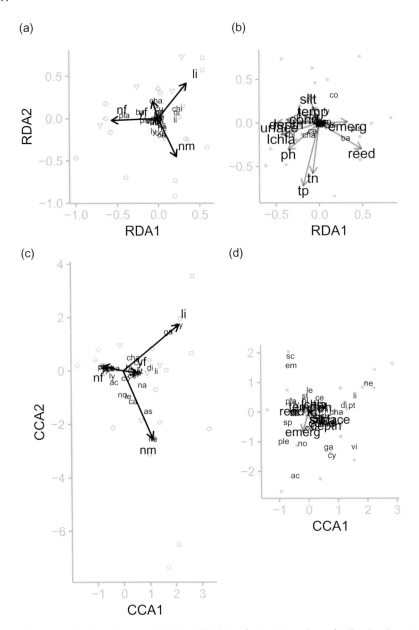

Figure 15.5 Triplots from constrained ordinations of macroinvertebrate family abundances from 23 ponds from Lemmens et al. (2015). Panels (a) and (b) show RDA constrained by four management categories (a) and 11 environmental variables (b); (c) and (d) show the equivalent triplots for CCA. Different symbols indicate different management types on the management plots and invertebrate families are indicated by two- or three-letter codes and shown in smaller fonts (as for Figure 15.4). Management and environmental arrows show centroids.

The RDA starts with a multiple linear regression model relating each response variable to the set of predictor variables and an n objects by p variables matrix of predicted Y-values for the response variables determined. This matrix of predicted Y-values is then subjected to a PCA using its covariance or correlation matrix.

The PCA step produces the canonical or constrained eigenvectors and their eigenvalues. This represents the component of the variance in our response variable matrix explained by the selected predictors. The sum of the eigenvalues for these constrained eigenvectors is the proportion of variance explained by the predictors (i.e. r^2), although an adjusted r^2 measure (Legendre & Legendre 2012) is available in most software that accounts for the number of predictors.

These constrained eigenvectors can be used to calculate scores for each object and are used as axes for the scaling/ordination of the objects. We can also derive the coefficients for each predictor on each constrained axis from the multiple regression component; even when using standardized predictors, care must be taken when comparing the sizes of these coefficients. Evaluating the overall null hypothesis that none of the predictors explain the variance in our response matrix and the null hypotheses for individual predictors can be carried out using a randomization/permutation version of an F statistic (Legendre & Anderson 1999a, b; Legendre & Legendre 2012). The variance not explained by the predictors is the unconstrained or residual variance, and there are also eigenvectors and eigenvalues associated with this residual component.

In PCA, we produced biplots where the objects are represented by points and the variables by vectors. With RDA, we not only have objects but two sets of variables: the response variables (e.g. taxon abundances) and predictor variables (e.g. environmental characteristics). Our plots become triplots (Figure 15.5), with objects represented by points and usually both types of variables by vectors; categorical predictors are typically represented on these plots by their group centroids. Triplots can be messy if there are many response or predictor variables. As with PCA biplots, there are two scaling types: distance scaling, where the distances between objects represent their true Euclidean (or modified Euclidean, such as Hellinger) distances, and correlation scaling, where the angle between two vectors (e.g. response and predictor) indicates the correlation between those variables.

A couple of comments about RDA, especially with ecological data:
• The predictor variables are usually standardized to zero mean and unit variance. This is often the default for RDA software.

• Whether the response variables are standardized depends on the data. For example, taxon abundances are not usually standardized to zero mean and unit variance. They may, however, be transformed or standardized in some other way to reduce the influence of very abundant taxa, e.g. fourth root or Hellinger transformations.
• A modification of RDA that uses a distance matrix for the response component is called distance-based RDA (db-RDA) and will be described in Chapter 16.

15.4.2 Canonical Correspondence Analysis

As we could extend PCA by including additional predictor variables in an RDA, we can also extend CA to canonical correspondence analysis (CCA). Here, the principal axes are extracted not only so they explain most of the total inertia but also so their correlation with the additional variables is maximized (Jongman et al. 1995; Legendre & Legendre 2012; ter Braak 1986). Canonical correspondence analysis is commonly used in ecological settings where objects are sampling units or sites, variables are taxa abundances, and the predictors are environmental variables associated with each unit. As with RDA, the predictors are usually standardized unless the variables are measured on the same scale/units, and differences in their variances are of interest. The abundances are transformed to their contribution to the χ^2 statistic, so additional transformation or standardization is usually unnecessary. We illustrate a simple CCA using the same data we used for RDA (Box 15.3).

While CCA can be done using CA's reciprocal averaging (RA) algorithm, it is best considered an RDA (i.e. a constrained eigenanalysis). Canonical correspondence analysis and the RDA described above differ in two ways (Greenacre 2017b; Legendre & Legendre 2012). First, the CCA is performed on the transformed (relativized or profiled) data matrix $\mathbf{Q}$ of a CA, representing each count's contribution to the χ^2 statistic. Second, the multiple regression component relating the response to the explanatory variables are weighted regression models, with the weights being each object's total as a proportion of the grand total. The matrix of residuals from these weighted regression models is converted into an SSCP, in contrast to RDA, which is usually based on a true covariance matrix. An eigenanalysis of this SSCP matrix results in eigenvectors and their eigenvalues, just like with RDA. Legendre and Legendre (2012) provide a detailed description.

Much of the CCA output resembles that from an RDA, remembering that we are dealing with inertia (total χ^2) rather than variances. The results include the constrained and unconstrained proportions of the total inertia (total χ^2), indicating how much of the independence of objects and variables is related to the predictors and how much is unrelated. We also get the eigenvalues for each constrained and unconstrained eigenvector and principal axis scores for each sampling unit and each taxon, again for constrained and unconstrained axes. As with RDA, tests of relevant null hypotheses use permutation/randomization methods.

Like RDA, we now have three components to include in our ordination plot – units, taxa, and predictor variables – so we have a triplot instead of the usual CA biplot (Figure 15.5). The units and taxa are usually represented by points and the predictors by vectors if continuous, or centroids if categorical. The two common scaling methods for a CCA triplot are similar to those described for CA; scaling 1 focusing on patterns among units and scaling 2 on patterns among taxa, with the predictors overlain on both types of plots. As with RDA, triplots can get very messy with lots of taxa or predictor variables.

15.5 Linear Discriminant Function Analysis

Discriminant function analysis, also termed linear discriminant analysis (LDA), is a "classification" technique, introduced by Fisher early last century, and is used when we have observations from predetermined factor groups and multiple variables recorded for each observation. Like PCA, LDA generates linear combinations of variables (termed discriminant functions), but now these functions are derived so they discriminate among groups the best and maximize the probability of correctly assigning observations to their predetermined groups. Linear discriminant analysis is essentially a constrained ordination method with the ordination constrained to maximize group differences. The discriminant functions can also be used to classify new observations into one of the groups and plot the observations in multivariate space. For example, Feinberg et al. (2014) used LDA to identify a new species of leopard frog based on morphological, acoustic, and genetic variables, where the groups were different frog species (including the potential new one). We present the LDA for these data in Box 15.4.

Linear discriminant analysis is mathematically equivalent to multivariate analysis of variance (MANOVA), where we model two or more response variables together against one or more categorical predictors or factors. LDA is simply a MANOVA where the response variables become the predictors, and the groups become the response; we are now modeling or predicting group membership based on multiple continuous variables.

15.5.1 Deriving Discriminant Functions

Linear discriminant analysis uses an eigenanalysis (like PCA) to extract successive discriminant functions and uses the logic of an ANOVA to partition the total variation between and within groups. The first discriminant function is the linear combination of variables that maximizes the ratio of between-groups to within-groups variance (i.e. maximizes the differences between groups). The second discriminant function is independent of (uncorrelated with) the first and best separates groups using the variation remaining after the first discriminant function has been determined, and so on. The number of discriminant functions that can be extracted depends on the number of groups and variables – it is the lesser of the degrees of freedom for groups (number of groups minus one) and the number of variables. With the acoustic data from Feinberg et al. (2014), with five species and six variables, there can be at most four discriminant functions. Like with the PCA and CA described earlier in this chapter, the first one or two usually explain most of the between-group variance.

Linear discriminant analysis is commonly based on standardized variables, although if the variables are measured in similar units and any differences in variance of the variables are of biological interest, we might just center them. We first calculate SSCP matrices (see Box 14.1), one matrix for between groups (sometimes termed the hypothesis or effect matrix, $\mathbf{H}$), one for within groups (the error or residual matrix, $\mathbf{E}$), and one for total (the total matrix, $\mathbf{T}$). The values in the main diagonal of these matrices are the univariate SS for each variable, either between-group means ($\mathbf{H}$) or pooled across observations within groups ($\mathbf{E}$). The other elements are the sums-of-cross-products between any two of the variables. For example, the cross-product for the between-groups matrix for two variables is the sum of the product of the differences between each group mean and the overall mean for one variable and the equivalent differences for the second variable. We could also base the analysis on variance–covariance matrices rather than SSCP matrices, where the former considers the degrees of freedom.

With an ANOVA, we would divide the between-groups SS by the within-groups SS. The matrix equivalent is to multiply $\mathbf{H}$ by the inverse of $\mathbf{E}$ (i.e. $\mathbf{HE}^{-1}$). We then decompose the resulting matrix product to calculate characteristic roots or eigenvalues of each linear combination or eigenvector. The eigenvalues measure how much of the between-group variance ("trace") in the variables (the sum of the between-group variances of each variable) is explained by each linear combination or eigenvector. The eigenvectors contain the coefficients or loadings for each

Box 15.4 🅡 🅔 Worked Example of Linear Discriminant (Function) Analysis: Cryptic Diversity in Leopard Frogs

Feinberg et al. (2014) examined morphological, genetic, and acoustic (call) criteria in four species of leopard frogs (*Rana sphenocephala*, *R. pipiens*, *R. palustris*, and a new species named *R. kauffeldi*) and an acoustically similar congener *R. sylvatica* (acoustic criteria only). For 283 museum specimens across the first four species, they measured size (snout–vent length) and 12 other morphological characteristics: head length, head width, eye diameter, tympanum diameter, foot length, eye to naris distance, naris to snout distance, thigh length, internarial distance, interorbital distance, shank length, and dorsal snout angle. Foot length was not recorded for 19 specimens, so these were excluded from the analysis. They also recorded seven call characteristics (call length, call rate, call rise time, call duty cycle, pulse number, pulse rate, and dominant frequency) from 45 frogs in the field across the five species. Call rate and call length were both adjusted based on regressions against temperature to a standard 14 °C. We will not cover the genetic analysis here, and we will use LDA to examine the different species in multivariate space and see if the proposed new species was distinguishable morphologically or acoustically.

Morphological Data

To account for body size differences, Feinberg et al. (2014) fitted regressions of each character against snout–vent length. They used the residuals from those models as variables in their LDA. We will use a different approach and standardize each variable to zero mean and unit variance, although the results of the two approaches are very similar. There was no evidence for skewed distributions or heterogeneity of the variance–covariance matrices across the four species (permutation $F_{3,260} = 0.863, P = 0.461$), so no transformation was applied.

We first ran an LDA with jackknife (leave-one-out) cross-validation to check our discriminant functions' success in predicting new observations. The overall accuracy was 0.795, and the table of predicted numbers versus observed numbers was as shown.

Actual	Predicted			
	R. kauffeldi	*R. sphenocephala*	*R. pipiens*	*R. palustris*
R. kauffeldi	147 (93.6%)	3	2	5
R. sphenocephala	16	30 (65.2%)	0	0
R. pipiens	4	1	23 (74.2%)	3
R. palustris	15	0	5	10 (33.3%)

These results suggested our discriminant functions could provide reasonable predictions for the new species, but were not very accurate for *R. palustris*.

We then ran the LDA without cross-validation. The first two discriminant functions explained 60% and 30% of the total variation, respectively. The predictions from this LDA were slightly more accurate than from the jackknife approach (*R. kauffeldi*: 94.3%; *R. sphenocephala*: 69.6%; *R. pipiens*: 80.6%; *R. palustris*: 36.7%), but the jackknife predictions are more realistic. The coefficients for the first two discriminant functions are as follows.

	DF 1	DF 2
Head length	−1.392	−3.194
Head width	0.005	3.220
Eye diameter	0.707	−0.136
Tympanum diameter	−0.885	−0.304
Foot length	−0.282	−0.511
Eye to naris distance	0.036	−0.885
Naris to snout distance	0.458	0.147
Thigh length	0.164	−0.148

Box 15.4 (cont.)

(cont.)

	DF 1	DF 2
Internarial distance	−0.061	0.534
Interorbital distance	−0.451	−0.006
Shank length	2.016	1.230
Dorsal snout angle	−0.255	−0.897

The plot of the first two discriminant functions is shown in Figure 15.6. There was considerable overlap between species, although the first function, most strongly representing shank length to the right and head length to the left, separated *R. kauffeldi* and *R. pipiens*. The second function, most strongly representing head width to the top and head length to the bottom, separated *R. sphenocephala* from the others.

Acoustic Data

Pulse rate was excluded from the analysis as *R. kauffeldi* has only one pulse in a call. The remaining six variables were standardized to zero mean and unit variance. There was some skewness for call rate and some evidence of heterogeneity in variance–covariance matrices between species (permutation $F_{4,40} = 4.024$, $P = 0.008$), but to be consistent with Feinberg et al. (2014) we did not transform any variables.

The jackknife predictions were as follows.

Actual	Predicted				
	R. kauffeldi	*R. spheno.*	*R. pipiens*	*R. palustris*	*R. sylvatica*
R. kauffeldi	13 (100 %)	0	0	0	0
R. sphenocephala	0	8 (100%)	0	0	0
R. pipiens	0	0	4 (100%)	0	0
R. palustris	0	0	0	11 (100%)	0
R. sylvatica	2	0	0	0	7 (77.8%)

Overall accuracy was 0.956, and all species were predicted perfectly except for *R. sylvatica*, where two frogs' calls were mistaken for *R. kauffeldi*.

The first two functions from the non-jackknife LDA models explained 61% and 24% of the variation, respectively. The predictions were identical to the jackknife model. The coefficients of the first two functions were as shown.

	DF 1	DF 2
Call length	−1.153	−5.497
Call rate	0.820	1.065
Call rise time	1.181	0.261
Call duty cycle	−0.278	−1.104
Pulse number	−4.996	5.147
Dominant frequency	0.050	0.525

The plot of the first two discriminant functions is shown in Figure 15.6(b). The five species formed distinct groups in the plot. The first function, representing mostly pulse number, separated *R. palustris* and *R. pipiens* from the rest. The second function represented pulse number again and call length, and separated *R. pipiens* from the rest.

Finally, we can reverse the LDA and do a MANOVA on both sets of data. There was strong evidence against the null hypothesis of no difference between species for both the morphological variables (Pillai = 1.198, approx. $F_{36,753} = 13.903$, $P < 0.001$) and acoustic variables (Pillai = 3.441, approx. $F_{24,152} = 38.976$, $P < 0.001$).

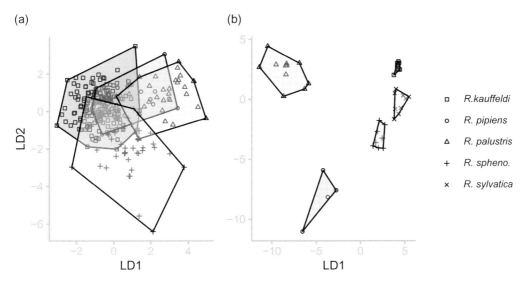

Figure 15.6 Plot of the scores for the first two discriminant functions from LDAs on 264 frogs based on 12 morphological variables (a) and 45 frogs based on six acoustic characteristics (b). Different symbols indicate species, and shaded areas show convex hulls for each species.

linear combination. The linear combination producing the largest eigenvalue is the linear combination that maximizes the ratio of between-group and within-group variance – that is, maximizes the explained variance between groups.

The eigenvectors are not usually orthogonal to each other, but can be made so by rescaling them, analogous to how we rescaled eigenvectors in PCA. A common rescaling maximizes the variance among group centroids (Legendre & Legendre 2012). Other rescaling options, such as using the square root of the relevant eigenvalue, are also possible. Determining the variables that contribute most to discriminant functions, and therefore to group separation, is usually based on the relative sizes of the scaled coefficients for each discriminant function (Box 15.4).

We can calculate discriminant function scores for each observation on each function by simply substituting the relevant observation (centered or standardized as appropriate) into each discriminant function, just as we did when determining PCA scores. These scores can be used in an LDA plot, with the first discriminant function scores on one axis and the second discriminant function scores on the other axis (Figure 15.5). Individual observations or centroids can be plotted. The distances between objects on the plot are Mahalanobis distances. These plots indicate subjectively how similar our different groups are in terms of the discriminant functions. For example, there

was a clear separation between species based on standardized acoustic variables from Feinberg et al. (2014) – see Figure 15.6.

The test of the H_0 of no difference between group centroids (i.e. a MANOVA) is sometimes recommended as the first step in an LDA because if there is little evidence against this null hypothesis, the discriminant functions will probably not be very useful for separating groups and therefore classifying observations.

15.5.2 Classification and Prediction

The second purpose of an LDA is to classify each observation into one of the groups and assess the success of the classification. A classification equation is derived for each group and is a linear combination of variables, like a discriminant function, including an intercept. A classification score for each observation for each group is then calculated by using the observed values for each variable to solve the classification equation for that group. Each observation is formally classified into the group for which it has the highest score. This may or may not be the actual group from which the observation came.

Linear discriminant analysis routines in most statistical software provide classification matrices (sometimes termed confusion matrices) that indicate the group to which each observation was classified and whether that classification was correct. The success of classifications would be greater if the groups clearly differed on the first discriminant function. For example, classification success from Feinberg et al. (2014) was higher for the acoustic data where there was a clear separation between species than for the morphological data where separation was less marked (Box 15.4; Figure 15.6).

One difficulty with the classification methodology is that we are classifying each observation with an equation derived using that observation. One way of avoiding the resulting inherent bias is to use a jackknife procedure. The classification of each observation is based on group classification functions determined when the observation is omitted (leave-one-out classification or cross-validation; Section 9.2.5). Jackknife classifications are generally less successful, but probably more realistic than the usual classifications using all observations (Box 15.4).

An even better cross-validation approach is to split the data into training and testing (or validation) sets. The LDA is developed on the training set, and then the classification is evaluated on the testing set. A constraint is that the original sample size needs to be large enough so the two datasets after splitting are adequate for model-building and testing classifications.

15.5.3 Assumptions of Discriminant Function Analysis

A key assumption for inference and classification in LDA (and MANOVA) is the homogeneity of the within-group variance–covariance matrices. Tabachnick and Fidell (2019) suggested plotting the scores for each observation for the first two discriminant functions (e.g. Figure 15.6) and checking if the spread of points is similar among the groups. Box's M test is available in many software packages, although it is very sensitive to nonnormality and considered too conservative. The best test is probably the multivariate equivalent of Levene's test for homogeneity of variances (Section 5.3.2) developed by Anderson (2006) for permutation testing of homogeneity of multivariate dispersions (Box 15.4). Strong collinearity among the variables can also be an issue.

If there is clear heterogeneity across the within-group variance–covariance matrices, transformations that remove some relationships between means and variances can help. Alternatively, you can try fitting quadratic functions instead of the usual linear ones. Quadratic functions include coefficients for squares of the variables and do not assume equal within-group covariances; statistical software usually offers quadratic discriminant analysis

(QDA) as an option. Quadratic terms are usually highly correlated with the linear term for the same variable, although standardizing the variables will usually minimize this risk.

15.5.4 Multivariate Analysis of Variance

We have already emphasized that LDA is mathematically identical to a related analysis called multivariate analysis of variance (MANOVA), which is used when we fit a linear model, usually with categorical predictors, to two or more response variables. If each response variable is of inherent biological interest, our research questions might be whether there are group or treatment effects on each variable separately and fitting separate univariate linear models is appropriate. However, we might be more interested in whether there are group differences on all the response variables considered simultaneously. Our hypothesis is now about group effects on a combination of the response variables, and instead of comparing group means on a single variable, we now compare group centroids for two or more variables. In the Feinberg et al. (2014) example we could test whether there is a difference between species for a combination of the 6 acoustic variables or the 12 morphological variables.

The MANOVA is based on the first discriminant function from an LDA, the one that best separates the groups. The null hypothesis of no difference between population group centroids can be evaluated with several test statistics (e.g. Wilk's lambda, Hotelling–Lawley trace). However, the most reliable statistic is the Pillai trace statistic (Johnson & Field 1993). All the MANOVA test statistics are usually converted to approximate F statistics. Wilk's, Hotelling's, and Pillai's statistics produce identical F statistics when there are only two groups and become Hotelling's T^2 statistic, the multivariate extension of the t statistic for comparing two group means. The remaining interpretation of a MANOVA is the same as interpreting the first discriminant function in LDA, and the assumptions are also the same.

15.6 Key Points

- Principal components analysis extracts summary variables, termed principal components, from a covariance or correlation matrix between the original variables; the first component explains most of the original variation between objects, the second is uncorrelated with the first and explains the next highest amount of variation, etc. If there are reasonable correlations among the original variables, the first few components should summarize most of the original variability.

• Each component is a weighted linear combination of the original variables. Loadings for each variable indicate how much each contributes to (is correlated with) each component; the explained variance for each component is measured by its eigenvalue.

• A component score can be calculated for each object. These object scores can be used to plot the objects on a two-dimensional (based on components 1 and 2) or three-dimensional (based on components 1 to 3) plot, termed an ordination plot. The loadings for each variable can also be included as vectors on the plot, which now becomes a biplot; different scaling methods can be used to emphasize patterns among objects or among variables.

• If our data are counts, such as abundances of different taxa (variables) for different sampling units (objects) in ecology, we can treat the data as a contingency table and use CA to extract components that summarize the lack of independence (contribution to the χ^2 statistic).

• Components are extracted jointly for variables and objects, and both can be represented on a CA biplot as points; as with PCA biplots, different scaling options are available depending on whether the focus is on objects, variables, or both.

• If we have two sets of variables, one a set of responses and the other a set of predictors, we can constrain the derivation of components from the response matrix to be a linear combination of the set of predictors. Constrained PCA is sometimes termed redundancy analysis and constrained CA is canonical corres-

pondence analysis. The ordination plots are usually presented as triplots with object scores, the response variables and the predictor variables all represented on the constrained ordination.

• When our multivariate data occur in predefined groups, we can extract our components in a way that maximizes the discrimination between groups by maximizing the ratio of between- and within-group variances; this is basically a constrained ordination based on groups being the predictor variable.

• Linear discriminant analysis is also used for classification and prediction, and we can use jackknife or cross-validation methods to determine the probability that we can correctly predict group membership for each object.

Further Reading

Principal components analysis: Jackson (2003) and Tabachnick and Fidell (2019) cover PCA in detail, with Legendre and Legendre (2012) providing an ecological perspective and Borcard et al. (2018) giving practical examples.

Correspondence analysis: Greenacre (2017a) is the standard text on CA, with Legendre and Legendre (2012) also providing a good explanation for ecologists.

Constrained analyses: Legendre and Legendre (2012) cover redundancy and canonical correspondence analyses.

Linear discriminant analysis: Huberty and Olejnik (2006) and Tabachnick and Fidell (2019) provide detailed overviews, and Legendre and Legendre (2012) put LDA in an ecological context with examples.

16 Multivariate Analyses Based on (Dis)similarities or Distances

In the previous chapter, we were mainly interested in analyses based on associations between variables. Objects were plotted as points where the distances between objects reflected their Euclidean or chi-square distances. Variables could be included on these plots as vectors or additional points. The plots could be constrained by a second set of variables, such as an ordination of taxon abundances being constrained by a set of environmental variables.

In the first part of this chapter, we focus on analyses that examine the relationship between objects directly, based on their similarities or dissimilarities. Objects are plotted on an ordination plot so the distances between objects represent their dissimilarities. Alternatively, they can be positioned on a tree-like diagram (e.g. a dendrogram) using cluster analysis, so the objects most dissimilar separate at the top of the tree and objects close together at the bottom branches are more similar. The second part of the chapter will look at the contribution of variables to the positioning of objects, or the relationships with a second set of variables (e.g. environmental variables or factors) determined by subsequent analyses after the initial ordination or clustering.

Some of the dissimilarity measures for dichotomous and continuous variables were outlined in Chapter 15; also see Legendre and Legendre (2012) for a complete treatment. One of the main advantages of the methods we describe in this chapter is that the user chooses the dissimilarity measure. That choice is critical, and different dissimilarities can cause very different patterns and interpretations when analyzed. The transformation and standardization of variables and objects, combined with the dissimilarity measure, can also be influential.

16.1 Multidimensional Scaling or Ordination

Multidimensional scaling (MDS) refers to a broad class of procedures that scale objects based on a reduced set of new variables derived from the original variables (Cox & Cox 2000). As the name suggests, MDS is specifically designed to represent relationships between objects graphically in multidimensional space as an ordination plot so the distances between objects represent the underlying dissimilarities. The plots look like those derived from association matrices via eigenanalysis in PCA and correspondence analysis (CA), especially if the same distance measures are used. One advantage of the MDS methods is that any dissimilarity or distance measure can be used, providing more flexibility when dealing with abundances in ecological research.

The basic data structure we will use is a data matrix of n objects by p variables. Any two objects will be identified as h and i (following Legendre & Legendre 2012). Their dissimilarity is d_{hi}. The distance between any two objects in the scaling (configuration) plot is $d_{\hat{h}i}$, and it is usually measured as simple Euclidean distance. Unfortunately, there is some inconsistency in the symbols used for dissimilarity and inter-object distance in the literature, with δ commonly used for dissimilarity. This seems inappropriate as Greek letters are usually reserved for unknown parameters.

Multidimensional scaling can be based on any of the dissimilarity measures described in Section 14.3, but is not restricted to them, and measures such as genetic dissimilarities (e.g. Nei's) are not uncommon.

16.1.1 Metric (Classical) Scaling: Principal Coordinates Analysis

Principal coordinates analysis (PCoA) is closely related to PCA and is sometimes called classical scaling. It is not used that much anymore in biology, so we'll only introduce it briefly; Legendre and Legendre (2012) provide details. We will illustrate PCoA using Lemmens et al.'s (2015) pond invertebrates example (see Box 16.1).

The steps in PCoA are:

• Create an $n \times n$ matrix of dissimilarities between objects (d_{hi}), using your preferred dissimilarity measure.
• Transform these dissimilarities to $-0.5d_{hi}^2$. This transformation maintains the original dissimilarities during subsequent calculations (Legendre & Legendre 2012).
• These transformed dissimilarities are double centered by subtracting the means for the relevant row and column and adding the overall mean from the dissimilarity matrix. This centering removes the first, and trivial, eigenvector in the next step. The relative positions of the objects in the final configuration won't be affected by the double centering.
• This symmetric $n \times n$ matrix of transformed dissimilarities is then subjected to an eigenanalysis to obtain the eigenvectors and their eigenvalues, in the same way as we did in an R-mode PCA. Hopefully, most of the information (as measured by the eigenvalues) in the dissimilarity matrix will be in the first few eigenvectors.
• As with PCA (Chapter 15), the eigenvectors are scaled, usually by the square roots of the eigenvalues (Legendre & Legendre 2012).
• The coefficients of these eigenvectors are then used to position the objects on the scaling plot (Figure 16.1).

If the original data were centered by variable means and Euclidean distance was used to create the matrix of object dissimilarities, the relative positions of objects in the PCoA ordination will be similar to those from a PCA based on a covariance matrix of variables. If the original data were double transformed by row and column totals so chi-square distance was used for dissimilarity, the objects' positions in the PCoA ordination will be like those from a CA. Principal coordinates analysis can be viewed as a generalization of PCA that allows a much wider range of dissimilarity measures to be used.

Principal coordinates analysis can also be viewed as a translation of inter-object dissimilarities between objects into Euclidean distances (Legendre & Anderson 1999a). If the dissimilarities were metric (e.g. Euclidean or chi-square), and all eigenvectors retained, the distances in principal coordinate space are the same as the original dissimilarities because all the variance in the original dissimilarity matrix is retained in the principal coordinates. However, biologists often use nonmetric dissimilarities, like Bray–Curtis for taxon abundance data, and the principal coordinates represent only part of the variation in the dissimilarities. Unfortunately, the remainder may be represented by negative eigenvalues, which are very difficult to interpret. This may not be a problem if we use PCoA as a variable-reduction technique because the first few eigenvalues will be positive. However, if we wish to use all the principal coordinates derived from a nonmetric dissimilarity matrix, such as in distance-based redundancy analysis (db-RDA; see Section 16.3.3), we usually have to correct for the negative eigenvalues. These corrections are somewhat technical (Legendre & Anderson 1999a; Legendre & Legendre 2012) and may cause conservative tests of complex hypotheses (McArdle & Anderson 2001).

16.1.2 Nonmetric (Enhanced) Multidimensional Scaling

Methods for MDS more familiar to biologists involve additional steps, beyond an initial scaling by PCoA, to improve the fit between the observed dissimilarities between objects and the inter-object distances in the configuration. Jackson (2003) termed these methods "enhanced multidimensional scaling," although most biologists know them as nonmetric MDS (NMDS). Basically, these methods iteratively reposition the objects in the configuration using an algorithm that improves the fit between the dissimilarities and the inter-object distances, the latter measured by a form of Minkowski metric such as Euclidean distance.

We present the NMDS analysis of the pond invertebrate data from (Lemmens et al. 2015) in Box 16.2. A second example is from Parkinson et al. (2020), who assessed the effect of artificial light at night on invertebrates living in experimental ponds (Box 16.3).

16.1.2.1 Deriving the Ordination

The steps for NMDS are as follows (Figure 16.2):

1. Set up a data matrix and decide about transformations or standardizations of the data.
2. Calculate a matrix of dissimilarities between objects (d_{hi}) using the preferred dissimilarity. Similarities could also be used; it makes no difference in the next steps.
3. Decide on the number (k) of dimensions (i.e. axes) for the scaling, which will be a compromise between the need to get the fit between dissimilarities and inter-object distances as good as possible and minimizing the number of scaling dimensions for simple plotting and interpretation.
4. Arrange the objects in a starting configuration in the k-dimensional space, at random or more commonly using coordinates from a PCoA or even a PCA.

Box 16.1 ® Worked Example of PCoA: Invertebrates in Artificial Ponds

We introduced the study by Lemmens et al. (2015), who sampled macroinvertebrate families in artificial ponds classified into one of four management types, to illustrate CA and redundancy analysis (RDA) in Boxes 15.2 and 15.3. We will use these data to produce a PCoA ordination based on Bray–Curtis dissimilarities between the 23 ponds using the abundances of 29 invertebrate families.

The first three eigenvectors had eigenvalues greater than predicted by the broken-stick model. For simplicity of plotting, we will just use the first two, which had eigenvalues of 1.426 and 1.266, representing 50% of the total variance. The last five eigenvalues were zero or negative.

The PCoA ordination plot of the 23 ponds is shown in Figure 16.1. It did not show a marked separation of the management types, although the no-management and no-fish ponds broadly separated as did the no-management and light-management ponds.

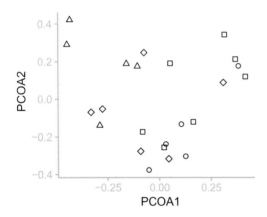

Figure 16.1 PCoA ordination plot of the 23 ponds from Lemmens et al. (2015), based on a Bray–Curtis dissimilarity matrix. Management types are represented by different symbols.

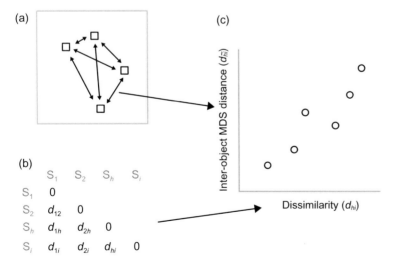

Figure 16.2 Illustration of the links between (a) configuration (ordination) plot, (b) dissimilarity matrix, and (c) Shepard plot in NMDS. S_1, S_2, etc., are objects (e.g. sampling units), shown as squares on the configuration plot.

Box 16.2 ® ⓔ Worked Example of NMDS with ANOSIM, PERMANOVA, MV-ABUND, BIO-ENV: Invertebrates in Artificial Ponds

We continue with the pond invertebrate example from Box 16.1. We will now use these data to produce an NMDS ordination plot showing the relationship among ponds based on invertebrate family abundances. We will compare this analysis to one where the abundances are converted to proportions of pond totals and where abundances are converted to presence–absence data. We will also compare the four management groups using ANOSIM, PERMANOVA, and MV-ABUND, and examine relationships between the patterns among ponds based on invertebrate families and patterns based on environmental variables.

We constructed three dissimilarity matrices: the first used Bray–Curtis dissimilarity based on original abundances; the second also used Bray–Curtis with abundances standardized to proportions of ponds totals to reduce the influence of the most abundant taxa; and the final one used Jaccard dissimilarity based on presence–absence of the families to remove the abundance component altogether. We then ran NMDS on these dissimilarity matrices and plotted the ponds in two-dimensional space in an ordination diagram. In the previous chapter, we included vectors representing the 11 environmental variables used for the RDA and CCA analyses.

All three MDS analyses converged to two-dimensional solutions with stress < 0.2 (BC-raw $= 0.115$, BC-prop $= 0.165$, Jaccard $= 0.168$). The ordination plots (Figure 16.3) did not show a marked separation of the management types, although the no-management and no-fish ponds broadly separated using original abundances (less marked using proportional abundances). One no-management pond differed greatly from the remaining ponds when using presence–absence data; this pond was missing some of the more abundant families.

To be consistent with the RDA and CCA on these data (Box 15.3), we did separate analyses relating the biotic dissimilarities to the categorical predictor management type and the 11 continuous predictors (some of the latter varied with management type). There was no evidence for heterogeneity of dispersions (variance–covariance matrices) between the four management types, even for the presence–absence data where the ordination showed one unusual pond. The ANOSIM, PERMANOVA, and MV-ABUND results suggested evidence against the null hypothesis of no difference between management types, especially for the raw abundance data.

	Homog dispersions		ANOSIM		PERMANOVA		MV-ABUND	
	$F_{3,19}$	P	R	P	$F_{3,19}$	P	Deviance	P
Raw	1.061	0.389	0.173	0.018	2.124	0.002	232.1	0.002
							Neg binomial (3 df)	
Prop	0.519	0.674	0.118	0.075	1.967	0.036	8.3	0.001[*]
							Neg binomial (3 df)	
Pres-abs	2.173	0.125	0.102	0.066	1.742	0.013	127.5	0.021
							Binomial (3 df)	

* Note that the proportional abundances were noninteger.

PERMANOVA and MV-ABUND can also be used with continuous predictors, although we can't easily check for homogeneity of dispersions. To illustrate, we will just use the raw biotic data and the 11 uncorrelated and standardized predictors we used in the RDA and CCA in Box 15.3. As both analyses are exploratory – that is, there was no biological reason for terms to be entered in a specific order – the PERMANOVA is based on marginal (Type III) tests and the MV-ABUND based on Wald tests.

Box 16.2 (cont.)

Predictor	PERMANOVA		MV-ABUND	
	$F_{1,11}$	P	Wald[*]	P
Depth	1.533	0.147	6.181	0.400
Silt thickness	0.842	0.546	6.815	0.281
pH	1.192	0.311	6.419	0.367
Temperature	1.050	0.396	6.454	0.351
Surface area	1.155	0.330	7.055	0.224
Conductivity	0.266	0.988	4.694	0.774
Log chlorophyll a	1.441	0.174	7.419	0.211
Total N	0.576	0.821	5.298	0.652
Total P	1.590	0.132	6.962	0.317
Percentage reeds	1.082	0.365	6.079	0.463
Percentage emergent	1.778	0.098	5.459	0.598

[*] Negative binomial, df = 1, 11

While the relative P-values differed, sometimes markedly (e.g. percentage emergent), between the two approaches, neither suggested that any of the predictors had a strong relationship with the raw abundances.

We also used the BIO-ENV procedure to select which subset of the 11 standardized predictors provided the highest rank correlation with the biotic dissimilarities based on raw abundances. The best subset included depth, total P and percentage emergent vegetation, but the Spearman correlation was only 0.105; these predictors also had the highest F statistics from the PERMANOVA.

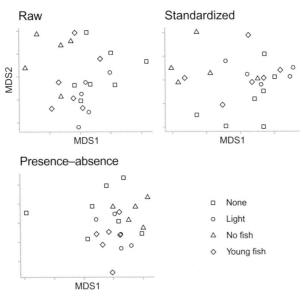

Figure 16.3 NMDS ordination plot of the 23 ponds from Lemmens et al. (2015), based on a Bray–Curtis matrix of dissimilarities between ponds using raw abundances, abundances standardized by pond totals, and presence–absence (Jaccard). Different symbols indicate the management types.

5. Move the location of objects in the k-dimensional space iteratively so the match between the inter-object distances in the configuration ($d_{\hat{h}i}$) and the actual dissimilarities (d_{hi}) improves at each step.

6. The final object positions and therefore the final ordination plot is achieved when further moving objects no longer improves the match between plot distances and actual dissimilarities.

Box 16.3 ℝ 🄴 Worked Example of NMDS with PERMANOVA, MV-ABUND, SIMPER: Artificial Light and Aquatic and Terrestrial Invertebrates

Parkinson et al. (2020) tested experimentally the effect of artificial light at night (ALAN) and fish presence on abundances of terrestrial and aquatic invertebrates in the littoral zone of a small lake. They established 20 mesocosms (1 m² plastic mesh walls and bottom submerged, with part of the walls above the water surface), with 10 provided with ALAN (solar lights in each corner) and 10 controls. Half the mesocosms in each group had fish added, but there were no effects of fish on any variables, so Parkinson et al. ignored fish in the subsequent analyses. After about six weeks, a single pan trap was placed in each mesocosm overnight, and invertebrates were identified to family and counted. We will use these data to do an ordination of mesocosms based on invertebrate family abundances and then fit models to examine differences between the two treatment groups (ALAN vs. no-ALAN).

We calculated a Bray–Curtis matrix of dissimilarities between mesocosms based on original abundances, but also on proportional abundances within each mesocosm to reduce the effects of the dominant taxa. The NMDS on the original abundances converged quickly in two dimensions with a stress of 0.031. The ordination plot showed a clear difference between the two groups except for two ALAN mesocosms that were grouped with the no-ALAN ones (Figure 16.4). The SIMPER analysis showed the two families most driving the difference were Caenidae (average/SD ratio: 2.558) and Chironomidae (2.004), followed by Baetidae (1.154) and Dolichopodidae (1.006). The NMDS on proportional abundance had a stress of 0.144 in two dimensions; the groups did not separate quite as clearly as previously (Figure 16.4). The SIMPER analysis showed three of the same families as before most contributing to the difference between groups were Caenidae (1.918), Baetidae (1.617), and Chironomidae (1.477); the contributions of Caenidae and Chironomidae were not just due to their higher overall abundance.

We evaluated heterogeneity of dispersions and used both ANOSIM and PERMANOVA to formally assess the difference between groups.

	Homog. dispersions		ANOSIM		PERMANOVA	
	$F_{1,18}$	P	R	P	$F_{1,18}$	P
Raw	0.456	0.508	0.678	0.001	15.415	0.001
Prop	3.518	0.077	0.299	0.001	2.549	0.001

There was no evidence for unequal dispersions for the raw abundances, and both ANOSIM and PERMANOVA indicated a clear difference between the two groups. There was a difference in dispersions for the proportional abundances, and this was evident in the ordination plot (Figure 16.4). This means the ANOSIM and PERMANOVA analyses are less reliable, and although both showed evidence for differences between the groups, this may reflect a difference in dispersions, not just locations. We could have tried transforming the abundances (e.g. to fourth roots) to make the dispersions more similar, but we will illustrate fitting multivariate generalized linear models (MV-ABUND) to both raw and proportional abundances using a negative binomial distribution. There was evidence for a difference between the groups for the raw abundances (df = 1,18, deviance = 106.7, $P = 0.001$). Abundances standardized to proportions of totals are noninteger, so fitting a negative binomial (or Poisson) model is not strictly valid as it requires data to be counts. These results should be interpreted cautiously, but nonetheless indicated a difference between groups (df = 1,18, deviance = 10.3, $P = 0.001$).

We can show the relationship between inter-object distances and dissimilarities for all pairs of objects in a Shepard diagram, which is a scatterplot with dissimilarity on the horizontal axis and inter-object distance on the vertical axis (Figure 16.2). Now consider a regression model relating inter-object Euclidean distance as the response variable to dissimilarity as the predictor. The differences between the observed inter-object distances and those predicted by the regression model ($\hat{d}_{hi}$), sometimes termed "disparities" in the NMDS literature, are the residuals from the regression

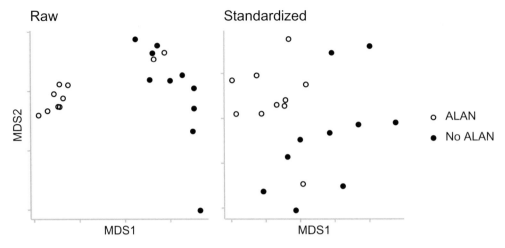

Figure 16.4 NMDS ordination plots of the 20 mesocosms from ALAN and control groups from Parkinson et al. (2020) based on Bray–Curtis dissimilarities calculated from raw abundances of invertebrate families and abundances standardized to proportions of mesocosm totals.

model. These residuals can measure the match between the dissimilarities and the inter-object distances.

One measure of fit is Kruskal's stress:

$$\sqrt{\frac{\sum\left(d_{\tilde{h}i} - \hat{d}_{\tilde{h}i}\right)^2}{\sum d_{\tilde{h}i}^2}}.$$

The summation is over all $n\,(n-1)/2$ object pairs. If there is a perfect match between inter-object distance and dissimilarity (i.e. they are directly proportional to each other), residuals and stress will be zero. The lower the stress value, the better the match. Other versions of stress are also used to measure fit (e.g. see Jackson 2003; Legendre & Legendre 2012). It is important to know which your software uses because they are scaled, and therefore interpreted, differently. The version shown is most used by biologists.

Shepard plots often show a nonlinear relationship between inter-object distance and dissimilarity (Figure 16.2). While this might suggest that a nonlinear model relating inter-object distance and dissimilarity is most appropriate, a more robust approach fits a monotonic regression. This is a form of nonparametric regression that relates the rank orders of the two variables (Section 6.6). Stress now measures the concordance in the ranks of the observed and predicted inter-object distances.

The algorithms for enhanced MDS are iterative, and the number of iterations depends on the complexity of the data. More rapid convergence can be achieved if the coordinates from an initial PCA or PCoA scaling are used as the starting configuration, and some software defaults to a preliminary PCoA. The iterative nature means that the iterations can converge to a "local" rather than global (i.e. best) solution. This problem can be avoided by repeating the MDS many times with different random starting configurations and comparing the stress and axis coordinates of the different configurations. We can only be confident of the final configuration if it occurs from most random starts. Configurations can be compared using Procrustes analysis, where one configuration is rotated and rescaled to most closely match a second. The fit is measured by the sum of squared distances between the corresponding objects in the two configurations.

16.1.2.2 Interpretation of Ordination Plot

The ordination plots (i.e. the final configurations) for our two worked examples are shown in Figures 16.3 and 16.4. Interpreting these plots depends on how well they represent the actual dissimilarities – that is, the stress value. Clarke (1993) suggested for ecological (taxon abundance) data that stress values >0.3 indicate the configuration is no better than arbitrary, and we should only try to interpret configurations with stress values <0.2, and ideally <0.15. We can always reduce the stress value by adding dimensions to the scaling. However, the more dimensions we use, the harder it is to display and interpret the final configuration, so we are trying to compromise between minimizing stress and minimizing the number of dimensions. Our experience with most ecological data is that two or three dimensions will usually produce adequate configurations.

The final orientation of the configuration is arbitrary, and only the relative distances between objects are relevant to interpretation. It is preferable to rotate the final configuration so the first axis lies along the direction of maximum variation. This can be achieved by a PCA on the NMDS axis scores (Clarke et al. 2014) and may be the default for some NMDS software. Note that actual object scores are also arbitrary, and they can be scaled in several ways. Only the relative distances between objects are important. Plots of the final configuration do not need scales on the axes, as long as the axes are scaled identically.

As with ordination plots in Chapter 15, interpreting the final scaling NMDS plot is subjective. Objects closer together are more similar (e.g. in taxonomic composition) than those farther apart. A useful addition to the plot is a minimum spanning tree, where the objects are joined by lines so the sum of line lengths is the smallest possible and there are no closed loops. Minimum spanning trees can be applied to any scatterplot of points. For NMDS configurations, objects joined by the shortest spans are closest on the plot and those separated by longest spans are furthest apart. Minimum spanning trees can be plotted in three dimensions, although they become ugly to interpret.

16.2 Cluster Analysis

Another way to represent dissimilarities among objects is to use cluster analysis to combine similar objects into groups or clusters that can usually be displayed in a tree-like diagram, called a dendrogram (Figure 16.5). Fielding (2006) distinguished cluster analysis, which places objects into natural groups, from classification, which places objects into predefined groups (e.g. linear discriminant analysis [LDA], Section 15.5). More sophisticated methods include decision trees, of which regression trees (Chapter 9) are an example, and artificial neural networks. Our focus will be on clustering. Probably the most important use of cluster analysis in biology is taxonomic and phylogenetic research, where the dissimilarity measures are often morphological or genetic/molecular differences between organisms, species, etc., and the dendrogram represents a possible evolutionary sequence.

16.2.1 Agglomerative Hierarchical Clustering

Agglomerative methods start with individual objects and join objects and then objects and clusters together until they all form one big group. This form of cluster analysis is familiar to most biologists. Objects are usually clustered, but sometimes you may wish to cluster variables (e.g. taxa). Most algorithms for agglomerative cluster analysis start with a matrix of pairwise similarities or dissimilarities between the objects, and the steps are:

1. Calculate a matrix of dissimilarities (d_{hi}) between all pairs of objects.
2. The first cluster is formed between the two objects with the smallest dissimilarity.
3. The dissimilarities between this cluster and the remaining objects are then recalculated.
4. A second cluster is formed between cluster 1 and the most similar object.
5. The procedure continues until all objects are linked in clusters.

The graphical representation of the cluster analysis is a dendrogram (Figure 16.5), showing the links between groups of objects with the line lengths representing dissimilarity. Like ordination plots, interpreting the groupings in the dendrogram is subjective. The decision about which groups to report is usually based on some arbitrary cut-off value for dissimilarity or a scree plot (as used for PCA) to look for marked discontinuities in the relationship between dissimilarity and cluster fusion (group formation) number starting at the bottom (Fielding 2006). If there are many objects, the standard dendrogram can be long and difficult to represent on a single page or screen.

The major difference between the available hierarchical agglomerative clustering methods is how the dissimilarities between clusters and between clusters and objects (step 3) are recalculated. These are termed linkage methods, and three common ones are:

• Single linkage (nearest neighbor), where the dissimilarity between two clusters is measured by the minimum dissimilarity between all combinations of two objects, one from each cluster.
• Complete linkage (furthest neighbor), where the dissimilarity between two clusters is measured by the maximum dissimilarity between all combinations of two objects, one from each cluster.
• Average linkage (group average or mean), where the dissimilarity between two clusters is measured by the average dissimilarity between all combinations of two objects, one from each cluster. The group mean (or average) linkage strategy, commonly called unweighted pairgroups method using arithmetic averages (UPGMA), is often recommended. There is a weighted version of UPGMA (WPGMA), which weights the original dissimilarities differently, and unweighted clustering based on centroids (UPGMC), which is equivalent to UPGMA except that we use centroids instead of means.

Legendre and Legendre (2012) discuss the pros and cons of these different linkage methods, and other clustering types. If there are distinct groups in your data, the different methods produce similar dendrograms, but less so when groups are unclear.

Box 16.4 ⓡ ⓔ Worked Example of Cluster Analysis: Cryptic Diversity in Leopard Frogs

We introduced in Box 15.4 the study of Feinberg et al. (2014), who used morphological, genetic, and acoustic (call) criteria to discriminate five species (one new) of congeneric frogs. We will use their acoustic data (six variables) to hierarchically cluster their specimens based on a Euclidean dissimilarity matrix calculated from standardized call variables. We used UPGMA, and the dendrogram is shown in Figure 16.5. The species segregated out strongly in the dendrogram. There was some intermingling of *R. pipiens* and *R. palustris* individuals, and the new species, *R. kauffeldi*, had acoustic characteristics in common with *R. sylvatica*. One *R. sylvatica* individual separated out early from the *R. sphenocephala* group. Interestingly, this slight intermingling of two *R. sylvatica* individuals was not evident from the LDA plot (Figure 15.6). However, this was the one species for which jackknife predictions from the LDA were <100%.

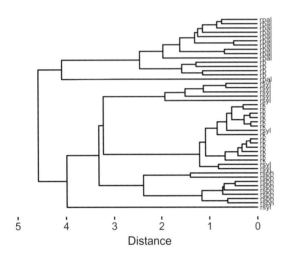

Figure 16.5 Dendrogram from hierarchical UPGMA cluster analysis of individual frogs from five species based on Euclidean distances calculated from standardized acoustic variables from Feinberg et al. (2014). Species codes are abbreviations of the names in Box 15.4.

Agglomerative cluster analysis has some disadvantages, primarily when interpreting the dendrogram. The hierarchical approach means that once a group or cluster is formed from two or more objects, it cannot be broken. Consequently, the dendrogram does not represent all pairwise dissimilarities between objects like in NMDS. A misleading cluster formed early in the process will influence the remaining clusters. The analysis also forces objects into clusters, and it would be easy for naïve biologists to place too much emphasis on these clusters without examining the actual dissimilarities. We prefer NMDS to represent relationships between objects based on dissimilarities graphically.

We illustrate UPGMA hierarchical cluster analysis for the acoustic variables for five species of frogs from Feinberg et al. (2014) in Box 16.4.

16.2.2 Divisive Hierarchical Clustering

Divisive methods have a long history for clustering ecological data. They start with the objects in a single group and split them into smaller and smaller groups until each

is a single object. One method once popular with ecologists is two-way indicator species analysis (TWINSPAN). This complex procedure uses the reciprocal averaging algorithm of CA (Chapter 15) to successively divide the first axis for both sampling units and species into smaller groups. The output includes a two-way table that orders the sampling units and species and shows the groupings and the relative abundances of species for each sampling unit. Our experience is that TWINSPAN is no longer common in ecological research.

16.2.3 Nonhierarchical Clustering

Nonhierarchical methods do not represent the relationship between objects in hierarchical form. They start with a single object and cluster other objects similar to the first one. In contrast to hierarchical clustering, objects can be reassigned to clusters during the clustering process. One method common in statistical software is *K*-means clustering – see Legendre and Legendre (2012) for a good description. *K*-means works by splitting the objects into a predefined number (*K*) of clusters. Then cluster membership of objects is iteratively re-evaluated by some criterion, such as to maximize the ratio of between-cluster to within-cluster variance.

16.3 Analyses Based on Dissimilarities

Ordination plots and dendrograms are imperfect representations of actual dissimilarities. Several statistical methods analyze the dissimilarities themselves, including using them as responses in a linear model. For example, we might wish to relate dissimilarities between units based on taxon abundances to dissimilarities based on environmental variables or genetic differences between populations to their geographic separation. Fortin and Gurevitch (2001) emphasized the importance of examining spatial structure in field experiments, where one matrix might be differences in responses of experimental units, and the other might be the actual physical distances between the units.

16.3.1 Contributions of Original Variables to Ordination

One disadvantage of methods like NMDS is that it can be difficult to determine how the variables influence the pattern on the ordination plot. We don't have the equivalent of variable loadings as we did with PCA, and biplots in the sense of joint scaling of objects and variables as with CA are impossible. We could correlate each variable with the NMDS axis scores, but recall that the axes are just used to define objects' positions in multidimensional space.

Clarke et al. (2014) described a procedure termed SIMPER (similarity percentages) for ecological data. This technique is designed for situations where the objects fall into predefined groups, such as a single-factor experimental design. It determines the taxa contributing most to group differences. For example, the Bray–Curtis dissimilarity for a pair of sampling units is basically the differences between the units in the abundance of each taxon, summed over all the taxa. SIMPER computes the percentage contribution of each taxon to the dissimilarities between all pairs of sampling units in different groups and the percentage contribution of each taxon to the similarities between all pairs of sampling units within each group. It then calculates the average of these percentage contributions, with its SD. Taxa with a large ratio of average/SD percentage contribution to the dissimilarity between sampling units in different groups best discriminate between the groups. There are no formal tests of hypotheses with SIMPER, just a list of taxa in order of their percentage contributions. With many taxa, the output from a SIMPER analysis can be voluminous and difficult to interpret. We illustrate a SIMPER analysis for the artificial light data from Parkinson et al. (2020) in Box 16.3.

16.3.2 Relating Dissimilarities to Other Variables

In Chapter 15, we described constrained ordination methods that incorporated additional variables (e.g. environmental variables) as part of the ordination of taxon abundances. Ordinations like NMDS are unconstrained, and describing the relationship between a set of dissimilarities, such as those based on taxon abundances, and other variables recorded on the same objects requires different methods. Some of these methods have been developed for specific cases where we have one or more grouping variables.

16.3.2.1 Mantel Test

The simple Mantel test tests hypotheses about correlations between two distance matrices. It uses a randomization procedure to test whether there is a linear (based on correlation coefficient) or monotonic (based on Spearman correlation coefficient) relationship between the two sets of distances (Legendre & Legendre 2012). The permutation component is because the elements in the two matrices are not independent of each other (the dissimilarity between objects 1 and 2 uses some of the same information as the dissimilarity between objects 1 and 3, etc.). Fortin and Gurevitch (2001) phrased the null hypothesis as: the observed correlation between the elements of the two matrices could be obtained by a random arrangement of objects. Other statistics equivalent to the correlation coefficient in Mantel's test include Z (the sum of the products of the corresponding elements in the two matrices) and the regression coefficient (slope) for elements in one matrix regressed against elements in the other matrix. If the distances in the two matrices are standardized to zero mean and unit variance, the values of the correlation coefficient, the regression slope, and Z/m, where m is the number of elements in each matrix, will be the same (Manly & Navarro Alberto 2022). Legendre and Fortin (2010) emphasized that Mantel tests can only address biological questions framed in terms of distances and not about the raw data that generated those distances. Guillot and Rousset (2013) were critical of the widespread use of Mantel tests in evolutionary biology, especially where spatial autocorrelation is likely.

To illustrate a Mantel test, we will use the pond invertebrate data from Box 16.2, where we have a matrix of Bray–Curtis similarities for the invertebrates and a similar matrix of Euclidean distances for the environmental variables. The correlation coefficient between the two matrices was -0.015 with a randomization P-value of 0.485 suggesting no detectable relationship between fauna and environmental variables. The Spearman correlation was similar (-0.044), with a more conservative P-value (0.585).

The Mantel test can be extended to more than two matrices. For example, we might have a matrix of genetic distances between populations, a matrix of dissimilarities based on environmental characteristics associated with each population, and a matrix of geographic distances. The partial Mantel test is an analog of the partial correlation coefficient, the correlation between matrices 1 and 2 controlling for matrix 3. Legendre and Legendre (2012) provide a worked example. Guillot and Rousset's (2013) criticism of Mantel tests extended to partial Mantel tests, especially the problems with these tests under conditions of spatial autocorrelation.

16.3.2.2 *BIO-ENV*

Clarke and Ainsworth (1993) proposed a procedure for relating dissimilarities from biotic data to other variables based on the Mantel test principles. Their procedure, BIO-ENV, measures the correlation between dissimilarities (e.g. Bray–Curtis) between units based on taxon composition (presence–absence or abundances) and the dissimilarities (e.g. Euclidean) between the same units based on environmental variables, usually standardized to unit variance. It then determines the Spearman rank correlation coefficient between the rank orders of these two matrices. Each pair of observations for the correlation will be the rank of the Bray–Curtis dissimilarity (from species abundances) between objects h and i and the rank of their Euclidean distance (from environmental variables). Note that Pearson's correlation could be used, but Clarke and Ainsworth (1993) and Clarke and Gorley (2015) emphasized using ranks with methods like BIO-ENV that are commonly used with NMDS.

BIO-ENV can be used in a stepwise fashion to find the combinations of environmental variables that produce dissimilarities between units with the highest correlations with dissimilarities based on taxa presence–absence or abundances (Clarke et al. 2014; see also Clarke & Gorley 2015). This is a descriptive approach, as they argued that their implementation of the Mantel test is not suitable for hypothesis testing because their stepwise procedure produces numerous correlation statistics and associated P-values that would be difficult to interpret (see also Section 9.2.2). We will illustrate the BIO-ENV procedure in Box 16.2 as part of an NMDS.

16.3.2.3 *Matrix Regression*

Legendre and Legendre (2012) briefly describe an extension of the Mantel test principles to fit multiple regression models to distance matrices. Matrix regression has been applied to general approaches for fitting linear models relating dissimilarities based on taxon composition/abundances to environmental or other predictors. These methods include db-RDA and PERMANOVA, which we discuss in Section 16.3.3. Ferrier et al. (2007) introduced generalized dissimilarity modeling (GDM) for predicting beta-diversity. Generalized dissimilarity modeling extends matrix regression to deal with nonlinearity by using generalized linear models and spline functions.

Tests based on dissimilarities are not straightforward for two reasons. First, the dissimilarities between objects are not always independent of each other, so randomization (permutation) procedures are required. Second, if we wish to use the dissimilarities in linear models, a limitation of some approaches is their inability to deal with nonmetric dissimilarities (see Anderson 2001).

16.3.2.4 *Analysis of Similarities*

Another method related to the Mantel test is the analysis of similarities (ANOSIM; Clarke 1993, Clarke et al. 2014) for testing for differences in dissimilarities between predefined groups of objects. It is commonly used with Bray–Curtis dissimilarities, although it could use any dissimilarity measure. This procedure uses a test statistic (R) based on the difference between the average of all the rank dissimilarities between objects between groups ($\bar{r}_B$) and the average of all the rank dissimilarities between objects within groups ($\bar{r}_W$):

$$R = \frac{\bar{r}_B - \bar{r}_W}{n(n-1)/4}.$$

This is equivalent to the linear regression slope of the ranked dissimilarities between objects and the rank distances from a model representing the groups (Somerfield et al. 2021a) – that is, a Mantel test where the second matrix represents the groups. Using rank rather than actual dissimilarities is in keeping with the spirit of NMDS.

The null hypothesis being tested by ANOSIM is that the average of the rank dissimilarities between all possible pairs of objects in different groups is the same as the average of the rank dissimilarities between pairs of objects in the same groups. R is scaled to be within the range $+1$ to -1, and Somerfield et al. (2021b) recommended that the size of R provides a useful relative measure of effect size. Differences between groups would be suggested by R values >0, where objects are more dissimilar between groups than within groups. R values of zero indicate that the null hypothesis is retained. Negative R values indicate that dissimilarities within groups are greater than dissimilarities between groups, an outcome considered unlikely (Clarke et al. 2014). However, Chapman and Underwood (1999) showed that negative R values can occur, especially with high levels of within-group variability that were similar between groups and when outliers were present. They argued that negative R values could be a useful diagnostic, indicating an inappropriate completely random sampling design when stratified sampling would be more appropriate.

Analysis of similarities uses a randomization procedure to allocate objects to groups to generate the distribution of R under the null hypothesis that all random allocations are equally likely. Clarke et al. (2014) described ANOSIM procedures for nested designs where averaging over the subsampling levels produces single-factor tests for each factor. They also proposed ANOSIM for testing main effects in factorial designs by treating each main effect as a single-factor test, averaging over the other factor. Somerfield et al. (2021a, b) described the ANOSIM test for two- and three-factor designs in more detail, including handling ordered and unordered factors.

One limitation of the ANOSIM procedure is that it is not a linear model method, so partitioning the variation in dissimilarities among terms in complex designs, such as those including interactions, is impossible. Although Somerfield et al. (2021a) argued that ANOSIM tests of main effects in these analyses have biological value despite the presence of interactions, most biologists want a method analogous to fitting linear models. This is challenging when dealing with nonmetric dissimilarities like Bray–Curtis, but two options for fitting linear models to dissimilarities, db-RDA and PERMANOVA, will be described in Section 16.3.3.

We illustrate the ANOSIM procedure with the pond invertebrate surveys of four management types (Lemmens et al. 2015) in Box 16.2 and the invertebrate family data from the artificial light treatments (Parkinson et al. 2020) in Box 16.3.

16.3.2.5 Multi-Response Permutation Procedures
Mielke (1976, 2006; see also Mielke & Berry 2007) proposed multi-response permutation procedures (MRPPs) that test hypotheses about group differences in Euclidean distances. Basically, the MRPP determines the mean of the Euclidean distances between objects within each group and calculates an MRPP statistic (*delta*) that is a linear combination of these mean within-group Euclidean distances. The statistic produces a weighted (by sample size) average of the within-group mean Euclidean distances. Small values of the statistic indicate that objects tend to be found in groups. The probability distribution of the MRPP statistic is determined by randomizing the

allocation of all objects to the groups, keeping the original sample sizes, with the null hypothesis being that all random allocations are equally likely. We compare our observed value of the MRPP statistic to the probability distribution generated under randomization to get the probability of obtaining the observed value of the statistic or one smaller under the null hypothesis. An alternative statistic for larger sample sizes is the Pearson type III T statistic that considers the skewness in the distribution of *delta* under the null hypothesis. The MRPP can be used for a range of hypotheses, including those associated with paired comparisons and randomized block designs. See Mielke and Berry (2007) for further extensions.

Multi-response permutation procedures have been traditionally based on Euclidean distance. Their use with more robust nonmetric dissimilarities might be tricky because of the difficulty of defining the centroid and calculating the mean within-group dissimilarity. Nonetheless, McCune and Mefford (1999) have suggested that MRPPs might work well with other dissimilarity measures, such as Bray–Curtis. They can also be based on rank dissimilarities, making it similar to ANOSIM (McCune & Grace 2002). Since Euclidean distance is not a particularly appropriate measure of dissimilarity for some biological data (Section 14.3.4), we could use the inter-object distances from PCoA or NMDS in an MRPP. This is not an ideal solution because we know these distances imperfectly represent the actual dissimilarities, and correction for negative eigenvalues would be required for PCoA. This approach is used, although not for MRPP, in db-RDA, described in Section 16.3.3.

16.3.2.6 Comparing Dispersions
An assumption underlying ANOSIM comparing average within and between-group dissimilarities (and MRPP focusing on within-group dissimilarities) is that the groups also should have similar dispersions (Warton et al. 2012), although this is an issue of some debate – see Roberts (2017) and response by Warton and Hui (2017). This is analogous to the univariate homogeneity of variance assumption underlying normal-based linear models, and applies to multivariate linear model methods like PERMANOVA, discussed in Section 16.3.3.2. Comparing multivariate dispersions may also be of intrinsic biological interest, irrespective of location differences.

Permutational comparison of dispersion (PERMDISP) (Anderson 2006; Anderson et al. 2006) provides a test of the null hypothesis of equal multivariate dispersions based on using Euclidean distances between each observation and its group centroid in a multivariate extension of Levene's test of homogeneity of variances (see Section

5.3.2). When using nonmetric dissimilarities like Bray–Curtis, their method does a PCoA on the B–C dissimilarities. It then uses all the axes to determine centroids (or spatial medians) and Euclidean distances between observations and their group centroids (or medians) as the input into Levene's test. *P*-values are calculated by permuting the residuals after accounting for location differences, so the hypothesis is only about dispersions.

Graphical methods are more useful in checking univariate assumptions than relying just on statistical tests. With multivariate dispersions, we can plot the Euclidean distances within each group or use the NMDS on the original Bray–Curtis dissimilarities or the PCoA plot of Euclidean distances to visualize the dispersions. We illustrate PERMDISP for the worked examples in Boxes 16.2 and 16.3.

16.3.3 Multivariate Linear Models

Multi-response permutation procedure or ANOSIM tests are not suitable for complex designs with interactions and/or continuous predictors (covariates). Another approach is to fit linear models or GLMs to multivariate responses by either fitting a model to the dissimilarities or fitting a model to the individual responses and combining the models in some way.

16.3.3.1 *Distance-Based Redundancy Analysis*

Distance-based redundancy analysis was proposed for testing group differences in dissimilarities (Legendre & Anderson 1999a, and 1999b for minor correction; McArdle & Anderson 2001). It uses PCoA to convert the original dissimilarities (e.g. Bray–Curtis) into their equivalent Euclidean distances, correcting for negative eigenvalues. The matrix of *n* objects by *p* principal coordinates is then related to continuous or categorical predictors using RDA. The predictors can be set up as a linear model. The SS is calculated as sum-of-squared deviations between observations and their centroid, partitioned among the model terms and the residual in the usual linear models framework, and model terms are evaluated using permutation *F* tests. The tests can be constructed in several ways – permuting the original data or the residuals from the reduced model or from the full model. These will be discussed further as part of PERMANOVA in Section 16.3.3.2. The model becomes a multiple linear regression model, and any combination of crossed and nested factors and continuous covariates can be included.

It turns out that we can get similar results by just doing a MANOVA test on the corrected principal coordinates. However, Legendre and Anderson (1999a) argued that db-RDA has more robust randomization tests and

does not require more objects than variables in the original data matrix. The latter advantage is important because ecological datasets nearly always have more taxa (variables) than sampling units (objects).

16.3.3.2 *Permutational Multivariate Analysis of Variance*

The partitioning of SS and variances used for testing linear models can also be applied directly to dissimilarities, even nonmetric ones like Bray–Curtis (Anderson 2001, 2017; McArdle & Anderson 2001). While we can't use distances from each object to its group centroid to calculate SS with nonmetric dissimilarities, we can get the same information by using the sum of squared dissimilarities between objects within each group (Anderson et al. 2008). This method means that using PCoA on the original dissimilarities is unnecessary, and the negative eigenvalues produced by db-RDA correspond to negative SS. The correction for negative eigenvalues in db-RDA described by Legendre and Anderson (1999a) actually produces overly conservative tests when random factors are included in the design (McArdle & Anderson 2001).

The PERMANOVA is elegantly simple and can be applied to any design structure (Anderson 2017). Consider a single-factor design with *p* groups and *n* objects in each group so there are $N = pn$ total objects. From an $N \times N$ matrix of dissimilarities between all pairs of objects, we calculate three SS:

- The SS dissimilarities between all pairs of objects divided by *N*; this is the SS_{Total}. Note that only the lower (or upper) diagonal of the dissimilarity matrix is used. The dissimilarity between objects *a* and *b* is the same as between *b* and *a*.
- The within-groups SS or $SS_{Residual}$, which is the SS dissimilarities between objects within each group, summed over the groups.
- The between-groups SS or SS_{Groups} is determined from the additive nature of partitioning SS in linear models: $SS_{Total} - SS_{Residual}$.

The pseudo *F* statistic for evaluating the H_0 that all allocations of objects, and therefore dissimilarities between objects, between groups are equally likely is:

$$F = \frac{SS_{Groups}/(p-1)}{SS_{Residual}/(N-p)}.$$

It is analogous to the *F* statistic for a single-factor linear model. The pseudo *F* is compared to a distribution generated using permutation methods.

Permutational analysis of variance can be extended to more complex designs with nesting and/or interactions, plus designs with continuous covariates. The main challenge is deriving a method of permutation when interactions are present. Anderson et al. (2008) proposed three methods:

- unrestricted permutation of the raw data;
- permutation of residuals from a reduced model;
- permutation of residuals from the full model.

The first is appropriate for single-factor designs, but the second is more generally recommended for complex designs except for small sample sizes (when the first method is better).

Random effects can also be included in PERMANOVA models and expected values of mean squares are used to determine the appropriate denominator for the pseudo F statistic as described in Section 10.2.1 for univariate models. For complex models with interaction terms, we also have to decide which type of SS to use with unbalanced designs (see Section 7.1.7). Type III are most commonly used but are not necessarily the default in some software. When fitting PERMANOVA models with multiple continuous variables, some software will fit the different predictors in the order in which they appear in the model, whereas assessing marginal effects (comparing the full model to a model omitting the term of interest) might be more appropriate.

We illustrate PERMANOVA for the two worked examples in Boxes 16.2 and 16.3.

16.3.3.3 *MV-ABUND*

One of the potential problems with methods like ANOSIM and PERMANOVA is that they can confound location and dispersion effects (Warton et al. 2012; see also Warton & Hui's 2017 response to Roberts 2017). While Anderson (2017) argued that PERMANOVA is robust to heterogeneity of dispersions with balanced designs, in contrast to ANOSIM, this may not be the case for unbalanced designs, just as was true for univariate linear models with unequal sample sizes. While there has been recent work developing pseudo F statistics robust to heterogeneous dispersions (Anderson et al. 2017), the usual approach to dealing with this problem has been to rely on univariate transformations of abundances to logs or, more commonly, fourth roots (so zero abundances are not a problem). The argument is that removing the univariate mean–variance relationship should also reduce the multivariate heterogeneity of dispersions (but see Warton et al.'s [2012] criticisms). Such a transformation will also reduce the influence of the most abundant taxa in calculating dissimilarity between objects, and transformations are often used for both purposes.

When fitting univariate linear models to count data, we have already recommended GLMs based on a Poisson distribution, or more realistically based on a negative binomial distribution for ecological data, which are likely to be overdispersed. Warton et al. (2012) introduced a multivariate extension of these GLMs, and suitable software (MV-ABUND) was described by Wang et al. (2012). Essentially, their method fits univariate GLMs relating the abundances of each taxon to the same set of predictors. These models are based on a binomial distribution for presence–absence data and Poisson or negative binomial distributions for count data. The multivariate aspect is that the relevant test statistics (e.g. likelihood ratio [LR] statistics) from the univariate models are summed to produce a "sum of LR" statistic. Permutations are then used to generate a distribution for this statistic under the null hypothesis of no effect of categorical predictors (or relationships with continuous predictors).

Warton et al. (2012) proposed that the relative contribution of each taxon to the overall "sum of LR" statistic can be assessed by the univariate LR statistics for each taxon, a measure of the between-group effect if we are dealing with a categorical predictor. They argued this is more reliable than assessing taxon contributions using SIMPER. Another advantage is that all the usual diagnostics available for assessing fits of linear and generalized linear models can be used; this is important to ensure that an appropriate distribution is used for the univariate GLMs.

The MV-ABUND approach deals directly with the mean–variance relationship by using an appropriate distribution for fitting univariate models. However, it is not a dissimilarity- or distance-based method, and there is no direct link to an ordination plot based on a specific dissimilarity. We illustrate MV-ABUND for worked examples in Boxes 16.2 and 16.3.

16.4 Key Points

- A common method of representing the relationships between objects based on their dissimilarities in terms of variable values is MDS. From a suitable dissimilarity matrix, MDS plots the objects in multidimensional space so that their relative interpoint distances on the plot most closely match their actual relative dissimilarities.
- Another way to represent a set of dissimilarities between objects is with agglomerative cluster analysis that produces a tree-like dendrogram, with all the objects in one group at the top and groups of objects most similar to each other at the bottom.
- If we have a set of response variables and a set of predictor variables for each object, and therefore a dissimilarity matrix of responses and a dissimilarity matrix

of predictors, we can test hypotheses about group differences in dissimilarities. Because dissimilarities are not independent from each other, these analyses are often based on randomization (permutation) tests:

– A Mantel test can be used to measure the correlation between the two matrices or the BIO-ENV procedure to find the subset of predictors that is best correlated to the predictors.

– An analysis of similarities (ANOSIM) can be used to compare average rank dissimilarities between predefined groups in simple (single-factor, two-factor without interaction) designs.

• Linear models can be fitted to dissimilarities using db-RDA or PERMANOVA, combining the usual ANOVA partitioning of SS within and between groups with randomization pseudo F tests. Both methods can be extended to complex designs and those that include random factors, by equating the MS to their expected values as described in Chapter 10.

• An alternative approach to analyzing taxon abundances against a set of predictors is MV-ABUND, which fits univariate GLMs based on a negative binomial distribution for each taxon with resulting LR statistics summed for the total dataset.

Further Reading

Multidimensional scaling: Legendre and Legendre (2012) and Clarke et al. (2014) describe MDS from an ecological perspective.

Cluster analysis: Fielding (2006) and Legendre and Legendre (2012) provide thorough discussions of clustering methods.

Analyses based on dissimilarities: Legendre and Legendre (2012) and Clarke et al. (2014) cover procedures such as Mantel test, BIO-ENV, ANOSIM, etc.

Linear models for multivariate data: Legendre and Legendre (2012) briefly describe db-RDA; few books cover PERMANOVA, but see Anderson et al. (2008). Warton (2022) describes the MV-ABUND procedure and associated analyses.

17 Telling Stories with Data

17.1 Research Doesn't Exist Until You Tell Someone

Collecting biological data is not an end in itself. We collect data to answer questions that build our understanding of the natural world, and we use this increased understanding to

- contribute to broad scientific knowledge and influence research directions;
- gain a qualification;
- establish and build our reputation among scientific peers;
- influence those responsible for deciding about environmental management, health, etc.;
- engage the community in science, and justify any community support we may have received;
- attract funding to sustain our work and justify funding already received;
- give us personal satisfaction.

To meet these objectives (besides the last one!) we must communicate our results. We are trying to engage a time-poor audience, who we must convince to keep reading or listening or to assess our proposal favorably. Often, the quality of our communication affects that decision.

Scientific communication does not have a glorious history. There are good examples of where the influence of science has been less than many scientists hoped, such as relatively unchanging public views of evolution in the United States and the hijacking of debate about climate change by discussion about post-1988 air temperatures when >90% of heat is absorbed in oceans, which provide a less noisy signal (see, e.g. Pew Research Center 2015). There are positive examples, such as how scientific evidence was trusted in responding to COVID-19 (at least in many places). At a more personal scale, most grant proposals fail, with many funding agencies supporting <20%

of proposals, and prestigious journals reject most papers submitted to them. Anything we can do to improve our chances is good.

We can lose audiences because of our **content** or **presentation.** So far in this book, we've tried to make sure that you collect data that will answer your questions unambiguously, but whether those questions are interesting or important is beyond our influence. The remaining component is whether you present what you found in the most engaging way.

We aren't usually taught much about writing after high school – we are trained as scientists, and within that, biologists. We are now aware of the gap between the writing skills many of us picked up more or less accidentally and those we can develop with some practice and expert advice. The same is true for analysis – we might become statistically competent, but communicating results uses graphic design, which we also were not taught.

As this chapter's title suggests, we believe we all should pay more attention to writing and illustrating. Data analysis and presentation should be a form of storytelling – putting information together logically, clearly, and engagingly.

17.1.1 Telling Better Stories: The Importance of Narrative

There is a long tradition of convincing scientists to write more clearly, with several excellent and essential guides (e.g. Lamott 2007; Pechenik 2015; Pinker 2014; Schimel 2012; Strunk & White 1999; Williams 2007). There's an increasing set of resources covering oral and written communication (e.g. Duarte 2008; Reynolds 2011), including to a broader range of audiences (Dean 2009; Greene 2013). One important view is that it is not helpful to distinguish between scientific and other kinds of communication

(Olson 2015), but simply to recognize good and bad communication. Good communication provides a clear, strong message between scientists and our target audience; poor communication obscures that message.

A second important suggestion offered frequently (Knaflic 2015; Bayer & Hettinger 2019; Dahlstrom 2014; Krzywinski & Cairo 2013; Morris et al. 2019; National Academies of Sciences 2017; Olson 2015) is that the most effective communication generally involves a story or narrative through which we lead the audience. Narrative is not new and can be traced back many centuries, but it is not a common feature of scientific communication. Narrative structure, or storytelling, has been developed in many other areas (e.g. Robert McKee's [1977] *Story*). Indeed, Randy Olson, a biologist turned film-maker turned communicator, argues that film, particularly the Hollywood industry, provides a model for communication, perhaps largely because the rewards (and penalties) surrounding effective communication are so substantial (Olson 2015).

Many of us now think carefully about our writing, but there is not such a long history of thinking about how to present the data. Because the data and analyses determine whether a scientific audience finds your story credible, you must present those results clearly – draw attention to the most important features of the results, rather than submerging them in a sea of extraneous material. In this chapter, we present simple ways to present analytical results and display results graphically – and highlight poor practice. Our aim is not to be prescriptive but to encourage you to think more about how you report your work.

We use the storytelling framework to think about how to extract the critical information from what are often very extensive outputs from statistical packages and then how to report on data and analyses. We discuss general principles behind summarizing data and also consider how our reporting of data and analyses should change with the target audience.

As for writing, there is a history of thinking about how best to summarize what is generally quantitative information, including the classic work of Cleveland (1993, 1994) and Tufte (1983, 1990, 2006). This literature has broadened recently to include the challenges provided by communicating complex experiments and also big data, and some of the best advice comes from outside science (e.g. Cairo 2016; Knaflic 2015; Reynolds 2011; Schwabish 2021).

> *In our story, data are evidence for credibility, and illustrations lead the audience to our key points*

Some of the essential parts of using data to enhance our biological stories are summarized by Knaflic (2015), who identified six steps:

1. Understand the context.
2. Choose an appropriate visual display.
3. Eliminate clutter.
4. Focus attention where you want it.
5. Think like a designer.
6. Tell a story.

Knaflic's (2015) emphasis was on data visualization, but we share her focus on these steps, although we treat them in a different order. We assume that the story is the starting point because the data were only collected to play a role in a biological story. The biological story affects the context of the data, but so too does the audience. We emphasize how to focus the reader on the most important parts of the story, choose appropriate illustrations, and make them effective. Our starting point is to make sense of the often voluminous output produced by most software packages.

17.2 Summarizing Data Analyses

We summarize analyses to help us digest the results, particularly with complex models. Many scientific journals now encourage authors to summarize analyses briefly in the main paper and move the analysis details to supplementary appendices, generally only available online. More generally, summarizing the analysis helps us focus on features of our results we want to highlight for our audience. We will deal with some of the most common analyses, although many of these concerns and suggestions apply to a range of other statistical analyses.

17.2.1 Linear Models
17.2.1.1 *Continuous Predictors*

Analyses of linear regression models are a clear example of where most statistics packages generate extensive output, but much of the information can be omitted. With a simple linear regression with a single predictor variable, you will get a result like Table 17.1. We use the data from Box 6.1, linking aphid density to the proportion of soldiers in a colony. The response variable is the proportion of soldiers, and the predictor variable is *aphiddens*.

The linear model also includes an intercept. The output gives us the estimated regression line, some measures of how precisely the parameters – slope and intercept – have been estimated, and tests of hypotheses about the slope and intercept (by default, that each = 0). We also get an idea of the distribution of residuals (range and

Table 17.1 Standard regression output from a major statistics package. The example is from *R*, produced by the *lm* function, but most packages produce similar levels of detail

Call:

lm(formula = soldiers ~ aphiddens, data = shibao)

Residuals:

Min	1Q	Median	3Q	Max
–14.299	–4.942	1.407	5.614	13.923

Coefficients:

	Estimate	Std. Error	t value	Pr(>|t|)	
Intercept	27.776	3.546	7.833	1.75e-06	***
aphiddens	29.297	4.684	6.255	2.11e-05	***

—

Signif. codes: 0 '***' 0.001 '**' 0.01 '*' 0.05 '.' 0.1 ' ' 1

Residual SE: 8.71 on 14 degrees of freedom
Multiple R-squared: 0.7365, Adjusted R-squared: 0.7176
F-statistic: 39.12 on 1 and 14 DF, p-value: 2.11e-05

quartiles). Some of this material could be added to a table, but most is of little interest to our audience, and we can present most of the useful information in the text in standardized form.

There is considerable redundancy for a simple model like this. Most statistics packages are written to deal with complex models, and a simple regression is treated as a special case. The last of the output is an ANOVA, testing whether the regression model explains more variation in the dependent variable than the model omitting the predictor. The top section of the table also shows tests of hypotheses – t tests for the slope and intercept. With only one predictor, the F and t tests for the effect of aphid density are identical: $F = t^2$, with identical P-values. There is no point in reporting both. Other parts of this output, such as adjusted multiple r^2, are only relevant with multiple predictors.

We are usually interested only in whether we can see a signal from the predictor against background (residual) noise and, if so, its strength. Tests for the effects of the predictor variable are a way of detecting a signal. For its strength, we need estimates of the model parameters to tell us if the relationship is biologically important and some measure of how well the model fits the data. The intercept and slope are listed under "Estimate" in the output table, and the simplest measure of fit is the r^2, provided at the bottom of the output. We could reduce that output to a single sentence in the text:

The percentage of soldiers as a colony rose <add your own adverb based on aphid biology> as aphid density increased, and density explained 72% of the variation in soldiers (equation: percentage of soldiers = 27.8 + 29.3 (aphid density), $F_{1,14} = 39.12, P < 0.0005$).

This format is standard, and you could expect a reader to be familiar with the estimates of the model parameters, etc. – assuming that you've mentioned it's a linear regression! Again, if we wished to be true minimalists, we could omit the r^2 or even the F. As discussed below, the df and P let you back-calculate the F and then r^2. This is overkill – most readers are comfortable with the information in the previous paragraph. The only additional information might be measures of uncertainty, such as CIs, for the parameter estimates.

The information from more complex regressions can also generally be compressed, although not to the same degree, and most complex regressions are presented in tables.

17.2.1.2 *Categorical Predictors*

The simplest way of presenting the results of a linear model with categorical predictor variables (i.e. a classical ANOVA model) is to display the complete ANOVA table, but we can present results more efficiently without sacrificing information.

We suggest:

- The df should always be presented, as they indicate the sample size.
- We do not need both SS and MS, as one can be calculated from the other.
- As long as the $MS_{Residual}$ and Fs are provided, we don't need the MS for groups or specific contrasts, as we can calculate them from the F, df, and $MS_{Residual}$. This step does require that you have described the statistical model adequately.
- If significance tests are used, we prefer P-values to be presented to allow readers to use their own level for deciding.

With a single predictor, there is generally no need to report your findings in a table; there is only one way to calculate the F. Report your analysis in the text, using a standardized form:

Attending a design and analysis course by the authors of this book did not markedly improve the quality of students' analyses ($F_{1,4} = 1.23, P = 0.546$).

The information in parentheses tells a reader that the conclusion is based on an F test with numerator df = 1, denominator df = 4, that the F is 1.23 (and hence the

Box 17.1 Different Arrangements of a Table

The tables present the results of testing for the effects of existing ascidians on settlement of marine invertebrate larvae. The analyses are single-factor linear models. The table shows the P-value from each analysis, with an estimate of the residual variance and power values (to detect a 50% change in settlement rate). The effect sizes were presented separately. We had already decided to omit SS, MS, and Fs from the ANOVA. The top table is laid out simply, with no unusual formatting. The degrees of freedom (df) were constant across species and were detailed separately and effects shown graphically.

We could improve the readability of the table by a few changes, shown in the middle panel. If our focus was on statistical significance, we could use boldface to highlight these results. The table shows results from two polychaetes, a species of barnacle, and a few bryozoans. If we want a reader to see them in their natural groups we could put faint lines or extra space between the groups. A reader then sees that the statistically significant results fall in the same taxonomic group. The lower panel shows one of the worst formatting styles; we have buried the important information behind unnecessary grid lines. Here, there is nothing to draw a reader's attention.

Taxon	P	$\sqrt{MS}_{Residual}$	Power
Serpulids	0.348	20.32	100
Spirorbids	0.455	2.60	47
Elminius	0.531	24.89	71
Cryptosula	0.025	1.90	48
Scruparia	0.789	0.62	61
Tricellaria	0.017	4.72	98
Watersipora	0.525	3.45	94
Bugula neritina	0.118	10.36	69
Bugula stolonifera	0.042	18.60	100

Taxon	P	$\sqrt{MS}_{Residual}$	Power
Serpulids	0.348	20.32	100
Spirorbids	0.455	2.60	47
Elminius	0.531	24.89	71
Cryptosula	**0.025**	1.90	48
Scruparia	0.789	0.62	61
Tricellaria	**0.017**	4.72	98
Watersipora	0.525	3.45	94
Bugula neritina	0.118	10.36	69
Bugula stolonifera	**0.042**	18.60	100

Taxon	P	$\sqrt{MS}_{Residual}$	Power
Serpulids	0.348	20.32	100
Spirorbids	0.455	2.60	47
Elminius	0.531	24.89	71
Cryptosula	0.025	1.90	48
Scruparia	0.789	0.62	61
Tricellaria	**0.017**	4.72	98
Watersipora	0.525	3.45	94
Bugula neritina	0.118	10.36	69
Bugula stolonifera	**0.042**	18.60	100

ratio of $\text{MS}_{\text{Groups}}$ to $\text{MS}_{\text{Residual}}$ is 1.23, so our "treatment" isn't adding much value), and gives the probability of this value of F or greater if the H_0 is true. There is no need for further information (except, perhaps, why this null hypothesis is retained or why so few participants were surveyed!). Of course, a minimalist might argue that the F is unnecessary; it follows automatically from the P-value and the two df.

Most analyses of this type will include a few planned contrasts or unplanned pairwise comparisons of the group means. Each contrast can be presented in the text as for the overall ANOVA or as part of the ANOVA table if it is presented since contrasts represent partitioning of the SS (see Chapter 6). If you have listed the $\text{df}_{\text{Residual}}$ and $\text{MS}_{\text{Residual}}$ already, all you need for most planned comparisons is the P-value since the vast majority have numerator df $= 1$.

While planned contrasts are much preferred to unplanned multiple comparisons, the latter are common in biological research with a focus on statistical significance adjusted for multiple testing. The results are commonly presented in two ways: (1) labeling means in graphs and tables with the same letter or symbol if they are not statistically significantly different; and (2) listing the means (or group labels) in order and joining those not different with an underline (see Chapter 6). The results can also be presented in the text – for example: "A Tukey's test (with $\alpha = 0.05$) showed that the two highest densities had slower growth rates than the two lowest densities."

This information only gives the reader our evidence for having detected a signal. For a simple OLS categorical predictor model, the nature of the signal is in the pattern of means, so we need to show our audience the group means or differences between means, with uncertainty.

With more complex models with multiple factors, plus interactions, it is more common to present the full ANOVA table, including any contrasts. Effect sizes are more challenging for multifactor designs, especially with interactions, but mean differences from planned contrasts should be provided if relevant. Two other issues are relevant for mixed model ANOVAs:

• When random factors are included, it is often useful to indicate the different error terms used to derive Fs for fixed effects in the ANOVA table (e.g. see Keough & Quinn 1998 for a very complex example and see also Box 17.2).

• Measures of explained variance (e.g. variance components [VCs]) are often incorporated into ANOVA tables when random factors are in the model.

The guiding principles here are that we must include sufficient information somewhere for the analysis to be checked for correctness, and we need to clearly describe the size of the effects. As we saw in Chapters 7–12, the technical details can be complex, as can the description of the effects.

17.2.2 Other Analyses

Many other statistical methods also produce voluminous output with lots of redundancy. The website for this book shows the code and output for the worked examples, and you can see the contrast between output and the content of the boxes. The worked example boxes themselves have more detail than expected in, for example, a scientific paper.

You can reduce the volume of analytical results without sacrificing information. There are often conventions about the critical pieces of information to assure a reader you know what you're talking about and that you have credible results. We will not detail these other analyses here, but in the earlier chapters the published examples we cite can also guide how to report those analyses. Unfortunately, few conventions exist for reporting some analyses that are only now making their way into the biological literature, such as generalized linear mixed model (GLMMs), generalized additive model (GAMs), etc. There are generally fewer conventions associated with ML estimation than for models fitted using OLS.

17.3 Visualizing Data

The best way to visualize data and results depends on the data and analyses to be reported, the purpose, and the target audience. We humans can quickly link a printed number to the value of a quantitative variable, so actual numbers can be sufficient, but graphs or pictures are effective for demonstrating relationships. As a rule of thumb, report your data or analyses in tabular form whenever you think the reader might want to know the precise value of individual data points or compare a few individual values, and use a graphical display when patterns or relationships are the main focus.

While the decision to report tables (or text) or graphs may seem dichotomous, the two options are complementary. As Gelman and Unwin (2013) point out, a picture may be worth 1,000 words (or numbers!), but a picture and some words may be worth much more than two pictures or 2,000 words. Recent trends in scientific publishing make it much easier to include graphic summaries, complete summaries of data analyses, and raw data, with the latter two being available online.

Box 17.2 Displaying Complex Analyses in a More Compact Form

The table below, taken from Wootton and Keough (2016) and reprinted with permission, shows the results of analyzing a five-factor experiment with eight different response variables. In the original paper, full analytical results accompany the table as electronic supplementary material.

Wootton and Keough measured the resilience of algae and invertebrates living on rocky shores to disturbance. The question was how different organisms responded to removing an algal canopy and how this response varied with the severity and spatial patterns of canopy removal. The experiment was done at two sites, and experimental plots were monitored through time. To complicate things further, experimental units were arranged in blocks at each site to remove some small-scale spatial noise. The net result was a complicated, five-factor linear mixed model. Only some interactions were relevant to the biological questions. In this table, the authors grouped model terms that were biologically related, then used squares to indicate statistically significant effects. They also used a bold typeface and filled symbols to highlight model terms of greatest interest to the biological question. The different response variables appeared as columns in the table.

Table 1. Summary of repeated measures ANOVAs on algal cover and invertebrate abundance (organized into functional groups) as well as rock and sediment cover in the disturbance experiment at *Hormosira*-dominated reef sites on the southeastern Australian coast. (A) Results from an analysis comparing all treatments, (B) results of a further analysis where the effects of severity (0 to 100% removal of *Hormosira* cover) and type of disturbance (cropped, clumped, haphazard removal; see 'Materials and methods') were separated (see Tables S3 & S4 in the Supplement at www.int-res.com/articles/suppl/m560p121_supp.pdf for full details). Bold face effect-titles denote effects of interest (discussed in main text). Open square: significant effect (at $p \leq 0.05$). Filled square: significant effect of interest (at $p \leq 0.05$)

Source	*Hormosira* cover	Sediment depth	Sediment cover	Rock cover	Ephemeral algae	Dead algae	Carnivorous whelks	Grazing snails
(A) All treatments together								
Time-averaged effects								
Site						□	□	□
Block(Site)	□	□	□		□		□	□
Treatment	□	□	□	□	□	□	□	□
Treatment × Site	□	□	□			□		□
Recovery profiles								
Time since disturbance	□		□	□		□	□	□
Site × Time						□		□
Treatment × Time	■		■	■	■	■		
Site-dependent effects of disturbance								
Treatment × Site × Time						□		
Block(Site) × Time	□	□	□	□		□	□	□
(B) Type and severity of disturbance								
Time-averaged effects								
Site						□	□	□
Block(Site)		□	□					□
Type	□		□	□	□	□	□	
Severity	□	□	□	□	□	□		□
Type × Site	□						□	
Severity × Site						□		□
Severity × Type	□		□					

Box 17.2 (cont.)

Table 1. (cont.)

Source	*Hormosira* cover	Sediment depth	Sediment cover	Rock cover	Ephemeral algae	Dead algae	Carnivorous whelks	Grazing snails
Severity × Type × Site	□	□		□			□	
Type × Block(Site)	□	□	□		□			
Severity × Block(Site)	□		□			□		
Recovery profiles								
Time since disturbance	□			□	□	□	□	□
Site × Time	□					□		
Type × Time	■	■	■	■	■	■		
Severity × Time	■	■	■	■	■	■		
Severity × Type × Site	■					■		
Site-dependent effects of disturbance								
Type × Site × Time	■							
Severity × Site × Time						■		
Severity × Type × Site × Time								
Block(Site) × Time	□		□	□		□	□	□

17.3.1 Just Show Us the Numbers!

If you have only two or three numbers to report, such as the means of a couple of groups, you don't need anything fancy – report the numbers.

17.3.2 Tables

Once you have decided which information to incorporate into a table, there is the matter of the table layout. Many software packages allow a wide range of formatting options. The table layout should make the reader's job as easy as possible. Look at the examples in Box 17.1, and see which table provides the clearest layout of simple information.

Sometimes, complex sets of results can best be displayed using nonstandard table designs. For example, if there are many analyses of the same kind, such as analyses on many species, the point of interest may be the patterns of *P*-values or statistical significance, power, etc. For example, Box 17.2 shows a highly reduced table, taken from Wootton and Keough (2016), summarizing results from a complex experiment with several factors and several response variables.

These examples aren't an exhaustive list, and there may be better ways to present the information. Still, they emphasize that there are alternatives to tedious standard ANOVA tables from complex models.

Most scientific journals use straightforward, uncluttered table layouts. However, when giving a talk or writing for a different target audience, you can change the presentation of the table to highlight the essential features. Simple changes to font size and emphasis (e.g. bold), and the use of color (e.g. in heat map tables) can be helpful (Figure 17.1). Few (2012) provides good suggestions for tables.

17.4 Graphical Summaries of the Data

The greatest value of a picture is when it forces us to notice what we never expected to see. (John Tukey 1977)

(a)

	Control		Agricultural				
	1	2	3	4	5	6	7
Achnanthes	2.7	1.0	1.0	0.7	0.3	0.3	0.7
Actinocyclus	2.0	2.7	0.3	0	0	0.7	0
Amphora	0	0	0.3	1.3	3.3	4.3	1.3
Bacillaria	13.0	6.7	0.7	1.7	1.3	3.3	4.0
Chroococcales	12.7	14.3	37.7	10.3	24	8.7	39.3
Cocconeis	0.3	0.3	0.7	0	0	0	0
Cymbella	2.3	1.7	0	0.3	1.0	2.0	1.7
Diploneis	3.7	5.3	0	0.7	3.7	1.0	1.3
Encyonema	1.3	0.7	0	0.3	1.0	3.0	0.3
Gomphonema	8.3	8.3	0.3	0	2.0	2.7	4.0
Melosira	10.3	8.3	19.7	59.3	3.0	27	4.7
Microcystis	0	0	0.3	0.3	0.7	0	0.3
Navicula	11.3	10.3	3.7	9.0	11.7	8.3	6.7
Nitzschia	15.3	19.0	2.7	3.7	18.7	4.3	3.3
Oedogonium	0	0	0.7	1.0	1.7	7.7	0
Oscillatoria	19.7	23.3	41.7	20.3	36	34.7	38.3
Pleurosigma	2	2.7	0	0	0.7	1.7	1.0
Synedra	2.7	5.0	0.3	1.0	0.3	0.3	2.7
Tabellaria	2.3	0.3	0	0	0.7	0	0.3

(b)

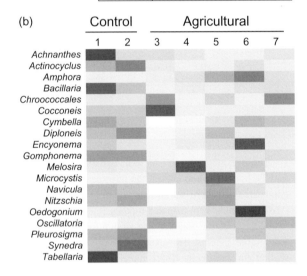

Figure 17.1 Heat map table, using shading to highlight aspects of the table. Heat maps can be useful to summarize data with several sampling units, with each unit having a range of measurements associated with it. Common examples might be measures of gene expression, enzyme activity, and protein synthesis that have been measured across several biological units, or environmental units where a substantial number of contaminants is measured and assessed against levels set by environmental regulation. In some cases, a reader might wish to access the actual data, but in others it may be more relevant to see which genes are up- or downregulated, or which contaminants exceed various screening thresholds. The figure shows a hypothetical environmental dataset comparing abundances of diatoms, microalgae and cyanobacteria in streams subject to two different classes of land use. Panel (a) shows means for each taxon across seven sites. In Panel (b), data for each taxon were standardized and color coded according to abundance, with darker colors representing greater relative abundance for a taxon.

17.4.1 Some Basic Principles for Visualizing Data

Most biologists use sophisticated graphics packages, but lack training in graphic design, and a substantial number of us have poor taste. These factors can combine to produce distracting visual displays. There is a substantial literature on the graphical display of quantitative information, including several classics by Tufte (1983, 1990), Tukey (1977), and Cleveland (1993, 1994). This literature also includes a formal theory of data graphics (Wilkinson 2013), many aspects of design (Tufte), and a recent explosion of books on data visualization, many of which also incorporate design principles.

The guiding principle in constructing graphs (data visualizations, in the current jargon or data graphics in "classic" terminology) is to produce clear, unambiguous representations of your results. These representations should draw a reader to what you consider the most important aspects and be free of distracting elements. This will usually mean simple, clean graphics rather than the ornate productions possible in many graphics packages. For complex experiments or sampling programs, we'll need to decide which factors to include, highlight, etc.

Your graphics must remain true to your analyses. Graphics are summaries of your results that highlight important features. You must not use design elements to (mis)lead your audience toward results you might wish you'd obtained.

A data graphic or visualization has many elements, such as the plot area, titles and labels, and legends. We need to decide what to include for a particular graph, how these elements are formatted, and where to place them.

Often, the centerpiece of a graphic will be a display summarizing the values of one or more quantitative measures, often to compare them under different conditions. The first decision to make is how to represent them, and we have several options, including length, area, and color of particular chart elements. These different ways of encoding information are not equivalent. As part of his pioneering work on visualizing data, Cleveland and his colleague McGill (Cleveland & McGill 1984) used experiments to compare different ways of encoding information. The results were formalized by Cleveland (1993, 1994) into a hierarchy, which has been extended (see, e.g. Cairo 2016; Knaflic 2015) to cover aspects such as color (intensity and hue) that have become widely available (Figure 17.2).

We are most effective at interpreting data encoded as length or position against a common scale, find it harder to extract information when scales are not aligned, struggle more to decode angles, area, and volume, and have great difficulty when quantitative information is encoded using color hues or shadings. Figure 17.3 contrasts a simple two-factor dataset for a continuous variable using area or length.

Encoding information is only one part of communication; your audience needs to decode it. We need to consider how that happens.

17.4.1.1 *Focus Attention Where You Want It*

Not all of the information on a graph is central to your point. Some is not needed at all (see Section 17.4.1.2), but other information might be necessary but not central. For example, axes, and sometimes axis scales, need to be there, but we want the audience to focus on the lines, bars, or data points. It is possible to highlight the data by using line thickness and color combinations to move less essential elements to the background (Figure 17.4).

Schwabish (2021) interestingly suggests starting with gray everywhere and using blacks and colors to move elements into the foreground. His suggestion forces us to think about the important message from the graph.

Using Colors

Color can be a powerful way of encoding information, but should be used carefully. There are several ways of describing color (RGB, CMYK, etc.), but the most widely

used is hue, saturation, and luminosity (or lightness, HSL). This is considered more intuitive than RGB, which describes colors by scores along three orthogonal axes of red, green, and blue. It is used in many image-processing packages. Broadly, hue specifies the "base" color, expressed by a position around a circle, with Red at $0°$. Saturation describes the expression of that color, from white to fully expressed, and can be visualized as the radius of the circle describing hue. Luminosity/lightness describes the brightness of the color. These aspects of color vary in their ability to convey information to the reader. Hue can be very helpful for identifying separate groups of data or for drawing attention to particular components of a graph. Saturation is not as effective at identifying groups, but can convey some quantitative information because saturation values can be a linear scale. However, saturation requires much more work to decode the information than measures such as length, so it should not generally be used. Saturation and luminosity can also discriminate groups when restricted to gray-scale images (as is commonly the case for scientific journals).

Some excellent general texts cover some basics of color perception (Duarte 2008; Few 2012; Tufte 1990). You may want to read some of that literature.

Some color combinations offered by many graphics packages, particularly red and green, will be indistinguishable to as many as 20% of your audience (especially if you are in a male-dominated forum). Red and green appear in default color palettes surprisingly often and you should always consider the different forms of colorblindness.

Colors can also reinforce stereotypes – think about the message you might convey to some readers by, for example, using pink and blue to represent female and male, or the common red/yellow/green/black "stoplight" showing a scale of good to bad. Even nonstereotypical color choices can vary in how they are perceived across cultures.

As a rule, graphics packages offer sets of colors recommended for producing a particular overall look for your presentation. These palettes are often, but not always, effective. We suggest that you choose a particular set, rather than designing your own.

Fortunately, we have help. Many color palettes have been assembled for specific tasks. Some provide appropriate contrast, are compatible with various forms of colorblindness, and, ideally, work when displayed in color and gray-scale. There are also palettes that offer colors effective for representing data along a gradient (sequential), for emphasizing divergence from a midpoint (diverging), and where categories are not ordered (qualitative). A large proportion of data visualization by biologists now takes place within *R* with publication quality

Easier

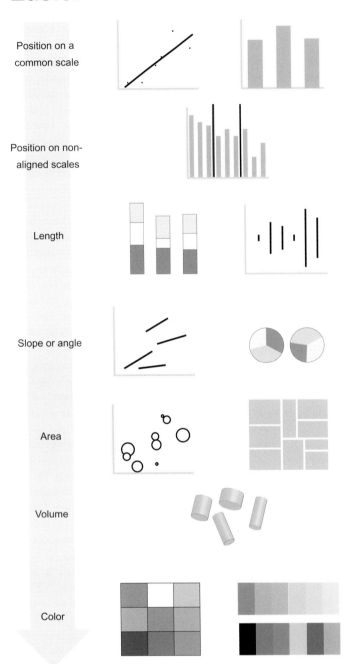

Position on a
common scale

Position on non-
aligned scales

Length

Slope or angle

Area

Volume

Color

Harder

Figure 17.2 Hierarchy of ways of encoding quantitative information, based on Cleveland (1994) and illustrated using common graphical types. The hierarchy reflects the accuracy with which quantitative information can be decoded by a reader. You should be aware that Cleveland did not see this as a strict hierarchy, and some of the ways may overlap in their efficiency. The hierarchy was derived from experimental studies, so it reflects the particular scenarios presented to the subjects. Use this as a guideline for what is most effective, but be prepared to disregard it if a particular situation suggests a specific graphical type would be most effective.

graphics from the package *ggplot2*, and several packages provide suitable palettes.

Colors and Fill Patterns

How the information is presented will also vary with your target audience – a figure in a paper can be more complex

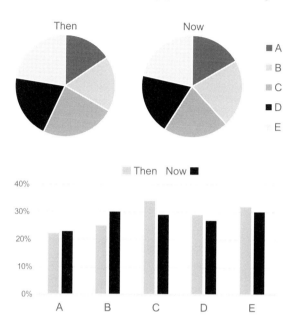

Figure 17.3 Comparison of angles/areas from a pie chart as a way of providing information, contrasted with the same data presented using a bar chart (length as the encoding). The data are from five groups (A–E), measured at two times (Then and Now).

than one you might show at a conference because the reader can sit and digest the information. Careful or thoughtful use of color can help an oral presentation. For written publication, most journals either don't permit colored graphs in their printed version or impose an extra charge for colored figures. Encouragingly, online journals and similar outlets happily include color. Other formats, such as reports that require fewer copies, can use color to great advantage.

The technical limitations of the medium will also influence how you construct graphs. For example, printers do not always reproduce solid colors well, and rather than solid black as a fill pattern, you may be better using hatched or stippled fill patterns. The same cross-hatched patterns may look awful as part of a digital presentation, when solid colors work much better (Figure 17.5). Tufte refers to some of the unfortunate choices of fill patterns as "unintentional optical art"! Many of these cross-hatching patterns are a legacy of journals that printed only in black and white.

17.4.1.2 Eliminate Clutter

Tufte coined phrases for other problems, including graphic elements that conveyed little information (and were, therefore, redundant) and elements that distracted the reader.

Data:Ink Ratio

This value originates in printed data graphics and reflects the ink used to present a given amount of data. High

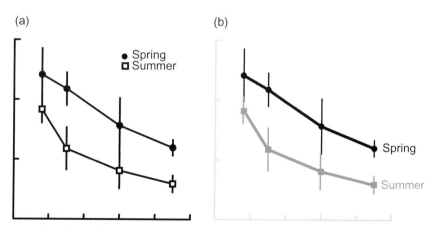

Figure 17.4 Line weight and shading used to focus attention on patterns in data. The two panels show a line graph from our first edition, where all lines are black and line weights identical. The two groups in the data are separated by solid vs. dashed lines. Panel (b) has the axes moved to the background by reducing line weight and changing color to 50% gray. The second group is now marked with a solid line of a different color. As an added step, group labels are placed alongside the lines, using a matching color. This step means the reader does not need to go first to the legend and then to the actual data. The groups would be even clearer if colored output is an option.

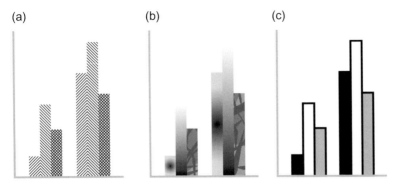

Figure 17.5 Fill patterns for solid areas can help or hinder our ability to identify patterns. Panel (a) shows poorly chosen cross-hatching, which creates Moiré effects. Panel (b) shows fill patterns with gradients in shading. The shading is a good example of chartjunk, because the actual shading level does not encode any information, but draws the reader's attention to it. Other patterns, such as gradient fills that blend into the background toward the end, can essentially move an object down the hierarchy shown in Figure 17.2 (e.g. from position along a common scale to simple length or area). Panel (c) shows simple solid fills, with a contrasting color in the middle of each group of bars

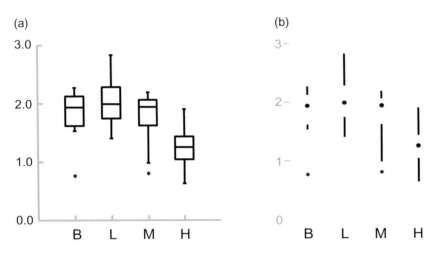

Figure 17.6 Increasing data:ink ratios by removing all lines that do not encode additional information from a box plot, using the Medley dataset from earlier chapters. Panel (b) is a rather extreme and probably less effective option.

values are desirable. Tufte offered very strong views on the steps that could maximize this ratio, including the minimalist box plot in Figure 17.6.

Data Density

Data density is similar to the data:ink ratio but reflects the space taken rather than the ink used.

Chartjunk

Chartjunk is extraneous ornamentation that puts fancy things all around but doesn't help explain your results. This is a particular problem in many graphics packages used to prepare talks. These days, it often includes a proliferation of corporate logos and watermarks.

17.4.1.3 White (Blank) Space Is Good

One common mistake is to cram lots of information into available space. That "space" is most often a poster for a conference or a slide for a talk, where we are trying to maximize the information that we provide. This can cause cluttered images where key points are hard to identify. This problem is less likely to occur in scientific papers and reports, where the convention is to distribute the graphics as smaller, separate figures.

Designers and experts in visual perception emphasize the value of open space as part of graphics. Open space can help draw a reader's attention to particular parts of the visualization, in part by setting it apart from other elements on the page. For a poster, this might mean putting less information on the poster, and for a talk, breaking up our information into a series of slides, each with less information.

The same consideration applies to individual graphics, where allowing some space around a graphic might help, rather than filling each slide completely.

This recommendation is somewhat at odds with our previous comment on data density, but Tufte, an advocate of high data density, also emphasized the value of white space.

17.4.1.4 The Nitty-Gritty: Scales, Ticks, Labels, and Legends

Scales

Our broad aim is to display quantitative information so it maximizes the use of the data area and shows patterns clearly, and we need some scale to describe the range of values shown. It is preferable to plot the data on the same scale on which they were measured in most cases, because the reader will have the easiest task of decoding that information. Sometimes, this will not be possible or desirable:

- Data may extend over several orders of magnitude. When plotted on a linear scale, points at the lower end of the scale may be clumped together, making relationships hard to discern.
- We are expecting our response variable to change proportionally rather than by fixed increments.
- The underlying relationship is nonlinear, and visualizing that relationship along part of the data range is difficult.
- We transformed variables before fitting our model, so the analysis shows patterns in the transformed variables. Under these circumstances, patterns in the data may be clearer if one or more variables is plotted on a transformed scale. We should make sure that the graphs match the analysis and make the visual decoding of the information easy and clear.

For example, when a log transformation has been done:

- The axis scale should represent actual numbers rather than logged numbers (Figure 17.7). For example, values of 10 and 100 should be labeled as such, and not as 1 and 2. Scientific readers can generally decode logged values

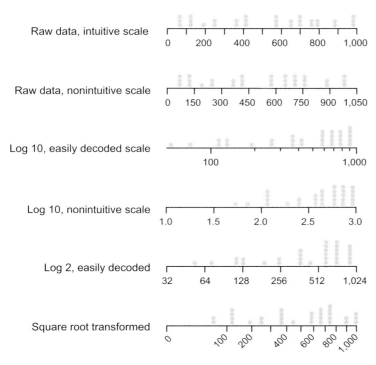

Figure 17.7 Choice of tick marks and labels along axes can help a reader decode the information. The figure shows dot plots of an example dataset, showing scales for raw data and transformed data. For the transformed data scales (log 10, log 2, and square root), ticks and labels are placed at spots on the axes corresponding to natural or intuitive values that a reader can easily decode. Minor ticks are unevenly spaced in some cases to reinforce the underlying transformation. A second log 10 panel shows data plotted with labels that are at nonintuitive spots along the axis.

in base 10 or 2, but most probably need lots of work to decode e^4 or e^7. The raw numbers are easiest to decode.

• Minor ticks should be unevenly spaced to reinforce the logarithmic scale.

• As discussed in Section 5.4, data with zeroes need to be incremented by some small amount to allow the log transformation. This can make axis labels awkward, and some researchers use square or fourth root transformations. These scales can also be plotted using regular, intuitive sequences.

Other transformations, particularly the arcsine (which we do not recommend), do not lead to axis scaling that a reader easily decodes.

Legends

Data displays that incorporate multiple predictors will need some explanation of what different symbols, fill patterns, etc. represent. The reader's task is easiest if the predictors or groups can be labeled directly, such as the example in Figure 17.4. The labeling is even clearer if you can match colors of lines or symbols and their labeling. While this might be desirable, it is not always possible – it is often easy for bar graphs and line graphs, but hard for scatterplots, and hard for bar graphs with multiple groups. In these cases, the legend needs to be where it does not distract the reader from the patterns you highlight in the data.

17.4.2 An Appropriate Visual Display

Scientific publications use relatively few graph types, although big data and the emergence of the various "–omics" are increasing the diversity. Here we outline the main features of the commonest scientific graphs (Figure 17.8). Some basic graphs, such as dotplots, boxplots, and histograms, have been introduced in Chapter 5, and we won't deal with them here. They are often used for essential data exploration but tend not to appear in publications and talks. Our focus in this section is on graphics that are part of a story. This is not to downplay exploratory graphics, because they are an essential part of our workflow. Their audience, however, is usually the research team, who are likely to be engaged and familiar with this biological system, so we don't need to polish the graphics to focus their attention. Gelman and Unwin (2013) make the point that we prepare two kinds of graphics, those to negotiate our way through our data and those for presentation and publication. They suggest a (very rough!) ratio of 1,000 graphs for one person at processing and one graph for 1,000 people at publication.

17.4.2.1 Bar Graph

A bar graph displays a quantitative response variable on the y-axis against a variable on the x-axis that is categorical or has been aggregated into groups or bins. The value of the response variable for each category is represented by the height of a rectangular bar (Figure 17.9). The bar width can be varied to improve aesthetics. The top of the bar may represent a single value or a summary statistic, such as a mean. In the latter case, some measure of variation or precision should be provided.

We can also improve the graph's readability by changing its orientation so the bars are horizontal rather than vertical (Figure 17.9).

A second grouping variable can be represented by adjacent bars, with different fill patterns or colors, at each category of the first grouping variable (Figure 17.3).

One fault made more common by the availability of graphics software designed for business presentations is the three-dimensional representation of two-dimensional data. This is particularly noticeable for bar graphs and pie charts. A "three-dimensional bar" graph is shown in Figure 17.10. There are many problems with this graph, the most serious is that it is hard to tell what value along the y-axis is displayed by the top of the bars.

Note this is not a three-dimensional graph – only the bars are three-dimensional. In Figure 17.10 the third dimension adds no new information. Note also that we haven't even tried to include error bars on this graph – the error bars would start somewhere on the tops of each bar below, and it would be very hard to see exactly how much overlap there is between our groups.

There may, however, be occasions when we have at least three grouping variables. Even then, several two-dimensional plots may be more effective than a single three-dimensional one.

If we were concerned about the waste of space or ink, we could reduce the simple bar graph even further – a minimalist might argue that we could replace the bar by a single point, losing no information). In Tufte's terminology, we'd be improving the data:ink ratio. We wouldn't go that far, because one important aspect of a graphic is the level of familiarity in the audience. Most biologists, and many members of the broader community, will be familiar with a bar graph, whereas dotplots are less familiar, so the "boring" bar graph might be easier for the reader to decode. We do note a version of these graphs that is still sparing with ink – the lollipop chart (Figure 17.11), in which a single point (or symbol) shows the value of the response variable, and a single line connects that point back to the base axis.

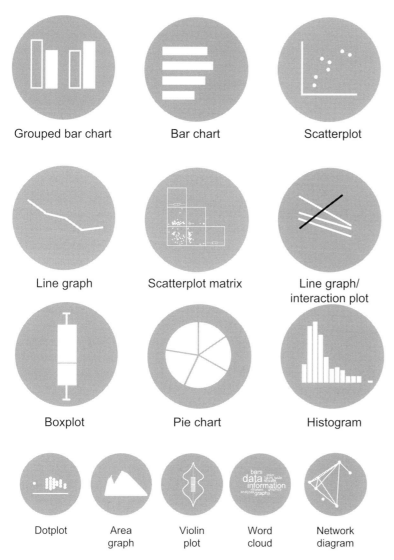

Figure 17.8 Gallery of common scientific graph types.

Guidelines

• As a broad rule, the y-axis should start at 0 for a bar graph (and several other types); starting at other values changes a reader's perception of the difference between values. Axes that start well away from 0 are often associated with attempts to exaggerate relatively small differences. Like most rules, this one can be broken (but always deliberately). If the biological variable being displayed could not plausibly take a value of 0, then the y-axis scale need not start at 0. However, you might be better with a different graph.

• Consider plotting the bar chart horizontally rather than the more common vertical orientation. The horizontal arrangement puts the bar labels to the left of the bars, with the text horizontal. The whole graph is easy to read.

• Consider the thickness of the bars. If the bars represent discrete categories, they should not be too thick, drawing attention to the area rather than the length of the bar, nor should they be too thin, leaving too much unused space. If the bars represent bins from a continuous histogram (e.g. in a size–frequency histogram), the bars may be adjoining.

• When the bars are in clusters, make sure that the clusters are separated. Within a cluster, individual bars need not be separated, but they should never overlap.

• Choose colors of adjacent bars carefully, so they can be readily distinguished. A colored outline is a good

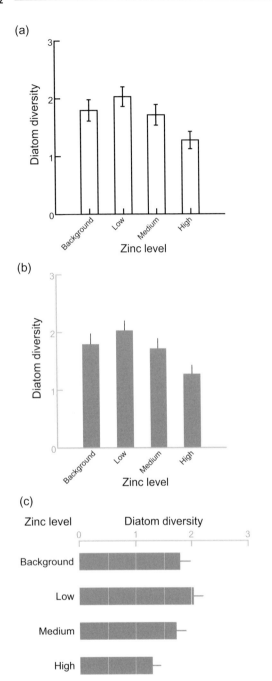

Figure 17.9 Bar chart for the dataset from Medley and Clements (1998) used in Chapter 6 where the mean of the response variable is displayed for each of four groups. (a) A "vanilla" graph, including standard errors on each bar. (b) Some improvement, where fill patterns are used to draw attention to the actual measurements, and the axes are pushed into the background by reducing line weights and using mid-level gray. In (c) the graph has been rotated, so the bars run across the page from left to right. This arrangement allows the group labels to be placed to the left of each bar (where the reader naturally starts), and the labels, being horizontal, are easier to read. The alignment of bars on the left makes the (old) x-axis redundant, and the addition of white gridlines, placed in front of the bars, provides an easy way to compare the heights of each bar.

example of chartjunk – it encodes no information not already provided by the bar itself. The exception is when one or a few bars are given a strong outline to draw the reader to them.

- Different colors should be used only to distinguish bars that are part of a cluster (e.g. to separate factors). When the graph features a series of individual bars (i.e. no clusters), the individual bars should only vary in color if we want to draw attention to one or two particular bars.

17.4.2.2 Line Graph

Line graphs are like bar graphs, except the top of the bar is replaced by a symbol and the adjacent symbols are joined by straight line segments (Figure 17.4). They are used when the x-axis can be ordered or is quantitative. The symbol can represent a single value or a summary statistic.

Suppose we wish to include a second grouping variable. In that case, it can be represented by an additional series of points, with different symbols, colors (see figure 1 in Cleveland 1994), or line styles (Figure 17.4).

These plots are also often used to show interactions (Chapter 7), and they work very well. It is very important to appreciate that the lines may simply indicate changes from group to group in the (mean) values, rather than a continuous trend. This is particularly the case for inter-action plots for fixed categorical effects in linear models because there can be no interpolation. The line connecting the symbols represents no formal relationship between Y and X and could be omitted (but see Box 17.3).

These graphs can be hard to interpret as the number of points and the number of groups increases, leading to what has been termed spaghetti graphs; the clutter can get worse if error bars are also included. Knaflic (2015) has good suggestions for making the reader's task easier, including "micro" graphs, each highlighting one group. Cleveland (1994, see his figure 2.17) also suggested this approach, and he showed a graph with each line on a single panel, with a composite panel at the end. Figure 17.13 shows another approach, with several "micro" graphs, with each graph "anchored" by two treatments that are shown on each panel. This approach is also termed small multiples; it may be more effective for some data, less effective for others. Try small multiples and compare them to single plots that use colors, line weights, and line styles to focus the reader's attention.

Banking to 45

Most software packages default to graphs where the area in which the data are enclosed (the data rectangle) is a square, or sometimes a wider-than-tall rectangle. The ratio of the height:width of the data rectangle affects how we perceive patterns in the data (Cleveland 1994). When the graph consists of line segments (e.g. in a line

(a) (b)

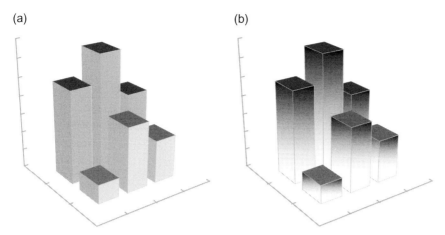

Figure 17.10 How to ruin a perfectly reasonable, if a little boring, data display. Panel (a) uses a third dimension in this case to display data with two predictor variables, but the visualization makes the data harder to discern by placing the bars in rows. Panel (b) confuses matters further by adding a distracting fill pattern, using a gradient that again encodes no additional information. These figures show the same data as were used in Figure 17.5, but with several distractions, each of which move the figure down Cleveland's hierarchy – the three-dimensional nature of the bars creates doubt about whether the quantitative measure is length, area, or volume, and the rotation of axes needed to see the bars clearly makes it hard for the reader to see a common base (or to be sure that all the measures are against the common scale). Some more enlightened software packages make it difficult or impossible to produce these kinds of graphics, but others make it unfortunately easy.

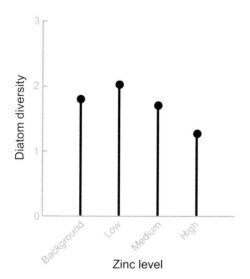

Figure 17.11 "Lollipop" chart, which encodes a similar amount of information to a bar chart, using instead a symbol to mark what was the top of each bar, and a drop line down to the base axis. Like a bar graph, it can be rotated through 90°, possibly improving the ease with which the reader decodes the information.

graph or a scatterplot with a nonlinear smoother), and the orientation of these segments informs the reader about rates of change, Cleveland argued that the aspect ratio of the graph should be adjusted so these line segments have orientations centered on 45°.

Guidelines

• Be wary of having too many series on the same figure because it may be hard for a reader to distinguish the lines.

• Try to label each line with text at one end, rather than using a separate legend or providing a key in the figure caption. This will be possible only if there aren't too many lines and not too many lines converging at the end.

• Different colors are often easier to separate than different line types.

• Consider varying the aspect ratios of plots, attempting to bank to 45, when aiming to show trends.

17.4.2.3 *Scatterplots*

We discussed scatterplots as an exploratory tool in Chapter 5, along with the related scatterplot matrices. They can also be very effective ways of presenting a bivariate relationship. For example, the scatterplot can include a line that represents a regression or smoothing function fitted to the observations (Figure 17.14). Note that the fitted line in this figure extends only across the range of X-values. Many computer graphics packages default to drawing the fitted curve across the entire axis. This is inappropriate, as we have no information about the relationship beyond our sample's largest and smallest X-values – even a simple linear relationship might change shape outside our range (Figure 17.14; see also Section 6.1.6).

Box 17.3 The Special Case of Interaction Plots

Interaction plots are a special graphical display showing the nonadditive effects of multiple factors. They are inherently complex, and common ways of displaying them are perhaps less than ideal.

Two Factors

With two factors, the most common visualization is to use the x-axis to show variation across groups of one factor and to separate groups of the second factor within the graph itself. In perhaps the commonest approach, a simple bar graph is used, with clusters of bars where each bar in the cluster is a group of factor B and a cluster at each factor A group (Figure 17.12).

This figure can be effective if B has relatively few groups (so clusters are small) and the bars are narrow. When either condition is not met, the data are effectively graphed on nonaligned scales rather than a common scale, making the reader's task harder (see Figure 17.12 for bars too wide).

We can prevent this problem in several ways:

- If the factor that appears on the x-axis is quantitative or can be ordered, consider a line or slope graph. These graphical types allow the groups of the second factor to be aligned, making comparisons easier. The line also creates a shape and enables the reader to just compare shapes.
- If the factor destined for the x-axis can't be ordered sensibly, consider a lollipop chart or something similar (see Figure 17.12). It may be necessary to offset the different groups of the second factor slightly so the drop lines are visible. The lollipops can easily accommodate error bars if desired.

  In this case, a line graph might be visually compelling, and we might still use it, but there will be a temptation for the reader to infer some more sophisticated relationship than exists because of how we interpret linear relationships. Where we can't order the factor, most software will default to alphabetical ordering of groups, but the order of groups is arbitrary. We should use the order that makes the most sense biologically. Figure 17.12 shows the same data as a line graph, but the groups have been reordered on the x-axis. The panel on the right suggests a divergent trend, while the one on the left does not. A particularly diligent (or time-rich) reader might recognize the data as coming from Figure 7.2, where the x-axis represented several sites that would not be ordered. The reordering in Figure 17.12 suggests some nonexistent spatial pattern.
- A factor that is not ordered, but has only two groups, lends itself to a slope graph.

More Than Two Factors

Depicting interactions between more than two factors is difficult. When there are few groups of most factors, a single panel may suffice. In this case, we could visualize the interaction between A and B using the technique discussed above, then show that interaction separately for each group of C. Where there are few C groups, this display might work, but otherwise we may need a separate two-factor interaction plot for each group of the third factor (or all combinations of the third, fourth factors, etc.).

The Order of Factors Matters

We might label our factors A, B, C, etc., but they are not ordered. How we describe the interaction will depend on the biological context of each factor. We might see A as a well-known effect, where we are asking questions about whether it varies with the group of factor B, or we could ask if B has an effect and, if so, does it change with A. The way we frame the biological question should guide how we describe the results, and we should construct our graphs to match the emphasis of the rest of the writing. This amounts to a decision about which factor will sit on the x-axis as the "main" factor and which one(s) are the modifying factors. This decision affects the graph, and the two versions may look quite different from each other (Figure 17.12).

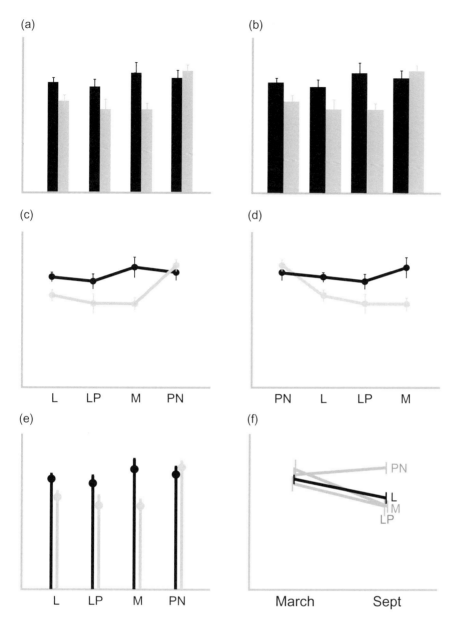

Figure 17.12 Options for displaying interaction plots. Panels (a) and (b) show a bar graph representation of the interaction between two factors. As the width of the bars increases, the black and gray sets (representing two seasons) become offset, making the scales less well aligned. Panels (c) and (d) show the same data as a line graph. This works fine for variables where the levels of a factor on the x-axis are ordered, but in this particular example the levels of this factor are sites, and were ordered alphabetically. The reordering of the sites creates a different impression of the interaction. Panel (e) shows a lollipop chart that overcomes the problem of bar width. Panel (f) shows the slope graph that could be used if times were used on the x-axis, and each line shows the seasonal change at a site. This figure, compared to (c), shows more clearly what is going on. The control, or unpolluted site, is highlighted, emphasizing the conclusion from the example that differences among sites are not closely linked to anthropogenic activity.

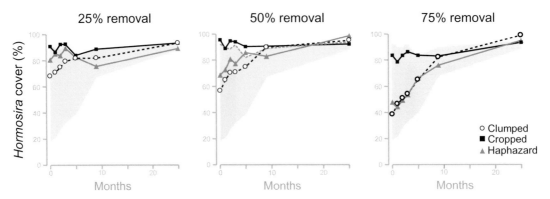

Figure 17.13 Simplifying a complex line graph. This figure is plotted from Wootton and Keough (2016), where they broke down time series from an experiment examining the effects of intensity and type of disturbance on a canopy-forming alga, with 11 treatments (a 3 × 3 factorial plus two "anchor" treatments – no disturbance and complete removal – which cannot be divided into different types) into three panels, each with three lines, along with the two anchor treatments in the background.

We could also plot CIs about the regression lines or confidence ellipses (Sokal & Rohlf 2012), and nonparametric confidence kernels (Silverman 1986) can be included to indicate our level of confidence in the centroid (the mean of the two variables in multidimensional space).

Multiple groups can be indicated on the scatterplot by simply using different symbols (or fill patterns or colors) for each group (Figure 17.15).

Guidelines

- Choose symbols carefully and use combinations of symbol type, color, and fill patterns to maximize the visibility of the data points.
- Consider using color shades and line weights to emphasize individual data points or smoothers, confidence, intervals, etc., depending on the main point to be made from the plot.
- When you show a trend line, restrict the line to the actual range of *X*-values in the data.

17.4.2.4 *Pie Charts*

A pie chart is a circle (or a "pie") where each category's value is represented by the size of its section or slice of the circle (Figure 17.3). The different sections can be further emphasized by different fill patterns or colors.

Pie charts are commonly used in business graphics (hence their presence in most presentation graphics software), but have a much-reduced role in scientific graphics and almost none in statistical graphics. Tufte (1983) argued that they should never be used because their "data density" is low, and they fail to order numbers along a visual dimension. He also asserted that the only thing worse than a pie chart is several pie charts! A reader can't

be sure whether to look at the angle or the area to get an idea of how big each group is, and it is hard to distinguish between broadly similar slices. Contrast that with a bar or line chart, where there is only one interpretation of the height of the bar or point. Figure 17.3 shows a comparison between two categorical factors, where the data are plotted as multiple pie charts, a simple bar chart, or a slope chart (see Section 17.4.2.5). The limitations of the pie chart become very clear. A pie chart can be made worse by allowing the software to produce a three-dimensional aspect to each slice.

The pie chart is essentially a bar chart with the data plotted on a different coordinate system, so bar and pie charts encode the same information.

Pie charts are, however, sometimes useful. They can work for visualizations of genetic data when the goal is to display geographic variation in genetic composition. There, the aim is to make broad comparisons of proportions, and each pie is placed on a map, representing the spatial arrangement of samples. This display can be useful for clear, large differences, but for more subtle differences bar charts may be more effective.

17.4.2.5 *Slope Charts*

Slope charts are an interesting relation to the line chart. They are used to display change in a quantitative variable between two (or occasionally more) groups (Figure 17.16) and can function as an alternative to bar charts with clustered bars.

In their simplest form, as shown in Figure 17.3, which shows data with two categorical grouping variables, they use lines to connect different groups of one variable, with different lines representing groups of the second variable.

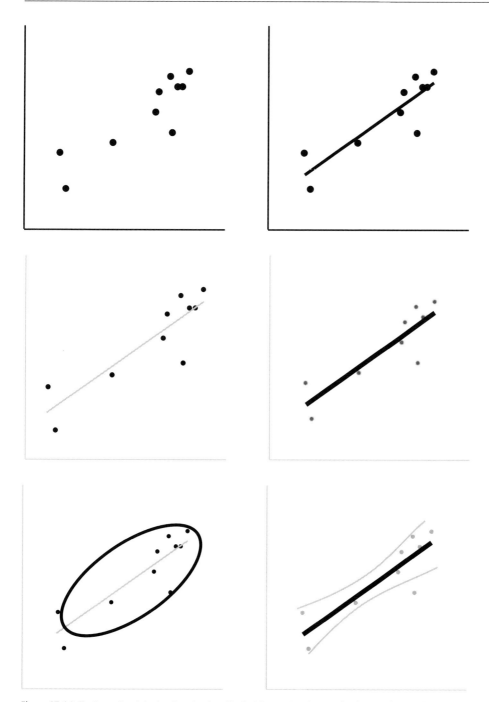

Figure 17.14 Basic scatterplot, showing simple edits that focus attention on the data, and to emphasize data more than the trend line (and vice-versa). Scatterplots can also be augmented with CIs around trend lines or confidence ellipses around data points.

This display can be effective when one of the grouping variables describes change, such as before–after or treatment–control. It is generally easy to label the groups by placing labels next to each line, particularly if they can be placed at the left. Some authors omit axes completely, adding a number by each point to indicate its value. This graph style allows the reader to determine values of each point accurately but does result in a more cluttered presentation, and a simple axis in the background may be more effective if exact values need not be determined.

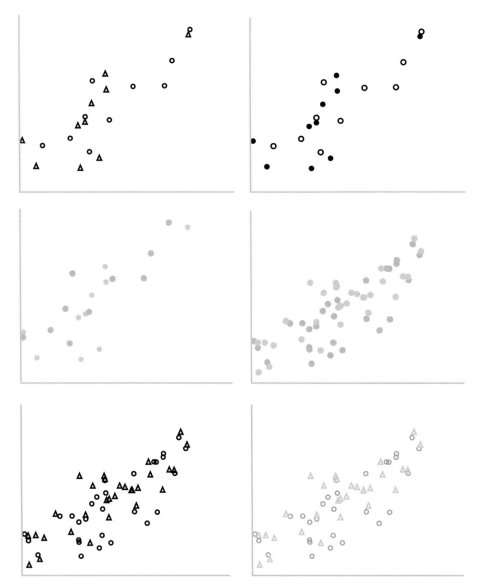

Figure 17.15 The choice of symbols is important for identifying patterns. In the example, filled symbols stand out, but when there are many data points or lots of overlap, unfilled symbols allow individual data points to be identified more easily. When multiple groups are involved, the choice of symbols can affect our ability to distinguish between groups. A combination of symbol type, fill pattern, and color can work well when we are trying to distinguish many groups or when there is considerable overlap between groups.

Color hues or saturation can highlight particular groups in the data, such as a group that shows a trend different from most others (Figure 17.16).

Slope charts are a special case of a parallel coordinates plot. The lines connect two response variables, one plotted on a left vertical axis and the other on the right vertical axis. Cleveland (1994) and others suggest these plots are less effective than a scatterplot when showing relationships between a pair of continuous variables.

Slope charts are very effective when showing change in one variable between different conditions. They can be a good way to visualize interactions (but see also Box 17.3).

17.4.2.6 *The Challenge of –Omics and Other Big Data*

Visualizing relationships becomes challenging as the number of variables increases. Visualization is straightforward when those multiple predictors act independently,

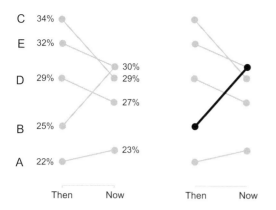

Figure 17.16 Slope chart using the data plotted in Figure 17.3, showing change in a quantitative variable for five groups between two times. The lines can be labeled in most cases by text next to one end of the line. There is a choice between adding a vertical scale and showing the value for each data point.

but gets more difficult when there are complex dependencies. Pairwise interactions are easy to visualize, but anything above that becomes challenging.

Multiple response variables can also prove challenging, and their visualization and analysis often require some simplification or summarization through techniques such as those discussed in Chapters 14–16. This is particularly the case when some responses are correlated, so they cannot be treated independently. Often the data cannot be reduced to a single summary measure, so the response variables need at least two dimensions. The grouping variables need to be represented as different symbols, or relationships indicated by connecting points or groups of points (e.g. network graphs). Simplifying these complex relationships into a graphic whose main point is clear, without sacrificing the underlying complexity, is a serious challenge. Often that challenge is not met well, particularly when multivariate data are part of oral presentations.

New technologies, particularly genomics, proteomics, metabolomics, etc., provide spectacularly rich data in large volumes and will also present communications challenges, but this complexity is beyond the scope of this book.

17.5 Error Bars: Visualizing Variation and Precision

Any representation of effects (i.e. most parameter estimates) should include some measure of uncertainty or precision, such as SDs, SEs, and CIs (Chapter 2). Error bars on graphs are usually represented by a straight line on either side of the mean. If we are using a bar graph

with filled bars, one-sided error bars can be used. The length of the line in each direction typically indicates one SD or one SE, so the total error bar is two SEs. Alternatively, bars might show the CI.

One problem with error bars on complex graphs with many plotting symbols is that the bars overlap with each other and other plotting symbols, making the graph messy and hard to read. In such cases, one alternative might be to present the largest and smallest error bars only in one section of the plot to indicate the range of variability or precision in the data.

Where a plot of means relates to a simple design and analysis, such as a single categorical predictor with OLS estimation, illustrating individual SDs or SEs may not even be crucial. In fitting this model, you have probably assumed that the variances of the different groups are similar and have compared the groups using a pooled within-groups variance. A single error bar may represent the variation used in the analysis more accurately, but individual group bars may indicate whether the assumption of homogeneous variance is appropriate. Software packages vary in whether they use pooled or individual error terms.

Simple error bars may sometimes be inappropriate, and they can even be misleading. The first situation is where we fit models without underlying normal distributions. With a normal distribution, the mean and symmetrical error bars represent the response variable accurately. When using models with different response variable distributions, the effects and their uncertainty are calculated on a measurement scale determined by the link function. If we report effects and uncertainty on original data scales, we should calculate our measures of uncertainty from the model and back-transform the intervals. Taking this approach often generates asymmetrical measures of uncertainty, in contrast to patterns generated by simple summary plots of means and SEs or CIs.

The second, tricker, situation involves mixed models, particularly when nesting is involved. In these cases, a simple SE or the $MS_{Residual}$ may be misleading. As discussed in Chapters 10–12, many hypotheses are tested in complex mixed models, often using different error terms. The graph is used to visualize differences between groups, and a scientific audience uses the error bars to assess the credibility of claims of differences. In the reader's mind, the error bars are linked to the background variation against which patterns have been detected. In complex mixed models there may not be a single measure of background variation, but instead several, each appropriate for one particular hypothesis or pattern in the data. The question is what to show an audience. Without intervention, most software defaults to a simple measure of variation like the $MS_{Residual}$. This measure may not have been used to assess the effect in question.

We illustrate this problem using a mixed model with three factors, A, B, and C, with B random and nested within A, and crossed with C – that is, a split-plot or partly nested design (Section 11.1 and see, e.g. Box 11.1). We generated two datasets with identical means for each combination of A and C, but different patterns within factor B. Analysis of the two datasets leads to very different conclusions, but the "standard" interaction plot, with default error bars, is identical for the two datasets (Box 17.4). One dataset suggests a strong AC interaction, and the other has little evidence for such an effect.

17.5.1 Possible Solutions

We recommend thinking about your message for the reader and then identifying the measure of variance appropriate for this message. With many linear models, the best indication of an appropriate variance often comes from the error term used to test this hypothesis. This variance *will not* be the residual term for many mixed models, and there may be little relationship between it and the overall model residual.

The best visualization is not obvious, and we consider several options:

1. The $\sqrt{MS_{Residual}}$ from the ANOVA.
2. Convert the model mean square used to assess the particular effect into a "pseudo" error bar that considers the df concerned. In some cases, this can be simple, such as a simple nested design with two factors, where factor A is tested using nested factor B. A plot of means and errors can be constructed by generating a summary file of means for each group of B, creating a file with a single factor (A). The resulting error bars reflect variation at the scale of B. The option may not be simple with a very complicated linear model.
3. Instead of using the mean square, extract the VC associated with a particular random effect – for example, in a nested design, calculate σ_B^2 instead of MS_B.
4. Produce a clean graph devoid of error bars. Use the graph to visualize the size and nature of differences, and allow the reader to use the model summary to see where variation occurs.

Option 1 is only appropriate when the model residual was used to assess the effect in question and is misleading otherwise. Options 2 and 3 arguably use appropriate measures of variation, but they can be hard to explain, particularly if the "denominator" term is itself an inter-action. Option 4 presents those patterns clearly and emphasizes the size of particular effects. It is our preferred

option, but there are two pitfalls. First, the visualization must show effects for which there is convincing evidence, and we should not use the removal of error bars as a device to discuss "effects" that are indistinguishable from noise. Second, readers, and more important, reviewers are trained/conditioned to expect error bars, and they may complain at their absence, despite clear arguments for the irrelevance. We have experience with reviewers requiring option 1, despite its irrelevance.

Our advice is:
- See if a "conventional" error bar (or CI) is appropriate, and if so, use it.
- Where the mixed models are complex and no simple error bar is appropriate, don't show an error bar, and point the reader to the analysis summary. Be careful not to mislead the reader when describing the patterns.
- If you believe it appropriate not to show an error bar, be prepared to argue your case (repeatedly).

17.6 Horses for Courses: What You Present Depends on Who's Listening

Our data summary and visualizations vary with our target audience and the mode of delivery. In a printed paper, we can place large amounts of information on a figure, giving the reader time to digest that information. When speaking, there's less time for the audience to assimilate the infor-mation. You are speaking more or less continuously. If you show a visual with lots of information, a large part of the audience immediately shifts their focus away from you to concentrate on reading. You've now lost control of the audience, and they aren't listening closely. They may also not be getting the information you want them to.

Your audience's expectations might vary widely. The readers of a scientific paper are mostly peers in your research field who share much of your biological and statistical background knowledge. A report prepared for a government agency may be read by people who may have been trained in your broad discipline, but these days they are likely to have backgrounds far from your topic. They may be experienced and proficient in using scien-tific information to help make decisions, but they may not have much specific technical knowledge or any interest in the mechanics of your data analysis. If you let the broader community know what you've discovered, the statistical details will probably be of even less interest.

The combination of audience and mode of delivery may require you to have several ways of communicating a particular outcome. The important first step is to be clear about the audience and delivery.

Box 17.4 ⓡ Worked Example of Ambiguous Error Bars

We illustrate how "default" error bars produced by most software packages can be misleading when summarizing mixed models. We generated two simple datasets for a three-factor mixed model design. Factor A has two groups and is fixed. Factor B is a random effect with four groups and is nested in A. Factor C is crossed with A and C and is fixed with three groups. In the first dataset (response variable $Y1$) there are strong effects of the random factor, and effects of factor C are consistent across that factor. The second dataset ($Y2$) has identical observations for each combination of A and C, but data were permuted so the effects of C vary across groups of B.

A	B	C	$Y1$	$Y2$
A	1	c1	1.42	1.42
A	1	c2	1.43	1.83
A	1	c3	1.44	1.89
A	2	c1	1.88	1.88
A	2	c2	1.88	1.43
A	2	c3	1.89	1.05
A	3	c1	1.05	1.05
A	3	c2	1.05	1.88
A	3	c3	1.05	1.84
A	4	c1	1.82	1.82
A	4	c2	1.83	1.05
A	4	c3	1.84	1.44
a	5	c1	1.18	1.18
a	5	c2	1.16	1.72
a	5	c3	1.14	1.01
a	6	c1	1.02	1.02
a	6	c2	1.01	1.16
a	6	c3	1.01	1.00
a	7	c1	1.02	1.02
a	7	c2	1.01	1.01
a	7	c3	1.00	1.72
a	8	c1	1.73	1.73
a	8	c2	1.72	1.01
a	8	c3	1.72	1.14

The linear model for this sampling design is $Y = \mu + A + C + B(A) + A \times C + \varepsilon$. There is only a single value for each BC combination, so $B(A) \times C$ is omitted. We described this design in Chapter 11.

The results for fitting this model to $Y1$ and $Y2$ are shown below.

Source	Num. df	Denom.df	$Y1$ MS	$Y1$ P	$Y2$ MS	$Y2$ P
A	1	6	0.6208	0.259	0.6208	0.010
C	2	12	0.0000	0.409	0.0000	1.000
$A \times C$	2	12	0.0005	0.001	0.0005	0.997
$B(A)$	6	12	0.3987	<0.0005	0.0446	0.949
Residual	12		0.0000		0.1771	

Box 17.4 (cont.)

The two analyses suggest different patterns in the data. For $Y1$, we detect an interaction between the two fixed factors, but there is no hint of such interaction for $Y2$, despite the identical cell means.

The reason for this is clear from Figure 17.17. For $Y1$, the consistent effects of C across each of the groups of B (the "plots" or "subjects") leads to a large amount of variation attributed to B and little to the residual (which includes $B(A) \times C$). In contrast, for $Y2$, effects of factor C vary widely across groups of B, giving a much larger residual variance and correspondingly smaller variance associated with B.

When we produce a standard interaction plot, we are surprised that the two datasets give the same interaction plot with identical standard errors. An inspection of that plot is not informative for explaining the $A \times C$ interaction. What's going on here is that the standard errors are being produced by examining the variation within each combination of A and C without taking factor B into account, so the errors calculated this way do not match the patterns of variance in the model. The extent to which the error terms can mislead a reader depends on the specific dataset and model. Still, the potential for a disconnect between the results and their visualization is not trivial.

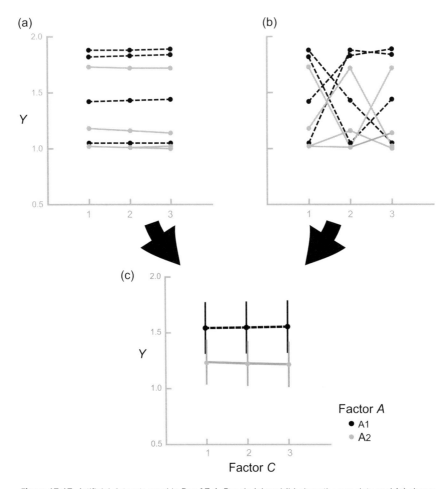

Figure 17.17 Artificial datasets used in Box 17.4. Panels (a) and (b) show the raw data and (c) shows the identical means and standard errors produced when the correlated nature of the data is not considered.

17.6.1 Know Your Audience as Well as Possible

The first question to ask is who the target audience is? Who are you trying to influence, inform, or inspire? How much background information about your topic can you assume? How much technical detail will be of interest or expected?

If it's an oral presentation, the overall structure of your talk will be influenced by its purpose. Are you issuing a call to arms, trying to inspire the audience to take some action in response to what you say? Are you providing concrete, practical information, or are you more interested in exciting concepts and theory rather than advice? Are you speaking as a way of showcasing your skills?

Ask these questions before preparing all talks or any writing, as they aren't specific to presentation of data. Still, they are essential for determining just how to present your data.

17.6.1.1 *Talks*

For an oral presentation, the important features are that
- **you** are the focus;
- **you** control the delivery of the material; and
- illustrations don't work without **you**

The focus is on you as the presenter, and your data graphics shouldn't be a distraction. In general, remove all extraneous information from the figures. As part of your talk, guide the audience through any particular figure – show them the key patterns, explain what the different symbols represent, and so on. That way, you control the emphasis placed on the information, and the audience feels that they are getting a scientist's view of some information rather than reading another paper. It is generally better to use more slides (or simple animations in which chart elements appear sequentially) to build a story rather than show a complex graphic that the audience will immediately focus on.

You probably need not show results of statistical tests on the figure. For example, a regression equation with Fs and P-values adds unnecessary clutter to a scatterplot. There is often a collective groan in the audience when the next slide is an ANOVA table or some similarly complex table of numbers. Omitting analyses from a talk starts with ethical behavior. If you talk about a pattern – a difference in groups, a correlation, etc. – you are describing the results of an analysis that separated an effect from background noise. The audience trust that you have this evidence. During the talk, there's little chance to scrutinize your experimental design or check the analysis; adding some statistical details (which may be right or wrong) doesn't change that. This reinforces the central idea that data graphics should emphasize important results but should not be used to mislead the reader.

17.6.1.2 *Conferences and Seminars*

Many talks are given to an audience with only a title and abstract for information. The data are provided only through the graphics you provide, accompanied by your narration. This interaction with your audience will require simple graphics, often with the material presented sequentially. Negotiating complex stories will be a challenge, but the advantage is that you are there to highlight the important aspects of the data.

17.6.1.3 *Talks with Handouts (Lecture, Briefing, etc.)*

Sometimes the audience has written material, such as lecture handouts or briefing notes, to go with your talk. When the audience is small, there is often an opportunity to work through more complex graphics. The audience members have information they can return to for clarification or digest in more detail. The printed material you provide tells part of the story, along with your narration. This presentation type involves some loss of control, as the audience may read ahead. You will need to use graphics on the screen to keep the audience focused, and those graphics may be different from those taken away. Take-away figures may require captions or explanatory notes, which will not be needed for the slides you lead the audience through.

17.6.1.4 *Written Stories (Papers, Theses, etc.)*

When you write the story down, several things change. For most of us, written stories serve different purposes. They can be scholarly publications directed at our peers (papers and theses), technical reports for nonspecialists, proposals to funding agencies, information for the community. Different outlets will require different details of the analyses, but the data graphics share several features.

First, and most important, the audience is on their own with the document, without you there to draw attention to aspects you feel are important. You also lose control over the order in which the material is read.

You can focus the reader's attention by your language and by careful preparation of graphics. All your data graphics will require detailed explanations, which are traditionally placed into figure captions. Those captions must be close to self-sufficient, so a casual reader can move through the figures and tables and get a picture of your important results.

The detail required for the statistical analysis will vary between outlets. Theses and papers typically require full details of statistical models fitted to the data, a brief statement of the extent to which any assumptions were satisfied, and comprehensive results. With many journals, the trend is for much of this detail to be placed into appendices, which are often available only online. Reports and proposals typically omit much statistical detail, and it is common for proposals to have very abbreviated analytical results. Community reports, particularly through social media, may lack this detail completely.

17.7 Software and Other Sources

The principles behind effective data visualization are increasingly incorporated into statistical software. Usually the default graphics will be fine for a first look at the data, with the graphics being polished for publication or presentation.

The development of scientific graphics has been formalized, where some principles espoused by Cleveland and others were placed into a formal "Grammar of Graphics" (Wilkinson 2013), and this grammar is embedded into some software packages, most notably the *R* package *ggplot2* (Wickham 2009), and previously SYSTAT (developed by Wilkinson, who embedded his grammar of graphics into the software; www.systat.com).

Other software packages were developed to aid data visualization but without an emphasis on scientific presentation. Some of the most widely known can produce compelling illustrations but need to be coerced into providing the measures of background variation that scientific audiences generally expect. It is also possible to develop effective graphics through standard office software such as Microsoft Excel, although considerable tweaking from the default settings is generally required.

There is also lots of good advice available online. You can find many sites devoted to advice about preparing effective illustrations, galleries of effective graphics, and types of graphics, and collections of bad examples. There are also sites where you can post your visualizations and receive helpful feedback. We also recommend keeping an idea on some mainstream media noted for the high quality of their data visualizations, such as the *New York Times* (www.nytimes.com) and the *Guardian* (www.guardian.co.uk). Online material is very fluid, so rather than listing individual sites we maintain a list of sites we find useful at our website (www.qkstats.com).

17.8 Key Points

• As biologists, we are telling biological stories.
• Data analysis is there to help tell the story; don't let it get in the way.

• Analysis helps us separate what we perceive as patterns from noise. Finding a signal is only the start when we fit a model to data. The bigger role is to tell us how strong the predictors' effects are. These effects and their biological meanings form the basis of our story and our illustrations.
• Most statistical packages produce considerable redundancy in their output. Omitting redundant elements produces cleaner, more concise descriptions of results.
• Some graphics packages readily produce graphs that obscure, rather than clarify, results.
• Tailor graphical illustration of results to the audience. Optimal use of colors, fill patterns, and explanatory text will be very different for published scientific papers and oral presentations.
• In preparing illustrations, decide which pattern in the data you wish to illustrate, then identify the variation that was the background against which that pattern was assessed. This variation is an appropriate candidate for error bars.
• There is lots of excellent communication advice outside of science. You can learn by reading advice from different fields and exposing yourself to effective communication from these fields, by reading, listening, and watching widely outside your discipline.
• As with many things, you'll find lots of people willing to give you rules for visualization. Don't follow them slavishly, but read the advice (particularly when it's evidence-based), think about it, and then decide what's best for the task in hand.

Further Reading

There's been a refreshing rise in scientists' commitment to improving how we communicate to peers and to the broader community, and it's reflected in many books of advice and dedicated scientific communications courses at many universities. We've highlighted our favorite broader guides.

Writing clearly: Pechenik (2015) and Schimel (2012) are extensive and targeted at biologists, but don't stay within biology. Pinker (2014) offers lots of advice on writing technique.

Producing clear graphics: Guides to visualizing data have grown the fastest. Most aren't targeted at biologists (or even scientists), but the principles don't change. We would start with Knaflic (2015), Cairo (2016), and Few (2012), but keep an eye on these authors' websites for new material.

A deeper dive into graphics: Many current principles come from some classic references, particularly Tukey's (1977) pioneering work on exploratory data analysis, Cleveland's (1993, 1994) ideas on visualization for scientists, along with Tufte's (1983) broad exploration of visualization principles. Those wanting to immerse themselves in *ggplot* could go to Wilkinson (2013) for its underlying principles.

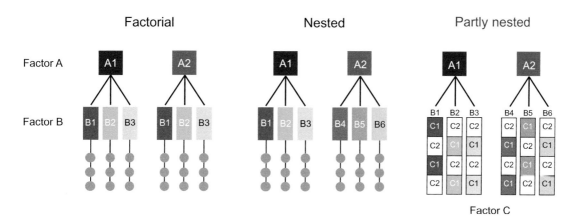

Figure 10.2 Layout of three common multifactor designs, showing differences between factorial, nested, and split-plot (partly nested) arrangements of factors.

Easier

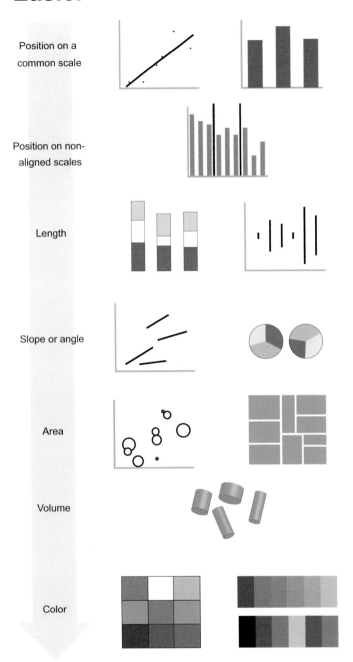

Position on a common scale

Position on non-aligned scales

Length

Slope or angle

Area

Volume

Color

Harder

Figure 17.2 Hierarchy of ways of encoding quantitative information, based on Cleveland (1994) and illustrated using common graphical types. The hierarchy reflects the accuracy with which quantitative information can be decoded by a reader. You should be aware that Cleveland did not see this as a strict hierarchy, and some of the ways may overlap in their efficiency. The hierarchy was derived from experimental studies, so it reflects the particular scenarios presented to the subjects. Use this as a guideline for what is most effective, but be prepared to disregard it if a particular situation suggests a specific graphical type would be most effective.

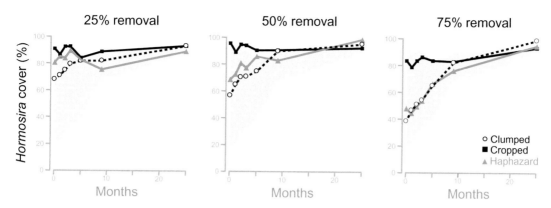

Figure 17.13 Simplifying a complex line graph. This figure is plotted from Wootton and Keough (2016), where they broke down time series from an experiment examining the effects of intensity and type of disturbance on a canopy-forming alga, with 11 treatments (a 3 × 3 factorial plus two "anchor" treatments – no disturbance and complete removal – which cannot be divided into different types) into three panels, each with three lines, along with the two anchor treatments in the background.

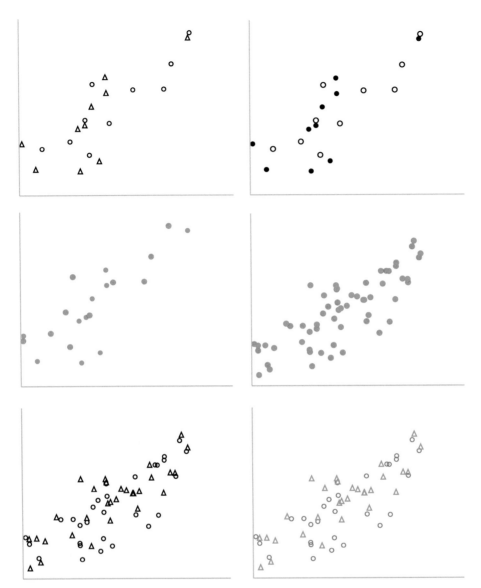

Figure 17.15 The choice of symbols is important for identifying patterns. In the example, filled symbols stand out, but when there are many data points or lots of overlap, unfilled symbols allow individual data points to be identified more easily. When multiple groups are involved, the choice of symbols can affect our ability to distinguish between groups. A combination of symbol type, fill pattern, and color can work well when we are trying to distinguish many groups or when there is considerable overlap between groups.

Glossary

Akaike's information criterion (AIC) – Measure of model fit that incorporates a penalty for too many parameters. Can be modified for small sample sizes.

analysis of covariance (ANCOVA) – Modification of ANOVA with one or more factors to account for the relationship between the response variable and a continuous covariate.

analysis of variance (ANOVA) – Partitioning of variance into amounts due to each term and the residual in a linear model fitted with OLS.

bagging and boosting – Methods to improve the predictive capacity of regression trees, often through resampling.

block design – Experimental design where experimental units are arranged in spatial blocks with one or more units for each factor group within each block.

bootstrap – Resampling with replacement to estimate a population parameter.

completely randomized design – Experimental design where experimental units are randomly assigned to factor groups.

compound symmetry – Covariance structure with equal variances and equal covariances.

confidence interval – One interval from a set of intervals that will include a parameter a given percentage of times (i.e. confidence) under repeated sampling.

confounding – Effects of one or more factors cannot be separated from other extraneous factors, commonly due to inadequate experimental design.

constrained ordination – Method applied mainly to principal components and correspondence analyses where derived components or ordination plot axes are constrained by linear relationships with additional covariates.

covariance structure – Pattern of variances and covariances in a dataset.

cross-validation – Method to assess the predictive capacity of a model by applying it to new data or a separate subset of original data.

deviance – Difference in fit, based on (log)likelihoods, between a specified model and the saturated model.

discriminant function – Linear combination of variables that maximizes the difference between groups of objects.

dissimilarity – Multivariate measure of difference between two objects based on their variable values.

distance – Measure of how far apart two objects are in multidimensional space.

dummy variable – Categorical variable recoded into binary (0,1) variables for regression models.

eigenanalysis – Method for decomposing a matrix of variable associations into a new set of derived variables or components.

experimental unit – Smallest unit to which an experimental treatment group is applied.

exponential family – Family of probability distributions used in linear models defined by their mean and variance.

fixed predictor – Predictor whose values are specifically chosen and inference is restricted to those values.

(general) linear model (LM) – Model relating a response variable to a linear combination of predictor variables; fitted using OLS.

generalized least squares (GLS) – Modification of OLS that allows unequal variances and nonzero covariances.

generalized linear mixed model (GLMM) – GLM that includes random effects.

generalized linear model (GLM) – Linear model where response variable or model error terms follow a probability distribution from the exponential family; fitted using ML.

heteroskedasticity – Variance of model error terms or responses is unequal across values of the predictors.

homoskedasticity – Variance of model error terms or responses is equal across values of the predictors.

imputation – Replacement of missing value of a variable by summary value from remaining data.

influential observation – Observation that has more influence on parameter estimates than other observations.

interaction – Effect of one predictor depends on other predictors.

jackknife – Resampling by omitting each observation in turn to estimate a parameter.

link function – Function relating a response variable to linear combination of predictors in a GLM; link function transforms response to turn a nonlinear relationship into a linear one.

maximum likelihood (ML) – Method for estimating parameters (e.g. for linear models) based on maximizing the likelihood of observing sample data.

mixed model – Linear model including both fixed and random predictors and their effects.

model averaging – Method to produce a single model by averaging (weighted by model fit) parameter estimates from all models.

model selection – Method for choosing one model over others based on measures of model fit.

multidimensional scaling – General procedure for deriving a plot of objects in multidimensional space where the distance between objects matches their dissimilarities as closely as possible.

multilevel regression – Term for a mixed model commonly used when a regression model relating a response to a continuous fixed predictor could vary across random groups.

nonparametric – Statistical procedure that does not assume a specific underlying probability distribution of data or model error terms.

observational unit – Unit at the scale that observations (e.g. variable values) are made; may be smaller than the experimental unit (subsampling).

ordinary least squares (OLS) – Method for fitting linear models and estimating their parameters based on minimizing the differences between observed and predicted values.

ordination/biplot/triplot – Encompassing term, mainly in ecology, for methods that produce a plot of objects in multidimensional space based on variable values; biplots include variables on the same plot; triplots include variables and covariates on the same plot.

outlier – Unusual value that is different from the rest of the sample or from the fitted model.

overdispersion – Data where the variance is greater than expected from the presumed probability distribution.

P-*value* – Probability under repeated sampling of obtaining our sample data or data more extreme if a null hypothesis is true.

parametric – Statistical procedure that assumes a specific underlying probability distribution of data or model error terms.

power of test – Measure of ability to statistically detect an effect if it occurs in the population(s).

pseudoreplication – Encompassing term to describe a range of experimental design errors, usually due to misunderstanding of true experiment units.

random predictor – Predictor whose values are a random sample from a population of possible values and inference is about the variance of associated random effects.

randomization test – Hypothesis-testing procedure that generates sampling distribution of a statistic via random reassignment of observations.

residual – Difference between observed value and that predicted by a given model; estimate of model error term.

restricted maximum likelihood – Modification of ML that provides unbiased estimates of variances.

split-plot design – Experimental design where factors are applied to experimental units at different scales of plots and sub-plots.

statistical significance – Rejection of a null hypothesis at some *a priori* defined *P*-value cut-off, conventionally 0.05.

statistical test – Procedure to assess a null hypothesis by comparing a sample statistic to its sampling distribution, thereby generating a *P*-value.

transformation – Change in the scale of a continuous variable, usually to meet distributional assumptions of an analysis.

Type I, II III SS – Different sums-of-squares for factorial ANOVA with unequal sample sizes.

variance component – Variance associated with a particular random effect in a linear model.

zero inflation – Data with more zeroes than expected from the presumed probability distribution.

References

Aarts, E., Verhage, M., Veenvliet, J. V., Dolan, C. V. & van der Sluis, S. (2014). A solution to dependency: using multilevel analysis to accommodate nested data. *Nature Neuroscience*, 17, 491–6.

Abdi, H. (2003). Partial least squares (PLS) regression. In *Encyclopedia of social sciences research methods*, eds. M. Lewis-Beck, A. Bryman, & T. Futing, pp. 792–5. Sage.

Abegaz, F., Chaichoompu, K., Genin, E., et al. (2019). Principals about principal components in statistical genetics. *Briefings in Bioinformatics*, 20, 2200–16.

Agresti, A. (2013). *Categorical data analysis*, 2nd ed. Wiley.

Agresti, A. (2015). *Foundations of linear and generalized linear models*. Wiley.

Agresti, A. (2019). *An introduction to categorical data analysis*, 3rd ed. Wiley.

Aguinis, H. & Gottfredson, R. K. (2010). Best-practice recommendations for estimating interaction effects using moderated multiple regression. *Journal of Organizational Behavior*, 31, 776–86.

Aiken, L. S. & West, S. G. (1991). *Multiple regression: testing and interpreting interactions*. Sage.

Akaike, H. (1978). Bayesian-analysis of minimum AIC procedure. *Annals of the Institute of Statistical Mathematics*, 30, 9–14.

Al-Janabi, B., Kruse, I., Graiff, A., et al. (2016). Buffering and amplifying interactions among OAW (Ocean Acidification & Warming) and nutrient enrichment on early life-stage *Fucus vesiculosus* L. (Phaeophyceae) and their carry over effects to hypoxia impact. *PLoS One*, 11, e0152948.

Alf, C. & Lohr, S. (2007). Sampling assumptions in introductory statistics classes. *The American Statistician*, 61, 71–77.

Allen, R. M. & Marshall, D. (2014). Egg size effects across multiple life-history stages in the marine annelid *Hydroides diramphus*. *PLoS One*, 9, e102253.

Allison, T. & Cicchetti, D. V. (1976). Sleep in mammals: ecological and constitutional correlates. *Science*, 194, 732–4.

Anderson, D. R., Burnham, K. P., & Thompson, W. L. (2000). Null hypothesis testing: problems, prevalence, and an alternative. *Journal of Wildlife Management*, 64, 912–23.

Anderson, M. J. (2001). Permutation tests for univariate or multivariate analysis of variance and regression. *Canadian Journal of Fisheries and Aquatic Sciences*, 58, 626–39.

Anderson, M. J. (2006). Distance-based tests for homogeneity of multivariate dispersions. *Biometrics*, 62, 245–53.

Anderson, M. J. (2017). Permutational Multivariate Analysis of Variance (PERMANOVA). In *Wiley StatsRef: statistics reference online*, pp. 1–15. Wiley.

Anderson, M. J. & Legendre, P. (1999). An empirical comparison of permutation methods for tests of partial regression coefficients in a linear model. *Journal of Statistical Computation and Simulation*, 62, 271–303.

Anderson, M. J. & Robinson, J. (2001). Permutation tests for linear models. *Australian and New Zealand Journal of Statistics*, 43, 75–88.

Anderson, M. J. & Ter Braak, C. J. F. (2003). Permutation tests for multi-factorial analysis of variance. *Journal of Statistical Computation and Simulation*, 73, 85–113.

Anderson, M. J. & Walsh, C. I. (2013). PERMANOVA, ANOSIM, and the Mantel test in the face of heterogeneous dispersions: what null hypothesis are you testing? *Ecological Monographs*, 83, 557–74.

Anderson, M. J., Ellingsen, K. E. & McArdle, B. H. (2006). Multivariate dispersion as a measure of beta diversity. *Ecology Letters*, 9, 683–93.

Anderson, M. J., Gorley, R. N. & Clarke, K. R. (2008). *PERMANOVA+ for PRIMER: guide to software and statistical methods*. PRIMER-E.

Anderson, M. J., Walsh, D. C. I., Clarke, K. R., Gorley, R. N. & Guerra-Castro, E. (2017). Some solutions to the multivariate Behrens–Fisher problem for dissimilarity-based analyses. *Australian & New Zealand Journal of Statistics*, 59, 57–79.

Anderson-Sprecher, R. (1994). Model comparisons and R^2. *The American Statistician*, 48, 113.

Andrew, N. L. & Mapstone, B. D. (1987). Sampling and the description of spatial pattern in marine ecology. *Oceanography and Marine Biology*, 25, 39–90.

Anscombe, F. J. (1973). Graphs in statistical analysis. *The American Statistician*, 27, 17–21.

Antelman, G. (1997). *Elementary Bayesian statistics*. Edward Elgar.

Aranda, S. C., Gabriel, R., Borges, P. A., et al. (2014). Geographical, temporal and environmental determinants of bryophyte species richness in the Macaronesian islands. *PLoS One*, 9, e101786.

Aune-Lundberg, L. & Strand, G.-H. (2014). Comparison of variance estimation methods for use with two-dimensional systematic sampling of land use/land cover data. *Environmental Modeling & Software*, 61, 87–97.

Austin, M. P. (2013). Inconsistencies between theory and methodology: a recurrent problem in ordination studies. *Journal of Vegetation Science*, 24, 251–68.

Ayres, M. P. & Thomas, D. L. (1990). Alternative formulations of the mixed-model ANOVA applied to quantitative genetics. *Evolution*, 44, 221–26.

Babyak, M. A. (2004). What you see may not be what you get: a brief, nontechnical introduction to overfitting in regression-type models. *Psychomatic Medicine*, 66, 411–21.

Banner, K. M., Irvine, K. M., Rodhouse, T. J. & O'Hara, R. B. (2020). The use of Bayesian priors in ecology: the good, the bad and the not great. *Methods in Ecology and Evolution*, 11, 882–89.

Barber, R. F., Candes, E. J., Ramdas, A. & Tibshirani, R. J. (2021). Predictive inference with the jackknife. *Annals of Statistics*, 49, 486–507.

Bates, D., Machler, M., Bolker, B. M. & Walker, S. C. (2015). Fitting linear mixed-effects models using lme4. *Journal of Statistical Software*, 67, 1–48.

Bayer, S. & Hettinger, A. (2019). Storytelling: a natural tool to weave the threads of science and community together. *Bulletin of the Ecological Society of America*, 100, 1–6.

Berk, R. A. (1990). A primer on robust regression. In *Modern methods of data analysis*, eds. J. Fox & J. S. Long, pp. 292–334. Sage.

Bi, J. (2012). A review of statistical methods for determination of relative importance of correlated predictors and identification of drivers of consumer liking. *Journal of Sensory Studies*, 27, 87–101.

Bjorklund, M. (2019). Be careful with your principal components. *Evolution*, 73, 2151–8.

Bland, J. M., Altman, D. G. & Rohlf, F. J. (2013). In defence of logarithmic transformations. *Statistics in Medicine*, 32, 3766–9.

Blasco-Moreno, A., Pérez-Casany, M., Puig, P., et al. (2019). What does a zero mean? Understanding false, random and structural zeros in ecology. *Methods in Ecology and Evolution*, 10, 949–59.

Bleeker, K., de Jong, K., van Kessel, N., Hinde, C. A. & Nagelkerke, L. A. J. (2017). Evidence for ontogenetically and morphologically distinct alternative reproductive tactics in the invasive round goby *Neogobius melanostomus*. *PLoS One*, 12, e0174828.

Bliss, C. I. (1934). The method of probits. *Science*, 79, 38–9.

Boehmke, B. & Greenwell, B. M. (2019). *Hands-on machine learning with R*. Chapman and Hall/CRC.

Boldina, I. & Beninger, P. G. (2016). Strengthening statistical usage in marine ecology: linear regression. *Journal of Experimental Marine Biology and Ecology*, 474, 81–91.

Bolker, B. M. (2008). *Ecological models and data in R*. Princeton University Press.

Bolker, B. M. (2015). Linear and generalized linear mixed models. In *Ecological statistics: contemporary theory and application*, eds. G. A. Fox, S. Negrete-Yankelevich, & V. J. Sosa, pp. 309–34. Oxford University Press.

Bolker, B. M., Brooks, M. E., Clark, C. J., et al. (2009). Generalized linear mixed models: a practical guide for ecology and evolution. *Trends in Ecology & Evolution*, 24, 127–35.

Booth, D. T. & Evans, A. (2011). Warm water and cool nests are best: how global warming might influence hatchling green turtle swimming performance. *PLoS One*, 6, e23162.

Borcard, D., Gillet, F. & Legendre, P. (2018). *Numerical ecology with R*, 2nd ed. Springer.

Breiman, L. (1996). Bagging predictors. *Machine Learning*, 24, 123–40.

Breiman, L., Friedman, J. H., Olshen, R. A. & Stone, C. J. (1984). *Classification and regression trees*. Wadsworth.

Breitwieser, M., Viricel, A., Graber, M., et al. (2016). Short-term and long-term biological effects of chronic chemical contamination on natural populations of a marine bivalve. *PLoS One*, 11, e0150184.

Bring, J. (1994). How to standardise regression coefficients. *The American Statistician*, 48, 209–13.

Buonaccorsi, J. P. (2010). *Measurement error: models, methods, and applications*. CRC Press.

Burnham, K. P. & Anderson, D. R. (2002). *Model selection and multimodel inference*. Springer-Verlag.

Burnham, K. P. & Anderson, D. R. (2004). Multimodel inference: understanding AIC and BIC in model selection. *Sociological Methods & Research*, 33, 261–304.

Burnham, K. P., Anderson, D. R. & Huyvaert, K. P. (2011). AIC model selection and multimodel inference in behavioral ecology: some background, observations, and comparisons. *Behavioral Ecology and Sociobiology*, 65, 23–35.

Caballes, C. F., Pratchett, M. S., Kerr, A. M., & Rivera-Posada, J. A. (2016). The role of maternal nutrition on oocyte size and quality, with respect to early larval development in the coral-eating starfish, *Acanthaster planci*. *PLoS One*, 11, e0158007.

Cabanellas-Reboredo, M., Vázquez-Luis, M., Mourre, B., et al. (2019). Tracking a mass mortality outbreak of pen shell *Pinna nobilis* populations: a collaborative effort of scientists and citizens. *Scientific Reports*, 9, 13355.

Cade, B. S. (2015). Model averaging and muddled multimodel inferences. *Ecology*, 96, 2370–82.

Cairo, A. (2016). *The truthful art: data, charts, and maps for communication*. Pearson Education.

Caley, M. J. & Schluter, D. (1997). The relationship between local and regional diversity. *Ecology*, 78, 70–80.

Carey, J. M. & Keough, M. J. (2002). Compositing and subsampling to reduce costs and improve power in benthic infaunal monitoring programs. *Estuaries*, 25, 1052–60.

Carpenter, S. R., Chisholm, S. W., Krebs, C. J., Schindler, D. W. & Wright, R. F. (1995). Ecosystem experiments. *Science*, 269, 324–7.

Carter, N. J., Schwertman, N. C. & Kiser, T. L. (2009). A comparison of two boxplot methods for detecting univariate outliers which adjust for sample size and asymmetry. *Statistical Methodology*, 6, 604–21.

Casler, M. D. (2015). Fundamentals of experimental design: guidelines for designing successful experiments. *Agronomy Journal*, 107, 692–705.

Chadha, A., Florentine, S., Chauhan, B. S., Long, B. & Jayasundera, M. (2019). Influence of soil moisture regimes on growth, photosynthetic capacity, leaf biochemistry and reproductive capabilities of the invasive agronomic weed; Lactuca serriola. *PLoS One*, 14, e0218191.

Chapman, M. G. & Underwood, A. J. (1999). Ecological patterns in multivariate assemblages: information and interpretation of negative values in ANOSIM tests. *Marine Ecology Progress Series*, 180, 257–65.

Chatterjee, S. & Hadi, A. S. (2012). *Regression analysis by example*, 5th ed. Wiley.

Choi, L., Blume, J. D. & Dupont, W. D. (2015). Elucidating the foundations of statistical inference with 2 × 2 tables. *PLoS One*, 10, e0121263.

Christensen, D. L., Herwig, B. R., Schindler, D. E. & Carpenter, S. R. (1996). Impacts of lakeshore residential development on coarse woody debris in north temperate lakes. *Ecological Applications*, 6, 1143–49.

Claesen, A., Gomes, S., Tuerlinckx, F. & Vanpaemel, W. (2021). Comparing dream to reality: an assessment of adherence of the first generation of preregistered studies. *Royal Society Open Science*, 8, 211037.

Clarke, K. R. (1993). Non-parametric multivariate analyses of changes in community structure. *Australian Journal of Ecology*, 18, 117–43.

Clarke, K. R. & Ainsworth, M. (1993). A method of linking multivariate community structure to environmental variables. *Marine Ecology Progress Series*, 92, 205–19.

Clarke, K. R. & Gorley, R. N. (2015). *PRIMER v7: user manual/tutorial*. PRIMER-E.

Clarke, K. R., Somerfield, P. J. & Chapman, M. G. (2006). On resemblance measures for ecological studies, including taxonomic dissimilarities and a zero-adjusted Bray–Curtis coefficient for denuded assemblages. *Journal of Experimental Marine Biology and Ecology*, 330, 55–80.

Clarke, K. R., Gorley, R. N., Somerfield, P. J., & Warwick, R. M. (2014). *Change in marine communities: an approach to statistical analysis and interpretation*, 3rd ed. PRIMER-E.

Clavel, J., Merceron, G., & Escarguel, G. (2014). Missing data estimation in morphometrics: how much is too much? *Systematic Biology*, 63, 203–18.

Cleveland, W. S. (1993). *Visualizing data*. Hobart Press.

Cleveland, W. S. (1994). *The elements of graphing data*, 2nd ed. Hobart Press.

Cleveland, W. S. & McGill, R. (1984). Graphical perception: theory, experimentation, and application to the development of graphical methods. *Journal of the American Statistical Association*, 79, 531–54.

Clevenger, A. P. & Waltho, N. (2000). Factors influencing the effectiveness of wildlife underpasses in Banff National Park, Alberta, Canada. *Conservation Biology*, 14, 47–56.

Cochran, W. G. & Cox, G. M. (1957). *Experimental designs*, 2nd ed. Wiley.

Cohen, J. (1988). *Statistical power analysis for the behavioral sciences*, 2nd ed. Lawrence Erlbaum.

Cohen, J., Cohen, P., West, S. G. & Aiken, L. S. (2003). *Applied multiple regression/correlation analysis for the behavioral sciences*, 3rd ed. Routledge.

Colegrave, N. & Ruxton, G. D. (2017). Statistical model specification and power: recommendations on the use of test-qualified pooling in analysis of experimental data. *Proceedings of the Royal Society B: Biological Sciences*, 284, 20161850.

Constable, A. J. (1993). The role of sutures in shrinking of the test in Heliocidaris erythrogramma (Echinoidea: Echinometridae). *Marine Biology*, 117, 423–30.

Cox, D. R. (2020). Statistical significance. *Annual Review of Statistics and Its Application*, 7, 1–10.

Cox, T. & Cox, M. (2000). *Multidimensional scaling*, 2nd ed. Chapman and Hall/CRC.

Crawley, M. J. (2014). *Statistics: an introduction using R*, 2nd ed. Wiley.

Dahlstrom, M. F. (2014). Using narratives and storytelling to communicate science with nonexpert audiences. *Proceedings of the National Academy of Sciences*, 111, 13614–20.

Dalal, D. K. & Zickar, M. J. (2012). Some common myths about centering predictor variables in moderated multiple regression and polynomial regression. *Organizational Research Methods*, 15, 339–62.

Dale, M. R. T. & Fortin, M.-J. (2009). Spatial autocorrelation and statistical tests: some solutions. *Journal of Agricultural, Biological, and Environmental Statistics*, 14, 188–206.

Day, R. W. & Quinn, G. P. (1989). Comparisons of treatments after an analysis of variance in ecology. *Ecological Monographs*, 59, 433–63.

De'ath, G. (2002). Multivariate regression trees: a new technique for modeling species–environment relationships. *Ecology*, 83, 1105–17.

De'ath, G. & Fabricius, K. E. (2000). Classification and regression trees: a powerful yet simple technique for ecological data analysis. *Ecology*, 81, 3178–92.

Dean, C. (2009). *Am I making myself clear? A scientist's guide to talking to the public*. Harvard University Press.

Dennis, B. (1996). Discussion: should ecologists become Bayesians? *Ecological Applications*, 6, 1095–103.

Dong, Y. & Peng, C.-Y. J. (2013). Principled missing data methods for researchers. *Springerplus*, 2, 222.

Dormann, C. F., Calabrese, J. M., Guillera-Arroita, G., et al. (2018). Model averaging in ecology: a review of Bayesian, information-theoretic, and tactical approaches for predictive inference. *Ecological Monographs*, 88, 485–504.

Dormann, C. F., Elith, J., Bacher, S., et al. (2013). Collinearity: a review of methods to deal with it and a simulation study evaluating their performance. *Ecography*, 34, 27–46.

Downes, B. J., Barmuta, L. A., Fairweather, P. G., et al. (2002). *Assessing ecological impacts: concepts and applications in flowing waters*. Cambridge University Press.

Downes, B. J., Lake, P. S., & Schreiber, E. S. G. (1993). Spatial variation in the distribution of stream invertebrates: implications of patchiness for models of community organization. *Freshwater Biology*, 30, 119–32.

Duarte, N. (2008). *Slide:ology: the art and science of creating great presentations*. O'Reilly Media.

Dunn, P. & Smyth, G. (2018). *Generalized linear models with examples in R*. Springer-Verlag.

Echambadi, R. & Hess, J. D. (2007). Mean-centering does not alleviate collinearity problems in moderated multiple regression models. *Marketing Science*, 26, 438–45.

Edgington, E. & Onghena, P. (2007). *Randomization tests*, 4th ed. Taylor & Francis.

Efron, B. & Hastie, T. (2016). *Computer age statistical inference: algorithms, evidence, and data science*. Cambridge University Press.

Elith, J., Leathwick, J. R., & Hastie, T. (2008). A working guide to boosted regression trees. *Journal of Animal Ecology*, 77, 802–13.

Ellison, A. M., Gotelli, N. J., Inouye, B. D., & Strong, D. R. (2014). P values, hypothesis testing, and model selection: it's deja vu all over again. *Ecology*, 95, 609–10.

Ercit, K., Martinez-Novoa, A. & Gwynne, D. T. (2014). Egg load decreases mobility and increases predation risk in female black-horned tree crickets (*Oecanthus nigricornis*). *PLoS One*, 9, e110298.

Faith, D. P., Minchin, P. R. & Belbin, L. (1987). Compositional dissimilarity as a robust measure of ecological distance. *Vegetatio*, 69, 57–68.

Faraway, J. J. (2014). *Linear models with R*, 2nd ed. Chapman & Hall.

Feinberg, J. A., Newman, C. E., Watkins-Colwell, G. J., et al. (2014). Cryptic diversity in metropolis: confirmation of a new leopard frog species (Anura: Ranidae) from New York City and surrounding Atlantic coast regions. *PLoS One*, 9, e108213.

Feng, C., Wang, H., Lu, N., & Tu, X. M. (2013a). Log transformation: application and interpretation in biomedical research. *Statistics in Medicine*, 32, 230–9.

Feng, C., Wang, H., Lu, N., & Tu, X. M. (2013b). Response to comments on 'Log transformation: Application and interpretation in biomedical research'. *Statistics in Medicine*, 32, 3772–4.

Ferrarezi, R. S., Qureshi, J. A., Wright, A. L., Ritenour, M. A. & Macan, N. P. F. (2019). Citrus production under screen as a strategy to protect grapefruit trees from Huanglongbing disease. *Frontiers in Plant Science*, 10, 1598.

Ferrier, S., Manion, G., Elith, J. & Richardson, K. (2007). Using generalized dissimilarity modelling to analyse and predict patterns of beta diversity in regional biodiversity assessment. *Diversity and Distributions*, 13, 252–64.

Few, S. (2012). *Show me the numbers: designing tables and graphs to enlighten*. Analytics Press.

Fewster, R. M. (2011). Variance estimation for systematic designs in spatial surveys. *Biometrics*, 67, 1518–31.

Fidler, F., Chee, Y. E., Wintle, B. C., et al. (2017). Metaresearch for evaluating reproducibility in ecology and evolution. *BioScience*, 67, 282–89.

Fielding, A. H. (2006). *Cluster and classification techniques for the biosciences*. Cambridge University Press.

Fill, J. M., Zamora, C., Baruzzi, C., Salazar-Castro, J., & Crandall, R. M. (2021). Wiregrass (*Aristida beyrichiana*) survival and reproduction after fire in a long-unburned pine savanna. *PLoS One*, 16, e0247159.

Finch, W. H., Holden, J. E., & Kelley, K. (2014). *Multilevel modeling using R*. CRC Press.

Finney, D. J. (1971). *Probit analysis*, 3rd ed. Cambridge University Press.

Fitzmaurice, G. M., Laird, N. M., & Ware, J. H. (2011). *Applied longitudinal analysis*, 2nd ed. Wiley.

Flack, V. F. & Chang, P. C. (1987). Frequency of selecting noise variables in subset regression-analysis: a simulation study. *American Statistician*, 41, 84–6.

Fortin, M.-J. & Gurevitch, J. (2001). Mantel tests: spatial structure in field experiments. In *Design and analysis of ecological experiments*, eds. S. M. Scheiner & J. Gurevitch, pp. 308–26. Oxford University Press.

Fox, J. & Weisberg, S. (2018). *An R companion to applied regression*, 3rd ed. Sage Publications.

Freckleton, R. P. (2011). Dealing with collinearity in behavioural and ecological data: model averaging and the problems of measurement error. *Behavioral Ecology and Sociobiology*, 65, 91–101.

Fry, J. D. (1992). The mixed-model analysis of variance applied to quantitative genetics: biological meaning of the parameters. *Evolution*, 46, 540–50.

Galipaud, M., Gillingham, M. A. F., David, M., & Dechaume-Moncharmont, F.-X. (2014). Ecologists overestimate the importance of predictor variables in model averaging: a plea for cautious interpretations. *Methods in Ecology and Evolution*, 5, 983–91.

Galipaud, M., Gillingham, M. A. F. & Dechaume-Moncharmont, F. X. (2017). A farewell to the sum of Akaike weights: the benefits of alternative metrics for variable importance estimations in model selection. *Methods in Ecology and Evolution*, 8, 1668–78.

Galway, N. W. (2014). *Introduction to mixed modelling: beyond regression and analysis of variance*, 2nd ed. Wiley.

Garamszegi, L. Z., Calhim, S., Dochtermann, N., et al. (2009). Changing philosophies and tools for statistical inferences in behavioral ecology. *Behavioral Ecology*, 20, 1363–75.

Garcia, N. W., Pfennig, K. S., & Burmeister, S. S. (2015). Leptin manipulation reduces appetite and causes a switch in mating preference in the Plains spadefoot toad (*Spea bombifrons*). *PLoS One*, 10, e0125981.

Gelman, A. (2016). Analysis of variance. In *New Palgrave Dictionary of Economics*, 2nd ed., eds. S. N. Durlauf & L. E. Blume, pp. 1–7. Palgrave Macmillan UK.

Gelman, A. & Hill, J. (2006). *Data analysis using regression and multilevel/hierarchical models*. Cambridge University Press.

Gelman, A. & Loken, E. (2014). The statistical crisis in science. *American Scientist*, 102, 460–5.

Gelman, A. & Shalizi, C. (2013a). Rejoinder to discussion of philosophy and the practice of Bayesian statistics. *British Journal of Mathematical & Statistical Psychology*, 66, 76–80.

Gelman, A. & Shalizi, C. R. (2013b). Philosophy and the practice of Bayesian statistics. *British Journal of Mathematical & Statistical Psychology*, 66, 8–38.

Gelman, A. & Unwin, A. (2013). Infovis and statistical graphics: different goals, different looks. *Journal of Computational and Graphical Statistics*, 22, 2–28.

Gelman, A., Simpson, D., & Betancourt, M. (2017). The prior can often only be understood in the context of the likelihood. *Entropy*, 19, e19100555.

Gelman, A., Hill, J., & Vehtari, A. (2020). *Regression and other stories*. Cambridge University Press.

Giam, X. & Olden, J. D. (2016). Quantifying variable importance in a multimodel inference framework. *Methods in Ecology and Evolution*, 7, 388–97.

Giri, S. S., Graham, J., Hamid, N. K. A., Donald, J. A. & Turchini, G. M. (2016). Dietary micronutrients and in vivo n-3 LC-PUFA biosynthesis in Atlantic salmon. *Aquaculture*, 452, 416–25.

Goldacre, B., Drysdale, H., Dale, A., et al. (2019). COMPare: a prospective cohort study correcting and monitoring 58 misreported trials in real time. *Trials*, 20, 118.

Goodrich, B., Gabry, J., Ali, I. & Brilleman, S. (2022) *rstanarm: Bayesian applied regression modeling via Stan. R package version 2.23.0*, 2022.

Green, S. B. (1991). How many subjects does it take to do regression analysis? *Multivariate Behavioral Research*, 26, 499–510.

Greenacre, M. (2010). Correspondence analysis of raw data. *Ecology*, 91, 958–63.

Greenacre, M. (2013a). Contribution biplots. *Journal of Computational and Graphical Statistics*, 22, 107–22.

Greenacre, M. (2013b). The contributions of rare objects in correspondence analysis. *Ecology*, 94, 241–9.

Greenacre, M. (2017a). *Correspondence analysis in practice*, 3rd ed. Chapman and Hall/CRC.

Greenacre, M. (2017b). 'Size' and 'shape' in the measurement of multivariate proximity. *Methods in Ecology and Evolution*, 8, 1415–24.

Greene, A. E. (2013). *Writing science in plain English*. University of Chicago Press.

Greenland, S. (2012). Nonsignificance plus high power does not imply support for the null over the alternative. *Annals of Epidemiology*, 22, 364–68.

Greenland, S. (2014). Confounding. *Encyclopedia of Statistical Sciences*, 2, 1–9.

Grömping, U. (2006). Relative importance for linear regression in R: the package relaimpo. *Journal of Statistical Software*, 17, 1–27.

Grömping, U. (2007). Estimators of relative importance in linear regression based on variance decomposition. *The American Statistician*, 61, 139–47.

Grueber, C. E., Nakagawa, S., Laws, R. J., & Jamieson, I. G. (2011). Multimodel inference in ecology and evolution: challenges and solutions. *Journal of Evolutionary Biology*, 24, 699–711.

Gruitjer, J., Brus, D., Bierkens, M., & Knotters, M. (2006). *Sampling for natural resource monitoring.* Springer.

Gueorguieva, R. (2017). *Statistical methods in psychiatry and related fields: longitudinal, clustered and other repeated measures data.* Chapman & Hall/CRC.

Guillot, G. & Rousset, F. (2013). Dismantling the Mantel tests. *Methods in Ecology and Evolution*, 4, 336–44.

Gunst, R. F. & Mason, R. L. (2009). Fractional factorial design. *Wiley Interdisciplinary Reviews: Computational Statistics*, 1, 234–44.

Gustavsson, S., Fagerberg, B., Sallsten, G., & Andersson, E. N. (2014). Regression models for log-normal data: comparing different methods for quantifying the association between abdominal adiposity and biomarkers of inflammation and insulin resistance. *International Journal of Environmental Research and Public Health*, 11, 3521–39.

Hadi, A. S. & Ling, R. F. (1998). Some cautionary notes on the use of principal components regression. *The American Statistician*, 52, 15–19.

Hairston, N. G. (1989). *Ecological experiments. purpose, design and execution.* Cambridge University Press.

Hand, D. J. (2021). Trustworthiness of statistical inference. *Journal of the Royal Statistical Society Series A (Statistics in Society)*, 185, 329–47.

Hanley, J. A. & Shapiro, S. H. (1994). Sexual activity and the lifespan of male fruitflies: a dataset that gets attention. *Journal of Statistics Education*, 2.

Harrison, F. (2011). Getting started with meta-analysis. *Methods in Ecology and Evolution*, 2, 1–10.

Harrison, X. A. (2014). Using observation-level random effects to model overdispersion in count data in ecology and evolution. *Peerj*, 2, e616.

Harrison, X. A. (2015). A comparison of observation-level random effect and beta-binomial models for modelling overdispersion in binomial data in ecology & evolution. *Peerj*, 3, e1114.

Hastie, T. J. & Tibshirani, R. J. (1990). *Generalized additive models.* Chapman & Hall.

Hays, W. L. (1994). *Statistics.* Holt, Rinehart and Winston.

Head, M. L., Holman, L., Lanfear, R., Kahn, A. T., & Jennions, M. D. (2015). The extent and consequences of p-hacking in science. *PLoS Biology*, 13, e1002106.

Hector, A. (2015). *The new statistics with R.* Oxford University Press.

Hector, A., Bell, T., Hautier, Y., et al. (2011). BUGS in the analysis of biodiversity experiments: species richness and composition are of similar importance for grassland productivity. *PLoS One*, 6, e17434.

Hector, A., von Felten, S., & Schmid, B. (2010). Analysis of variance with unbalanced data: an update for ecology & evolution. *Journal of Animal Ecology*, 79, 308–16.

Heinze, G., Wallisch, C. & Dunkler, D. (2018). Variable selection: a review and recommendations for the practicing statistician. *Biometrical Journal*, 60, 431–49.

Hey, M. H., DiBiase, E., Roach, D. A., Carr, D. E., & Haynes, K. J. (2020). Interactions between artificial light at night, soil moisture, and plant density affect the growth of a perennial wildflower. *Oecologia*, 193, 503–10.

Hilbe, J. M. (2007). *Negative binomial regression.* Cambridge University Press.

Hilbe, J. M. & Robinson, A. P. (2013). *Methods of statistical model estimation.* Chapman & Hall/CRC.

Hilborn, R. & Mangel, M. (1997). *The ecological detective: confronting models with data.* Princeton University Press.

Hill, M. O. & Gauch, H. G. (1980). Detrended correspondence analysis, an improved ordination technique. *Vegetatio*, 42, 47–58.

Hines, W. G. S. (1996). Pragmatics of pooling in ANOVA tables. *American Statistician*, 50, 127–39.

Hintze, J. L. & Nelson, R. D. (1998). Violin plots: a box plot-density trace synergism. *The American Statistician*, 52, 181.

Hoaglin, D. C. & Welsch, R. E. (1978). Hat matrix in regression and anova. *American Statistician*, 32, 17–22.

Hoaglin, D. C., Mosteller, F. & Tukey, J. W. (1983). *Understanding robust and exploratory data analysis.* Wiley.

Hocking, R. R. (2013). *Methods and applications of linear models: regression and the analysis of variance*, 3rd ed. Wiley.

Hoenig, J. M. & Heisey, D. M. (2001). The abuse of power: the pervasive fallacy of power calculations for data analysis. *The American Statistician*, 55, 19–24.

Hoeting, J. A., Madigan, D., Raftery, A. E. & Volinsky, C. T. (1999). Bayesian model averaging: a tutorial. *Statistical Science*, 14, 382–417.

Hollander, M., Wolfe, D. A., & Chicken, E. (2013). *Nonparametric statistical methods*. Wiley.

Hooten, M. B. & Hobbs, N. T. (2015). A guide to Bayesian model selection for ecologists. *Ecological Monographs*, 85, 3–28.

Hosmer, D. W. & Lemeshow, S. (1989). *Applied logistic regression*. Wiley.

Hosmer, D. W., Lemeshow, S., & Sturdivant, R. X. (2013). *Applied logistic regression*, 3rd ed. Wiley.

Hostetler, C. M., Phillips, T. J. & Ryabinin, A. E. (2016). Methamphetamine consumption inhibits pair bonding and hypothalamic oxytocin in prairie voles. *PLoS One*, 11, e0158178.

Huberty, C. J. & Olejnik, S. (2006). *Applied MANOVA and discriminant analysis*, 2nd ed. Wiley.

Huitema, B. E. (2011). *The analysis of covariance and alternatives: statistical methods for experiments, quasi-experiments, and single-case studies*, 2nd ed. Wiley.

Hurlbert, S. H. (1984). Pseudoreplication and the design of ecological field experiments. *Ecological Monographs*, 54, 187–211.

Hurlbert, S. H. (2009). The ancient black art and transdisciplinary extent of pseudoreplication. *Journal of Comparative Psychology*, 123, 434–43.

Ialongo, C. (2016). Understanding the effect size and its measures. *Biochemia Medica (Zagreb)*, 26, 150–63.

Ieno, E. N. & Zuur, A. F. (2015). *A beginner's guide to data exploration and visualisation with R*. Highland Statistics.

Inoue, L. Y. T., Berry, D. A., & Parmigiani, G. (2005). Relationship between Bayesian and frequentist sample size determination. *The American Statistician*, 59, 79–87.

Ives, A. R. (2015). For testing the significance of regression coefficients, go ahead and log-transform count data. *Methods in Ecology and Evolution*, 6, 828–35.

Ives, A. R. (2019). R^2s for correlated data: phylogenetic models, LMMs, and GLMMs. *Systematic Biology*, 68, 234–51.

Jackson, D. A. (1993). Stopping rules in principal components-analysis: a comparison of heuristic and statistical approaches. *Ecology*, 74, 2204–14.

Jackson, D. A. & Somers, K. M. (1991). Putting things in order: the ups and downs of detrended correspondence analysis. *American Naturalist*, 137, 704–12.

Jackson, J. E. (2003). *A user's guide to principal components*. Wiley.

Janky, D. G. (2000). Sometimes pooling for analysis of variance hypothesis tests: a review and study of a split-plot model. *American Statistician*, 54, 269–79.

Johnson, C. R. & Field, C. A. (1993). Using fixed-effects model multivariate analysis of variance in marine biology and ecology. *Oceanography and Marine Biology Annual Review*, 31, 177–221.

Johnson, J. W. & LeBreton, J. M. (2004). History and use of relative importance indices in organizational research. *Organizational Research Methods*, 7, 238–57.

Jolliffe, I. T. & Cadima, J. (2016). Principal component analysis: a review and recent developments. *Philosophical Transactions of the Royal Society A*, 374, 20150202.

Jones, E., Harden, S., & Crawley, M. J. (2022). *The R book*, 3rd ed. Wiley.

Jongman, R. H. G., ter Braak, C. J. F., & Tongeren, O. F. R. (1995). *Data analysis in community and landscape ecology*. Cambridge University Press.

Kampstra, P. (2008). Beanplot: a boxplot alternative for visual comparison of distributions. *Journal of Statistical Software*, 28, 1–9.

Kass, R. E. & Raftery, A. E. (1995). Bayes factors. *Journal of the American Statistical Association*, 90, 773–95.

Kenny, D. A. & Judd, C. M. (1986). Consequences of violating the independence assumption in analysis of variance. *Psychological Bulletin*, 99, 422–31.

Keough, M. J. & Quinn, G. P. (1998). Effects of periodic disturbances from trampling on rocky intertidal algal beds. *Ecological Applications*, 8, 141–61.

Keough, M. J. & Raimondi, P. T. (1995). Responses of settling invertebrate larvae to bioorganic films: effects of different types of films. *Journal of Experimental Marine Biology and Ecology*, 185, 235–53.

Keough, M. J., Quinn, G. P., & King, A. (1993). Correlations between human collecting and intertidal mollusc populations on rocky shores. *Conservation Biology*, 7, 378–91.

Kery, M. (2010). *Introduction to WinBUGS for ecologists*. Academic Press.

Kery, M. & Hatfield, J. S. (2003). Normality of raw data in general linear models: the most widespread myth in statistics. *Bulletin of the Ecological Society of America*, 84, 92–4.

Kirk, R. E. (2012). *Experimental design: procedures for the behavioral sciences*, 4th ed. Sage.

Kleinbaum, D. G., Kupper, L. L., Nizam, A., & Rosenberg, G. B. (2013). *Applied regression analysis and other multivariable methods*, 5th ed. Cengage Learning.

Knaflic, C. N. (2015). *Storytelling with data: a data visualization guide for business professionals*. Wiley.

Kramer, M. & Font, E. (2017). Reducing sample size in experiments with animals: historical controls and related strategies. *Biological Reviews of the Cambridge Philosophical Society*, 92, 431–45.

Krause, W. C., Rodriguez, R., Gegenhuber, B., et al. (2021). Oestrogen engages brain MC4R signalling to drive physical activity in female mice. *Nature*, 599, 131–5.

Kruschke, J. K. (2011). Bayesian assessment of null values via parameter estimation and model comparison. *Perspectives in Psychological Science*, 6, 299–312.

Krzywinski, M. & Cairo, A. (2013). Storytelling. *Nature Methods*, 10, 687.

Kuehl, R. O. (2000). *Design of experiments: statistical principles of research design and analysis*. Duxbury/Thomson Learning.

Kuhn, M. & Johnson, K. (2013). *Applied predictive modeling*. Springer.

Kutner, M. H., Nachtsheim, C. J., Neter, J., & Li, W. (2005). *Applied linear statistical models*, 5th ed. McGraw-Hill/Richard D. Irwin.

Lakens, D. (2013). Calculating and reporting effect sizes to facilitate cumulative science: a practical primer for *t*-tests and ANOVAs. *Frontiers in Psychology*, 4, 1–12.

Lamott, A. (2007). *Bird by bird: some instructions on writing and life*. Knopf Doubleday.

Legault, R., Zogg, G. P., & Travis, S. E. (2018). Competitive interactions between native *Spartina alterniflora* and non-native *Phragmites australis* depend on nutrient loading and temperature. *PLoS One*, 13, e0192234.

Legendre, P. & Anderson, M. J. (1999a). Distance-based redundancy analysis: testing multispecies responses in multifactorial ecological experiments. *Ecological Monographs*, 69, 1–24.

Legendre, P. & Anderson, M. J. (1999b). Distance-based redundancy analysis: testing multispecies responses in multifactorial ecological experiments (vol 69, pg 1, 1999). *Ecological Monographs*, 69, 512.

Legendre, P. & De Caceres, M. (2013). Beta diversity as the variance of community data: dissimilarity coefficients and partitioning. *Ecology Letters*, 16, 951–63.

Legendre, P. & Fortin, M. J. (2010). Comparison of the Mantel test and alternative approaches for detecting complex multivariate relationships in the spatial analysis of genetic data. *Molecular Ecology Resources*, 10, 831–44.

Legendre, P. & Legendre, L. F. J. (2012). *Numerical ecology*. Elsevier Science.

Lemmens, P., Mergeay, J., Van Wichelen, J., De Meester, L. & Declerck, S. A. (2015). The impact of conservation management on the community composition of multiple organism groups in eutrophic interconnected man-made ponds. *PLoS One*, 10, e0139371.

Lentner, M., Arnold, J. C. & Hinkelmann, K. (1989). The efficiency of blocking: how to use MS(blocks) MS(error) correctly. *The American Statistician*, 43, 106–8.

Leonard, G. H., Bertness, M. D. & Yund, P. O. (1999). Crab predation, waterborne cues, and inducible defenses in the blue mussel, *Mytilus edulis*. *Ecology*, 80, 1–14.

Levy, P. & Lemeshow, S. (2008). *Sampling of populations: methods and applications*, 4th ed. Wiley.

Link, W. & Barker, R. (2009). *Bayesian inference with ecological applications*. Academic Press.

Linton, S., Barrow, L., Davies, C., & Harman, L. (2009). Potential endocrine disruption of ovary synthesis in the Christmas Island red crab *Gecarcoidea natalis* by the insecticide pyriproxyfen. *Comparative Biochemistry and Physiology, Part A*, 154, 289–97.

Little, R. J. A. & Rubin, D. B. (2019). *Statistical analysis with missing data*, 3rd ed. Wiley.

Long, J. D. & Porturas, L. D. (2014). Herbivore impacts on marsh production depend upon a compensatory continuum mediated by salinity stress. *PLoS One*, 9, e110419.

Low, L., Bauer, L., & Klaunberg, B. (2016). Comparing the effects of isoflurane and alpha chloralose upon mouse physiology. *PLoS One*, 11, e0154936.

Loyn, R. H. (1987). Effects of patch area and habitat on bird abundances, species numbers and tree health in fragmented Victorian forests. In *Nature conservation: the role of remnants of native vegetation*, eds. D. A. Saunders, G. W. Arnold, A. A. Burbidge, & A. J. M. Hopkins, pp. 65–77. Surrey Beatty and Sons.

Lozada-Misa, P., Kerr, A. M., & Raymundo, L. (2015). Contrasting lesion dynamics of white syndrome among the scleractinian corals *Porites* spp. *PLoS One*, 10, e0129841.

Lukacs, P. M., Burnham, K. P., & Anderson, D. R. (2010). Model selection bias and Freedman's paradox. *Annals of the Institute of Statistical Mathematics*, 62, 117–25.

Mac Nally, R. (1996). Hierarchical partitioning as an interpretative tool in multivariate inference. *Australian Journal of Ecology*, 21, 224–28.

Mac Nally, R. (2000). Regression and model-building in conservation biology, biogeography and ecology: the distinction between, and reconciliation of, predictive and exploratory models. *Biodiversity and Conservation*, 9, 655–71.

Mac Nally, R., Duncan, R., Thomson, J. & Yen, J. (2017). Model selection using information criteria, but is the 'best' model any good? *Journal of Applied Ecology*, 55, 1441–4.

Manly, B. & Navarro, J. (2014). *Introduction to ecological sampling*. Chapman and Hall/CRC.

Manly, B. F. J. & Navarro Alberto, J. A. (2022). *Randomization, bootstrap, and Monte Carlo methods in biology*, 4th ed. Chapman and Hall/CRC.

Mapstone, B. D. (1995). Scalable decision rules for environmental impact studies: effect size, Type I, and Type II errors. *Ecological Applications*, 5, 401–10.

Marshall, D. J. & Steinberg, P. D. (2014). Larval size and age affect colonization in a marine invertebrate. *Journal of Experimental Biology*, 217, 3981–87.

Martin, J. T. (1942). The problem of the evaluation of rotenone-containing plants. VI. The toxicity of *l*-ellptone and of poisons applied jointly, with further observations on the rotenone equivalent method of assessing the toxicity of Derris root. *Annals of Applied Biology*, 29, 69–81.

Maxwell, S. E., Delaney, H. D. & Kelley, K. (2018). *Designing experiments and analyzing data: a model comparison perspective*. Routledge.

Mayekar, H. V. & Kodandaramaiah, U. (2017). Pupal colour plasticity in a tropical butterfly, *Mycalesis mineus* (Nymphalidae: Satyrinae). *PLoS One*, 12, e0171482.

Mayo, D. G. (1996). *Error and the growth of experimental knowledge*. University of Chicago Press.

Mayo, D. G. (2018). *Statistical inference as severe testing: how to get beyond the statistics wars*. Cambridge University Press.

Mayo, D. G. (2019). P-value thresholds: forfeit at your peril. *European Journal of Clinical Investigation*, 49.

Mayo, D. G. (2021). Significance tests: vitiated or vindicated by the replication crisis in psychology? *Review of Philosophy and Psychology*, 12, 101–20.

Mayo, D. G. & Hand, D. (2022). Statistical significance and its critics: practicing damaging science, or damaging scientific practice? *Synthese*, 200, 220.

McArdle, B. H. (1988). The structural relationship: regression in biology. *Canadian Journal of Zoology/ Revue Canadienne De Zoologie*, 66, 2329–39.

McArdle, B. H. & Anderson, M. J. (2001). Fitting multivariate models to community data: a comment on distance-based redundancy analysis. *Ecology*, 82, 290–7.

McCarthy, M. A. (2007). *Bayesian methods for ecology*. Cambridge University Press.

McCarthy, M. A. (2015). Approaches to statistical inference. In *Ecological statistics: contemporary theory and application*, eds. G. A. Fox, S. Negrete-Yankelevich & V. J. Sosa, pp. 15–43. Oxford University Press.

McCulloch, C. E., Searle, S. R. & Neuhaus, J. M. (2008). *Generalized, linear, and mixed models*, 2nd ed. Wiley.

McCune, B. & Grace, J. B. (2002). *Analysis of ecological communities*. MjM Software.

McCune, B. & Mefford, M. J. M. (1999). *PC-ORD: multivariate analysis of ecological data*. Version 4 for Windows. MjM Software.

McDonald, S., Cresswell, T., Hassell, K. & Keough, M. (2021). Experimental design and statistical analysis in aquatic live animal radiotracing studies: a systematic review. *Critical Reviews in Environmental Science and Technology*, 52, 2772–801.

McGowen, M. R., Tsagkogeorga, G., Williamson, J., Morin, P. A. & Rossiter, A. S. J. (2020). Positive selection and inactivation in the vision and hearing genes of cetaceans. *Molecular Biology and Evolution*, 37, 2069–83.

McKean, J. W. (2004). Robust analysis of linear models. *Statistical Science*, 19, 562–70.

McKee, R. (1997). *Story: substance, structure, style, and the principles of screenwriting*. Harper-Collins.

McKillup, S. C. (2012). *Statistics explained: an introductory guide for life scientists*, 2nd ed. Cambridge University Press.

McKnight, P. E., McKnight, K. M., Sidani, S., & Figueredo, A. J. (2007). *Missing data: a gentle introduction*. Guilford Press.

Mead, R., Gilmour, S. G., & Mead, A. (2012). *Statistical principles for the design of experiments: applications to real experiments.*. Cambridge University Press.

Medley, C. N. & Clements, W. H. (1998). Responses of diatom communities to heavy metals in streams: the influence of longitudinal variation. *Ecological Applications*, 8, 631–44.

Menard, S. (2000). Coefficients of determination for multiple logistic regression. *The American Statistician*, 54, 17–24.

Mielke, P. W. (2006) Multiresponse permutation procedures. *Encyclopedia of Statistical Sciences*. Available at: https://doi.org/10.1002/0471667196 .ess1714.pub2.

Mielke, P. W. & Berry, K. J. (2007). *Permutation Methods: A Distance Function Approach*, 2nd ed. Springer.

Mielke, P. W., Berry, K. J., & Johnson, E. S. (1976). Multi-response permutation procedures for *a priori* classifications. *Communications in Statistics Part A: Theory and Methods*, 5, 1409–24.

Milliken, G. A. & Johnson, D. E. (2001). *Analysis of messy data. Volume 3: Analysis of covariance*: CRC Press.

Milliken, G. A. & Johnson, D. E. (2009). *Analysis of messy data. Volume 1: designed experiments*, 2nd ed. Chapman and Hall/CRC.

Morehouse, A. T., Graves, T. A., Mikle, N., & Boyce, M. S. (2016). Nature vs. nurture: evidence for social learning of conflict behaviour in grizzly bears. *PLoS One*, 11, e0165425.

Morris, B. S., Chrysochou, P., Christensen, J. D., et al. (2019). Stories vs. facts: triggering emotion and action-taking on climate change. *Climatic Change*, 154, 19–36.

Morrissette-Boileau, C., Boudreau, S., Tremblay, J. P., & Côté, S. D. (2018). Simulated caribou browsing limits the effect of nutrient addition on the growth of *Betula glandulosa*, an expanding shrub species in Eastern Canada. *Journal of Ecology*, 106, 1256–65.

Morrissey, M. B., Ruxton, G. D., & Peres-Neto, P. (2020). Revisiting advice on the analysis of count data. *Methods in Ecology and Evolution*, 11, 1133–40.

Nakagawa, S. (2015). Missing data: mechanisms, methods, and messages. In *Ecological statistics: contemporary theory and application*, eds. G. A. Fox, S. NegreteYankelevich, & V. J. Sosa, pp. 81–105. Oxford University Press.

Nakagawa, S. & Cuthill, I. C. (2007). Effect size, confidence interval and statistical significance: a practical guide for biologists. *Biological Reviews*, 82, 591–605.

Nakagawa, S. & Freckleton, R. P. (2008). Missing inaction: the dangers of ignoring missing data. *Trends in Ecology and Evolution*, 23, 592–6.

Nakagawa, S. & Freckleton, R. P. (2011). Model averaging, missing data and multiple imputation: a case study for behavioural ecology. *Behavioral Ecology and Sociobiology*, 65, 103–16.

Nakagawa, S. & Schielzeth, H. (2013). A general and simple method for obtaining R^2 from generalized linear mixed-effects models. *Methods in Ecology and Evolution*, 4, 133–42.

Nakagawa, S., Johnson, P. C. D., & Schielzeth, H. (2017). The coefficient of determination R-2 and intra-class correlation coefficient from generalized linear mixed-effects models revisited and expanded. *Journal of the Royal Society Interface*, 14.

Nakajima, T. (2013). Probability in biology: overview of a comprehensive theory of probability in living systems.

Progress in Biophysics and Molecular Biology, 113, 67–79.

National Academies of Sciences (2017). *Communicating science effectively: a research agenda*. National Academies Press.

Nelder, J. A. & Lane, P. W. (1995). The computer analysis of factorial experiments: in memoriam Frank Yates. *The American Statistician*, 49, 382–85.

Newman, E., Manning, J., & Anderson, B. (2015). Local adaptation: mechanical fit between floral ecotypes of *Nerine humilis* (Amaryllidaceae) and pollinator communities. *Evolution*, 69, 2262–75.

Nimon, K. F. (2012). Statistical assumptions of substantive analyses across the general linear model: a mini-review. *Frontiers in Psychology*, 3, 322.

Nosek, B. A., Ebersole, C. R., DeHaven, A. C., & Mellor, D. T. (2018). The preregistration revolution. *Proceedings of the National Academy of Sciences of the United States of America*, 115, 2600–6.

O'Hara, R. B. & Kotze, D. J. (2010). Do not log-transform count data. *Methods in Ecology and Evolution*, 1, 118–22.

Olejnik, S. & Algina, J. (2003). Generalized Eta and Omega squared statistics: Measures of effect size for some common research designs. *Psychological Methods*, 8, 434–47.

Olson, R. (2015). *Houston, we have a narrative: why science needs story*. University of Chicago Press.

Open Science Collaboration (2015). Psychology: estimating the reproducibility of psychological science. *Science*, 349, aac4716.

Parkinson, A., Mac Nally, R., & Quinn, G. P. (2002). Differential macrohabitat use by birds on the unregulated Ovens River floodplain of southeastern Australia. *River Research and Applications*, 18, 495–506.

Parkinson, E., Lawson, J., & Tiegs, S. D. (2020). Artificial light at night at the terrestrial–aquatic interface: effects on predators and fluxes of insect prey. *PLoS One*, 15, e0240138.

Partridge, L. & Farquhar, M. (1981). Sexual activity and the lifespan of male fruitflies. *Nature*, 294, 580–1.

Paruelo, J. M. & Lauenroth, W. K. (1996). Relative abundance of plant functional types in grasslands and shrublands of North America. *Ecological Applications*, 6, 1212–24.

Peake, A. J. & Quinn, G. P. (1993). Temporal variation in species–area curves for invertebrates in clumps of an intertidal mussel. *Ecography*, 16, 269–77.

Pearl, D. (2014). Making the most of clustered data in laboratory animal research using multi-level models. *ILAR Journal*, 55, 486–92.

Pechenik, J. A. (2015). *A short guide to writing about biology*. Pearson Education.

Pekár, S. & Brabec, M. (2016). *Modern analysis of biological data: generalized linear models in R*. Masaryk University Press.

Peres-Neto, P. R., Jackson, D. A., & Somers, K. M. (2003). Giving meaningful interpretation to ordination axes: assessing loading significance in principal component analysis. *Ecology*, 84, 2347–63.

Peres-Neto, P. R., Jackson, D. A., & Somers, K. M. (2005). How many principal components? Stopping rules for determining the number of non-trivial axes revisited. *Computational Statistics & Data Analysis*, 49, 974–97.

Petraitis, P. S. (1998). How can we compare the importance of ecological processes if we never ask, "compared to what?" In *Experimental ecology: Issues and perspectives*, eds. W. J. Resetarits & J. Bernardo, pp. 183–201. Oxford University Press.

Pew Research Center (2015). Americans, politics and science issues. Available at: www.pewresearch.org/ science/2015/07/01/americans-politics-and-science-issues/.

Pinheiro, J. C. & Bates, D. M. (2000). *Mixed-effects models in S and S-Plus*. Springer-Verlag.

Pinker, S. (2014). *The sense of style: the thinking person's guide to writing in the 21st century*. Penguin.

Polis, G. A., Hurd, S. D., Jackson, C. T., & Sanchez-Pinero, F. (1998). Multifactor population limitation: variable spatial and temporal control of spiders on Gulf of California islands. *Ecology*, 79, 490–502.

Pursall, E. R. & Rolff, J. (2011). Immune responses accelerate ageing: proof-of-principle in an insect model. *PLoS One*, 6, e19972.

Raudenbush, S. & Bryk, A. (2002). *Hierarchical linear models: applications and data analysis methods*. Sage.

Rawlings, J. O., Pantula, S. G., & Dickey, D. A. (1998). *Applied regression analysis: a research tool*. Springer.

Resetarits, W. & Fauth, J. (1998). From cattle tanks to Carolina bays: the utility of model systems for understanding natural communities. In *Experimental ecology: issues and perspectives*, eds. W. Resetarits & J. Bernardo, pp. 133–51. Oxford University Press.

Reynolds, G. (2011). *Presentation zen: simple ideas on presentation design and delivery*. Pearson Education.

Rhodes, J. R. (2015). Mixture models for overdispersed data. In *Ecological statistics: contemporary theory and application*, eds. G. A. Fox, S. Negrete-Yankelevich, & V. J. Sosa, pp. 284–308. Cambridge University Press.

Richards, S. A. (2005). Testing ecological theory using the information-theoretic approach: examples and cautionary results. *Ecology*, 86, 2805–14.

Richards, S. A. (2015). *Likelihood and model selection*. Oxford University Press.

Richards, S. A., Whittingham, M. J., & Stephen, P. A. (2011). Model selection and model averaging in behavioural ecology: the utility of the IT-AIC framework. *Behavioral Ecology and Sociobiology*, 65, 77–89.

Roberts, D. W. (2017). Distance, dissimilarity, and mean–variance ratios in ordination. *Methods in Ecology and Evolution*, 8, 1398–407.

Rodgers, J. L. & Nicewander, W. A. (1988). 13 ways to look at the correlation-coefficient. *The American Statistician*, 42, 59–66.

Rousseeuw, P. J., Ruts, I., & Tukey, J. W. (1999). The bagplot: a bivariate boxplot. *The American Statistician*, 53, 382.

Rowland, J. A., Bland, L. M., James, S., & Nicholson, E. (2021). A guide to representing variability and uncertainty in biodiversity indicators. *Conservation Biology*, 35, 1669–82.

Rubin, D. B. (1987). *Multiple imputation for nonresponse in surveys*. Wiley.

Ruxton, G. D. & Neuhäuser, M. (2010). When should we use one-tailed hypothesis testing? *Methods in Ecology and Evolution*, 1, 114–17.

Ruxton, G. D. & Neuhäuser, M. (2012). Review of alternative approaches to calculation of a confidence interval for the odds ratio of a 2 x 2 contingency table. *Methods in Ecology and Evolution*, 4, 9–13.

Ryeland, J., Weston, M. A., Symonds, M. R. E., & Overgaard, J. (2017). Bill size mediates behavioural thermoregulation in birds. *Functional Ecology*, 31, 885–93.

Scheffé, H. (1959). *The analysis of variance*. Wiley.

Schielzeth, H. (2010). Simple means to improve the interpretability of regression coefficients. *Methods in Ecology and Evolution*, 1, 103–13.

Schielzeth, H. & Nakagawa, S. (2013). Nested by design: model fitting and interpretation in a mixed model era. *Methods in Ecology and Evolution*, 4, 14–24.

Schielzeth, H., Dingemanse, N. J., Nakagawa, S., et al. (2020). Robustness of linear mixed-effects models to violations of distributional assumptions. *Methods in Ecology and Evolution*, 11, 1141–52.

Schimel, J. (2012). *Writing science: how to write papers that get cited and proposals that get funded*. Oxford University Press.

Schlegel, P., Havenhand, J. N., Gillings, M. R. & Williamson, J. E. (2012). Individual variability in reproductive success determines winners and losers under ocean acidification: a case study with sea urchins. *PLoS One*, 7, e53118.

Schützenmeister, A., Jensen, U., & Piepho, H. P. (2012). Checking normality and homoscedasticity in the general linear model using diagnostic plots. *Communications in Statistics: Simulation and Computation*, 41, 141–54.

Schwabish, J. (2021). *Better data visualizations: a guide for scholars, researchers, and wonks*. Columbia University Press.

Schwarz, C. J. (1993). The mixed model ANOVA: the truth, the computer packages, the books. Part I: balanced data. *American Statistician*, 47, 48–59.

Schwarz, G. (1978). Estimating the dimension of a model. *Annals of Statistics*, 6, 461–64.

Scott, A. & Wild, C. (1991). Transformations and R^2. *The American Statistician*, 45, 127–9.

Searle, S. R. (1993). Unbalanced data and cell means models. In *Applied analysis of variance in behavioral science*, ed. L. K. Edwards, pp. 375–420. Marcel Dekker.

Searle, S. R. & Gruber, M. H. J. (2016). *Linear Models*, 2nd ed. Wiley.

Searle, S. R., Casella, G., & McCulloch, C. E. (1992). *Variance components*. Wiley.

Shaw, P. J. A. & Shackleton, K. (2010). Carnivory in the teasel *Dipsacus fullonum*: the effect of experimental feeding on growth and seed set. *PLoS One*, 6, e17935.

Shaw, R. G. & Mitchell-Olds, T. (1993). ANOVA for unbalanced data: an overview. *Ecology*, 74, 1638–45.

Shibao, H., Kutsukake, M. & Fukatsu, T. (2004). Density triggers soldier production in a social aphid. *Proceedings: Biological Sciences*, 271(Suppl. 3), S71–4.

Silverman, B. W. (1986). *Density estimation for statistics and data analysis*. Chapman & Hall.

Simmons, J. P., Nelson, L. D., & Simonsohn, U. (2012) A 21 word solution. Social Science Research Network. doi: 10.2139/ssrn.2160588.

Sinclair, A. R. E. & Arcese, P. (1995). Population consequences of predation-sensitive foraging: the Serengeti wildebeeste. *Ecology*, 76, 882–91.

Singh, K., Samant, M. A., Tom, M. T., & Prasad, N. G. (2016). Evolution of pre- and post-copulatory traits in male *Drosophila melanogaster* as a correlated response to selection for resistance to cold stress. *PLoS One*, 11, e0153629.

Skrip, M. M., Seeram, N. P., Yuan, T., Ma, H., & McWilliams, S. R. (2016). Dietary antioxidants and flight exercise in female birds affect allocation of nutrients to eggs: how carry-over effects work. *Journal of Experimental Biology*, 219, 2716–25.

Snedecor, G. W. & Cochran, W. G. (1989). *Statistical methods*, 8th ed. Iowa State University Press.

Snee, R. D. & Pfeifer, C. G. (1983). Graphical representation of data. In *Encyclopedia of statistical sciences*, eds. S. Kotz & N. L. Johnson, pp. 488–11. Wiley.

Snijders, T. A. B. & Bosker, R. J. (2012). *Multilevel analysis: an introduction to basic and advanced multilevel modeling*, 2nd ed. Sage.

Sokal, R. R. & Rohlf, F. J. (2012). *Biometry: the principles and practice of statistics in biological research*, 4th ed. W.H. Freeman.

Somerfield, P. J. (2008). Identification of the Bray–Curtis similarity index: comment on Yoshioka (2008). *Marine Ecology Progress Series*, 372, 303–6.

Somerfield, P. J., Clarke, K. R., & Gorley, R. N. (2021a). Analysis of similarities (ANOSIM) for 2-way layouts using a generalised ANOSIM statistic, with comparative notes on permutational multivariate analysis of variance (PERMANOVA). *Austral Ecology*, 46, 911–26.

Somerfield, P. J., Clarke, K. R., & Gorley, R. N. (2021b). Analysis of similarities (ANOSIM) for 3-way designs. *Austral Ecology*, 46, 927–41.

Spanos, A. (2019). *Probability theory and statistical inference: empirical modeling with observational data*, 2nd ed. Cambridge University Press.

Spiegelhalter, D. J., Myles, J. P., Jones, D. R., & Abrams, K. R. (2000). Bayesian methods in health technology assessment: a review. *Health Technology Assessment*, 4, 1–130.

Sprenger, J. (2011). Science without (parametric) models: the case of bootstrap resampling. *Synthese*, 180, 65–76.

Steiger, S., Franz, R., Eggert, A. K., & Muller, J. K. (2008). The Coolidge effect, individual recognition and selection for distinctive cuticular signatures in a burying beetle. *Proceedings of the Royal Society B*, 275, 1831–8.

Stokes, A. N., Ducey, P. K., Neuman-Lee, L., et al. (2014). Confirmation and distribution of tetrodotoxin for the first time in terrestrial invertebrates: two terrestrial flatworm species (*Bipalium adventitium* and *Bipalium kewense*). *PLoS One*, 9, e100718.

Strunk, W. & White, E. B. (1999). *Elements of style*, 4th ed. Pearson.

Symonds, M. R. E. & Moussalli, A. (2011). A brief guide to model selection, multimodel inference and model averaging in behavioural ecology using Akaike's information criterion. *Behavioral Ecology and Sociobiology*, 65, 13–21.

Tabachnick, B. G. & Fidell, L. S. (2019). *Using multivariate statistics*, 7th ed. Pearson.

Tartu, S., Bustamante, P., Angelier, F., et al. (2016). Mercury exposure, stress and prolactin secretion in an Arctic seabird: an experimental study. *Functional Ecology*, 30, 596–604.

Teng, K. T., McGreevy, P. D., Toribio, J., & Dhand, N. K. (2020). Positive attitudes towards feline obesity are strongly associated with ownership of obese cats. *PLoS One*, 15, e0234190.

ter Braak, C. J. F. (1986). Canonical correspondence-analysis: a new eigenvector technique for multivariate direct gradient analysis. *Ecology*, 67, 1167–79.

ter Braak, C. J. F. & Smilauer, P. (2015). Topics in constrained and unconstrained ordination. *Plant Ecology*, 216, 683–96.

Thompson, S. (2012). *Sampling*, 3rd ed. Wiley.

Thompson, S. K. & Seber, G. A. F. (1996). *Adaptive sampling*. Wiley.

Thorarensen, H., Kubiriza, G. K., & Imsland, A. K. (2015). Experimental design and statistical analyses of fish growth studies. *Aquaculture*, 448, 483–90.

Tjur, T. (2009). Coefficients of determination in logistic regression models: a new proposal – the coefficient of discrimination. *The American Statistician*, 63, 366–72.

Tufte, E. R. (1983). *The visual display of quantitative information*. Graphics Press.

Tufte, E. R. (1990). *Envisioning information*. Graphics Press.

Tufte, E. R. (2006). *Beautiful evidence*. Graphics Press.

Tukey, J. W. (1977). *Exploratory data analysis*. Addison-Wesley Publishing Company.

Underwood, A. J. (1997). *Experiments in ecology: their logical design and interpretation using analysis of variance*. Cambridge University Press.

van der Geest, M., van der Heide, T., Holmer, M., & de Wit, R. (2020). First field-based evidence that the seagrass-lucinid mutualism can mitigate sulfide stress in seagrasses. *Frontiers in Marine Science*, 7, 11.

Vatcheva, K. P., Lee, M., McCormick, J. B., & Rahbar, M. H. (2016). Multicollinearity in regression analyses conducted in epidemiologic studies. *Epidemiology*, 6, 1000227.

Vittinghoff, E., Glidden, D. V., Shiboski, S. C., & McCulloch, C. E. (2012). *Regression methods in biostatistics: linear, logistic, survival, and repeated measures models*, 2nd ed. Springer.

Voss, D. T. (1999). Resolving the mixed models controversy. *The American Statistician*, 53, 352–6.

Walter, D. E. & O'Dowd, D. J. (1992). Leaves with domatia have more mites. *Ecology*, 73, 1514–18.

Wamelink, G. W. W., Frissel, J. Y., Krijnen, W. H. J., Verwoert, M. R. & Goedhart, P. W. (2014). Can plants grow on Mars and the Moon: a growth experiment on Mars and Moon soil simulants. *PLoS One*, 9, e103138.

Wang, H., Chow, S. C., & Chen, M. (2005). A Bayesian approach on sample size calculation for comparing means. *Journal of Biopharmaceutical Statistics*, 15, 799–807.

Wang, Y., Naumann, U., Wright, S. T., & Warton, D. I. (2012). mvabund: an R package for model-based analysis of multivariate abundance data. *Methods in Ecology and Evolution*, 3, 471–4.

Wartenberg, D., Ferson, S., & Rohlf, F. J. (1987). Putting things in order: a critique of detrended correspondence analysis. *The American Naturalist*, 129, 434–48.

Warton, D. I. (2022). *Eco-stats: data analysis in ecology*. Springer.

Warton, D. I. & Hui, F. K. C. (2011). The arcsine is asinine: the analysis of proportions in ecology. *Ecology*, 92, 3–10.

Warton, D. I. & Hui, F. K. C. (2017). The central role of mean–variance relationships in the analysis of multivariate abundance data: a response to Roberts (2017). *Methods in Ecology and Evolution*, 8, 1408–14.

Warton, D. I., Wright, I. J., Falster, D. S., & Westoby, M. (2006). Bivariate line-fitting methods for allometry. *Biological Reviews*, 81, 259–91.

Warton, D. I., Wright, S. T., & Wang, Y. (2012). Distance-based multivariate analyses confound location and dispersion effects. *Methods in Ecology and Evolution*, 3, 89–101.

Warton, D. I., Lyons, M., Stoklosa, J., & Ives, A. R. (2016). Three points to consider when choosing a LM or GLM test for count data. *Methods in Ecology and Evolution*, 7, 882–90.

Wasserstein, R. L. & Lazar, N. A. (2016). The ASA's statement on p-values: context, process, and purpose. *The American Statistician*, 70, 129–33.

Wasserstein, R. L., Schirm, A. L., & Lazar, N. A. (2019). Moving to a world beyond "p < 0.05". *The American Statistician*, 73, 1–19.

Weisberg, S. (2013). *Applied linear regression*. Wiley.

West, B. T., Welch, K. B., & Galecki, A. T. (2015). *Linear mixed models: a practical guide using statistical software*, 2nd ed. Chapman and Hall/CRC.

Westfall, P. H. & Young, S. S. (1993a). On adjusting p-values for multiplicity. *Biometrics*, 49, 941–44.

Westfall, P. H. & Young, S. S. (1993b). *Resampling-based multiple testing: examples and methods for p-value adjustment*. Wiley.

Westley, L. C. (1993). The effect of inflorescence bud removal on tuber production in *Helianthus tuberosus* L (Asteraceae). *Ecology*, 74, 2136–44.

Whitlock, M. C. & Schluter, D. (2020). *The analysis of biological data*, 3rd ed. W.H. Freeman.

Whittingham, M. J., Stephens, P. A., Bradbury, R. B., & Freckleton, R. P. (2006). Why do we still use stepwise modelling in ecology and behaviour? *Journal of Animal Ecology*, 75, 1182–9.

Wickham, H. (2009). *ggplot2: elegant graphics for data analysis*. Springer.

Wilcox, R. R. (1987). Within row pairwise comparisons in a 2-way ANOVA design. *Communications in Statistics: Simulation and Computation*, 16, 939–55.

Wilcox, R. R. (2022). *Introduction to robust estimation and hypothesis testing*, 5th ed. Academic Press.

Wilkinson, L. (2013). *The grammar of graphics*, 2nd ed. Springer.

Williams, J. M. (2007). *Style: lessons in clarity and grace*. Pearson Longman.

Winer, B. J., Brown, D. R., & Michels, K. M. (1991). *Statistical principles in experimental design*. McGraw-Hill.

Winter, B. (2019). *Statistics for linguists: an introduction using R*. Routledge.

Wood, S. N. (2017). *Generalized additive models: an introduction with R*, 2nd ed. Chapman & Hall/CRC.

Wootton, H. F. & Keough, M. J. (2016). Disturbance type and intensity combine to affect resilience of an intertidal community. *Marine Ecology Progress Series*, 560, 121–33.

Wu, W., Li, Y., Yan, M., et al. (2021). Surface soil metal elements variability affected by environmental and soil properties. *PLoS One*, 16, e0254928.

Yandell, B. S. (1997). *Practical data analysis for designed experiments*. Chapman & Hall/CRC.

Yasuhara, M., Hunt, G., van Dijken, G., et al. (2012). Patterns and controlling factors of species diversity in the Arctic Ocean. *Journal of Biogeography*, 39, 2081–8.

Yavno, S. & Fox, M. G. (2013). Morphological change and phenotypic plasticity in native and non-native pumpkinseed sunfish in response to sustained water velocities. *Journal of Evolutionary Biology*, 26, 2383–95.

Yoshioka, P. M. (2008a). The Bray–Curtis similarity index remains misidentified: Reply to Somerfield (2008). *Marine Ecology Progress Series*, 372, 307–9.

Yoshioka, P. M. (2008b). Misidentification of the Bray–Curtis similarity index. *Marine Ecology Progress Series*, 368, 309–10.

Zuur, A. F. (2012). *Beginner's guide to generalized additive models with R*. Highland Statistics.

Zuur, A. F. & Ieno, E. N. (2016). *A beginner's guide to zero-inflated models with R*. Highland Statistics.

Zuur, A. F. & Ieno, E. N. (2022). *The world of zero-inflated models. Volume 1: Using GLM*. Highland Statistics.

Zuur, A., Ieno, E. N., & Smith, G. M. (2007). *Analyzing ecological data*. Springer-Verlag.

Zuur, A. F., Ieno, E. N., Walker, N., Saveliev, A. A., & Smith, G. M. (2009). *Mixed effects models and extensions in ecology with R*. Springer-Verlag.

Zuur, A., Saveliev, A. A., & Ieno, E. N. (2012). *Zero inflated models and generalized linear mixed models with R*. Highland Statistics.

Zuur, A. F., Hilbe, J. M., & Ieno, E. N. (2013). *Beginner's guide to GLM and GLMM with R*. Highland Statistics.

Index